Recent Developments in General Relativity, Genoa 2000

Springer-Verlag Italia Srl.

R. Cianci · R. Collina
M. Francaviglia · P. Fré (Eds)

Recent Developments in General Relativity, Genoa 2000

 Springer

R. CIANCI
Dipartimento di Metodi e Modelli
Matematici (DIMET),
Università di Genova
Genova, Italy

M. FRANCAVIGLIA
Dipartimento di Matematica,
Università di Torino
Torino, Italy

R. COLLINA
Dipartimento di Fisica,
Università di Genova and
Istituto Nazionale di Fisica Nucleare,
Sezione di Genova, Genova, Italy

P. FRÉ
Dipartimento di Fisica Teorica,
Università di Torino and
Istituto Nazionale di Fisica Nucleare,
Sezione di Torino, Torino, Italy

© Springer-Verlag Italia 2002
Originally published by Springer-Verlag Italia in 2002

MyCopy version of the original edition 2002

http://www.springer.de

DOI 10.1007/978-88-470-2101-3

Library of Congress Cataloging-in-Publication Data : Applied for

Cover design: Simona Colombo, Milan
Typesetting: Bürosoft/Text- und DTP-Service, Berlin/Heidelberg

Springer.com/mycopy

Preface

The 14th SIGRAV Conference on General Relativity and Gravitational Physics was held in Genova from September 18 to September 22, 2000. The SIGRAV Conference, which is held every two years, collects a highly selected list of speakers who discuss with the participants the most recent researche on General Relativity and Gravitational Physics that, in these years, are receiving a new impetus from astrophysical discoveries and technical progress in field and string theories.

The important experimental effort in gravitational wave detection has also attracted numerous experimental researchers to the sector of General Relativity and Gravitational Physics.

The Conference was structured into 22 talks given by invited speakers and four parallel sessions of short reports (subjected to referee) concerning Classical Relativity, Relativistic Astrophysics and Cosmology, Experimental Gravitation and Space physics, Quantum Relativity and strings. The topics dealt with mathematical, theoretical, experimental and applicative aspects.

This meeting, also called the "Sigrav 2000 Conference", attracted more than 70 participants from many Department of Astronomy and Astrophysics, Departments of Theoretical and Applied Mathematics, Departments of Experimental and Theoretical Physics and Experimental Laboratories spread all over the world.

The Conference saw also the award of the highly distinguished SIGRAV PRIZES, which are granted, every two years, to researchers of high international prestige and relatively young age, typically below forty, who made important contributions to the advance of knowledge in the field of Relativity and Gravity.

The Scientific Selection Committee for this awarding of SIGRAV PRIZES, was composed as follows: Prof. Sergio Ferrara (CERN) (President), Prof. Alfonso Cavaliere (Università di Roma Tor Vergata), Prof. Thibault Damour (Bur sur Yvette), Prof. Guido Pizzella (Università di Roma Torvergata), Prof. Bernard Schutz (Max Planck Institute, Potsdam), Prof. J. Ehlers (Max Planck Institute, Potsdam).

The Conference took place in the pleasant and atmospheric surroundings of the ancient "Magazzini del Cotone" in the Ancient Port of Genova, less than ten meters from the sea. The participants had then the possibility of enjoying both the historical center of Genova with its extremely impressive monuments, the international Aquarium, the wonderful landscape, and a well structured meeting place with all the modern facilities of an International Congress.

As Editors of the Proceedings Volume and Sigrav members, we wish to thank all our sponsors, without whose help this Conference would not have been possible. Their financial generosity was very important for our L.O.C. We wish to thank the City of Genova and the Fondazione Cassa di Risparmio di Genova e Imperia, (CARIGE), the University of Genova, the Department of Mathematical Models and

Methods (DIMET), the Department of Physics (DIFI), the Department of Mathematics (DIMA) and the Faculty of Engineering. Also, wish to thank the local section of the National Institute of Nuclear Physics (INFN) whose personnel provided important technical support in the organization of the conference, the National Group of Mathematical Physics (GNFN) and the "Istituto di Fisica dello Spazio interplanetario" (IFSI).

November 2000

R. Cianci
R. Collina
P. Fré
M. Francaviglia

Contents

CLASSICAL RELATIVITY

RELATIVISTIC ASTROPHYSICS
AND COSMOLOGY

EXPERIMENTAL GRAVITATION
AND SPACE PHYSICS

Recent Developments in General Relativity, Genoa 2000

SIGRAV Scientific Committee

U. Bruzzo (S.I.S.S.A., Trieste)
R. Cianci (Univ. of Genova) – Chairman
I. Ciufolini (Univ. of Lecce)
E. Coccia (Univ. of Roma II)
V. Ferrari (Univ. of Roma I)
M. Francaviglia (Univ. of Torino)
P. Fre' (Univ. of Torino) – SIGRAV President
V. Gorini (Univ. of Insubria, Como)
L. Lusanna (INFN of Firenze) – SIGRAV Secretary
G. Marmo (Univ. of Napoli)
A. Masiero (S.I.S.S.A., Trieste)
T. Regge (Univ. of Torino)
A. Treves (Univ. of Insubria, Como)

Local Organizing Committee

R. Cianci (Univ. of Genova) – Chairman
R. Collina (Univ. of Genova, INFN of Genova) – Co-Chairman
P. Fre' (Univ. of Torino)
L. Opisso (INFN of Genova) – Conference Secretary
C. Dellepiane(Univ. of Genova) – Conference Secretary

Workshop Coordinators

U. Bruzzo (S.I.S.S.A., Trieste)
E. Coccia (Univ. of Roma II)
M. Francaviglia (Univ. of Torino)
A. Treves (Univ. of Insubria, Como)

List of Contributors and Participants

Agnese A.G.
Dipartimento di Fisica, Università di Genova, and INFN, Via Dodecaneso 33, 16146
Genova, Italy
e-mail: agnese@ge.infn.it

Allemandi G.
Dipartimento di Matematica, Università di Torino, Via C. Alberto 10, 10123 Torino,
Italy
e-mail: allemandi@dm.unito.it

Andersson N.
Department of Mathematics, University of Southampton, Southampton SO17 1BJ,
UK
e-mail: N.Andersson@maths.soton.ac.uk

Astone P.
Istituto Nazionale di Fisica Nucleare, Sezione di Roma 1, Piazzale A. Moro 2,
00185 Roma, Italy

Braccini S.
Progetto VIRGO, Istituto Nazionale di Fisica Nucleare, Sezione di Pisa,
56010 S. Piero a Grado, Pisa, Italy

Bachas C.
LPT, Ecole Normale Supérieure, 24 rue Lhomond, 75231 Paris, France
e-mail: bachas@lpt.corto.ens.fr

Barrett J.
University of Nottingham, University Park, Mathematical Sciences,
NG7 2RD Nottingham, UK
e-mail: jwb@maths.nott.ac.uk

Bassan M.
Dipartimento di Fisica, Università di Roma "Tor Vergata", and INFN, Sezione di
Roma 2, Via della Ricerca Scientifica 1, 00133 Roma, and CNR, Istituto Fisica
Spazio Interplanetario, Via Fosso del Cavaliere, 00133 Roma, Italy

Bertotti B.
Dipartimento di Fisica Nucleare e Teorica, Università degli Studi di Pavia, and Istituto
Nazionale di Fisica Nucleare, Sezione di Pavia, Via A. Bassi 6, 27100 Pavia, Italy
e-mail: bruno.bertotti@pv.infn.it

Bičák J.
Institute of Theoretical Physics, Charles University, V Holesovickach 2, Prague,
Czech Republic
e-mail: bicak@mbox.troja.mff.cuni.cz

Bignotto M.
Dipartimento di Fisica, Università di Padova, and INFN Sezione di Padova, Via
Marzolo 8, 35100 Padova, Italy

Blasi A.
Dipartimento di Fisica, Università di Genova, Via Dodecaneso 33, 16146 Genova,
Italy. e-mail: blasi@ge.infn.it

Bonaldi M.
Dipartimento di Fisica, Università di Padova, and INFN Sezione di Padova, Via
Marzolo 8, 35100 Padova, Italy

Bonelli G.
Spinoza Institute, University of Utrecht, Leuvenlaan 4, 3584 CE Utrecht,
The Netherlands

Bonifazi P.
Consiglio Nazionale delle Ricerche, Istituto di Elettronica dello Stato Solido, Via
Cineto Romano 42, 00156 Roma, and Istituto Nazionale di Fisica Nucleare, Sezione
di Roma 1, Piazzale A. Moro 2, 00185 Roma, Italy
e-mail: Paolo.Bonifazi@Roma1.infn.it

Bonora L.
S.I.S.S.A., ISAS, Via Beirut 2–4, 34014 Trieste, and INFN, Sezione di Trieste,
Trieste, Italy
e-mail: bonora@sissa.it

Bruzzo U.
S.I.S.S.A., Via Beirut 2–4, 34014 Trieste, Italy
e-mail: bruzzo@sissa.it

Buonanno A.
Californian Institute of Technology 1200, E. California Boulv. Caltech 130-33,
91125 Pasadena, California, USA
e-mail: buonanno@newman.tapir.caltech.edu

Camacho Q.
Abel Astrophysikalisches Institut Potsdam, An der Sternwarte 16, 14482 Potsdam
Brandenburg, Germany
e-mail: acamacho@aip.de

Canarutto D.
Dipartimento di Matematica Applicata "G. Sansone", Via S. Marta 3, 50013 Firenze,
Italy. e-mail: canarutto@dma.unifi.it

Carbone G.
S.I.S.S.A., Via Beirut 2–4, 34014 Trieste, Italy, and Ecole Normale Supérieure de
Lyon, Laboratoire de Physique, Allee d'Italie 46, 6936 Lyon, Cedex 07, France
e-mail: carbone@sissa.it

Carelli P.
Consiglio Nazionale delle Ricerche, Istituto di Fisica dello Spazio Interplanetario, Via
del Fosso del Cavaliere 100, 00133 Roma, and Istituto Nazionale di Fisica Nucleare,
Sezione di Roma 2, Viale della Ricerca Scientifica 1, 00133 Roma, Italy

Carfora M.
Dipartimento di Fisica Nucleare e Teorica, Università degli Studi di Pavia, and Istituto
Nazionale di Fisica Nucleare, Sezione di Pavia, Via A. Bassi 6, 27100 Pavia, Italy
e-mail: mauro.carfora@pv.infn.it

Caricato G.
Università di Roma "La Sapienza", Piazzale A. Moro 2, 00185 Roma, Italy
Casteill P.Y.
Laboratoire de Physique Theorique et des Hautes Energies, Unite associé au CNRS
URA 280, University of Paris 7, 2 Place Jussieu, 75251 Paris Cedex 05, France
Castellano M.G.
Istituto Nazionale di Fisica Nucleare, Sezione di Roma 2, Viale della Ricerca Scientifica 1, 00133 Roma, Italy
Caponio E.
Dipartimento di Matematica "U. Dini", Università di Firenze, 50134 Firenze, Italy
Cavallari G.
CERN, Geneva, Switzerland
Cenni R.
INFN Sezione di Genova, Via Dodecaneso 33, 16146 Genova, Italy
e-mail: cenni@ge.infn.it
Cerdonio M.
Dipartimento di Fisica, Università di Padova, and INFN Sezione di Padova, Via
Marzolo 8, 35100 Padova, Italy
Chu K.C.
Department of Mathematics, University of Utah, Salt Lake City, Utah, USA, and Dipartimento di Fisica, Università di Roma "Tor Vergata", Via della Ricerca Scientifica 1, 00133 Roma, Italy
Cianci R.
DIMET, Università di Genova, P.le J.F. Kennedy, Padiglione D, 16129 Genova, Italy
e-mail: cianci@unige.it
Ciufolini I.
Dipartimento di Ingegneria Innovazione, Campus Universitario, Università di Lecce,
Via Arnesano, 23100 Lecce, Italy
e-mail: ciufoli@nero.ing.uniroma1.it
Coccia E.
Dipartimento di Fisica, Università degli Studi di Roma "Tor Vergata", Viale della
Ricerca Scientifica 1, and Istituto Nazionale di Fisica Nucleare, Sezione di Roma 2,
Via della Ricerca Scientifica 1, 00133 Roma, Italy
e-mail: coccia@roma2.infn.it
Collina R.
INFN, Università di Genova, Via Dodecaneso 33, 16146 Genova, Italy
e-mail: collina@ge.infn.it
Cosmelli C.
Dipartimento di Fisica, Università degli Studi di Roma "Tor Vergata", Viale della
Ricerca Scientifica 1, 00133 Roma, and Istituto Nazionale di Fisica Nucleare, Sezione
di Roma 1, Piazzale A. Moro 2, 00185 Roma, Italy
Conti L.
Dipartimento di Fisica, Università di Padova, and INFN Sezione di Padova, Via
Marzolo 8, 35100, Padova, Italy

Crivelli Visconti V.
Dipartimento di Fisica, Università di Padova, and INFN Sezione di Padova, Via Marzolo 8, 35100, Padova, Italy

Czapor S.R.
Department of Mathematics and Computer Science, Laurentian University, Sudbury, Ontario, Canada

D'Antonio S.
Istituto Nazionale di Fisica Nucleare, Laboratori Nazionali di Frascati, Viale E. Fermi, 00044 Frascati, Italy

D'Auria R.
Politecnico di Torino, C.so degli Abruzzi 24, 10129 Torino, Italy
e-mail: dauria@polito.it

De Rosa M.
Dipartimento di Fisica, Università di Firenze, and INFN Sezione di Firenze, L.go E. Fermi 2, 50125 Arcetri (Firenze), Italy

Di Virgilio A.
INFN Sezione di Pisa, Via Livornese 1291, 56010 S. Piero a Grado, Pisa, Italy
e-mail: angela.divirgilio@pi.infn.it

Dolesi R.
Dipartimento di Fisica, Università di Trento, Via Sommarive 38050, POVO (TN), Italy
e-mail: dolesi@science.unitn.it

Esposito G.
INFN, Sezione di Napoli, Complesso Universitario di Monte S. Angelo, and Dipartimento di Scienze Fisiche, Complesso Universitario di Monte S. Angelo, Via Cintia, 80126 Napoli, Italy
e-mail: giampiero.esposito@na.infn.it

Fafone V.
Istituto Nazionale di Fisica Nucleare, Laboratori Nazionali di Frascati, Viale E. Fermi, 00044 Frascati, Italy
e-mail: fafone@lnf.infn.it

Falferi P.
Centro di Fisica degli Stati Aggregati CNR-ITC, and INFN Gruppo Collegato di Trento, 38050 Povo (Trento), Italy

Fatibene L.
Dipartimento di Matematica, Università di Torino, Via C. Alberto 10, 10123 Torino, Italy
e-mail: fatibene@dm.unito.it

Federici G.
INFN, Sezione di Roma 1, Piazzale A. Moro 2, 00185 Roma, Italy

Ferrara S.
Theoretical Physics Division, CERN, 1211 Geneva 23, Switzerland
e-mail: sergio.ferrara@cern.ch

Focardi S.
Dipartimento di Fisica and INFN, Università di Bologna, Bologna, Italy
e-mail: focardi@bo.infn.it

Francaviglia M.
Dipartimento di Matematica, Università di Torino, Via C. Alberto 10, 10123 Torino,
Italy
e-mail: francaviglia@dm.unito.it

Fre' P.
Dipartimento di Fisica Torica, Via Giuria 1, 10125 Torino, Italy
e-mail: fre@to.infn.it

Frossati G.
Leiden University, Niels Bohrweg 2, 2333CA Leiden, The Netherlands
e-mail: giorgio@phys.leidenuniv.nl

Garay L.J.
Instituto de Matematicas y Fisica Fundamental, CSIC, Serrano 121, 28006 Madrid,
Spain

Gemme G.
INFN, Università di Genova, Via Dodecaneso 33, 16146 Genova, Italy
e-mail: gianluca.gemme@ge.infn.it

Giazotto A.
Progetto VIRGO, Istituto Nazionale di Fisica Nucleare, Sezione di Pisa, Traversa H
di Via Macerata, Santo Stefano a Macerata, 56021 Cascina (Pisa), Italy
e-mail: adalberto.giazotto@pi.infn.it

González-Díaz P.
Instituto de Matemáticas y Física Fundamental, CSIC, Serrano 121, 28006 Madrid,
Spain
e-mail: p.gonzalezdiaz@imaff.cfmac.csic.es

Gorini V.
Università dell'Insubria, Via Lucini 3, 22100 Como, Italy
e-mail: gorini@fis.unico.it

Gupta S.N.P.
Bhilai Steel Plant Bhilai m.p 1b, Street 57, sec 8, Bhilai 490006, India

Harpaz A.
Department of Physics, University of Haifa at Oranim, Tivon 36006, Israel
e-mail: phr89ah@tx.technion.ac.il

Janyška J.
Department of Mathematics, Masaryk University, Janáčkovo nám 2a, 662 95 Brno,
Czech Republic

Kokkotas K.
Department of Physics, Aristotle University of Thessaloniki, Thessaloniki 54006,
Greece, and Department of Mathematics, University of Southampton, Southampton
SO17 1BJ, UK
e-mail: kokkotas@astro.auth.gr

Kokorelis C.
CITY University, Frobisher Crescent EC2Y 8HB, London, UK
e-mail: c.kokorelis@sussex.ac.uk

Lusanna L.
INFN, Largo Fermi 2, 50125 Firenze, Italy
e-mail: lusanna@fi.infn.it

Magli G.
Dipartimento di Matematica, Politecnico di Milano, P.le Leonardo da Vinci 32,
20131 Milano, Italy
e-mail: giumag@mate.polimi.it

Maraschi L.
Osservatorio Astronomico di Brera, Via Brera 28, 20121 Milano, Italy
e-mail: maraschi@brera.mi.astro.it

Marin A.
Dipartimento di Fisica, Università di Padova, and INFN Sezione di Padova, Via
Marzolo 8, 35100 Padova, Italy

Marin F.
Dipartimento di Fisica, Università di Firenze, and INFN Sezione di Firenze, L.go E.
Fermi 2, 50125 Arcetri (Firenze), Italy

Marini A.
INFN, Laboratori Nazionali di Frascati, Via E. Fermi 40, 00044 Frascati, Italy

Marzuoli A.
Dipartimento di Fisica Nucleare e Teorica, Università degli Studi di Pavia, and Istituto
Nazionale di Fisica Nucleare, Sezione di Pavia, Via A. Bassi 6, 27100 Pavia, Italy
e-mail: annalisa.marzuoli@pv.infn.it

Masiello A.
Dipartimento di Matematica, Politecnico di Bari, 70125 Bari, Italy

Mbonye M.
Department of Physics, Randall Laboratory, 48109-1120, University of Michigan,
Ann Arbor, Michigan, USA
e-mail: mbonye@umich.edu

McLenaghan R.
Department of Applied Mathematics, University of Waterloo, 200 University Avenue
W. Waterloo Ontario, Canada
e-mail: rgmclena@sirius.uwaterloo.ca

Menotti P.
Dipartimento di Fisica dell'Università, and INFN Sezione di Pisa, Via F. Buonarroti
2, 56100 Pisa, Italy
e-mail: menotti@df.unipi.it

Mezzena R.
Dipartimento di Fisica, Università di Trento, and INFN Gruppo Collegato di Trento,
38050 Povo (Trento), Italy

Miller J.
S.I.S.S.A., Via Beirut 2–4, 34014 Trieste, Italy
e-mail: miller@sissa.it

Minenkov Y.
Istituto Nazionale di Fisica Nucleare, Sezione di Roma 2, Viale della Ricerca Scientifica 1, 00133 Roma, and CNR, Istituto Fisica Spazio Interplanetario, Via Fosso del Cavaliere, 00133 Roma, Italy

Modena I.
INFN, Laboratori Nazionali di Frascati, Via E. Fermi 40, 00044 Frascati, Roma, Italy
e-mail: modena@roma2.infn.it

Modestino G.
INFN, Laboratori Nazionali di Frascati, Viale E. Fermi 40, 00044 Frascati, Roma, Italy

Modugno M.
Dipartimento di Matematica Applicata, Via S. Marta 3, 50139 Firenze, Italy
e-mail: modugno@dma.unifi.it

Moleti A.
Dipartimento di Energetica, Università degli Studi dell'Aquila, Roio Poggio, and Istituto Nazionale di Fisica Nucleare, Sezione di Roma 2, Viale della Ricerca Scientifica 1, 00133 Roma, Italy

Moschella U.
Dipartimento di Scienze Matematiche Fisiche e Chimiche, Università dell'Insubria, Via Valleggio 11, 22100 Como, and INFN, Sezione di Milano, Milano, Italy
e-mail: moschell@fis.unico.it

Nesti F.
International School for Advanced Studies (S.I.S.S.A./ISAS), and INFN, Sezione di Trieste, Via Beirut 2–4, 34014 Trieste, Italy

Nobili A.
Gruppo di Meccanica Spaziale, Dipartimento di Matematica, Università di Pisa, Via Filippo Buonarroti 2, 56127 Pisa, Italy
e-mail: nobili@pi.infn.it

Pallottino G.V.
Dipartimento di Fisica, Università degli Studi di Roma "Tor Vergata", Viale della Ricerca Scientifica 1, 00133 Roma, and Istituto Nazionale di Fisica Nucleare, Sezione di Roma 1, Piazzale A. Moro 2, 00185 Roma, Italy and Istituto Nazionale di Fisica Nucleare, Sezione di Roma 1, Piazzale A. Moro 2, 00185 Roma, Italy

Panin A.
Department of Physics, Utah Valley State College, Orem, Utah, 84058, USA
e-mail: paninal@uvsc.edu

Pavlis E.
Joint Center for Earth Systems Technology, University of Maryland Baltimore County, NASA Goddard Space Flight Center, Greenbelt, MD, 20771-0001, USA
e-mail: epavlis@Helmert.gsfc.nasa.gov

Pedroni M.
Dipartimento di Matematica dell'Università di Genova, Via Dodecaneso 35, 16146 Genova, Italy
e-mail: pedroni@dima.unige.it

Pizzella G.
Dipartimento di Fisica, Università di Roma "Tor Vergata", Via della Ricerca Scientifica 1, 00133 Roma, and INFN, Laboratori Nazionali di Frascati, V. E. Fermi 40, 00044 Frascati, Italy

Prodi G.A.
Dipartimento di Fisica, Università di Trento, and INFN Gruppo Collegato di Trento, 38050 Povo (Trento), Italy

Quintieri L.
INFN, Laboratori Nazionali di Frascati, Via E. Fermi 40, 00044 Frascati, Italy

Raiteri M.
Dipartimento di Matematica, Università di Torino, Via C. Alberto 10, 10123 Torino, Italy
e-mail: raiteri@dm.unito.it

Rezzolla L.
S.I.S.S.A., Via Beirut 2–4, 34014 Trieste, Italy
e-mail: rezzolla@sissa.it

Righetti R.
European Patent Office Huis te Hoornkade 15 2282 JW, Rijswijk Netherlands
e-mail: rrighetti@epo.org

Rizzi G.
Dipartimento di Fisica, Politecnico di Torino, Corso Duca d'Abruzzi 24, 10129 Torino, Italy
e-mail: rizzi@polito.it

Rocchi A.
Diparimento Fisica, Università di Roma "Tor Vergata", Via della Ricerca Scientifica 1, 00133 Roma, and INFN, Laboratori Nazionali di Frascati, Via E. Fermi 40, 00044 Frascati, Italy

Ronga F.
INFN, Laboratori Nazionali di Frascati, Via E. Fermi 40, 00044 Frascati, Italy

Rubano C.
Dipartimento di Scienze Fisiche, Università Federico II di Napoli, and INFN, Sezione di Napoli, Complesso Universitario di M. Sant'Angelo, Via Cinthia, Ed. N, 80126 Napoli, Italy
e-mail: claudio.rubano@na.infn.it

Ruffini R.
ICRA, and Università di Roma "La Sapienza", P.le Aldo Moro 5, 00185 Roma, Italy
e-mail: ruffini@icra.it

Saller D.
Department of Mathematics, Mannheim University, 68131 Mannheim, Germany

Salviato M.
Dipartimento di Fisica, Università di Padova, and INFN Sezione Padova, Via Marzolo 8, 35100 Padova, Italy

Schneider R.
Università di Roma "La Sapienza", P.le Aldo Moro 2, 00185 Roma, Italy
e-mail: raffaella.schneider@roma1.infn.it

Slagter R.
University of Amsterdam, Physics Department, and ASFYON, Astronomisch Fysisch Onderzoek, The Nederlands
e-mail: rjs@asfyon.nl

Soranzo G.
Dipartimento di Fisica, Università di Padova, and INFN Sezione di Padova, Via Marzolo 8, 35100 Padova, Italy

Stella L.
Osservatorio Astronomico di Roma, Via Frascati 33, 00040 Monteporzio Catone (Roma), Italy
e-mail: stella@coma.mporzio.astro.it

Szocs H.L.
University Coll. J. Kodolanyi, Szabadsagharcos Str.59, 8000 Szekesfehervar, Hungary
e-mail: szh@uranos.kodolanyi.hu

Taffarello L.
Dipartimento di Fisica, Università di Padova, and INFN Sezione di Padova, Via Marzolo 8, 35100 Padova, Italy

Tartaglia A.
Dipartimento di Fisica, Politecnico di Torino, Corso Duca d'Abruzzi 24, 10129 Torino, Italy
e-mail: tartaglia@polito.it

Terenzi R.
Consiglio Nazionale delle Ricerche, Istituto di Elettronica dello Stato Solido, Via Cineto Romano 42, 00156 Roma, and Istituto Nazionale di Fisica Nucleare, Sezione di Roma 2, Viale della Ricerca Scientifica 1, 00133 Roma, Italy

Terna S.
International School for Advanced Studies (S.I.S.S.A./ISAS), Via Beirut 2–4, 34014 Trieste, and INFN, Sezione di Trieste, Trieste, Italy

Tomasiello A.
International School for Advanced Studies (S.I.S.S.A./ISAS), Via Beirut 2–4, 34014 Trieste, Italy

Torrioli G.
CERN, Geneva, Switzerland and Istituto Nazionale di Fisica Nucleare, Sezione di Roma 2, Viale della Ricerca Scientifica 1, 00133 Roma, Italy

Treves A.
Università dell'Insubria, Via Lucini 3, 22100 Como, Italy
e-mail: treves@fis.unico.it

Vetrano F.
Università di Urbino, Via S. Chiara 27, 61029 Urbino, Italy
e-mail: vetrano@fis.uniurb.it
Vietri M.
Dipartimento di Fisica, Università Roma 3, Via della Vasca Navale 84, 00147 Roma,
Italy
e-mail: vietri@fis.uniroma3.it
Villani P.
Dipartimento di Fisica Nucleare e Teorica, Università degli Studi di Pavia, and Istituto
Nazionale di Fisica Nucleare, Sezione di Pavia, Via A. Bassi 6, 27100 Pavia, Italy
Vinante A.
Dipartimento di Fisica, Università di Trento, and INFN Gruppo Collegato di Trento,
38050 Povo (Trento), Italy
Visco M.
Consiglio Nazionale delle Ricerche, Istituto di Elettronica dello Stato Solido, Via
Cineto Romano 42, 00156 Roma, and Istituto Nazionale di Fisica Nucleare, Sezione
di Roma 2, Viale della Ricerca Scientifica 1, 00133 Roma, Italy
e-mail: visco@roma2.infn.it
Vitale S.
Dipartimento di Fisica, Università di Trento, and INFN Gruppo Collegato di Trento,
38050 Povo (Trento), Italy
Votano L.
INFN, Laboratori Nazionali di Frascati, Via E. Fermi 40, 00044 Frascati, Italy
Will C.
Department of Physics, Washington University, St. Louis, Campus Box 1105, One
Brookings Drive 63130, St. Louis MO, USA
e-mail: cmw@wuphys.wustl.edu
Zaffaroni A.
INFN, and Università di Milano Bicocca, Piazza della Scienza 3, 20126 Milano, Italy
e-mail: alberto.zaffaroni@mi.infn.it
Luca Zampieri
INFN, and Dipartimento di Fisica, Università di Padova, Via F. Marzolo 8, 35131
Padova, Italy
e-mail: luca.zampieri@pd.infn.it
Zendri J.P.
Dipartimento di Fisica, Università di Padova, and INFN Sezione di Padova, Via
Marzolo 8, 35100 Padova, Italy
e-mail: zendri@lnl.infn.it

In Memory of Ruggiero de Ritis
July 1, 1943 – September 8, 2000

C. Rubano

1 Introduction

At the age of 57 years, Ruggiero de Ritis suddenly passed away on September 8, 2000. His premature departure has left a big void in those who had the luck to meet him, either as a coworker or a teacher.

He was scheduled to give an invited talk at the XIV National Conference on General Relativity and Gravitational Physics in Genova, September 2000. As a homage to his memory, we decided to participate in his place, commemorating his life and work with this brief talk. Later, a meeting in his memory has been rapidly organized at Vietri sul Mare (Salerno), on December 1–2, 2001[1]. In this sad situation, we also thought to publish a book in his memory, containing some contributions by

---

[1] The meeting has been entitled "General Relativity, Cosmology, and Gravitational Lensing" and organized by many local institutions – Dipartimento di Fisica "E. R. Caianiello" (Università di Salerno), Dipartimento di Scienze Fisiche (Università di Napoli), Osservatorio

Ruggiero's friends and colleagues, in order to give that kind of signal that Ruggiero would have liked best – science, as one of the responses to the absurdity of the unfathomable life project.

Something of these pages, here, is also dedicated to some aspects of his life, which may be unknown to many people.

2 Life and academy

Ruggiero de Ritis was born in Napoli during the II World War, and deeply shared the local atmospheres and trends. He certainly represented that rich kind of culture which could sometimes appear as a separated one in our town, so profoundly popular and noisy.

As a student, his approach to universitary life was in fact critical and intellectual. This led him to fully participate in the great social movements during 60's and 70's, joining the so-called "Sinistra Universitaria", a political movement with a deep influence not only in the universitary environment. He got the degree in physics in 1972, with a thesis entitled "The formalization of quantum mechanics" (under the guide of Prof. G. F. dell'Antonio), revealing his already deep interest in theoretical physics and philosophy of science.

Before becoming a researcher, he won some grants which allowed him to remain in the same places where he had been as a young man, now making his first teaching experiences, and, also, to go to the United States. In 1978 he began to teach physics at the first year of the course of geological studies, what he continued to do also when he got his permanent position as an Associate Professor. Nonetheless, he had the possibility (or, we must say, created it) to teach subjects much closer to his researches (as Field Theory, Spatial Physics, Relativity, Gravitational Physics, Cosmology), for senior and Ph.D. students.

In his career at the university he supervised 16 degree and 6 Ph.D. theses, clearly witnessing his incessant research for pupils and the exceptional care he dedicated to them. Mostly, those who had him as a supervisor have remained in contact with him, often as a friend. As a sign of the influential contact that Ruggiero had with them, most of them are now engaged in a research activity.

Warm and passionate, but also able to keep a deep self control, Ruggiero de Ritis was a very gentle man, always involved in the collective cultural life of his university and beyond. He had a big sense of responsability towards the community hosting him, and this, non occasionally, led him to be responsible of many activities in I.N.F.N. (Istituto Nazionale di Fisica Nucleare) and at the University physics department.

Skipping most of his many academic engagements, it seems important to remind the organization of the Peter G. Bergmann's Celebration and the IX National Conference on General Relativity and Gravitational Physics, both held in Capri (September 1990). In this occasion SIGRAV was conceived and organized, in a process which

Astronomico di Capodimonte (Napoli), Facoltà di Scienze MM. FF. NN. (Università di Napoli), Istituto Italiano di Studi Filosofici (Napoli), IIASS "E. R. Caianiello" (Vietri sul Mare, Salerno).

saw him as a very active participant and a promoter. Since then, he took part in all subsequent SIGRAV conferences, always bringing his cultural and organizative contributions.

The blind academic point of view denied Ruggiero de Ritis to become a Full Professor, something he had surely been worth of (as his scientific paper production shows, even if we do believe that such an element has to be considered only together with other not less important credits).

3　Science

Ruggiero de Ritis was interested in many and, sometimes, apparently divergent fields. Actually, such interests were constantly joined with a philosophical view of the world, mainly based on unity and search for it. To mention only his scientific work, it is possible to see that it covers a wide area; essentially, it can be roughly divided in four great domains: i) epistemology and philosophy, ii) theoretical cosmology, iii) gravitational lensing, and iv) astronomy. The first place in this list is not casual, since he always kept a philosophical attitude towards his scientific work.

With the same dedication he had towards teaching, he also shed his scientific views through many lectures and seminars, wherever he was invited to, in order to bring his peculiar insights. He collaborated with many researchers, often leaving in them memories of his kind way of approaching relationships and of his deep and sophisticated philosophical points of view.

3.1　Epistemology and philosophy

Since his thesis on axiomatization of quantum mechanics, dealing with the connection of some existing approaches (Birkhoff–von Neumann, Jauch, Segal), Ruggiero de Ritis was always interested in epistemology, seen as a natural bridge between science and philosophy.

Very briefly, his reflections have been above all developed in order to study some aspects of Special and General Relativity, mainly examining the scientific production of A. Einstein. As a matter of fact, all Einstein's works were deeply familiar to him, and he found there was a well precise continuity in the work of that famous scientist. Defining and analyzing his idea of completeness of a physical theory, de Ritis was so led to a general reconstruction of Einstein's epistemology, based on the four notions of monism, realism, completeness of a theory, and nonlinearity of field equations. In such a historical-cultural investigation, the distinction (made by Einstein) between a constructive theory and a theory based on principles was discussed, also clarifying and generalizing the models built by Popper and Lakatos.

But the general philosophical interests followed by Ruggiero de Ritis were wider. Mostly, they were private studies (for instance, those on Hegel) and appear to be relevant and of great importance to understand his character. It is natural to cite them here, since they were never separated from science in his mind. An important

4   C. Rubano

example was his interest in Freud, since, among other things, this led him to a book
on the notion of time in Freud.

With the same spirit, he intensively collaborated from the very beginning with the
activities of the Italian Institute of Philosophical Studies (directed by G. Marotta).
Many were the seminars and conferences, not only about physics, there organized
by de Ritis, who often personally intervened.

He also participated into the activities of Bibliopolis editions, directed by one
of his best old friends, F. del Franco. Since 1985 (and together with G. Marmo),
he has been the scientific coordinator of the series "Monographs and Textbooks in
Physical Science" (30 volumes). Since 1997 (together with M. Capaccioli and G.
Marmo), he has also been the scientific coordinator of the "Napoli Series on Physics
and Astrophysics" (3 volumes).

He contributed very much to open the traditional philosophical area in Napoli
(in which Institutions like Marotta's and del Franco's worked) towards sciences
in general and physics in particular, and to establish connections and interchanges
between these two fields of knowledge. We mention here also the fact that, in the
last period of his life, he founded (with some friends) a cultural association, "IL
MILLEPIEDI", sometimes more evidently connected with his scientific work and
always investigating the philosophical insights in all branches of knowledge.

3.2 Theoretical cosmology

He first began to be more directly involved in cosmology with his friend G. Platania,
in 1975. This is certainly the most important field of interests in which Ruggiero
de Ritis produced work and results, also due to its natural connections to so many
domains of science and philosophy.

First of all, he dedicated himself to the study of the general group of invariance in
Newtonian cosmology (Heckmann–Schucking group), constructing its Lie algebra
and giving the cosmological conditions to get the inhomogeneous Galilei group from
it. A variational formulation of such a cosmology was also produced.

Nextly, he investigated theories with torsion and General Relativity as a gauge
theory of Poincaré group. At that time only few groups in the world were interested
in torsion, and he greatly contributed to gather attention towards the group in Napoli.
Searching for a Lagrangian able to describe a self-gravitating fluid with spin, he
found that the material part of such a Lagrangian was nothing but the pressure of the
cosmological fluid. In the study of isotropization in cosmology and the relationship
between Big Bang and torsion singularities, the extensions of Bernoulli theorem and
geometrical optics to the case with torsion were considered.

Also, the inclusion of a quadratic curvature scalar term in cosmology allowed to
prove the validity of Birkhoff theorem and find some exact solutions, so giving new
insights in Big Bang singularities.

Throwing the seeds of future developments, there was the immersion of the concept of torsion in the scenario of inflationary cosmology, with a minimally or nonminimally coupling between the inflationary scalar field and the spacetime curvature.
Qualitative and approximated methods have been used to investigate both anisotropic

and isotropic cosmologies, also obtaining asymptotic behaviours of the scale factor with several scalar field potentials, or finding the conditions for isotropization.

Afterwards, in the context of torsionless inflationary cosmology, Ruggiero de Ritis has highly contributed to develop a new powerful approach to the study of the Universe, the so-called "Nöther cosmology". In such an approach, the Lagrangian can be seen as depending just on two coordinates, the scale factor and the scalar field, plus their time derivatives, and then considered as a point-like Lagrangian. Thus, Nöther's symmetries of this function can be used to search for exact solutions of Einstein's equations with a minimally or nonminimally coupled scalar field. Some different potentials have been used, finding asymptotically exponential or power-law inflationary solutions, also being able to reconstruct the inflationary potential from the equation of state. In particular, searching for Nöther's symmetries in the nonminimally coupled case, a connection between the potential and the coupling function was discovered.

The exact knowledge of the temporal behaviours of the scale factor and inflationary scalar field has allowed a better analysis of large scale structures in the Universe, also presenting an apparently periodic version. Due to the possibility of introducing an arbitrary parameter into this game, in fact, Ruggiero de Ritis has then also investigated cosmological perturbations in exact-Nöther background solutions.

In a more general context, but with similar techniques, he examined the string cosmology Lagrangian, connecting Nöther's and duality symmetries.

After such researches, two relevant problems in contemporary cosmology, i.e. dark matter and cosmological constant, have been considered. Firstly a new definition of a dynamical cosmological "constant" has been advanced, making a proposal about its origin in scalar-tensor theories of gravity. (This was done fixing its asymptotic behaviour through the generalization of the cosmic no-hair theorem). This work has also produced, as byproducts, an analysis of asymptotic freedom in gravitational physics and an analysis of cosmological conformal equivalence between the Einstein and Jordan frames.

Further investigations on the role of a cosmological term in cosmology with dark matter and energy have then led Ruggiero de Ritis to turn his attention more and more to phenomenology in cosmology. "Quintessence" fields and how such a kind of (minimally or nonminimally coupled) field could be connected to both a cosmological constant and string cosmology duality was one of the resulting fields of interest.

3.3 Gravitational lensing

In close relation with the two just above cited problems, Ruggiero de Ritis also began to study gravitational lensing.

General relativity predicts that the light rays are deflected by the gravitational field of a given mass distribution (the field itself working like a refraction index). The lensing effect takes into account singularities and caustics of the so-called "lens map", allowing to study what is connected to such a kind of "gravitational optics". It has become a very powerful tool to look for dark matter in, for instance, our obscure galactic halo (microlensing), and is still playing a role in determining the nature of such a matter.

Ruggiero de Ritis has analyzed many of such problems and has contributed to develop an original approach to the so-called "defocusing", in which light, instead of being focused, is defocused in certain peculiar situations. Due to the importance that de Ritis attributed to his relationship with students, we must here particularly mention the work for the elaboration of a textbook on basics of gravitational lensing, a book intentionally written in Italian for being used in the lectures on the introduction to lensing in the Napoli course of gravitational physics.

In cosmology, the lensing phenomenon is also affected by expansion and, therefore, can give information on the cosmological parameters. Starting from the analysis of the concept of distance induced by gravitational lensing in a clumpy universe, Ruggiero de Ritis has examined how such a distance is changed by means of a cosmological constant and how this can infer new insights on the time delay and the Sunyaev-Zel'dovich effect. Furthermore, the role of a clumpiness parameter in statistical lensing and, in general, with respect to exact solutions in cosmology has been considered.

3.4 Astronomy

In close connection with his cosmological and gravitational lensing studies, Ruggiero de Ritis was truely involved in astronomy only in the last part of his life, when he decided to consistently develop his wide interests for phenomenology in those fields. Due to the scientific contact with astronomers, he took part to the life of the local observatory (O.A.C.), and became the leader of its Laboratory of Gravitational Physics.

Both for theoretical and observative aims, he has been one of the promoters of the international group of microlensing research called SLOTT-AGAPE (including Lecce, Napoli, Paris, Pavia, Salerno, and Zürich). He was the local coordinator, of course, while working on the "inverse" problem in microlensing, to get informations on the Galaxy model from the optical depth, and on the pixel lensing search of extrasolar planets.

Anyway, Ruggiero de Ritis has passed away before he could form his own experience at a telescope and see the fulfilment of this other step of his greater dream of connection between phenomenology and theory, science and philosophy.

4 Conclusions

Even if Ruggiero de Ritis has disappeared, many traces of his huge work have remained, and can be followed and developed, if possible.

For example, among others, he left three smart Ph.D. students (G. Covone. E. Piedipalumbo, and M. Sereno), still working on his ideas, and still remembering the many passionate discussions they had together. It is as if such a situation possesses a kind of inner momentum, so keeping energy from the past.

Ruggiero de Ritis also left a great legacy of contacts with many researchers all around. They often were good friends with him and still continue in the contacts with the remaining research group in General Relativity, cosmology, and gravitational lensing in Napoli. For us all this has become a challenge, and we sincerely hope to be able to sustain it.

The Lectures

Radiative Spacetimes

J. Bičák

Abstract. The question of existence of general, asymptotically flat radiative spacetimes and examples of explicit classes of radiative solutions of Einstein's field equations are discussed in the light of some new developments. The examples are cylindrical waves, Robinson–Trautman and type N spacetimes, and especially boost- rotation symmetric spacetimes, representing uniformly accelerated particles or black holes.

1 Introduction

In physical theories on a fixed background spacetime, as in Newtonian theory or special relativity, it is not difficult to formulate asymptotic fall-off conditions on fields of spatially bounded systems. For example, the gravitational potential due to a Newtonian star is usually required to decay to zero at infinity of Euclidean space, with the decay rate being compatible with Laplace's equation. In general relativity no a priori given background space exists. The metric itself is both a dynamical field and a quantity which determines distances. One expects that in a suitable coordinate system far away from a system of bodies the metric should have a form $g_{\mu\nu} = \eta_{\mu\nu} +$ small quantities, where $\eta_{\mu\nu}$ is Minkowski metric. What, however, does it mean "far away", what is "infinity"? Can one formulate suitable boundary conditions in a coordinate-free manner? What is the decay of a *radiative* gravitational field?

After several important contributions to the gravitational radiation theory in the late 1950's and early 1960's by Pirani, Bondi, Robinson, Trautman and others, a landmark paper by Bondi et al. [1] appeared in which the radiative properties of isolated (spatially bounded) axisymmetric systems were studied along outgoing null hypersurfaces $u = $ constant, with u representing a retarded time function. An ansatz was made that the metric along $u = $ constant can be expanded in inverse powers of r,

$$g_{\mu\nu} = \eta_{\mu\nu} + h_{\mu\nu}(\theta)r^{-1} + f_{\mu\nu}(\theta)r^{-2} + \dots , \tag{1}$$

where r denotes a suitable parameter along null generators (parametrized by coordinates θ, φ) on the hypersurfaces $u = $ constant. Under the assumption (1) Einstein's vacuum equations were shown to determine uniquely formal power series solution of the form (1), provided that a free "news function" $c(u, \theta)$ is specified. The news function contains all the information about radiation at infinity ("$r = \infty$"). It enters the fundamental "Bondi mass-loss formula" for the total mass $M(u)$ of an isolated system at retarded time u. The field equations imply that $M(u)$ is a monotonically decreasing function of u if $\partial_u c \neq 0$. A natural interpretation is that gravitational waves carry away positive energy from the system and thus decrease its mass. In the work of Bondi et al. [1] as well as in the important generalizations by Sachs [2], Newman and Penrose [3], the decay of radiative fields was studied in preferred coordinate systems.

In 1963 Penrose[1] [4] formulated a beautiful *geometrical* framework for description of the "radiation zone" in general relativity in terms of conformal infinity.

Penrose's definition of asymptotically flat radiative spacetimes avoids such problems as "distances" or "suitable coordinates", and incorporates a clear definition of what is infinity. It is inspired by the work on radiation theory mentioned above, and by the properties of conformal infinity in Minkowski spacetime. In contrast to an Euclidean space, in Minkowski spacetime one can go to infinity in various directions: moving along timelike geodesics we come to the future (or past) timelike infinity I^+ (or I^-); along null geodesics (cf. Eq. (1)) we reach the future (past) null infinity $\mathcal{J}^+(\mathcal{J}^-)$; spacelike geodesics lead to spatial infinity i_0. Minkowski spacetime can be compactified and mapped into a finite region by an appropriate conformal transformation. Thus one obtains the well-known *Penrose diagram* in which the three types of infinities are mapped into the boundaries of the compactified spacetime – see Fig. 1.

It is generally accepted that Penrose's definition forms the only rigorous, geometrical basis for the discussion of gravitational radiation from isolated systems. It enables us to use techniques of local geometry "at infinity" and to define covariantly such fundamental quantities as the total (Bondi) mass of an isolated system.

2 Asymptotically flat radiative spacetimes: Existence

Despite its rigour and elegance, Penrose's *definition* might turn to be of a limited importance if no interesting radiative spacetimes *exist* which satisfy the definition. In Sect. 5 we shall describe special exact radiative spacetimes which represent "uniformly accelerated" sources in general relativity and admit $\mathcal{J}$ as required; however, at least four points on $\mathcal{J}$ – those in which worldlines of the sources start and end – are singular. There are no other *explicit* exact radiative solutions describing finite sources available at present and this situation will probably not change soon. Nevertheless, thanks to the work of Friedrich, reviewed in [7], and Christodoulou and Klainerman [5] we know that globally non-singular (including *all* $\mathcal{J}$) asymptotically flat exact solutions of Einstein's equations really exist.

The key idea of Friedrich is in realizing that Penrose's treatment of infinity not only permits to use the methods of local differential geometry at $\mathcal{J}$, but also to analyze global existence problems of solutions of Einstein's equations in the physical spacetime M by solving initial value problems in conformally related unphysical spacetime $\tilde{M}$.

By using this approach Friedrich established that formal Bondi-type expansions (1) converge locally at $\mathcal{J}^+$. He succeeded to show that one can formulate the "hy-

---

[1] How this framework became powerful is especially evident from the 2-volume monography by R. Penrose and W. Rindler (1986), *Spinor & space-time*, Cambridge University Press, Cambridge

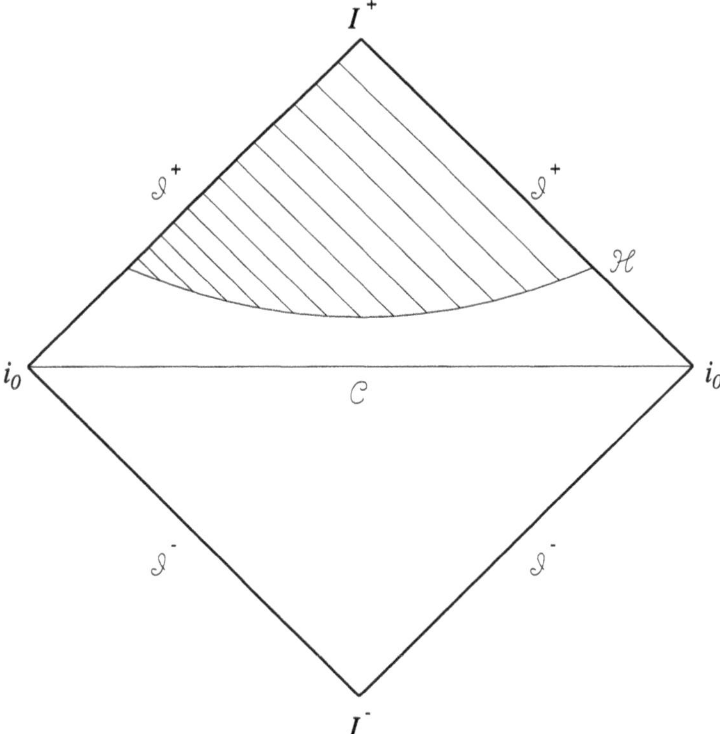

Fig. 1. The Penrose conformal diagram of an asymptotically flat spacetime. The Cauchy hypersurface and the hyperboloidal hypersurface are indicated

perboloidal initial value problem" for Einstein's vacuum equations in which initial data are given on a hyperboloidal spacelike hypersurface $\mathcal{H}$ which intersects $\mathcal{J}$ (see Fig. 1). It can then be proven that hyperboloidal initial data, which are sufficiently close to Minkowskian hyperboloidal data (i.e. to the metric induced on the hypersurface $\mathcal{H}$ in Minkowski spacetime by the standard Minkowski metric) evolve to a vacuum spacetime which is smooth on $\mathcal{J}^+$ and I^+ as required by Penrose's definition.

Despite the deep and complicated work by Friedrich, one still did not have what one really would wish. Let us first note that initial data *"sufficiently close"* to Minkowskian data in Friedrich's result we mentioned above, do not mean any approximation – "smallness" of data is understood in a functional sense in an appropriate Sobolev space. One cannot hope to prove global existence of smooth general solutions developed from general, arbitrarily *strong* data. Vacuum data representing strong gravitational waves could, due to nonlinearities, lead to the creation of black holes with singularities inside. Nevertheless, ultimately one would like to have also results on the evolution of strong data.

More importantly, however, we would like to have initial data given on a *standard* spacelike Cauchy hypersurface which does not intersect $\mathcal{J}$ but "ends" at spatial infinity (cf. hypersurface $\mathcal{C}$ in Fig. 1), rather than data given on a hyperboloidal initial hypersurface. It could well happen that a spacetime evolved from data on $\mathcal{H}$ is smooth "above" $\mathcal{H}$ (in the shaded region in Fig. 1), however, it does not satisfy Penrose's requirements of asymptotic flatness "below" $\mathcal{H}$.

A remarkable progress in proving rigorously the existence of general, asymptotically flat radiative spacetimes was achieved by Christodoulou and Klainerman. Their treatise [5] contains the first really global general existence statement for full, nonlinear Einstein's vacuum equations with vanishing cosmological constant: Any smooth asymptotically flat initial data set (determined by the first and the second fundamental form on a Cauchy hypersurface) which is "near flat (Minkowski) data" leads to a unique, smooth and geodesic complete development solution of Einstein's vacuum equations. This solution is "globally asymptotically flat" in the sense that the curvature tensor decays to zero at infinity in all directions. The Christodoulou–Klainerman theorem involves a "global smallness assumption" which requires appropriate (integral) norms of curvature tensor of initial data integrated over the initial Cauchy hypersurface to be small.

The theorem demonstrates the existence of singularity-free, asymptotically flat radiative vacuum spacetimes. In hydrodynamics, even for arbitrary small initial data (decaying at infinity), an analogous theorem is not true since shocks arise. In general relativity the effect of nonlinear terms, which could have led to formation of singularities, is excluded owing, apparently, to the covariance and algebraic properties of the Einstein vacuum equations. That singularities could have had well developed is evident: the collision of arbitrarily weak, vacuum plane gravitational waves leads (due to nonlinearities) to the formation of a singularity. This, of course, does not contradict the Christodoulou–Klainerman result because initial data for plane waves are not asymptotically flat.

The work of Friedrich, Christodoulou, Klainerman and others demonstrates rigorously that the general picture of null infinity is compatible with the vacuum Einstein field equations. However, important open questions remain. In classical papers by Bondi et al. [1], Sachs [2], Newman and Penrose [3] and others, the decay of the curvature (characterized by the Weyl tensor) along outgoing null geodesics at infinity exhibits "the peeling-off" property: the fall-off of various components of the Weyl tensor is related to their Petrov algebraic type. To be more specific, certain complex linear combinations of the Weyl tensor in the orthonormal frame, $\Psi_k (k = 0, 1, 2, 3, 4)$, behave as $\Psi_k = O(r^{k-5})$ as $r \to \infty$. (In particular, $\Psi_4 \sim r^{-1}$ has the same algebraic structure as the Weyl tensor of a plane wave – the radiative field of a bounded system resembles asymptotically that of a plane wave.) This decay of the curvature can be shown to follow from a sufficient differentiability (smoothness) of the conformally rescaled (unphysical) metric $\tilde{g}$. A sufficient smoothness of null infinity is thus commonly assumed. The results of Christodoulou and Klainerman [5], however, show a weaker peeling. They were only able to prove that the asymptotically flat vacuum initial data lead to $\Psi_0 \sim r^{-\frac{7}{2}}$ (not $\sim r^{-5}$) at null infinity.

An increasing evidence against the proposal of a smooth $\mathcal{J}$ (suggested also by some results obtained by approximation methods used, for example, in the studies of the gravitational scattering of two particles) has led Chruściel et al. [6] to introduce the concept of a *polyhomogeneous* $\mathcal{J}$. The metric is called polyhomogeneous if at large r it admits an expansion in terms of $r^{-j} \log^i r$ rather than r^{-j} (as it has been assumed in the works of Bondi and others [1] – cf. Eq. (1)). The hypothesis of polyhomogeneity of $\mathcal{J}$ has been shown to be formally consistent with Einstein's vacuum equations. Under appropriate assumptions on the asymptotic form of the polyhomogeneous metric, one can demonstrate that the Bondi mass-loss law can be formulated, and the peeling-off property of the curvature holds, with the first two terms identical to the standard peeling, the third term being $\sim r^{-3} \log r$. At null infinity the conformally rescaled (unphysical) metric is not smooth.

In more recent investigations, Friedrich (see also his review [7]) constructed the new – finite but "wider" than the point i_0 – representation of spacelike infinity. This construction enables one to make much deeper analysis of the initial data in the region where null infinity touches spacelike infinity. Good chances now exist to obtain clear criteria determining which data lead to the smooth and which just to the polyhomogeneous null infinity. In [7] such criteria are proposed.

Although a substantial progress in understanding the existence of radiative solutions of vacuum field equations and the asymptotic structure of corresponding radiative spacetimes has been achieved, we have seen that open problems remain. Curiously enough, in the case of vacuum Einstein's equations with a *non-vanishing cosmological constant* a more complete picture is known for some time already. By using his regular conformal field equations, Friedrich demonstrated [8] that initial data sufficiently close to de-Sitter data develop into solutions of Einstein's equations with a positive cosmological constant, which are asymptotically simple (with a smooth conformal infinity), as required in the original framework of Penrose. Later Friedrich [9] also discussed the existence of asymptotically simple solutions to the Einstein vacuum equations with a negative cosmological constant.

The ultimate goal of rigorous work on the existence and asymptotics of solutions of the Einstein equations is *physics*: one hopes to be able to consider astrophysical sources, to relate their behaviour to the characteristics of the far fields. One would like to have under control various (both analytical and numerical) approximation procedures. A still more ambitious program is to consider strong initial data so as to be able to analyze such issues as cosmic censorship.

In the following we shall briefly discuss three classes of *explicit* radiative solutions of Einstein's equations: cylindrical waves, Robinson–Trautman and type N spacetimes, and the boost-rotation symmetric spacetimes. The latter represent the only known examples describing moving, radiating objects; except for points of null infinity where particles start and end, the null infinity is smooth. We here closely follow our recent, more detailed reviews [10,11] in which also other classes of radiative spacetimes are analyzed, such as plane waves and gravitational waves representing inhomogeneous cosmological models.

3 Cylindrical waves

Despite the fact that cylindrically symmetric waves cannot describe exactly the radiation from bounded sources, they even recently played an important role in clarifying a number of complicated issues, such as testing the quasilocal mass-energy, testing codes in numerical relativity, investigation of the cosmic censorship, and quantum gravity.

With Ashtekar and Schmidt [12,13], we considered gravitational waves with a space-translation Killing field ("generalized Einstein–Rosen waves"). In the $(2 + 1)$-dimensional framework, the Einstein–Rosen subclass forms a simple instructive example of explicitly given spacetimes which admit a smooth global null (and timelike) infinity even for strong initial data.

The 4-dimensional vacuum gravity which admits a spacelike hypersurface Killing vector $\partial/\partial z$, the norm of which is $\exp(2\psi)$, is equivalent to 3-dimensional gravity coupled to the scalar field ψ. In 3 dimensions, there is no gravitational radiation. Hence, the local degrees of freedom are all contained in the scalar field. One therefore expects that Cauchy data for the scalar field will suffice to determine the solution. For data which fall off appropriately, we thus expect the 3-dimensional Lorentzian geometry to be asymptotically flat in the sense of Penrose, i.e. that there should exist a 2-dimensional boundary representing null infinity. In general cases, this is analyzed in [12].

Restricting ourselves to the Einstein–Rosen waves by assuming that there is a further spacelike, hypersurface orthogonal Killing vector $\partial/\partial\varphi$ which commutes with $\partial/\partial z$, we find the 3-metric given by

$$d\sigma^2 = g_{ab}dx^a dx^b = e^{2\gamma}(-dt^2 + d\rho^2) + \rho^2 d\varphi^2. \tag{2}$$

The field equations become

$$-\ddot{\psi} + \psi'' + \rho^{-1}\psi' = 0, \quad \gamma' = \rho(\dot{\psi}^2 + \psi'^2), \quad \dot{\gamma} = 2\rho\dot{\psi}\psi'. \tag{3}$$

Thus, we can first solve the axisymmetric wave equation for ψ on Minkowski space and then solve for γ – the only unknown metric coefficient – by quadratures.

By analyzing the asymptotic behavior of the solutions we find [12,13] that *cylindrical waves in (2+1)-dimensions give an explicit model of the Bondi-Penrose radiation theory which admits smooth null and timelike infinity for arbitrarily strong initial data.* There is no other such model available. The general results on the existence of $\mathcal{J}$ in 4 dimensions assume weak data.

4 Robinson–Trautman and type N twisting solutions

These vacuum spacetimes have attracted increased attention in the last decade – most notably in the work by Chruściel, and Chruściel and Singleton [14]. In these studies the Robinson–Trautman spacetimes have been shown to exist globally for all positive "times", and to converge asymptotically to a Schwarzschild metric. Interestingly, the extension of these spacetimes across the "Schwarzschild-like" event horizon can only be made with a finite degree of smoothness. These studies are based on the derivation and analysis of an asymptotic expansion describing the long-time behaviour of the solutions of the nonlinear parabolic Robinson–Trautman equation.

In our work [15,16] we analyzed Robinson–Trautman spacetimes with a positive cosmological constant Λ. The results proving the global existence and convergence of the solutions of the Robinson–Trautman equation can be taken over from the previous studies since Λ does not explicitly enter this equation. We have shown that, starting with arbitrary, smooth initial data at $u = u_0$, these cosmological Robinson–Trautman solutions converge exponentially fast to a Schwarzschild-de Sitter solution at large retarded times ($u \to \infty$). The interior of a Schwarzschild-de Sitter black hole can be joined to an "external" cosmological Robinson–Trautman spacetime across the horizon $\mathcal{H}^+$ with a higher degree of smoothness than in the corresponding case with $\Lambda = 0$. In particular, in the extreme case with $9\Lambda m^2 = 1$, in which the black hole and cosmological horizons coincide, the Robinson–Trautman spacetimes can be extended smoothly through $\mathcal{H}^+$ to the extreme Schwarzschild–de Sitter spacetime with the same values of Λ and m. However, such an extension is not analytic (and not unique).

We have also demonstrated that the cosmological Robinson–Trautman solutions represent explicit models exhibiting the cosmic no-hair conjecture. As far as we are aware, these models represent the only exact analytic demonstration of the cosmic no-hair conjecture under the presence of gravitational waves. They also appear to be the only exact examples of black hole formation in nonspherical spacetimes which are not asymptotically flat.

4.1 Type N twisting spacetimes

Since diverging, non-twisting Robinson–Trautman spacetimes of type N have singularities, there has been hope that if one admits a nonvanishing twist a more realistic radiative spacetime may exist.

Stephani [17], however, indicated, by constructing a general solution of the linearized equations, that singularities at infinity probably exist. Later, Finley et al. [18] found an approximative twisting type N solution up to the third order of iteration on the basis of which they suggested that it seems that the twisting, type N fields can describe a radiation field outside bounded sources. However, employing the Newman–Penrose formalism and MAPLE one succeeds in discovering a nonvanishing quartic invariant in the 2nd derivatives of the Riemann tensor [19], which shows that solutions of both Stephani and Finley et al. contain singularities at large r. Mac Alevey [20] argued that an approximate solution at any finite order can be calculated without occurrence of singularities. It is very likely, however, that a corresponding exact solution must contain singularities since Mason [21] proved that the only vacuum algebraically special spacetime that is asymptotically simple is the Minkowski space.

Even if an explicit radiative solution with complete smooth null infinity may be out of reach, it is of interest to construct radiative solutions which at least admit a global null infinity in the sense that its smooth cross sections exist although this null infinity is not necessarily complete. The only explicit examples of such solutions are spacetimes with boost-rotation symmetry.

5   Boost-rotation symmetric radiative spacetimes

I reviewed these spacetimes representing "uniformly accelerated objects" in various places (see e.g. [10,22]); here I shall just mention some new results. The "physical picture" is given in Fig. 2 and its caption; the Penrose diagram is schematically illustrated in Fig. 3.

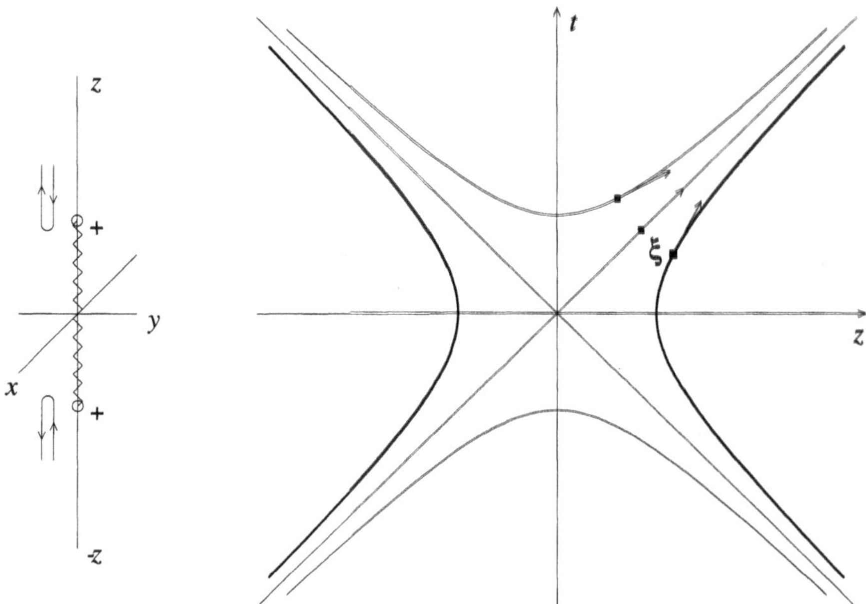

Fig. 2. Two particles uniformly accelerated in opposite directions. The nodal (conical) singularity ("a spring") between the particles gives them the accelerations. The nodal singularities may occur outside particles (two "strings" extending from each particle to infinity), the axis between particles being regular. There are also solutions in which particles are "self-accelerating" (due to the presence of a negative mass in their "inner structure"), and the axis is regular everywhere outside the particles

The unique role of the boost-rotation symmetric spacetimes is exhibited by a theorem [23] which roughly states that in axially symmetric, locally asymptotically flat electrovacuum spacetimes (in the sense that a null infinity satisfying Penrose's requirements exists, but it need not necessarily exist globally), the only additional symmetry that does not exclude radiation is the *boost* symmetry.

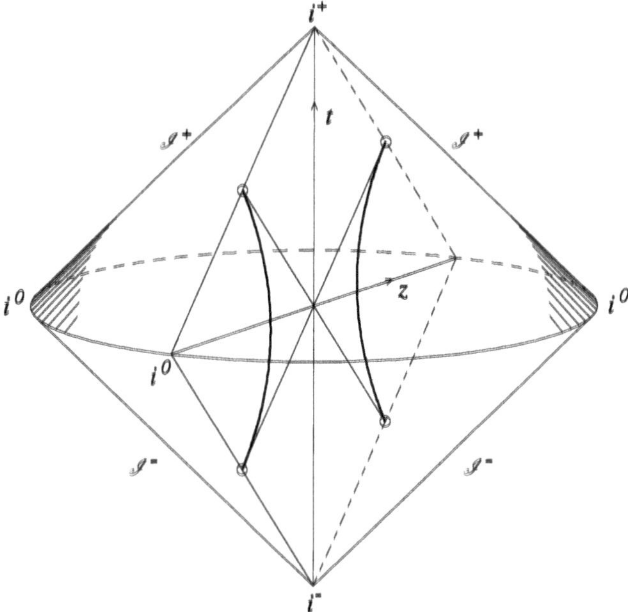

Fig. 3. The schematic illustration of the Penrose compactified diagram of a boost-rotation symmetric spacetime. Null infinity can admit smooth sections, with timelike and spacelike infinities being also smooth

To prove such result we start from the metric

$$ds^2 = \left(r^{-1} V e^{2\beta} - r^2 e^{2\gamma} U^2 \cosh 2\delta - r^2 e^{-2\gamma} W^2 \cosh 2\delta \right.$$

$$\left. - 2r^2 U W \sinh 2\delta \right) du^2$$

$$+ 2e^{2\beta} du dr + 2r^2 \left(e^{2\gamma} U \cosh 2\delta + W \sinh 2\delta \right) du\, d\theta$$

$$+ 2r^2 \left(e^{-2\gamma} W \cosh 2\delta + U \sinh 2\delta \right) \sin\theta\, du\, d\phi$$

$$- r^2 \left[\cosh 2\delta \left(e^{2\gamma} d\theta^2 + e^{-2\gamma} \sin^2\theta\, d\phi^2 \right) + 2\sinh 2\delta \sin\theta\, d\theta\, d\phi \right],$$

where all the functions describing metric and electromagnetic field tensor $F_{\mu\nu}$ are independent of ϕ. Assuming asymptotic expansions of these functions at large r with u, θ, ϕ fixed to guarantee asymptotic flatness, and using the outgoing radiation condition and the field equations, one finds these expansions to have specific forms. For example,

$$\gamma = \frac{c}{r} + (C - \tfrac{1}{6}c^3 - \tfrac{3}{2}cd^2)\frac{1}{r^3} + \cdots, \quad V = r - 2M + \cdots,$$

$$F_{02} = X + (\epsilon_{,\theta} - e_{,u})\frac{1}{r} + \cdots, \quad F_{03} = Y - \frac{f_{,u}}{r} + \cdots, \tag{4}$$

where the 'coefficients' $c, d, \cdots$ are functions of u and θ. The expansions are needed to further orders – see [23] for their forms. Let us only recall that the decrease of the Bondi mass, $m(u) = \frac{1}{2} \int_0^\pi M(u, \theta) \sin \theta d\theta$, is given by

$$m_{,u} = -\frac{1}{2} \int_0^\pi (c_{,u}^2 + d_{,u}^2 + X^2 + Y^2) \sin \theta d\theta \leq 0, \tag{5}$$

where $c_{,u}, d_{,u}, X, Y$ are the gravitational and electromagnetic news functions.

Now one writes down the Killing equations and solves them asymptotically in r^{-1}. One arrives at the following theorem [23]: Suppose that an axially symmetric electrovacuum spacetime admits a "piece" of $\mathcal{J}^+$ in the sense that the Bondi–Sachs coordinates can be introduced in which the metric takes the form (5), with the asymptotic form of the metric and electromagnetic field given by (4). If this spacetime admits an additional Killing vector forming with the axial Killing vector a two-dimensional Lie algebra then the additional Killing vector has asymptotically the form

$$\eta^\alpha = \left[-ku \cos \theta + \alpha(\theta), \ kr \cos \theta + \mathcal{O}(r^0), \ -k \sin \theta + \mathcal{O}(r^{-1}), \ \mathcal{O}(r^{-1}) \right], \tag{6}$$

where k is a constant. For $k = 0$ it generates asymptotically translations (the function α has then a specific form). For $k \neq 0$ it is the boost Killing field.

The case of translations is analyzed in detail in [24]. Theorem 1, precisely formulated and proved there, states that if the asymptotically translational Killing vector is spacelike, then null infinity is singular at some $\theta \neq 0, \pi$; if it is null, null infinity is singular at $\theta = 0$ or π. The first case corresponds to cylindrical waves, the second case to a plane wave propagating along the symmetry axis. We refer to [24] for the case when there is also a cosmic string present along the symmetry axis. The case of timelike Killing vector is described by Theorem 2 (proved also in [24]): "If an axisymmetric electrovacuum spacetime with a non-vanishing Bondi mass admits an asymptotically translational Killing vector and a complete cross section of $\mathcal{J}^+$, then the translational Killing vector is timelike and spacetime is thus stationary."

The case of the boost Killing vector ($k \neq 0$) is thoroughly analyzed in [23]. The general functional forms of the news functions (both gravitational and electromagnetic), and of the mass aspect and total Bondi mass of boost-rotation symmetric spacetimes are there given. Recently these results were obtained [25] by using the Newman–Penrose formalism and under more general assumptions (for example, $\mathcal{J}$ could in principle be polyhomogeneous).

The general structure of the boost-rotation symmetric spacetimes with hypersurface orthogonal Killing vector was analyzed in detail in [26]. Their radiative properties, including explicit construction of radiation patterns and of Bondi mass for the specific boost-rotation symmetric solutions were investigated in several works – we refer to the reviews [10,22] and [27] for details. In these reviews the role of the boost-rotation symmetric spacetimes in such diverse fields like numerical relativity and quantum production of black-hole pairs is also noticed and a number of references are given.

Here I would like to mention yet a recent progress in understanding specific boost-rotation symmetric spacetimes with Killing vectors which are *not* hypersurface orthogonal. This is the *spinning C*-metric (see e.g. [28]). It was discovered by Plebański and Demiański [29] as a generalization of the standard C-metric which is known to represent uniformly accelerated non-rotating black holes. In [30] we first transformed the metric into Weyl coordinates, and then found a transformation which brings it into the canonical form of the radiative spacetimes with the boost-rotation symmetry:

$$
\begin{aligned}
ds^2 = {}& e^\lambda d\rho^2 + \rho^2 e^{-\mu} d\phi^2 \\
& + (z^2 - t^2)^{-1}\left[(e^\lambda z^2 - e^\mu t^2)dz^2 - 2zt(e^\lambda - e^\mu)dz\,dt \right. \\
& \left. + (e^\lambda t^2 - e^\mu z^2)dt^2\right] - 2\mathcal{A}e^\mu(zdt - tdz)d\phi - \mathcal{A}^2 e^\mu(z^2 - t^2)d\phi^2,
\end{aligned}
$$

$$(7)$$

where the functions e^μ, e^λ and $\mathcal{A}$ are given in terms of (t, ρ, z) in a somewhat complicated but explicit manner. This metric can represent two uniformly accelerated, spinning black holes, either connected by a conical singularity, or with conical singularities extending from each of them to infinity. The behaviour of the curvature invariants clearly indicates the presence of a non-vanishing radiation field (see Fig. 5 in [30]). The spinning C-metric is the only explicitly known example with two Killing vectors which are not hypersurface orthogonal, in which one can give arbitrarily strong initial data on a hyperboloid "above the roof" ($t > |z|$) which evolve into the radiative spacetime with smooth $\mathcal{J}^+$.

Acknowledgement. I am grateful to the organizers of the SIGRAV 2000 congress, in particular to Prof. L. Lusanna, for the invitation to Genoa and hospitality. I thank Tomáš Ledvinka and Martin Žofka for their help with the manuscript. Support from the grants GACR 202/99/0261 and J13/98:113200004 is also acknowledged.

References

1. Bondi, H., van der Burg, M.G.J., Metzner, A.W.K. (1962): Gravitational waves in general relativity. VII. Waves from Axi-symmetric isolated systems. Proc. Roy. Soc. Lond. A **269**, 21

2. Sachs, R.K. (1962): Gravitational waves in general relativity VIII. Waves in asymptotically flat space-time. Proc. Roy. Soc. Lond. A **270**, 103

3. Newman, E.T., Penrose, R. (1962): An approach to gravitational radiation by a method of spin coefficients. J. Math. Phys. **3**, 566

4. Penrose, R. (1963): Asymptotic properties of fields and space-times. Phys. Rev. Lett. **10**, 66

5. Christodoulou, D., Klainerman, S. (1994): The nonlinear stability of the Minkowski space-time. Princeton University Press, Princeton

6. Chruściel, P., MacCallum, M.A.H., Singleton, P.B. (1995): Gravitational waves in general relativity XIV. Bondi expansions and the "polyhomogeneity" of $\mathcal{J}$. Phil. Trans. Roy. Soc. Lond. **A350**, 113

7. Friedrich, H. (1998): Einstein's equation and geometric asymptotics, in *Gravitation and relativity: At the turn of the millenium*, ed. by N. Dadhich, J. Narlikar, Proceedings of the GR-15 conference, Inter-University Centre for Astronomy and Astrophysics Press, Pune

8. Friedrich, H. (1986): On the existence of n-geodesically complete or future complete solutions of Einstein's field equations with smooth asymptotic structure. Commun. Math. Phys. **107**, 587

9. Friedrich, H. (1995): Einstein equations and conformal structure: Existence of anti-de Sitter-type space-times. J. Geom. Phys. **17**, 125

10. Bičák, J. (2000): Selected solutions of Einstein's field equations: their role in general relativity and astrophysics, in *Einstein's Field Equations and Their Physical Meaning*, ed. by B.G. Schmidt, Springer Verlag Berlin Heidelberg New York

11. Bičák, J. (2000): Exact ratiative spacetimes: Some recent developments. Annalen Phys. **9**, 207–216

12. Ashtekar, A., Bičák, J., Schmidt, B.G. (1997): Asymptotic structure of symmetry-reduced general relativity. Phys. Rev. D **55**, 669

13. Ashtekar, A., Bičák, J., Schmidt, B.G. (1997): Behaviour of Einstein–Rosen waves at null infinity. Phys. Rev. D **55**, 687

14. Chruściel, P.T. (1992): On the global structure of Robinson–Trautman space-times. Proc. Roy. Soc. Lond. A **436**, 299;
Chruściel, P.T., Singleton, D.B. (1992): Non-smoothness of event horizons of Robinson–Trautman black holes. Commun. Math. Phys. **147**, 137

15. Bičák, J., Podolský, J. (1997): The global structure of Robinson–Trautman radiative space-times with cosmological constant. Phys. Rev. D **55**, 1985

16. Bičák, J., Podolský, J. (1995): Cosmic no-hair conjecture and black-hole formation: An exact model with gravitational radiation. Phys. Rev. D **52**, 887

17. Stephani, H. (1993): A note on the solutions of the diverging twisting, type N, vacuum field equations. Class. Quantum Grav. **10**, 2187

18. Finley, J.D., Plebański, J.F., Przanowski, M. (1997): An iterative approach to twisting and diverging, type N vacuum Einstein equations: A (third order) resolution of Stephani's paradox. Class. Quant. Grav. **14**, 489

19. Bičák, J., Pravda, V. (1998): Curvature invariants in type N spacetimes. Class. Quantum Grav. **15**, 1539

20. MacAlevey, P. (1999): Approximate solutions of Einstein's vacuum field equations in the type N, twisting and diverging case. Class. Quantum Grav. **16**, 2259

21. Mason, L. (1998): The asymptotic structure of algebraically special spacetimes. Class. Quantum Grav. **15**, 1019

22. Bičák, J. (1997): Radiative spacetimes: Exact approaches, in *Relativistic Gravitation and Gravitational Radiation* Proceedings of the Les Houches School of Physics, ed. by J.-A. Marck, J.-P. Lasota, Cambridge University Press, Cambridge

23. Bičák, J., Pravdová, A. (1998): Symmetries of asymptotically flat electrovacuum space-times and radiation. J. Math. Phys. **39**, 6011

24. Bičák, J., Pravdová, A. (1999): Axisymmetric electrovacuum spacetimes with a translational Killing vector at null infinity. Class. Quantum Grav. **16**, 2023

25. Valiente-Kroon, J.A. (2000): On Killing vector fields and Newman–Penrose constants. J. Math. Phys. **41**, 898

26. Bičák, J., Schmidt, B.G. (1989): Asymptotically flat radiative space-times with boost-rotation symmetry: The general structure. Phys. Rev. D **40**, 1827

27. Pravda, V., Pravdová, A. (2000): Boost-rotation symmetric spacetimes – review. Czech. J. Physics **50**, 333
28. Kramer, D., Stephani, H., Herlt, E., MacCallum, M.A.H. (1980): Exact solutions of Einstein's field equations. Cambridge University Press, Cambridge
29. Plebański, J., Demiański, M. (1976): Rotating, charged and uniformly accelerating mass in general relativity. Ann. Phys. (N.Y.) **98**, 98
30. Bičák, J., Pravda, V. (1999): Spinning C-metric: Radiative spacetime with accelerating, rotating black holes. Phys. Rev. D **60**, 044004

Instantons and Scattering

G. Bonelli, L. Bonora, F. Nesti, A. Tomasiello, S. Terna

Abstract. This is a review of some recent developments in the study of classical solutions of Yang–Mills theories in various dimensions and their significance in the path integral of the corresponding theories. These particular solutions are called instantons because of their kinship with ordinary instantons. Just as ordinary instantons interpolate between different vacua, the new instantons interpolate between different asymptotic states. Therefore they represent scattering phenomena. Here we review the two dimensional and four dimensional Yang-Mills case.

1 Introduction

The rich structure of non-Abelian gauge theories is at the core of their successful employment as theories of the elementary particles. It is this complicated structure that allows us to claim to be able, at least potentially, to describe in a consistent way by the same theory both confinement and asymptotic freedom, quark, gluons, mesons and hadrons. Classical and quantum non-Abelian gauge theories have been analyzed in countless papers in the past, but they do not cease to surprise us by revealing from time to time new structure. This has happened also recently in connection with the so-called second string revolution. Gauge theories have been naturally associated with branes and this association has revealed new, previously unsuspected features. This review is devoted to one such new development: the connection between Yang–Mills theories, strings and branes via classical configurations called new or interacting instantons.

The (up to now) clearest example is based on 2D SYM theories with $\mathcal{N} = (8, 8)$ supersymmetry through the emergence in the latter of classical solutions modeled over Riemann surfaces, which naturally lend themselves to a string interpretation. We refer to them as *interacting* or *Riemannian instantons*. They expose a string-like nature of Yang–Mills theories, which could actually have been found long ago, but in fact was brought to light only after the proposal of Matrix Theory [1]. The latter, in the large N limit, is expected to describe M theory. Therefore upon compactifying it on a circle one should end up, in the appropriate limit, with type IIA superstring theory [2–5] . Now, upon compactifying Matrix Theory on a circle we obtain $\mathcal{N} = (8, 8)$ super-Yang–Mills (SYM) on a cylindrical 2D space-time with gauge group $U(N)$. Therefore the conjecture, supported with various arguments [2,3,5], is that this theory represents in the strong Yang–Mills coupling limit a theory of type II superstrings (see also [5,6]). Hereafter we refer to this theory as Matrix String Theory (MST).

The role of Riemannian instantons, [7–9], in this correspondence between MST at strong coupling and type IIA superstring theory, is fundamental. One can show that any Riemann instanton represents a type IIA closed string interaction, the string interaction based on the corresponding Riemannian surface. Evidence of this result was collected in a series of papers, [10–12]. One can even establish a precise correspondence between moduli space of Riemann instantons and moduli space of closed superstring theory.

A quite similar analysis can be carried out for the 4D SYM theory with $\mathcal{N} = 4$ supersymmetry and gauge group $U(N)$. We find interacting instantons as well, but their interpretation is different than in the 2d case. They can only represent scattering of D3 branes. A preliminary study of this case has been carried out in [13].

The paper is organized as follows. In the next section a general characterization of interacting instantons is given. In section 3 the 2d case (Riemann instantons) is described in some detail, since it is the most analyzed one up to now and also the most pedagogical. Section 4 contains a summery of interacting instantons in 4d.

2   Characterization of interacting instantons

An interacting instanton in dimension $d < 4$ *satisfies a set of differential equations which are obtained as a dimensional reduction of the (anti)self-duality equations in 4D*. Let us consider, for definiteness, a Euclidean Yang–Mills theory in 4D with gauge group $U(N)$. Let A be the gauge connection form with curvature F. We use the hermitean convention for the Lie algebra-valued matrices so that the covariant derivative is $D = d + igA$, where g is the Yang–Mills coupling. The self-duality condition for the Yang–Mills field strength is

$$F_{\mu\nu} = \frac{1}{2}\epsilon_{\mu\nu\lambda\rho}F^{\lambda\rho}. \tag{1}$$

Dimensionally reducing these equations to one dimensions, we obtain the so-called Nahm equations. Reducing to three dimensions one gets the monopole equations. These are all well-known and well studied equations. Let us concentrate in particular on the self-duality equations reduced to 2D. Rewritten in complex coordinates $w = x^1 + ix^2, \bar{w} = x^1 - ix^2, y = x^3 + ix^4, \bar{y} = x^3 - ix^4$, eqs.(1) become

$$F_{w\bar{w}} = F_{y\bar{y}}, \qquad F_{w\bar{y}} = 0 = F_{\bar{w}y}. \tag{2}$$

We want to single out solutions of these equations which are independent of $y, \bar{y}$. Then, introducing the notation $X = A_y, \overline{X} = A_{\bar{y}} = X^\dagger$, (2) becomes

$$\begin{aligned}
&F_{w\bar{w}} - ig^2[X, \bar{X}] = 0, \\
&D_w \bar{X} = 0, \qquad D_{\bar{w}} X = 0.
\end{aligned} \tag{3}$$

We refer to these equations as *Riemannian instanton equations* or *Hitchin equations*, [14].

Interacting instanton equations for $d \geq 4$ can be obtained from dimensional reduction of (anti)selfduality equations in 8D. The latter are written down explicitly in [13]. One reduces this system to 4D by keeping the dependence on $x^1, \ldots, x^4$ and dropping the dependence on the remaining coordinates. Let us introduce the complex coordinates $v = \frac{1}{2}(x^1 + ix^2)$, $w = \frac{1}{2}(x^3 + ix^4)$, set $A_7 = A_8 = 0$ and call $X = A_5 - iA_6$. Then the system becomes:

$$F_{v\bar{v}} + F_{w\bar{w}} + i[X, \bar{X}] = 0,$$
$$F_{vw} = 0, \quad F_{\bar{v}\bar{w}} = 0, \quad D_{\bar{v}}X = 0 = D_v\bar{X}, \quad D_{\bar{w}}X = 0 = D_w\bar{X}. \tag{4}$$

These are the interacting instanton equations we will analyse later on.

Analogous equations can be obtained for $d > 4$, but no analysis of such reduced equations has been carried out so far.

Up to now the derivation of instanton equations has been formal (dimensional reduction), but it shows how close is the connection between ordinary instantons (solutions of (anti)self-duality equations) and interacting instantons, which are solutions of dimensionally reduced (anti)self-duality equations.

Much more important is the next characterization of interacting instantons: *they are BPS solutions of some SYM theory.* In other words, they preserve a fraction of supersymmetry in the SYM theory. We will see this in some examples.

Finally a much more substantial characterization is the following one. Among the solutions of the above mentioned dimensionally reduced instanton equations, we will call *interacting instantons those that describe scattering of physical objects*; this is possible because the geometry underlying such instantons represent the merging of and the splitting into distinct geometrical objects, which can be interpreted as the carriers of physical objects such as strings and branes.

3   A pedagogical example

Let us consider a Riemannian instanton solution of Eqs. (3) in the simplest nontrivial case (gauge group $U(2)$). To conform with the convention of MST, which will be used in the following sections, we suppose that the coordinate x^1, x^2 span an infinite cylinder (precisely x^2 is the periodic coordinate).

We look for a couple (A, X) that satisfies Eqs. (3). To this end we choose the following ansatz

$$X = Y^{-1}MY, \qquad A_w = i\partial_w Y^\dagger (Y^{-1})^\dagger, \tag{5}$$

where Y is a suitable matrix $\in SL(2, \mathbb{C})$, and M is the following 2×2 matrix

$$M = \begin{pmatrix} 0 & a \\ 1 & 0 \end{pmatrix}, \tag{6}$$

where a is a function of the point in the w cylinder. As a consequence of the equation $D_{\bar{w}}X = 0$, it follows that $\partial_{\bar{w}}a = 0$, i.e. a is holomorphic in w (except perhaps at infinity on the cylinder). Now, given such a holomorphic a we want to find Y so that (3) is satisfied. We parametrize Y as follows:

$$Y = \begin{pmatrix} e^p & 0 \\ 0 & e^{-p} \end{pmatrix} = KL, \quad L = \begin{pmatrix} e^{\frac{u}{2}} & 0 \\ 0 & e^{-\frac{u}{2}} \end{pmatrix}, \quad K = \begin{pmatrix} |a|^{\frac{1}{4}} & 0 \\ 0 & |a|^{-\frac{1}{4}} \end{pmatrix}, \tag{7}$$

where u is a function to be determined and $p = \frac{u}{2} + \frac{1}{4}ln|a|$. Then using (5) we find

$$X = \begin{pmatrix} 0 & ae^{-2p} \\ e^{2p} & 0 \end{pmatrix}, \quad A_w = i\partial_w p \begin{pmatrix} 1 & 0 \\ 0 & -1 \end{pmatrix}. \tag{8}$$

Now it is easy to verify that the first equation in (3) implies

$$2\partial_w \partial_{\bar{w}} p - g^2\left(e^{4p} - |a|^2 e^{-4p}\right) = 0. \tag{9}$$

Inserting the explicit form of p and the change of variable $w \to \zeta$, s.t. $\frac{\partial \zeta}{\partial w} = \sqrt{a}$, one can rewrite (9) as

$$\partial_\zeta \partial_{\bar{\zeta}} u - 2g^2 \sinh u = -\frac{\pi}{4}\delta(a)(\partial_\zeta a)(\partial_{\bar{\zeta}} \bar{a}), \tag{10}$$

where the derivatives are understood in the sense of complex distribution theory. This is the sinh-Gordon equation with delta-function-like boundary conditions at the points where a vanishes. If u is a smooth solution of this equation, the couple (X, A) is a solution of (3) which is smooth everywhere except perhaps at infinity on the cylinder. As we will see later such solutions exist, therefore solutions of the instanton equations do exist. However before we go to more details on the existence of solutions, it is important to discuss their meaning.

There are two distinct ingredients in (5): one is the matrix M (the *core*) and the other is the *group theoretical factor* Y. We discuss them in turn.

3.1 Branched coverings

The matrix M represents a branched covering of the cylinder spanned by the coordinate w. In order to see this we diagonalize it by means of a matrix in $SL(2, \mathbb{C})$:

$$M = S\hat{M}S^{-1}, \quad \hat{M} = \begin{pmatrix} \sqrt{a} & 0 \\ 0 & -\sqrt{a} \end{pmatrix}, \quad S = \frac{i}{\sqrt{2}}\begin{pmatrix} a^{\frac{1}{4}} & a^{\frac{1}{4}} \\ a^{-\frac{1}{4}} & -a^{-\frac{1}{4}} \end{pmatrix}. \tag{11}$$

It is convenient now to pass to a new coordinate $z = e^w$, which maps the cylinder into the complex z–plane with two punctures at $z = 0$ and $z = \infty$. The two eigenvalues of M (or, equivalently, of X), which are the two roots of the algebraic equation $X^2 = a$, can be thought of as the sheets of a double covering of the cylinder. Each sheet is a copy of the complex z–plane. Each point on each sheet projects to the corresponding point on the z–plane. Such projection will be denoted π. Suppose, for simplicity, that

$a = z - z_0$. Then we have a branch point at $z = z_0$ and another at $z = \infty$. We can draw a cut between these two branch points (for definiteness the cut will lie in the region $|z| > |z_0|$). The two sheets are connected through the cut. What we mean by this, as is well known, is that if we consider going around the origin in the z–plane along a small circle of radius $< |z_0|$, we produce an inverse image (under π) on the covering formed by two small circles around the origin, one for each sheet. But if we do the same operation for a circle with radius $> |z_0|$, we are bound to cross the cut: crossing the cut multiplies the root by a phase $e^{i\pi}$, so that $\sqrt{a}$ goes over to $-\sqrt{a}$ and viceversa, i.e. by crossing the cut we pass from one sheet to the other. This means that the counterimage of the circle on the covering this time is a long circle that extends over both sheets.

This result can be interpreted in two ways: a geometrical and a string theoretical one. Geometrically what we have just described is a Riemann surface represented by a branched covering of the complex z plane. The Riemann surface in question has genus 0 and three punctures (see Fig. 1).

The string interpretation is rather obvious. The coordinate x_1 is taken to be the Euclidean time. Therefore $z = 0$ corresponds to time $-\infty$ and $z = \infty$ to time $+\infty$. It is natural to interpret the counterimages of the small circle around the origin as two incoming strings. This configuration does not change if we enlarge the circle around $z = 0$ as long as its radius remains smaller than $|z_0|$. We say that the two strings propagate in time without interacting. If the radius of the circle in the z plane becomes larger than $|z_0|$, the counterimage becomes a unique closed curve extending over both sheets, and this configuration propagates unchanged as far as ∞. Therefore

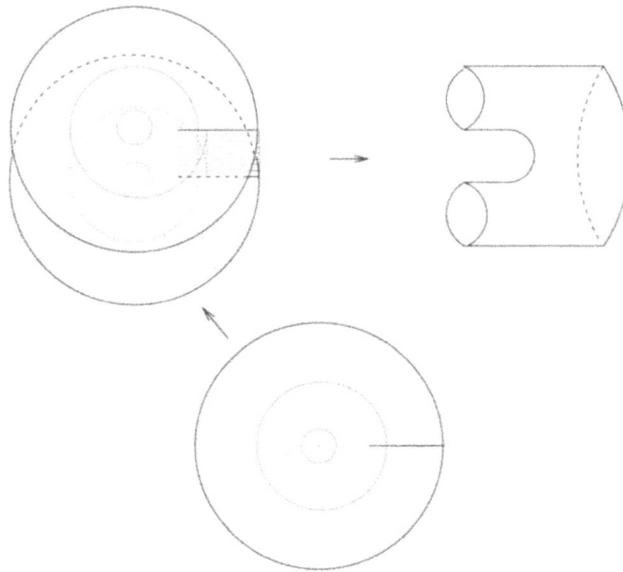

Fig. 1. Simple branch

the instanton we are considering represents the joining of two strings to form one long string of double length. The joining interaction takes place at $z = z_0$. The inverse images under π of $z = 0$ and $z = \infty$ correspond to the points of the Riemann surface where the incoming strings enter and the outgoing string exits, respectively.

If there are more branch points besides the one considered above, it is not difficult to see how they may give rise to more complicated Riemann surfaces with handles (and more complicated string interactions). We will return later on to the problem of constructing more complicated Riemann surfaces and justifying the string interpretation outlined above. For the time being we would like to complete the description of our simple example. We notice that the passage through the cut can be described mathematically as the monodromy transformation

$$\hat{M} \to \Lambda \hat{M} \Lambda^{-1} = \begin{pmatrix} -\sqrt{a} & 0 \\ 0 & \sqrt{a} \end{pmatrix}, \qquad \Lambda = \begin{pmatrix} 0 & 1 \\ 1 & 0 \end{pmatrix}, \tag{12}$$

where M is monodromy invariant since simultaneously $S \to S\Lambda^{-1}$.

Another useful remark is the following: X can be rewritten as follows

$$X = Y^{-1}MY = L^{-1}U\hat{M}U^{-1}L, \tag{13}$$

where

$$U = \frac{i}{\sqrt{2}} \begin{pmatrix} \left(\frac{a}{\bar{a}}\right)^{\frac{1}{8}} & \left(\frac{a}{\bar{a}}\right)^{\frac{1}{8}} \\ \left(\frac{\bar{a}}{a}\right)^{\frac{1}{8}} & -\left(\frac{\bar{a}}{a}\right)^{\frac{1}{8}} \end{pmatrix}. \tag{14}$$

It is a remarkable (and intentionally looked for) fact that U is a unitary matrix. U has the same monodromy as S: $U \to U\Lambda^{-1}$.

3.2 The group theoretical factor

Let us now pass to the analysis of the group theoretical factor $Y = KL$. We have just seen that K concurs with the diagonalizing matrix S to form a unitary matrix U. We notice however that K diverges at $z = z_0$. Let us analyze next the *dressing factor* L. This amounts to returning to the solutions of Eq. (10). The problem is to find a solution u that satisfies the sinh-Gordon equation with the boundary condition

$$u \sim -\frac{1}{2} \ln |a|, \qquad a \sim 0. \tag{15}$$

It is thanks to this logarithmic singularity that the factors $e^{\pm\frac{u}{2}}$ and K, contained in Y, compensate for the the singularities of each other, so that the resulting solution (X, A) is smooth. Let us suppose again, for simplicity, that $a = z - z_0$. We want to single out, for this simple case, solutions of (10) with the right asymptotic behavior, that is behaving like (15) at $z = z_0$ and vanishing at $z = 0, \infty$. First we rescale $\zeta \to \sqrt{2}g\zeta$, so that the sinh–Gordon equation takes the standard form

$$\partial_\zeta \partial_{\bar{\zeta}} u - \sinh u = 0. \tag{16}$$

We do not know an exact solution of this equation satisfying the boundary condition (15) and vanishing at the origin and at infinity of the z–plane. We can therefore proceed in two ways: find a numerical solution or an approximate analytical one. The latter was discussed in [10]. The conclusion there was that the solution u, in the strong coupling limit, shrinks around the branch point $z = z_0$, that is it is practically zero everywhere except in a small neighborhood of this point where it has a spike-like behaviour.

We can easily extend the previous analysis to the case in which a contains several distinct zeroes. In the following we will simply assume that this is always the case, namely that in the strong coupling limit the dressing factor L tends very rapidly to 1 outside the branch points of the covering surface.

From the example discussed in this section we learn how to proceed in order to find the general solution. A few things should be retained because they turn out to hold in general: *a Riemannian instanton* is made of two ingredients, the *core* constituted by a matrix M which defines a branched covering representing a Riemann surface with punctures, and a group theoretical factor. The latter splits into two factors: a dressing factor which tends to 1 in the strong coupling limit and the remnant, which concurs to form a unitary matrix when applied to M.

Our next step is to see the astonishing consequences of this fact at work in MST.

4 Matrix string theory

The MST is by definition $U(N)$ SYM in a 1+1 dimensions with $\mathcal{N} = (8, 8)$ supersymmetry. The world–sheet coordinates are σ and τ, where σ is periodic; they define an infinite cylinder. We are interested here in the Euclidean version of this theory. To this end we make a Wick rotation and introduce the complex coordinates

$$w = \frac{1}{2}(\tau + i\sigma), \quad \bar{w} = \frac{1}{2}(\tau - i\sigma), \quad A_w = A_0 - iA_1, \quad A_{\bar{w}} = A_0 + iA_1.$$

The action becomes

$$S = \frac{1}{\pi} \int_C d^2w \, \mathrm{Tr} \left(D_w X^i D_{\bar{w}} X^i - \frac{1}{4g^2} F^2_{w\bar{w}} - \frac{g^2}{2}[X^i, X^j]^2 \right.$$
$$\left. + i(\theta_s D_{\bar{w}}\theta_s + \theta_c D_w\theta_c) + ig\theta^T \Gamma_i[X^i, \theta] \right),$$

(17)

where C is the infinite Euclidean cylinder spanned by w.

In (17), g is the gauge coupling, X^i, with $i = 1, \dots, 8$, are hermitean $N \times N$ matrices and $D_w X^i = \partial_w X^i + i[A_w, X^i]$. $F_{w\bar{w}}$ is the curvature of $A_w, A_{\bar{w}}$. θ represents $2 N \times N$ matrices whose entries are simultaneously 2D and $SO(8)$ spinors. It can be written as $\theta^T = (\theta_s, \theta_c)$, where $\pm$ denotes the 2D chirality and θ_s, θ_c are spinors in the $\mathbf{8_s}$ and $\mathbf{8_c}$ representations of $SO(8)$, while ^{T} represents the 2D transposition. The matrices Γ_i are the 16×16 $SO(8)$ gamma matrices.

For definiteness we will write the matrices Γ_i in the form

$$\Gamma_i = \begin{pmatrix} 0 & \gamma_i \\ \tilde{\gamma}_i & 0 \end{pmatrix},$$

and γ_i, $\tilde{\gamma}_i$ are the same as in Appendix 5B of [15].

The action (17) is invariant under the supersymmetric transformations

$$\begin{aligned}
\delta X^i &= \frac{i}{g}(\epsilon_s \gamma^i \theta_c + \epsilon_c \tilde{\gamma}^i \theta_s), \\
\delta\theta_s &= (-\frac{i}{2g^2} F_{w\bar{w}} + \frac{1}{2}[X^i, X^j]\gamma_{ij})\epsilon_s - \frac{1}{g} D_w X^i \gamma_i \epsilon_c, \\
\delta\theta_c &= (\frac{i}{2g^2} F_{w\bar{w}} + \frac{1}{2}[X^i, X^j]\tilde{\gamma}_{ij})\epsilon_c - \frac{1}{g} D_{\bar{w}} X^i \tilde{\gamma}_i \epsilon_s, \\
\delta A_w &= -2\epsilon_s \theta_s, \qquad \delta A_{\bar{w}} = -2\epsilon_c \theta_c,
\end{aligned} \qquad (18)$$

where

$$\gamma_{ij} = \frac{1}{2}(\gamma_i \tilde{\gamma}_j - \gamma_j \tilde{\gamma}_i), \quad \gamma_{ij} = \frac{1}{2}(\tilde{\gamma}_i \gamma_j - \tilde{\gamma}_j \gamma_i).$$

A string interpretation, which is what we want to arrive at, is more natural after the coordinate transformation, already considered above, $w \to z = e^w$, i.e. after passing from the cylinder to the complex plane with the origin deleted, i.e. $\mathbb{C}^*$, or the Riemann sphere $\mathbb{C}P^1$ with two punctures.

The MST in the strong coupling limit admits a string interpretation in the following sense. Let us rescale $A \to \frac{1}{g} A$ and imagine that the fields represent small oscillations about a trivial background. The naive strong coupling limit ($g \to \infty$) in the action tells us that all fields commute, therefore they can be simultaneously diagonalized. The formal aspect of the resulting theory is that of the Green–Schwarz superstring theory in the light-cone approach; however there is a significant variant. The fields are the eigenvalues of X_i and θ. Therefore for each field of the Green–Schwarz theory we have here N fields. We will see later on a completely satisfactory interpretation of this fact. For the time being let us notice that the theory becomes a free theory of the diagonal degrees of freedom. The gauge freedom can be completely fixed up to a residual gauge invariance which takes the form of the Weyl group, i.e. the permutation group of N elements. We are therefore allowed to fix the boundary conditions for the diagonal degrees of freedom up to this residual gauge transformation. On this basis these degrees of freedom lend themselves naturally to an interpretation as free strings of various lengths. The naive strong coupling limit of (17) is therefore a free theory of closed superstring of various lengths. As we will see, there is much more than that: SYM theory contains any information needed to construct the full string perturbation theory. For the time being we note that, if the limit we have described is the true strong coupling limit, we can interpret it as the weak coupling limit of (type II) string theory, i.e. if g_s is the string coupling, $g_s \sim g^{-1}$ [2,5].

5   Riemannian instantons and their construction

The string interpretation of the previous subsection will be confirmed by an analysis of the theory in the background of its instantons. Before going into that, let us describe such classical solutions. As a first step we will show that, as anticipated above, they are BPS solutions, i.e. they preserve a fraction of the original supersymmetry. In fact, in MST, we will look for classical Euclidean supersymmetric configurations that preserve half supersymmetry. To this end we set $\theta = 0$ and look for solutions of the equations $\delta\theta^{\pm} = 0$, i.e. from (18),

$$
\left(\frac{i}{2g^2} F_{w\bar{w}} + \frac{1}{2}\left[X^i, X^j \right] \tilde{\gamma}_{ij} \right) \epsilon_c = 0, \quad D_w X^i \gamma_i \epsilon_c = 0,
$$
$$
\left(-\frac{i}{2g^2} F_{w\bar{w}} + \frac{1}{2}\left[X^i, X^j \right] \gamma_{ij} \right) \epsilon_s = 0, \quad D_{\bar{w}} X^i \tilde{\gamma}_i \epsilon_s = 0.
$$
$$\tag{19}$$

Solutions of these equations that preserve half supersymmetry are the following ones. Set $X^i = 0$ for all i except two, for definiteness $X^i \neq 0$ for $i = 1, 2$; note that γ_{12} is an antisymmetric 8×8 matrix, and $\gamma_{12}^2 = -1$ and therefore its eigenvalues are $\pm i$ (moreover $\tilde{\gamma}_{12} = \gamma_{12}$). It is easy to show that there exists ϵ^+ and ϵ^-, each with four independent components, such that

$$
\gamma_{12}\epsilon^{\pm} = \pm i\epsilon^{\pm}, \quad \gamma_1\epsilon^+ = -i\gamma_2\epsilon^+, \quad \tilde{\gamma}_1\epsilon^- = i\tilde{\gamma}_2\epsilon^-.
$$

Now it is convenient to introduce the complex notation $X = X^1 - iX^2$, $\bar{X} = X^1 + iX^2 = X^\dagger$. Then the conditions to be satisfied in order to preserve half supersymmetry are

$$
F_{w\bar{w}} - ig^2[X, \bar{X}] = 0, \tag{20}
$$
$$
D_{\bar{w}} X = 0, \quad D_w \bar{X} = 0. \tag{21}
$$

These are the same equations as (3), but they have been obtained in the context of MST. The solutions of these equations preserve half supersymmetry.

It is easy to verify that, if non-trivial solutions to such equations exist, they satisfy the equations of motion of the action (17). The instanton action is the action with $\theta = 0$, $X^i = 0$ for $i = 3, \ldots 8$

$$
S_{\text{inst}} = \frac{1}{2\pi} \int d^2w \, \text{Tr}\left(- X D_w D_{\bar{w}} \bar{X} - \bar{X} D_w D_{\bar{w}} X \right.
$$
$$
\left. - \frac{1}{2g^2} F_{w\bar{w}}^2 + \frac{g^2}{2}[X, \bar{X}]^2 \right). \tag{22}
$$

It is elementary to prove that S_{inst} vanishes in correspondence with smooth solutions of (20), (21).

5.1 General ansatz for Riemannian instantons

In order to find solutions (A, X) of (20), (21), we follow and generalize the example $N = 2$ presented in section 3. We start from the analogous ansatz

$$A_w = i\,dw\,Y^\dagger(Y^{-1})^\dagger, \qquad X = Y^{-1}MY, \tag{23}$$

where Y is a generic element in the complex group $SL(N, \mathbb{C})$ and M specifies a branched covering of the cylinder. As a consequence of (23) the equation $D_{\bar{w}}X = 0$ is equivalent to $\partial_{\bar{w}}M = 0$ or

$$\partial_{\bar{z}}M = 0. \tag{24}$$

This parametrization is defined therefore in terms of two factors Y and M. As above, Y will be referred to as the *group theoretical factor*, while M defines a general branched covering of the cylinder. These two factors will be discussed later. For the time being let us give some essential information. Let us consider the polynomial

$$P_X(y) = \det(y - X) = y^N + \sum_{i=0}^{N-1} y^i a_i,$$

where y is a complex indeterminate. The equation

$$P_X(y) = 0, \tag{25}$$

can also be written as the matrix equation

$$X^N + a_{N-1}X^{N-1} + \cdots + a_0 = 0. \tag{26}$$

A diagonalizable matrix, which is solution of Eq. (26), can always be cast in the canonical form

$$M = \begin{pmatrix} -a_{N-1} & -a_{N-2} & \cdots & \cdots & -a_0 \\ 1 & 0 & \cdots & \cdots & 0 \\ 0 & 1 & 0 & \cdots & 0 \\ \cdots & \cdots & \cdots & \cdots & \cdots \\ 0 & 0 & \cdots & 1 & 0 \end{pmatrix}. \tag{27}$$

The case considered in Sect. 3 is evidently a particular case of the above formulas.

Due to (24), we have $\partial_{\bar{z}}a_i = 0$, which means that the set of functions $\{a_i\}$ are analytic in the complex plane, although they are allowed to have poles at $z = 0$ and $z = \infty$. In fact, Eq. (25) identifies in the (z, y) space a Riemann surface Σ, which is an N-sheeted branched covering of the cylinder. Generalizing what has been said in Sect. 3, the explicit form of the covering is given by the set $\{x^{(1)}(z), \ldots, x^{(N)}(z)\}$ of eigenvalues of X. Each eigenvalue spans a sheet. The projection map to the base cylinder C will be denoted $\pi : \Sigma \to C$. The points where two or more eigenvalues coincide are called branch points. The identification cuts in the sheets start or end at these points (which may include 0 and ∞). We stress that the covering is independent of the coupling g.

The Riemann surface Σ will be characterized by the genus h and by a certain number n of punctures. Each puncture will come associated with an integer, the length of the string entering or exiting at that puncture. Moreover each surface is characterized by certain numbers, the moduli. All this information is stored in the a_i analytic functions. It has been worked out in [11] and will be summarized later.

5.2 Explicit construction of instanton solutions

Each solution of (20), (21) consists of two parts: a branched covering of the cylinder via the relative X characteristic polynomial and a group theoretical factor. The aim of the present subsection is to extend the construction of Sect. 3 to the general case.

Let us deal first with the group theoretical factor. In our ansatz (23) the group theoretical factor Y takes values in the complex group $SL(N, \mathbb{C})$, while the matrix M determines the branched covering. The dependence on the Yang–Mills coupling constant g is contained in the Y factor, while M does not depend on g. In Sect. 3 we have shown an example in which $Y = KL$ where L, *the dressing factor*, tends to 1 in the strong coupling limit outside the string interaction points, while K is a special matrix, independent of g, endowed with the property that $K^{-1}MK$ and $K^{\dagger}M^{\dagger}(K^{\dagger})^{-1}$ are simultaneously diagonalizable.

We proceed in a parallel way also in the general case. It is well-known, [7], that the matrix M can be diagonalized

$$M = S\hat{M}S^{-1}, \quad \hat{M} = \text{Diag}(\lambda_1, \dots, \lambda_N), \tag{28}$$

by means of a suitable matrix $S \in SL(N, \mathbb{C})$:

We notice first that the role of a in section 3 is played now by Δ, the discriminant of M, which vanishes whenever two eigenvalues coincide. Two coinciding eigenvalues define a branch point of the covering. Going around a branch point in the complex z–plane, produces a reshuffling of the eigenvalues that can be represented via a monodromy matrix Λ: $\hat{M} \to \Lambda\hat{M}\Lambda^{-1}$. Correspondingly, we have $S \to S\Lambda^{-1}$, so that the single-valuedness of M is preserved.

The construction of K and L in the general case is subtle and technically rather complicated. It has been given in [11]. On the basis of the example of Sect. 3, it is not difficult to understand the general features of these matrices.

First one introduces a monodromy–invariant K such that $K^{-1}S = U$ be unitary. As it turns out, K may have singularities at the points of C where any two eigenvalues of M coincide, i.e. at the branch points of the spectral covering. Therefore $K^{-1}MK$ is in general singular at these points. That is why we must introduce into the game a new monodromy invariant matrix L, with the purpose of canceling the singularities of $K^{-1}MK$ in such a way that $L^{-1}K^{-1}MKL$ be smooth and satisfy (20), (21). In the general case neither K nor L are diagonal. The entries of L can be taken to be generalizations of the u field of Sect. 3. Let us denote again by u any such field. For

(20), (21) to be satisfied these fields u must satisfy, [10], an equation of the WZNW type with the following general structure

$$\partial_w \partial_{\bar{w}} u + \cdots \sim \partial_w \partial_{\bar{w}} \ln |\Delta| = \pi \frac{\partial \Delta}{\partial w} \frac{\partial \bar{\Delta}}{\partial \bar{w}} \delta(\Delta), \tag{29}$$

where dots represent all the other terms, which are irrelevant in the cancellation of singularities. Let us refer to these equations as the "dressing equations". On the right-hand side we see the typical delta-function-type source which characterizes them. The sources are point-like and located at the zeroes of Δ, that is at the branch points of the covering.

By construction K is independent of g while L does depend on g. One can show that in fact $L \to 1$ as $g \to \infty$.

5.3 Branched coverings and plane curves

It is time to turn to a more careful description of the second ingredient of Riemannian instantons, i.e. to branched coverings of the cylinder (or the Riemann sphere with two punctures). To this end let us return to the definition in Sect. 3.5, in particular Eq. (25) or (26). Branched coverings make their appearance in MST as solutions of affine equations

$$P(y, z) \equiv \sum_{p,q} a_{p,q} y^q z^p = 0, \quad (y, z) \in \mathbb{C}^2, \tag{30}$$

where P is a polynomial of degree N. Actually, from Eq. (25) it follows that P_X has degree N in y, but the $a_i(z)$'s could be any analytic functions on the punctured Riemann sphere. In order to preserve the string interpretation we will limit ourselves to $a_i(z)$'s which are polynomials in z in such a way that $P(y, z)$ has overall degree N.

The locus in $\mathbb{C}^2$ of the solutions (y, z) of (30) is known as a *plane curve*. The study of plane curves has been carried out in [11]. Any plane curve is characterized by its topological type (h, n), where h is the genus and n is the number of punctures of the curve. Punctures are the sites on the embedded Riemann surface, that is on the corresponding plane curve, where the incoming strings enter and the outgoing strings exit. They are the counterimages by π of $z = 0$ and $z = \infty$, respectively. If any such point on the plane curve is a branch point of multiplicity $l - 1$, then the corresponding incoming or outgoing string has length l. On the other hand, in MST the length of an incoming or outgoing string is interpreted as the $+$ component of the momentum in the light-cone framework. Therefore the multiplicities of the branch points in the inverse image of $z = 0, \infty$ have a precise physical meaning.

Next, the independent non-vanishing coefficients $a_{p,q}$ can be varied without changing, in general, the topological type of the curve. They are the *moduli* of the plane curve. Counting them is necessary in order to see whether the moduli space of MST coincides with the moduli space of IIA superstring theory, or more realistically to what extent $\mathcal{M}_N^{(h,n)}$ approximates $\mathcal{M}^{(h,n)}$, where the former represents the moduli space of Riemannian instantons in MST and the second denotes the moduli space of Riemann surfaces that appear in string theory. Now $\mathcal{M}^{(h,n)}$ has complex dimension $3h - 3 + n$. Our analysis in [11] has shown that $\mathcal{M}_N^{(h,n)}$ has a stratified structure, where each continuous layer has complex dimensions $2h - 3 + n$ and the number of layers is parametrized by h discrete parameters. It is likely that when $N \to \infty$ the two moduli spaces tend to become identical.

Let us summarize the results described in this section. We have seen that *utilizing a very general recipe, an instanton solution can be constructed in correspondence with the most general branched covering of the base cylinder, i.e. to any Riemann surface with punctures, which can be represented as a branched covering of the cylinder, we can associate an instanton.* The coefficients of the algebraic equations that define such branched coverings are (up to symmetry identification) to be identified with moduli. *The moduli space of Riemannian instantons is only an approximated version of the moduli space of Riemann surfaces which appear in string interaction theory. In particular h among the former are a discrete version of h among the latter. It is however reasonable to assume that in the large N limit the two spaces tend to coincide.*

6   MST on a Riemannian instanton background

Let us go back to the path integral of MST. We would like to interpret an amplitude on a given instanton background as an amplitude for a string process characterized by the Riemann surface which underlies the instanton. It is in fact possible to show that MST in the strong coupling limit in the background of a given Riemannian instanton solution reduces to the Green–Schwarz superstring theory plus a decoupled Maxwell theory, and compute string interaction amplitudes in such background. Since the latter interpolates between an initial and a final string configuration via a punctured Riemann surface Σ (which represents a branched covering of the base cylinder), the amplitudes can be interpreted in a string theoretic way as the transition amplitudes between two such configurations. We will show below that their leading term is proportional to $g_s^{-\chi}$, where $\chi = 2 - 2h - n$ is the Euler characteristic of the Riemann surface of genus h with n punctures, which characterizes the given classical solution. This is the result one expects from perturbative string interaction theory and proves beyond any doubt the relation between MST at strong coupling and string interaction theory.

Let us sketch the relevant proof. To start with we split every field Φ in the action (17) into a classical instanton background part and a quantum fluctuating part

$$\Phi = \Phi^{(b)} + \phi \tag{31}$$

and expand the action. The instanton action vanishes, as we have seen above, and the piece of action linear in ϕ vanishes due to the equations of motion. We want to evaluate what remains in the strong coupling limit, by considering an expansion of the action in $1/g$.

As a first step let us analyze the background part. To this end let us recall a few facts from the previous section. The dependence on the coupling is entirely contained in the factor L. We have seen that in the strong coupling limit $L \rightarrow 1$ outside the branch points of the covering. Since here we are interested in expanding the action (17) in inverse powers of $1/g$, and actually in singling out the dominant term in this expansion, we will consider the action (17) around a given classical solution stripped of the above dressing factor, and exclude from the integration region the branch points on the cylinder. Said otherwise, we introduce in our integrated action a regulator (which will eventually be removed).

After getting rid of the dressing factor, the classical background configuration is specified by $X = K^{-1}MK$ and $A_{\bar{w}} = -iK^{-1}\partial_{\bar{w}}K$. In the strong coupling limit $X \rightarrow U\hat{M}U^{-1}$, where $U = K^{-1}S$ is a unitary matrix and therefore simultaneously diagonalizes X and $\bar{X}$. Corresponding to $\hat{X} = \hat{M}$ we have $\hat{A}_w$ which, as was shown above, vanishes everywhere, even at the branch points. What has happened is that the unitary transformation has swallowed entirely the connection, including the singularities. Now, with a gauge transformation, we can remove the unitary factor U. This leads us to the diagonal representation: $\hat{X}$ diagonal and $\hat{A} = 0$ for the classical background in the strong coupling limit.

Now let us insert this information into the action (17). To extract the strong coupling effective theory, we first rewrite the action in the following useful form

$$S = \frac{1}{\pi} \int d^2w \ \mathrm{Tr} \left(D_w X^I D_{\bar{w}} X^I - \frac{g^2}{2}[X^I, X^J]^2 - g^2[X^I, X][X^I, \overline{X}] \right.$$

$$- \frac{1}{2}\bar{X}D_w D_{\bar{w}}X - \frac{1}{2}X D_{\bar{w}} D_w \bar{X} - \frac{1}{4g^2}\left(F_{w\bar{w}} - ig^2[X, \overline{X}] \right)^2$$

$$\left. + i(\theta_s D_{\bar{w}}\theta_s + \theta_c D_w\theta_c) + ig\theta^T \Gamma_i[X^i, \theta] \right),$$

where $I = 3, 4, \ldots, 8$. Then we expand it around a generic instanton configuration as in (31), but we further split the quantum field ϕ

$$\Phi = \Phi^{(b)} + \phi^{\mathrm{t}} + \phi^{\mathrm{n}} \equiv \Phi^{(b)} + \phi \equiv \Phi^{\circ} + \phi^{\mathrm{n}}, \tag{32}$$

where $\Phi^{(b)}$ is the background value of the field at infinite coupling, ϕ^{t} are the fluctuations along the Cartan directions and ϕ^{n} are the fluctuations along the complementary directions in the Lie algebra $\mathfrak{u}(N)$.

Next, we fix the gauge as follows:

$$\mathcal{G}_{w\bar{w}} = D^{\circ}_w a_{\bar{w}} + D^{\circ}_{\bar{w}} a_w + ig^2([X^{\circ}, \bar{x}] + [\bar{X}^{\circ}, x]) + 2ig^2[X^{\circ I}, x^I] = 0, \tag{33}$$

where D° is the covariant derivative with respect to A°. Next, we introduce the appropriate Faddeev–Popov ghost and antighost fields c and $\bar{c}$ and expand them like all the other fields. Then we add to the action the gauge fixing term and Faddeev–Popov term.

At this point, to single out the strong coupling limit of the action, we rescale the fields in appropriate manner. Precisely, we redefine our fields as follows

$$A_w = g a_w^\mathrm{t} + a_w^\mathrm{n}, \qquad X = \hat{X} + x^\mathrm{t} + \frac{1}{g} x^\mathrm{n},$$

$$X^I = x^{It} + \frac{1}{g} x^{In}, \qquad \theta^\mathrm{n} = \theta^\mathrm{t} + \frac{1}{\sqrt{g}} \theta^\mathrm{n},$$

and likewise for the conjugate variables. For the ghosts we set

$$c = g c^\mathrm{t} + \sqrt{g} c^\mathrm{n}, \qquad \bar{c} = g \bar{c}^\mathrm{t} + \frac{1}{\sqrt{g}} \bar{c}^\mathrm{n}.$$

Finally the action becomes

$$S = S_{sc} + S_\mathrm{n} + O\left(\frac{1}{\sqrt{g}}\right),$$

where

$$S_{sc} = \frac{1}{\pi} \int_{C_0} d^2w \ \mathrm{Tr}\left[\partial_w x^{It} \partial_{\bar{w}} x^{It} + \partial_w x^\mathrm{t} \partial_{\bar{w}} \bar{x}^\mathrm{t} + i(\theta_s^\mathrm{t} \partial_{\bar{w}} \theta_s^\mathrm{t} + \theta_c^\mathrm{t} \partial_w \theta_c^\mathrm{t})\right. \tag{34}$$
$$\left. + \partial_w a_{\bar{w}}^\mathrm{t} \partial_{\bar{w}} a_w^\mathrm{t} + \partial_w \bar{c}^\mathrm{t} \partial_{\bar{w}} c^\mathrm{t}\right].$$

S_n is the purely quadratic term in the ϕ^n fluctuations. The latter can be easily integrated over and, since they do not involve zero modes, give exactly 1. Let us see this in detail. S_n has the form

$$S_\mathrm{n} = \frac{1}{\pi} \int d^2w$$

$$\mathrm{Tr}\left[\bar{x}^\mathrm{n} \mathcal{Q} x^\mathrm{n} + x^{In} \mathcal{Q} x^{In} + a_{\bar{w}}^\mathrm{n} \mathcal{Q} a_w^\mathrm{n} + \bar{c}^\mathrm{n} \mathcal{Q} c^\mathrm{n} + i(\theta_s^\mathrm{n}, \theta_c^\mathrm{n}) \mathcal{A} \begin{pmatrix} \theta_s^\mathrm{n} \\ \theta_c^\mathrm{n} \end{pmatrix}\right], \tag{35}$$

where

$$\mathcal{Q} = \mathrm{ad}_{\bar{X}^\circ} \cdot \mathrm{ad}_{X^\circ} + \mathrm{ad}_{a_{\bar{w}}^\mathrm{t}} \cdot \mathrm{ad}_{a_w^\mathrm{t}} + \mathrm{ad}_{x^{It}} \cdot \mathrm{ad}_{x^{It}}$$

and

$$\mathcal{A} = \begin{pmatrix} i\,\mathrm{ad}_{a_{\bar{w}}^\mathrm{t}} & \gamma_i\,\mathrm{ad}_{X^{\circ i}} \\ \tilde{\gamma}_i\,\mathrm{ad}_{X^{\circ i}} & i\,\mathrm{ad}_{a_w^\mathrm{t}} \end{pmatrix}.$$

In the path integral we can now integrate over the non-Cartan modes. The net result of integrating over the non-Cartan modes is 1. This is the result expected from supersymmetry in the absence of zero modes.

In conclusion, in the strong coupling limit we are left with the quadratic action (34) over the Cartan modes.

Now, since in (34) all the matrices involved are diagonal we can rewrite this action in terms of the diagonal modes $\phi^t = \phi_{(1)}, \ldots, \phi_{(N)}$

$$S_{sc} = \frac{1}{\pi} \int d^2w \sum_{n=1}^{N} \left[\partial_w x^i_{(n)} \partial_{\bar{w}} x^i_{(n)} + i(\theta_{s(n)} \partial_{\bar{w}} \theta_{s(n)} + \theta_{c(n)} \partial_w \theta_{c(n)}) \right.$$
$$\left. + \partial_w a_{\bar{w}(n)} \partial_{\bar{w}} a_{w(n)} + \partial_w \bar{c}_{(n)} \partial_{\bar{w}} c_{(n)} \right].$$

This is a theory of free fields on the base cylinder, excluding the punctures, and it is tempting to extend the action to $\mathcal{C}$ by just forgetting the punctures on the cylinder corresponding to the branch points. However this is not correct. The fields x^i are not single–valued on the cylinder. For example, upon going around a simple branch point, at least one x^i is mapped to another one, and this is precisely the way a joining or a splitting of two strings is represented in this formalism, [8,9]. The point is that the fields x^i (as well as the others) are not well defined on each sheet, but all together they form a well defined field on the covering surface. In fact let us look at the realization of local fields on a Riemann surface represented as a branched covering. If Σ is a branched covering of the cylinder $\mathcal{C}$, then, as we have seen, there is a projection map $\pi : \Sigma \to \mathcal{C}$ whose inverse image is N-valued. In our language this is simply

$$\pi^{-1} : w \to \left(x_{(1)}(w), \ldots, x_{(N)}(w) \right). \tag{36}$$

Suppose a local complex field $\tilde{\psi}$ is given on Σ; applying the above construction $\tilde{\psi}$ can be represented as an N-tuple $\left(\psi_{(1)}(w), \ldots, \psi_{(N)}(w) \right)$ representing the field on each copy of the cylinder $\mathcal{C}$ that composes the covering Σ; the $\psi_{(i)}(w)$'s are related by the appropriate monodromy properties along the cuts.

Going now back to the action (6), we see that we have to interpret any set of N fields $(\phi_{(1)}, \ldots, \phi_{(N)})$ in it as a unique field $\tilde{\phi}$ on the covering Σ. $\tilde{\phi}$ is locally a function of a coordinate z in Σ. From the point of view of Σ, the w coordinate is locally defined via an abelian differential $\omega = dw$ with imaginary periods, which is canonical, i.e. is fixed only by the complex structure of the surface.

Finally we can write the strong coupling action (6) as follows

$$S^{\Sigma}_{sc} = S^{\Sigma}_{GS} + S^{\Sigma}_{\text{Maxwell}}, \tag{37}$$

$$S^{\Sigma}_{GS} = \frac{1}{\pi} \int_{\Sigma} d^2z \left(\partial_z \tilde{x}^i \partial_{\bar{z}} \tilde{x}^i + i(\tilde{\theta}_s \partial_{\bar{z}} \tilde{\theta}_s + \tilde{\theta}_c \partial_z \tilde{\theta}_c) \right), \tag{38}$$

$$S^{\Sigma}_{\text{Maxwell}} = \frac{1}{\pi} \int_{\Sigma} d^2z \left(g^{z\bar{z}} \partial_z \tilde{a}_{\bar{z}} \partial_{\bar{z}} \tilde{a}_z + \partial_z \tilde{\bar{c}} \partial_{\bar{z}} \tilde{c} \right). \tag{39}$$

In (38), a $\sqrt{\omega_z}$ (resp. $\sqrt{\omega_{\bar{z}}}$) factor has been absorbed in $\tilde{\theta}_s$ (resp. $\tilde{\theta}_c$) which is a $(\frac{1}{2}, 0)$ (resp. $(0, \frac{1}{2})$) differential on Σ and the metric in the Maxwell term is $g_{z\bar{z}} = \omega_z \omega_{\bar{z}}$.

Summarizing, what we obtained above is that the strong coupling effective theory is given by the Green–Schwarz superstring action on the branched covering worldsheet plus a decoupled Maxwell theory on the same surface.

Let us concentrate now on the role of the Maxwell action (39) in the path integral. Our aim is to compute correlators that involve only $\tilde{x}$ and $\tilde{\theta}$ modes, therefore what we really need is the amplitude for the Maxwell and ghost modes on the background of a given instanton. As we have already pointed out several times, this amplitude has a string interpretation as the amplitude for the transition from the initial to the final string configuration described by the instanton. For this interpretation to be consistent, this amplitude, to the leading order, should be proportional to $g_s^{-\chi}$ where χ is the Euler characteristic of the Riemann surface Σ, i.e. the covering surface introduced above.

A necessary step in order to evaluate one such amplitude is to integrate over the Maxwell Cartan modes in the functional integral with action (38). Since the action is free, the integration produces a ratio of determinants, which turns out to be a constant. However, we have to take into account the zero modes for the fields that have been rescaled (the unrescaled zero modes are irrelevant in this argument). The rescaled fields in $\mathcal{C}$ are just the Maxwell and the ghost fields. The corresponding fields in Σ will be rescaled too

$$\tilde{a}_z \to g\,\tilde{a}_z, \quad \tilde{a}_{\bar{z}} \to g\,\tilde{a}_{\bar{z}}, \quad \tilde{c} \to g\,\tilde{c}, \quad \tilde{\bar{c}} \to g\,\tilde{\bar{c}}. \tag{40}$$

The counting of zero modes is rather standard, [10]. There is one zero mode for the ghost field and $2h + n - 1$ zero modes for the gauge field, with opposite sign. Therefore the overall number of zero modes is $2h + n - 2$. Finally, the factor in front of the vacuum to vacuum amplitude will be $g^{-2h-n+2} = g_s^{2h+n-2}$. The exponent of g is precisely the Euler characteristic of Σ, as we wanted to prove.

On this basis one can calculate string interaction amplitudes directly from MST. Of course real string amplitudes must contain vertex insertions, i.e. should be correlators of the vertex operators corresponding to the various in – and out – (super)strings. In this regard we simply remark that such vertex operators are constructed in terms of the string fields $\tilde{x}^i$ and $\tilde{\theta}$. These can be regarded as eigenvalues of the matrices X^i and θ. Therefore it is legitimate to construct asymptotic states and vertex operators out of them, and calculate correlators of such vertex operators in MST even at finite Yang–Mills coupling g. Of course they will recover their meaning of string correlators only when $g \to \infty$. s So let us introduce the vertex operators $V_1, \dots, V_n$ corresponding to n incoming and outgoing strings, expressed in terms of $\tilde{x}, \tilde{\theta}$, and of the string transverse momenta, and insert them into the path integral. The genus h amplitude (in the strong coupling limit) can be written symbolically:

$$\langle V_1, \dots, V_n \rangle_h = g_s^{-\chi} \int_{\mathcal{M}_N^{(h,n)}} dm \int \mathcal{D}[\tilde{x}, \tilde{\theta}, \tilde{a}, \tilde{c}] V_1 \dots V_n \, e^{-S_{GS} - S_{\text{Maxwell}}}.$$

We have singled out the integration over $\mathcal{M}_N^{(h,n)}$, namely over all distinct instantons which underlie the given string process for fixed N, that is to say with assigned incoming and outgoing strings and string interactions. We have already seen above that $\mathcal{M}_N^{(h,n)}$, is for finite N, an approximation of the actual string moduli space $\mathcal{M}^{(h,n)}$. The approximation is the better the larger N is.

6.1 Summary

We have just seen that by expanding the MST action about a Riemannian instanton one gets, in the strong coupling limit, the Green–Schwarz action plus the free Maxwell action over the Riemann surface supporting the instanton. If this Riemann surface has genus h and n punctures, the path integral is proportional to a factor $g^{-2h-n+2} = g_s^{2h+n-2}$. This is the correct factor one expects from string interaction theory for a string process mediated by such a Riemann surface. One can actually compute all the type IIA string amplitudes in the framework of MST. Our conclusion is therefore that the strong coupling limit of MST is the type IIA superstring theory.

Similar results have been seen to hold, also for the Heterotic Matrix String Theory (HMST), [16], i.e. for $\mathcal{N} = (8, 0)$ SYM in 2D with gauge group $O(N)$: in the strong Yang–Mills coupling limit one finds the heterotic superstring theory with suitably broken gauge group. It is very likely that also the type IA theory in 10d can be approximated in a similar way by an appropriate MST orbifold.

7 New instantons in 4D

In this section we sketch the discussion of interacting instantons in 4d. For a more complete treatment see [13,17]. The Minkowski action of $\mathcal{N} = 4$ SYM theory in 4D is

$$S = \int_{\mathcal{X}} d^4x \, \mathrm{Tr} \left(-\frac{1}{4g^2} F_{\mu\nu} F^{\mu\nu} - \frac{1}{2} D_\mu X^i D^\mu X_i + \frac{g^2}{4} \left[X^i, X^j \right]^2 \right.$$
$$\left. + \frac{i}{2} \bar{\lambda} \gamma^\mu D_\mu \lambda + \frac{g}{4} \left(\lambda^T C \gamma^{i\dagger} \left[X^i, \lambda \right] - \lambda^\dagger C \gamma^i \left[X^i, \lambda^* \right] \right) \right),$$

where $i = 1, \dots, 6$. $\mathcal{X}$ is a four dimensional manifold of the type $\mathcal{X} = \mathbb{R} \times M_3$, where M_3 is a three–dimensional compact manifold and $\mathbb{R}$ is the line $-\infty < x^0 < \infty$. Although the action (7) can be studied on more general manifolds, we will consider in the following essentially the example of $M_3 = \mathbb{T}^3$, the 3–torus defined by periodic x^1, x^2, x^3.

$F^{\mu\nu}$ is the field strength of the gauge field A_μ, the X^i are $N \times N$ hermitean matrices in the adjoint of $U(N)$. λ is an $N \times N$ matrix whose entries are both Weyl spinors of $SO(1, 3)$ and vectors in the fundamental of $SU(4)$: namely the γ^μ's will act on the $SO(1, 3)$ spinorial indices, while the γ^i's on the $SU(4)$ ones.

The action (7) is invariant under the supersymmetric transformations

$$\delta X^i = \frac{i}{g} \left(\epsilon^T C \gamma^{i\dagger} \lambda - \epsilon^\dagger C \gamma^i \lambda^* \right),$$
$$\delta A_\mu = -i \left(\bar{\epsilon} \gamma^\mu \lambda - \bar{\lambda} \gamma^\mu \epsilon \right),$$
$$\delta \lambda = -\frac{1}{g^2} F_{\mu\nu} \gamma^{\mu\nu} \epsilon - i \left[X_i, X_j \right] \gamma^{ij} \epsilon + \frac{2}{g} D_\mu X_i \gamma^\mu \gamma^0 C \gamma^i \epsilon^*.$$

After going to the Euclidean and introducing the complex coordinates $v = \frac{1}{2}(x^1 + ix^2)$, $w = \frac{1}{2}(x^3 + ix^4)$, we look for solutions that preserve $\frac{1}{4}$ supersymmetry. To this end we set all fermions and all X^i, with $i = 3, \ldots, 6$, to zero, and define $X = X^1 + iX^2$ and $\bar{X} = X^\dagger$. The equations that define such solutions are

$$F_{v\bar{v}} + F_{w\bar{w}} - ig^2[X, \bar{X}] = 0, \tag{41}$$

$$F_{vw} = 0, \quad F_{\bar{v}\bar{w}} = 0, \tag{42}$$

$$D_{\bar{v}}X = 0 = D_v\bar{X}, \quad D_w\bar{X} = 0 = D_{\bar{w}}X. \tag{43}$$

We will refer to the solutions of these equations as *interacting instantons*. Analogous equations for interacting anti-instantons can be obtained by an anti-holomorphic involution.

These equations can be obtained also from dimensional reduction of 8D self-duality or anti-self-duality equations. The solutions of (41–43) are constructed in a completely parallel way with respect to 2D. One introduces the ansatz

$$A_v = i\partial_v Y^\dagger (Y^{-1})^\dagger, \quad A_w = i\partial_w Y^\dagger (Y^{-1})^\dagger, \quad X = Y^{-1}MY, \tag{44}$$

where Y is a generic element in the complex group $SL(N, \mathbb{C})$ and M specifies a branched covering of the base manifold. A more general ansatz has been used in [13], but will not be discussed here. As a consequence of (44) the equations $D_{\bar{v}}X = 0 = D_v\bar{X}$ are equivalent to

$$\partial_{\bar{v}}M = 0 = \partial_{\bar{w}}M, \tag{45}$$

which means that the matrix M is holomorphic in v, w. Eq. (45) guarantees that Eqs. (43) are satisfied.

In the above ansatz Y is again referred to as the *group theoretical factor*. We split it into $Y = KL$, where the entries of L satisfy equations of the type (29) above, so that one can conclude that $L \to 1$ in the strong coupling limit, except at the locus where the discriminant of M vanishes. As for M we proceed as in 2D. We consider the polynomial equation

$$P_X(y) = \det(y - X) = y^N + \sum_{i=0}^{N-1} y^i a_i = 0,$$

where y is a complex indeterminate. A diagonalizable matrix, which is solution of Eq. (7), can always be cast in the canonical form (27), with the obvious change that due to (45), we have $\partial_{\bar{v}}a_i = 0 = \partial_{\bar{w}}a_i$. Therefore $\{a_i\}$ are holomorphic in v, w, although they are allowed to have poles at $z = 0$ and $z = \infty$.

All this means that Eq. (7) identifies in the (y, z, w) space a complex 2-manifold (a surface) Σ, which is an N-sheeted branched covering of the base manifold. The explicit form of the covering is given by the set $\{x^{(1)}(z, w), \ldots, x^{(N)}(z, w)\}$ of eigenvalues of X. Each eigenvalue spans a sheet. The projection map to the base cylinder $\mathcal{X}$ will be denoted $\pi : \Sigma \to \mathcal{X}$. The divisor (complex 1-submanifold) where two eigenvalues coincide is the branch locus.

Now, in parallel to what happens in MST, such coverings may describe scatterings of 3D objects (D3-branes). The idea is simple. The ic-instantons present in our theory describe four-manifolds of various topologies, which cover various base spaces; the latter are topologically of the form $\mathbb{R} \times M_3$, and so we may define slices of the covering at constant time. In general our interacting instantons at $t = -\infty$ are represented by a disjoint union of 3-manifolds of various topologies and at $t = +\infty$ by another (in general different) disjoint union. Now, we interpret the $t = -\infty$ configuration as a set of incoming 3-branes and the $t = \infty$ one as a set of outgoing 3-branes. Any interacting instanton interpolates between two such asymptotic configurations. If we want to describe a given scattering process we will choose, among all the interacting instantons, those with the given asymptotic structure, i.e. whose slices at $t \to \pm\infty$ correspond to the assigned unions of 3-manifolds.

Let us consider the simplest possible example: a scattering of 3-tori. In this case the base manifold $M_3 = \mathbb{T}^3$. We think of the base space as $(\mathbb{P}^1 - \{0, \infty\})$ times an elliptic curve $\mathcal{C}$. We call z the coordinate on the first factor (time is given by $e^t = |z|$) and w the one on $\mathcal{C}$. The a_i's as functions of z are just meromorphic functions on $\mathbb{P}^1$, with poles in the excluded points 0 and ∞; as functions of w, they are constant. The resulting coverings are very simple: for each fixed z, the covering space is nothing but a disjoint union of 2-tori (the eigenvalues are constant in w); so the process is of the type (scattering of strings) $\times T^2$, where the first factor is exactly what was already examined in MST [9–11]. This simply means that 3-tori are really scattering just along one of their dimensions. The surface Σ has Euler characteristic $\chi_\Sigma = 0$.

The expansion of the action around any such instanton solution is rather similar to the one carried out in 2D. There is no need to repeat it here. The final result is that in the strong coupling limit the action on the covering is a free $U(1)$ gauge theory with matter:

$$S = \frac{1}{2} \int_\Sigma d^4\xi \left(\frac{1}{2} \partial_\mu \tilde{x}^i \partial^\mu \tilde{x}^i + \frac{1}{2} \partial_\mu \tilde{a}_\nu \partial^\mu \tilde{a}^\nu + \frac{1}{2} \partial_\mu \tilde{\bar{c}} \partial^\mu \tilde{c} - \frac{1}{2} \tilde{\lambda}^\dagger \gamma^\mu \partial_\mu \tilde{\lambda} \right),$$

(46)

where ξ are local coordinates on Σ (for example, z and w). We remark that (46) contains the fields which are expected to live on a D3-brane and it is itself the low energy and low curvature action for a D3-brane. This is a strong reason to look at the process represented by the underlying instanton as a scattering of D3-branes.

Like in 2D this is not the end of the story. In the path integral there is a non-trivial dependence on the original gauge coupling g which is due to the integration over the zero modes. For our previous rescaling of the various fields by powers of g involves, in particular, a rescaling of both the gauge and ghost diagonal degrees of freedom. When defining the path integral we have to take this fact into account, which amounts to rescaling it by an overall factor for any given instanton. The calculation of this factor is not as straightforward as in 2D, in particular it requires a summation over the line bundles on Σ. The final result is that the corresponding amplitude in the path integral should be preceded by a factor $g^{-\chi_\Sigma}$ where χ_Σ is the Euler characteristic of the covering. This suggests that we interpret $g_3 = 1/g$ as the D3-branes perturbative interaction coupling, analogous to the string coupling of MST.

8   Comments

We have seen that in 2D and in 4D SYM theories contain classical solutions, which we have called interacting instantons, that describe scattering of strings and D3-branes, respectively. The analysis is by now fairly complete in 2D, while it is still at a preliminary stage in 4D. Concerning other dimensions nothing has been done yet. Many scattering calculations have been carried out in the context of Matrix Theory and its compactifications (i.e. SYM theories in 1+d dimensions). But they all concern scatterings of classical object (0-branes, membranes, etc.) on a trivial background, i.e. no instantonic solutions like the one discussed in this paper have been considered. Moreover they generally involves weak YM coupling calculations. A remarkable exception is the paper [18], where a scattering of membranes in the background of an instanton in $SYM_{1,2}$ (i.e. an ordinary monopole) has been calculated, allowing the authors to compute scattering with momentum transfer in the 11-th direction of M-theory.

One remarkable problem we have not considered here concerns the stability of the classical solutions we have called interacting instantons. In this paper we have only considered the leading term in the $1/g$ expansion about each classical solution. Upon integrating out the nondiagonal fluctuations we found a free theory (it only interacts with the background metric). The question is now how taking into account the full theory would correct this result (with the understanding that we only consider amplitudes involving diagonal degrees of freedom). There are two types of corrections: perturbative and nonperturbative ones. While the latter are certainly present but difficult to evaluate for the time being, the perturbative corrections can in principle be calculated order by order in $1/g$. This task was started by S.Terna, [17]. He argued that such corrections should identically vanish up to (and including) order $1/g^2$. These arguments unfortunately cannot be considered conclusive, the basic reason being that supersymmetry in theories with 16 supercharges can only be realized on a shell. One consequence is that gauge fixing breaks supersymmetry and many divergences which appear in perturbation theory are gauge fixing artifacts.

The problem of perturbative $1/g$ corrections is under study. It is clearly a crucial problem, since their nonvanishing would introduce an explicit cutoff in the theory and would lead us to riconsider the procedure of taking the $g \to \infty$ limit.

Apart from the problem of $1/g$ corrections a lot of work has still to be done in the context of interacting instantons: finding scattering instantons in dimensions different from 2 and 4, calculating the corresponding path integral, finding a representation of the asymptotic states, ... What appears to be sure from now is that interacting instantons exist, have a physical interpretation and, therefore, will stay with us.

Acknowledgements. This work was partially supported by the Italian MURST for the program "Fisica Teorica delle Interazioni Fondamentali".

References

1. Banks, T., Fischler, W., Shenker, S.H., Susskind, L. (1997): M Theory as a matrix model: A conjecture. Phys. Rev. D **55**, 5112; hepth/9610043]
2. Motl, L.: Proposals on nonperturbative superstring interactions; hepth/9701025
3. Banks, T., Seiberg, N. (1997): Strings from matrices. Nucl. Phys. B **497**, 41; hepth/9702187
4. Taylor, W. (1997): D-brane field theory on compact spaces. Phys. Lett. B **394**, 283; hepth/9611042
5. Dijkgraaf, R., Verlinde, E., Verlinde, H. (1997): Matrix string theory. Nucl. Phys. B **500**, 43; hepth/9703030
6. Bonora, L., Chu, C.S. (1997): On the string interpretation of M(atrix) Theory. Phys. Lett. B **410**, 142; hepth/9705137
7. Wynter, T. (1997): Gauge fields and interactions in matrix string theory. Phys. Lett. B **415**, 349; hepth/9709029]
8. Giddings, S.B., Hacquebord, F., Verlinde, H. (1999): High energy scattering of D-pair creation in matrix string theory. Nucl. Phys. B **537** 260; hepth/9804121
9. Bonelli, G., Bonora, L., Nesti, F. (1998): Matrix string theory, $2D$ instantons and affine Toda field theory. Phys. Lett. B **435**, 303; hepth/9805071]
10. Bonelli, G., Bonora, L., Nesti, F. (1999): String interactions from matrix string theory. Nucl. Phys. B **538**, 100; hepth/9807232
11. Bonelli, G., Bonora, L., Nesti, F., Tomasiello, A.: Matrix string theory and its moduli space; hepth/9901093, to be published in Nucl. Phys. B
12. Bonora, L. (2000): Yang–Mills Theory and matrix string theory, in *Quantum field theory. A 20th century profile*, ed. by A.N. Mitra, Industan Book Agency
13. Bonelli, G., Bonora, L., Terna, S., Tomasiello, A.: Instantons and scattering in $\mathcal{N} = 4$ SYM in 4D; hepth/9912227
14. Hitchin, N.J. (1987): The self-duality equations on a Riemann surface. Proc. London Math. Soc. **55**, 59;
   (1992): Lie groups and Teichmüller space. Topology **31**, 449
15. Green, M.B., Schwarz, J.H., Witten, E. (1987): Superstring theory. Cambridge University Press, Cambridge
16. Bonelli, G., Bonora, L., Nesti, F., Tomasiello, A.: Heterotic matrix string theory and Riemann surfaces; [hepth/9905092
17. Terna, S. (2000): PhD Thesis, Trieste
18. Polchinski, J., Pouliot, P.: Membrane scattering with M-momentum transfer; hepth/9704029

The Early Universe, the Present Universe

A. Buonanno

Abstract. Almost ten years ago Gasperini and Veneziano proposed a new picture of the very primordial Universe ($z \gg 1$) based on string theory, called the pre–big-bang scenario. Here we review the key ideas of this model, its main phenomenological consequences, and the most striking differences with respect to ordinary inflationary models. The second part of this Proceedings is concerned with cosmology at much lower redshifts, $z < 2$. We tackle the problem of the motion of inspiraling and merging black-hole binaries, which are among the most promising astrophysical sources of gravitational waves. We discuss a new approach for the two-body problem in general relativity, which makes it possible to study the transition between the adiabatic inspiral and the plunge, and which provides a first estimate of the gravitational waveform emitted during the late dynamical evolution of a binary black-hole system.

1 A scenario for the very primordial Universe from string theory

Friedman–Robertson–Walker (FRW) cosmological solutions diverge when extrapolated backward in time, raising the so-called singularity problem [1]. Moreover, if the entire history of the universe follows a FRW solution, it is impossible to explain the degree of homogeneity and flatness of our present visible universe. These conundrums are known as the standard cosmological problems [1]; until now, the most credible solution to them assumes that soon after its birth the universe underwent an inflationary phase [1]. Yet inflationary models, still have to deal with the initial-singularity problem, with the problem of the naturalness of initial conditions and the issue of describing the universe at high energy and/or strong coupling. If we had a particle theory capable of describing this initial cosmological phase, then some questions would come naturally: what was before such high-energy stage? Could the origin of time and the issue of initial conditions be decoupled from the singularity problem?

Superstring theory/M-theory [2] is currently considered the most promising extension of the standard model of particle physics. It is the only theory that can unify quantum mechanics and general relativity. If string theory describes our real world, then we should expect that it contains the solution to the cosmological conundrums that we just mentioned. Various attempts have been made in the literature to apply string theory to cosmology. Some results were obtained within M-cosmology, or within its much better understood low-energy limits, such as 11-D supergravity [3,4]. Progresses in nonperturbative string theory have been made but until now only with classical solutions that respect a large number of supersymmetries [4]. More recently, interesting new ideas came from models with extra dimensions [5].

At low energy, string theory does not give only Einstein general relativity. In the simplest case it leads to the four-dimensional action

$$\Gamma_{\text{eff}} = \frac{1}{\lambda_s^2} \int d^4x \sqrt{|g|} e^{-\varphi} \left[\mathcal{R} + g^{\mu\nu} \partial_\mu \varphi \partial_\nu \varphi - \frac{1}{12} (dB)^2 \right], \tag{1}$$

where φ is the dilaton field, related to the string coupling by $g^2 = e^\varphi$; $dB = \partial_\mu B_{\nu\rho} + \partial_\nu B_{\rho\mu} + \partial_\rho B_{\mu\nu}$, where $B_{\mu\nu}$ is the two-form gauge field or antisymmetric field; and where λ_s is the string scale. In writing Eq. (1) we disregard for simplicity the internal dimensions, whose dynamics can be described in terms of moduli fields [3]. [Henceforth, we pose $\hbar = 1$.]

It was realized long ago that potential-driven inflationary scenarios cannot be implemented in string theory if the dilaton field is simply identified with the inflaton field [6]. This result forced people to conceive new ways of reconciling inflation and string theory. Henceforth, we shall discuss one of those attempts, the so-called pre–big-bang (PBB) scenario, originally proposed by Gasperini and Veneziano [7,8].

In the homogeneous and isotropic limit with $B = 0$ [$ds^2 = -dt^2 + a^2(t)dx^2$, $\varphi = \varphi(t)$], the solution of the low-energy string-effective action (1) satisfies the scale-factor–duality (SFD) symmetry: $a(t) \to 1/a(t)$, $\varphi(t) \to \varphi(t) - 6\log a(t)$,[1] with $a(t) \sim t^{1/\sqrt{3}}$ and $\varphi(t) \sim -\log t$. Deducing this result Veneziano [7] conceived the idea, subsequently sharpened by Gasperini and Veneziano [8], of implementing the inflationary phase at times before the *would-be* big-bang singularity. Indeed, it is easily shown that for $t < 0$, $\dot{a} > 0$, $\ddot{a} > 0$; that is the universe undergoes a (super) inflationary phase! Two different but physically equivalent descriptions of the PBB phase exist: either in the string-frame picture given by Eq. (1), where the universe undergoes an accelerated expansion ($H > 0$, $\dot{H} > 0$, $\dot{\varphi} > 0$), or in the Einstein-frame picture, where the action has the standard Hilbert–Einstein form and the evolution of the universe is described by an accelerated contraction, or gravitational collapse ($H < 0$, $\dot{H} < 0$, $\dot{\varphi} > 0$).

This radically new kind of inflation is driven by the kinetic energy of the dilaton field and forces both the string coupling ($\dot{g} > 0$) and the spacetime curvature to grow toward the future. As a consequence, at least in the homogeneous case, the inflationary stage lasts for ever ($t \to -\infty$) and the initial state of the universe is nearly flat, cold and decoupled: $g \ll 1$, $\mathcal{R}\lambda_s^2 \ll 1$. At some later time, evolving toward the *would-be* big-bang singularity, the universe enters a phase of high curvature ($\mathcal{R}\lambda_s^2 \sim 1$) and/or strong coupling ($g \sim 1$), where the perturbative description of the PBB phase breaks down and higher order corrections to the low-energy string effective action (1) should be taken into account:

$$\Gamma_{\text{eff}} = \frac{1}{\lambda_s^2} \int d^4x \sqrt{|g|} e^{-\varphi} \left\{ \mathcal{R} + g^{\mu\nu} \partial_\mu \varphi \partial_\nu \varphi + \cdots + \alpha' [\mathcal{R}^2 + (\partial\varphi)^4 + \cdots] \right.$$

$$\left. + e^\varphi [\mathcal{R} + (\partial\varphi)^2 + \alpha'(\mathcal{R}^2 + (\partial\varphi)^4 + \cdots)] + \cdots \right\},$$

$$\tag{2}$$

---

[1] Here for convenience the origin of time has been fixed at $t = 0$.

where $\alpha' = \lambda_s^2$ governs the finite–string-size effects, and $g = e^\varphi$ governs the quantum loop corrections. The issue of connecting the perturbative and nonperturbative PBB phases to the FRW cosmologies of radiation and matter eras has been called the *graceful exit* problem. Various analytical and numerical investigations have been made including first order finite–string-size effects and/or quantum loop corrections [9]. A complete solution of the graceful exit problem is still not available.

In Fig. 1 we draw the evolution of the Hubble expansion rate $H = \dot{a}/a$ in the homogeneous PBB cosmology. The central blob refers to the high-curvature and/or strong-coupling phase.

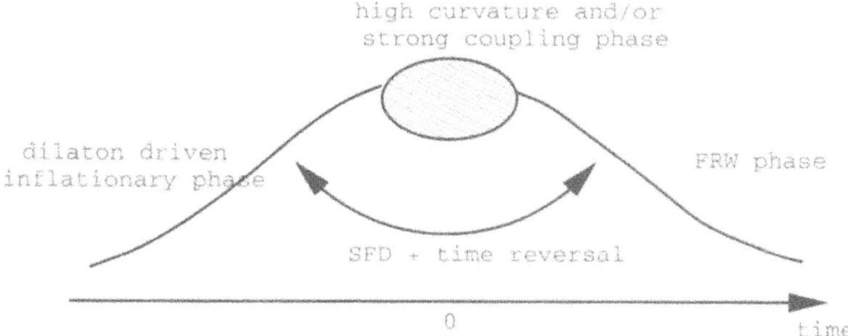

Fig. 1. Evolution in time of the Hubble expansion rate $H = \dot{a}/a$ in the homogeneous PBB cosmology

It was shown that to solve the standard cosmological problems of homogeneity and flatness in PBB cosmology, we must require that initially the string coupling $g_{\rm in} < 10^{-26}$ and that $H_{\rm in}\lambda_s < 10^{-19}$ [7,10,11]. Quasi-homogeneous cosmological solutions in the perturbative inflationary phase of the PBB scenario have been derived [11,12] by applying the spatial-gradient expansion technique. Because of the presence of the dilaton field in the action (1), the Belinskii–Khalatnikov–Lifshitz (BKL) oscillations can last at most for a finite time [13]. Afterwards, the universe enters an era where spatial gradients become less and less important as we move toward the singularity; that is $(\nabla_{\rm spatial}\varphi)^2/\dot{\varphi}^2$, $R/\dot{\varphi}^2 \to 0$ as $t \to 0^-$. However, recently it was pointed out [14] that this result is somewhat spoiled when other p-form fields (including the antisymmetric field B) are present in the low-energy string-effective action (1). Their presence leads to the generic appearance of an inhomogeneous chaos near the *would-be* big-bang singularity, ultimately leading to a string-scale foam [14].

As we anticipated, one of the most striking differences between the PBB scenario and the standard cosmological models is the description of the initial state of the universe. In the PBB model the universe is initially in a weakly-coupled, classical state, consisting of a stochastic bath of gravitational and dilatonic waves [15]. By the mechanism of gravitational instability, this state can give rise to our universe (modulo the assumption of a graceful exit from the PBB phase to the standard FRW era). Indeed, if those initial waves satisfy a certain strength criterion [15], viewed

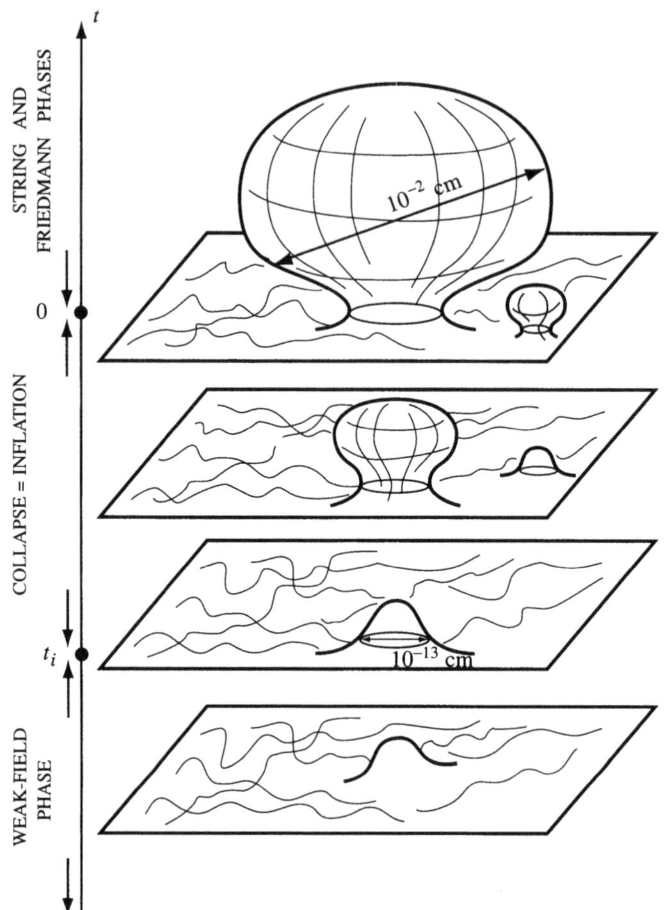

Fig. 2. Naive evolution of a PBB bubble of initial size 1fm, which originated from a classical fluctuation in the initial bath of gravitational and dilatonic waves. We follow the PBB bubble up to the *would-be* big-bang singularity hypersurface, where its size is on the order of 0.1mm (this is the right size to explain the present scale of homogeneity of the universe)

in the Einstein frame, they collapse but viewed in the physically more appropriate string frame, each gravitational collapse leads to the local birth of a baby inflationary universe. It was then claimed [15] that the occurrence of a PBB inflation period is as generic in string theory as the occurrence of gravitational collapse in general relativity. In Fig. 2 we draw the evolution of a PBB bubble, from its inception as classical fluctuation of the initial sea of dilatonic and gravitational waves, up to the beginning of the FRW era.

The stochastic version of PBB cosmology was originally intended to address some concerns about fine-tuning [10,16] in the PBB scenario. It can be shown [11,12,15] that the only condition needed for the birth of a PBB bubble of size H^{-1} (in string units), is similar to the corresponding condition in chaotic inflation [1]. Namely, the inhomogeneous contributions to the local Friedmann equation should be fractionally small (say by a factor of five) compared to the homogeneous contribution, $\dot{\varphi}^2 \sim H^2$. However, it has been recently realized that the PBB scenario is not very effective in smoothing out the initial tensor classical inhomogeneities [17]. As a consequence, if we wish that generic coarsely homogeneous bubbles evolve into our universe, we need to require that $g_{\text{in}} \lesssim 10^{-35}$, that is the initial value of the string coupling should be parametrically smaller than the minimal value $g_{\text{in}}^{\min} \simeq 10^{-26}$ needed to solve the standard cosmological conundrums mentioned earlier.

Significant effort has been spent on extracting the observational predictions of the PBB scenario. During the dilaton-driven inflationary phase, the kinetic energy of the dilaton field is converted into particles, by the well known mechanism of amplification of quantum vacuum fluctuations. This phenomenon is also present in ordinary inflationary models – for example it is supposedly responsible of the inhomogeneities in the Cosmic Microwave Background Radiation (CMBR). Due to the richness of the particle content of string theory (axions, scalar fields, gauge fields,...), and due to the nontrivial coupling (depending also on the compactification of internal dimensions) between those fields and the background fields (i.e., the dilaton, moduli and gravitational field), a much larger number of species can be produced out of vacuum in the PBB scenario than in ordinary inflationary models. It is interesting to note that at second order in perturbation theory the fluctuation ψ of whatever field is governed by the action [written in the conformal time η $(dt = ad\eta)$] [18]

$$\delta\Gamma_{\text{eff}} = \int d\eta d^3x \, \tilde{a}_\psi^2(\eta) \left[\left(\frac{\partial \psi}{\partial \eta} \right)^2 - (\nabla \psi)^2 \right], \tag{3}$$

where the only dependence on the specific field comes through the function $\tilde{a}_\psi$, the so-called *pumping field*. For example, for gravitons, for scalar fields and for moduli fields, we find $\tilde{a} = ae^{-\varphi/2}$ [18–20]. If we assume static internal dimensions, then gauge fields have $\tilde{a} = e^{-\varphi/2}$ [18,21], while for axion fields $\tilde{a} = ae^{\varphi/2}$ [18,22]. Contrary to what happens in potential-driven inflation, in PBB cosmology the spectrum of the energy density versus frequency can grow, decrease or be constant. More specifically the PBB model predicts a stochastic background of gravitational waves whose energy-density spectrum increases at very low frequencies [19], and whose amplitude might be well above that predicted by models of standard inflation at frequencies $\sim 100\,\text{Hz}$, just in the band of best sensitivity for future earth-based gravitational-wave interferometers, such as the Laser Inteferometric Gravitational Wave Observatory (LIGO) and VIRGO [23]. In certain regions of the free-parameter space, quantum vacuum electromagnetic fluctuations, amplified by the nonconformal coupling between gravitational and electromagnetic fields, could produce the primordial seeds responsible for the formation of galactic and extragalactic magnetic fields [21]. However, despite the efforts to investigate the amplification of axion quantum fluctuations [18,22,24] (which can produce a nearly constant energy-density spectrum versus frequency), the PBB scenario still lacks a mechanism for describing the inhomogeneities in the CMBR, and the formation of large scale structures.

The mechanism of reheating in PBB cosmology has also been investigated [20]. Let us first note that in standard cosmological models the inflationary era is dominated by potential energy; the post-inflationary phase is driven by inflaton condensates, that later on, decay into radiation (in the reheating process) giving rise to the birth of the *hot big bang*. By contrast, as discussed above, in the PBB scenario it is the kinetic energy of the dilaton field which drives the inflationary phase. Therefore, to explain the birth of the hot big-bang era in the PBB model, it was originally suggested [25] that the particles present at the very beginning of the radiation era could have originated from the mechanism of quantum-vacuum fluctuations during the PBB phase. However, it was found [20] that PBB models inevitably face a severe gravitino/moduli problem. Indeed, they predict quite generically that at the beginning of the radiation era, the moduli and gravitinos, produced gravitationally or from scattering processes of the thermal bath, will have a number-density to entropy-density ratio that is far in excess of the big-bang-nucleosynthesis bound. Hence, in the PBB scenario, reheating cannot be implemented solely by gravitational production. Depending on the details of the transition from PBB era to FRW phase, late-entropy production to the level of $\Delta s \gtrsim 10^5 - 10^{10}$ is mandatory to dilute those dangerous relics. This entropy production can be viewed as a period of secondary reheating; that is as the *real* birth of the hot big-bang era in the PBB scenario. Sufficient entropy can be produced by the domination and decay of the zero-mode of a modulus field with mass $\sim 10^6$ GeV; this could well be the dilaton field, initially displaced from the minimum of its potential by an amount on the order of the string mass $M_s \sim 10^{18}$ GeV [26]. Nevertheless, the above source of entropy comes with a bonus: baryogenesis can be implemented in a natural way via the so-called Affleck–Dine mechanism [27]. Finally, we notice that in PBB cosmology typical reheating temperatures vary in the range $T_{\mathrm{RH}} \sim 1 - 10^5$ GeV.

In conclusion, the PBB scenario is certainly an interesting attempt of reconciling string theory and cosmology. It has proposed a new, elegant way to implement inflation, which is based on a duality symmetry of string theory and uses the kinetic energy of the dilaton field; it has proposed the rather unconventional idea of decoupling the singularity problem from the issue of initial conditions, by assuming that the universe originated from a classical, weakly coupled state; it has pointed out the rich variety of energy-density spectra of particles produced out of vacuum during the PBB inflationary phase, whose details strongly depend on the background dynamics of dilaton, moduli and gravitational field.

2   Coalescing compact binaries: a new approach to the two-body problem in general relativity

Binary systems made of compact objects (neutron stars or black holes) that inspiral toward coalescence because of gravitational-radiation damping are among the most promising candidate sources for interferometric gravitational-wave (GW) detectors, such as the Laser Inteferometric Gravitational Wave Observatory (LIGO) and VIRGO [23]. One of the most important issue in gravitational-wave research is the *generation problem* [28]; that is, the link between the radiative transverse traceless (TT) gravitational field $h_{ij}^{\rm TT}$, far away from the source, and the motion of the source. This link is provided at lowest order in the post-Newtonian (PN) expansion by the Einstein's quadrupole formula [28], which gives for the radiative field:

$$h_{ij}^{TT}(T, D) = \frac{2G}{c^4 D} \mathcal{P}_{ijkm}(N) \frac{d^2}{dT^2} Q_{km} \left(T - \frac{D}{c} \right), \tag{4}$$

where Q_{ij} $(i, j = 1, 2, 3)$ is the tracefree quadrupole moment of the source; D is the distance from the source; $N = X/D$ is the unit vector from the source to the observer; $\mathcal{P}_{ijkm}(N)$ is the TT projection operator onto the plane orthogonal to N; G is the Newton constant; and c is the speed of light.

The inspiral waveform enters the detector band during the last few minutes of evolution of the binary. The LIGO/VIRGO community plans to track the signal phase and build up the signal-to-noise ratio by integrating the signal for the time during which it stays in the detector band. This is achieved by filtering the detector output with a template which is an (approximate) copy of the exact, observed signal. From Eq. (4) (and its extensions at higher PN orders) we see that the more precisely we know the two-body motion, the more accurately the PN template $h_{ij}^{\rm TT}$ will describe the exact gravitational waveform.

Henceforth, our analysis will be restricted to nonspinning black holes. In Fig. 3 we show a typical gravitational waveform. The part of the waveform drawn with a continuous line is emitted during the inspiral phase when the two black holes are largely separated ($r \geq 10M$). We denoted by r the radial separation and by M the total mass of the binary system. During the inspiral, the two black holes follow an adiabatic sequence of quasi-circular orbits. The equation of motion in the center-of-mass frame can be written schematically as [28]

$$\frac{d^2 x}{dt^2} = -\frac{GMx}{r^3}[1 + \mathcal{O}(\epsilon) + \mathcal{O}(\epsilon^2) + \mathcal{O}(\epsilon^{5/2}) + \cdots] \times [1 + \mathcal{O}(v) + \cdots],$$
$$\tag{5}$$

where x denotes the separation vector between the two bodies and $r = |x|$. Equation (5) is characterized by a double expansion: in the PN parameter $\epsilon \sim v^2/c^2 \sim M/r$, and in the parameter $v = m_1 m_2 / M^2$, where m_1 and m_2 are the masses of the two black holes. The parameter v ranges between 0 (test-mass limit) and 1/4 (equal-mass case).

The PN expansion converges badly: as the two bodies draw closer, it becomes more and more difficult to extract nonperturbative information from the PN series. Specifically, when the distance between the inspiraling black holes shrinks to $r \lesssim 10M$, the PN expansion can no longer be trusted [29]. The dashed line in Fig. 3 depicts the part of the waveform emitted during the final phase of evolution, when nonlinearities and strong curvature effects become important. During this stage the PN expansion fails and nonperturbative analytical and/or numerical techniques should be used. This final phase includes the transition from the adiabatic inspiral to the plunge, beyond

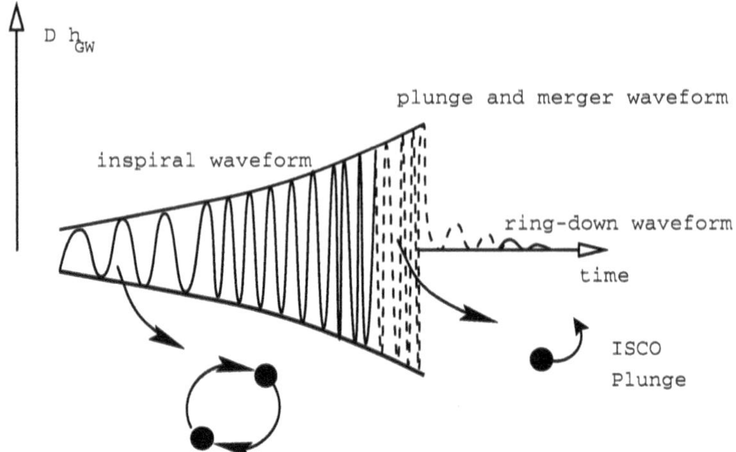

Fig. 3. Typical gravitational waveform emitted throughout the inspiral, plunge and ring-down phase

which the two-body motion is driven (almost) only by the conservative part of the dynamics. The plunge starts at the innermost stable circular orbit (ISCO) of the binary black holes. Due to the failure of the PN expansion, different predictions for the ISCO location have been provided so far in the literature [30,31]. Beyond the plunge the two black holes merge, forming a Kerr black hole. As the system reaches the stationary Kerr state, the nonlinear dynamics of the merger resemble more and more the oscillations of the black-hole quasi-normal modes [32]. During this phase, often called the ring-down phase, the gravitational signal will be a superposition of exponentially damped sinusoids.

It seems likely that the first detection of gravitational waves with LIGO and VIRGO interferometers will come from binary systems made of massive black holes of comparable masses, say with a total mass $M \simeq 15M_\odot + 15M_\odot$. If we restrict our attention to nonspinning black holes, it is easily shown that the gravitational-wave frequency at the ISCO for such massive systems is very close (not accidentally!) to the location of the minimum for the detector's noise spectral density. For example, for the first generation of LIGO interferometers, the maximum of the signal-to-noise ratio for nonspinning black holes of total mass $M \simeq 15M_\odot + 15M_\odot$ is reached at $f_{\text{detection}} \simeq 167\,\text{Hz}$, which is quite close to $f_{\text{GW}}^{\text{ISCO}} \simeq 180\,\text{Hz}$. Therefore, for data analysis purposes it is quite desirable to have a thorough knowledge of the late dynamical evolution of comparable-mass binaries.

Despite the progress made by the numerical relativity community during the recent years, an estimate of the complete waveform emitted by a black-hole binary with comparable masses has not yet been provided. Preliminary results for the plunge, merger and ring-down waveform were only recently obtained [33]. To tackle the delicate issue of the late dynamical evolution, Buonanno and Damour introduced a new *nonperturbative analytical* approach to study the motion of two nonspinning bodies in general relativity [31,34]. This approach should be able to capture the crucial features of the transition from the adiabatic inspiral to the plunge. Henceforth, we shall refer to this new technique as effective-one-body (EOB) approach.

The EOB approach combines two PN resummation techniques. The first method [31], inspired by an approach introduced by Brézin, Itzykson and Zinn-Justin [35] to study electromagnetically interacting two bodies, makes it possible to derive a nonperturbative estimate for the conservative part of the nonlinear force law that governs the motion of comparable-mass binaries. The basic idea [31], illustrated in Fig. 4, is to map the *real* conservative two-body dynamics up to 2PN order (see below for the extension at 3PN order) onto an *effective* one-body problem, where a test particle of mass m_0 moves in some effective background metric $g_{\mu\nu}^{\text{eff}}$. This mapping

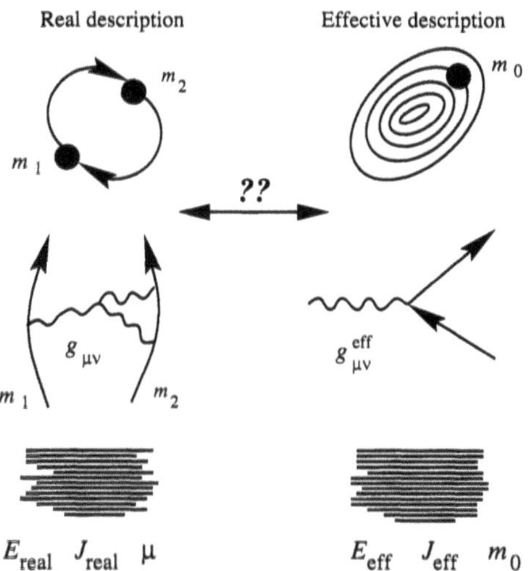

Fig. 4. How the EOB approach matches the real two-body problem (on the left) and the effective one-body problem (on the right) in general relativity

has been worked out within the Hamilton–Jacobi formalism, imposing that the adiabatic invariants of the real and effective description coincide $J_{\text{real}} = J_{\text{eff}}, \mathcal{I}_{\text{real}} = \mathcal{I}_{\text{eff}}$, where J denotes the total angular momentum, and $\mathcal{I}$ the radial action variable. While doing so, we allow a transformation of the energy axis, $E_{\text{real}} = f(E_{\text{eff}})$, where f is a generic function. The test mass m_0 in the effective description was assumed to be equal to the reduced mass $\mu = m_1 m_2 / M$ of the two-body system.

After applying the rules to define the mapping, we found that, as long as radiation-reaction effects are not taken into account, the effective metric is just a deformation of the Schwarzschild metric, with deformation parameter $\nu = \mu / M$. The effective metric reads [31]

$$ds_{\text{eff}}^2 = -A(R)c^2 dt^2 + \frac{D(R)}{A(R)} dR^2 + R^2 d\Omega^2,$$

$$A(R) = 1 - 2\frac{GM}{c^2 R} + 2v\left(\frac{GM}{c^2 R}\right)^3, \quad D(R) = 1 - 6v\left(\frac{GM}{c^2 R}\right)^2. \tag{6}$$

The effective and real (nonrelativistic) energies are related by [31]

$$\frac{E_{\text{eff}}^{\text{NR}}}{m_0 c^2} = \frac{E_{\text{real}}^{\text{NR}}}{\mu c^2}\left(1 + \frac{v}{2}\frac{E_{\text{real}}^{\text{NR}}}{\mu c^2}\right). \tag{7}$$

Remarkably, this mapping between the real and the effective nonrelativistic energies coincides with the mapping obtained by Brézin, Itzykson and Zinn-Justin [35] in the context of quantum electrodynamics, where these authors mapped the one-body relativistic Balmer formula onto the two-body energy formula.

The EOB approach provides a method to resum nonperturbatively the badly convergent PN-expanded dynamics of the real description. Indeed, it gives the following *improved* real Hamiltonian [34]:

$$H_{\text{real}}^{\text{improved}} = Mc^2 \sqrt{1 + 2v\left(\frac{H_{\text{eff}}^v - \mu c^2}{\mu c^2}\right)}, \tag{8}$$

where

$$H_{\text{eff}}(v, R, P_R, P_\varphi) = \mu c^2 \sqrt{A(R)\left(1 + \frac{A(R)P_R^2}{\mu^2 c^2 D(r)} + \frac{P_\varphi^2}{\mu^2 c^2 R^2}\right)}. \tag{9}$$

The basic idea that underlies the mapping of two-body general relativistic dynamics onto an effective one-body problem, was recently extended to classical electrodynamics to test its robustness. Reference [36] discussed the mapping of the conservative part of two-body electrodynamics (i.e., of a two-body system of charges e_1 and e_2 with $e_1 e_2 < 0$) onto the dynamics of a test particle of charge e_0 moving in some external electromagnetic field (see Fig. 5); the author took into account recoil effects and relativistic corrections up to second post-Coulombian order. In this case the expansion parameter is the classical radius $\alpha_0/m_0 c^2$, where $\alpha_0 = e_0^2$. Unlike the results obtained in general relativity, in classical electrodynamics it is not possible to implement the matching without introducing external parameters in the effective electromagnetic field. For example, it was found that the effective vector potential A_μ^{eff} must depend either on the energy or the angular momentum. However, if we relax the assumption that the effective test particle moves in a flat spacetime, then it is sufficient to introduce a scalar potential φ^{eff} to obtain the matching between real and effective descriptions. Let us finally note that even in classical electrodynamics the real and effective nonrelativistic energies are mapped through the same Eq. (6).

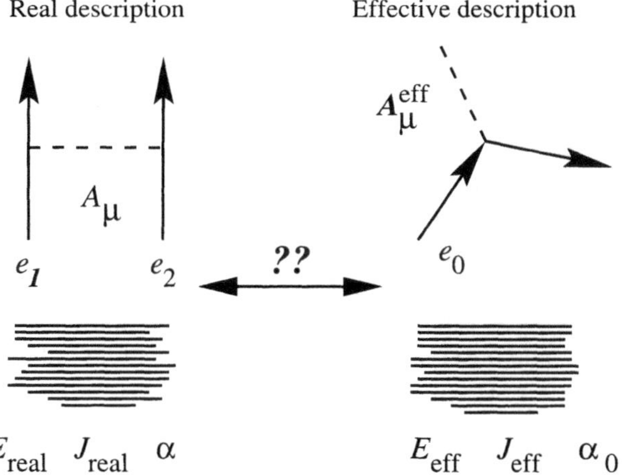

Fig. 5. Matching between the real two-body problem (on the left) and the effective one-body problem (on the right) in classical electrodynamics

We now go back to the general relativistic case. Earlier we only discussed the conservative part of the dynamics; now let us introduce radiation-reaction effects. Using Padé approximants, Damour, Iyer and Sathyaprakash [37] gave a resummed estimate of the energy-loss rate along circular orbits Φ_{circ}, up to 2.5 PN order. Buonanno and Damour [34] then combined this resummation method with the EOB approach, and deduced a system of ordinary differential equations which describe the late dynamical evolution of a binary–black-hole system. In spherical coordinates $(\varphi, R, P_\varphi, P_R)$, their relevant equations are [34]:

$$\frac{dR}{dt} = \frac{\partial H_{\text{real}}^{\text{impr}}}{\partial P_R}, \qquad \frac{dP_R}{dt} + \frac{\partial H_{\text{real}}^{\text{impr}}}{\partial R} = 0,$$

$$\frac{d\varphi}{dt} = \frac{\partial H_{\text{real}}^{\text{impr}}}{\partial P_\varphi}, \qquad \frac{dP_\varphi}{dt} = -\frac{\Phi_{\text{circ}}}{\dot\varphi}. \tag{10}$$

These equations can be used analytically or numerically to study the transition between the adiabatic inspiral and the plunge. Specifically, by a linear expansion in the radial velocity $\dot R$, they deduced the following characteristic equation [34]

$$\frac{d^3 R}{dt^3} + \omega_R^2(R)\frac{dR}{dt} = -B_R(R). \tag{11}$$

The quantity ω_R^2 plays the role of a *restoring force*. It is the square of the frequency of radial oscillations, and it is proportional to the curvature of the effective radial potential (it vanishes at the ISCO). The quantity $-B_R$ ($\propto \nu$) is a *driving force*, coming from gravitational radiation damping. The term $d^3 R/dt^3$ is an *inertia term*, which is neglected in the adiabatic approximation, but should be retained when describing the motion in proximity of the ISCO and beyond it. Ori and Thorne [38] independently derived an equation analogous to Eq. (11) for a test particle moving along quasi circular equatorial orbits in Kerr spacetime.

Let us discuss the main features of the transition from inspiral to plunge in the two extreme limits $\nu \ll 1$ and $\nu = 1/4$. The case $\nu \ll 1$ refers to binary–black-hole systems in which a very small black hole spirals around a supermassive black hole. These are typical GW sources for the future Laser Interferometer Space Antenna (LISA). In this case, the transition from adiabatic inspiral to plunge is sharply localized around the ISCO and various interesting quantities satisfy very simple scaling laws. For example the radial momentum at the ISCO scales like $\nu^{3/5}$, and the number of GW cycles left after the ISCO scales like $\nu^{-1/5}$ [34,38]. Ori and Thorne [38] pointed out that likely LISA could observe the transition from inspiral to plunge.

For equal-mass binaries ($\nu = 1/4$), we compare in Fig. 6 the "exact" gravitational waveform, obtained by solving Eqs. (11) numerically, with its adiabatic approximation. Contrary to the case $\nu \ll 1$, for equal-mass black holes the radiation damping effects become important in an extended region on the order of $\Delta(Rc^2/GM) \sim 1$ above the naive (Schwarzschild) ISCO $R = 6GM/c^2$. In Fig. 6 the naive ISCO is found at $t \sim 50M$. Hence, the transition from inspiral to plunge is rather blurred. Moreover, as Fig. 6 shows, the dephasing between the exact and the adiabatic waveform becomes visible somewhat before the naive ISCO. The plunge part of the exact waveform looks like a continuation of the inspiral part. This happens because the orbital motion remains quasi-circular throughout the plunge.

Recently Damour, Iyer and Sathyaprakash [39] investigated the consequences of the EOB waveform for LIGO/VIRGO data analysis. They found the interesting result that GW radiation coming from the plunge and merger can significantly enhance the signal-to-noise ratio for binaries of total mass $M \gtrsim 30M_\odot$.

In Fig. 7 we have blown up the plunge and merger part of the waveform shown in Fig. 6, and we have included the ring-down waveform [34]. The ringdown waveform contains only the mode that is damped more slowly, $l = 2$, $m = 2$ [32], at frequency $\omega_{qnm} \sim 1880(10M_\odot/M_{BH})$ Hz, where M_{BH} is the mass of the final hole formed. The dimensionless rotation parameter is $a_{BH} = J_{BH}/(GM_{BH}^2) = 0.795$, where we denoted the angular momentum of the final Kerr black hole by J_{BH}. The energy emitted during the plunge is $\sim 0.7\%$ of M, with a comparable energy loss $\sim 0.7\%$ of M during the ring-down phase. This gives a total energy released of $\sim 1.4\%$ of M to be contrasted with the much larger value 4–5 % of M recently estimated in [33].

Before closing, let us observe that the EOB approach was extended to the 3PN order by Damour, Schafer and Jaranowski [40]. They found that at the 3PN order the mapping between the effective and the real problem exists *only if* we abandon the hypothesis (used at 2PN order [31]) that the effective test-mass motion is geodesic. Note also that the relation (11) between the effective and real (nonrelativistic) energies survives at 3PN order.

In conclusion, we have discussed how analytical resummation techniques can cope with the final nonperturbative phase of binary–black-hole evolution. By reducing the two-body dynamics onto a simpler auxiliary one-body problem [31], and by resumming radiation-reaction effects with Padé approximants [37], we end up with

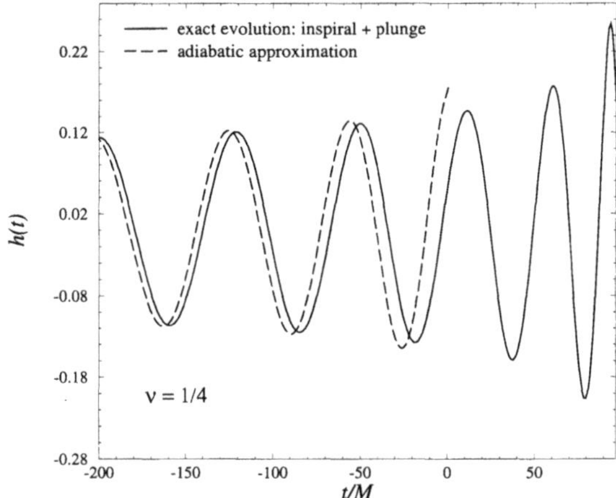

Fig. 6. Comparison between the exact gravitational waveform and the adiabatic waveform. Note that $t \sim -200M$ and $t \sim 90M$ corresponds to the radial separations $R \sim 8GM/c^2$ and $R \sim 2.8GM/c^2$, respectively. The naive (Schwarzschild) ISCO $R = 6GM/c^2$ is located at $t \sim 50M$

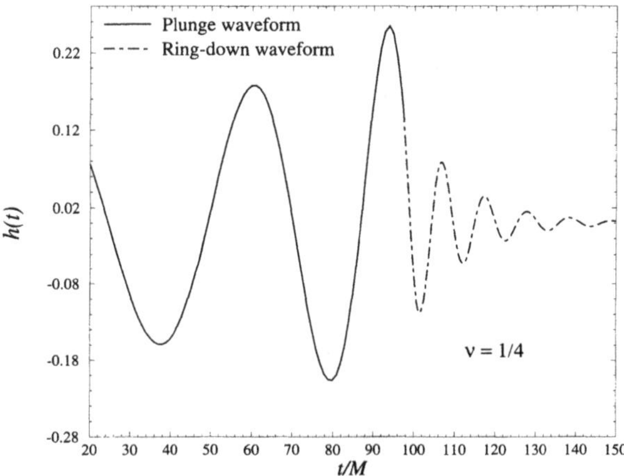

Fig. 7. Plunge and ring-down gravitational waveform obtained from the EOB approach

an explicit analytical system of ordinary differential equations that describes the transition from inspiral to plunge [34]. Beyond the estimation of the gravitational waveform, the most important and urgent application of this approach will be to provide initial dynamical data for numerical relativity investigations, of black holes that have just started their plunge motion.

Acknowledgements. The author wishes to thank all the people who collaborated with her on these research topics. They deserve much of the merit for the SIGRAV Prize the author received. They are: T. Damour, M. Gasperini, M. Lemoine, M. Maggiore, K. Meissner, K.A. Olive, C. Ungarelli and G. Veneziano.

References

1. Kolb, E.W., Turner, M.S. (1990): Early cosmology. Reading Massachusetts, Addison Wesley;
   Linde, A.D. (1990): Particle physics and inflationary cosmology. Harwood
2. Polchinski, J. (1998): String theory. Cambridge
3. Caroll, S.: Tasi lectures: Cosmology for string theorists; hep-th/0011110
4. Banks, T. (1999): M-theory and cosmology, Les Houches 1999 Summer School, *L'Univers Primordial*; hep-th/9911067
5. Antoniadis, I., Arkani-Hamed, N., Dimopoulos, S., Dvali, G. (1998): Phys. Lett. B **429**, 263; (1998): Phys. Lett. B **436**, 257;
   Randall, L., Sundrum, R. (1999): Phys. Rev. Lett. **83**, 3370;
   (1999): Phys. Rev. Lett. **83**, 4690
6. Campbell, B.A., Linde, A.D., Olive, K.A. (1991): Nucl. Phys. B **335**, 146;
   Brustein, R., Steinhardt, P.J. (1993): Phys. Lett. B **302** 196
7. Veneziano, G. (1991): Phys. Lett. B **265**, 287
8. Gasperini, M., Veneziano, G. (1993): Astropart. Phys. **1**, 317;
   (1993): Mod. Phys. Lett. A **8**, 3701
9. Gasperini, M., Maggiore, M., Veneziano, G. (1997): Nucl. Phys. B **494**, 315;
   Brustein, R., Madden, R. (1997): Phys. Lett. B **410**, 110;
   (1998): Phys. Rev. D **57**, 712;
   Foffa, S., Maggiore, M., Sturani, R. (1999): Nucl. Phys. B **552**, 395
10. Turner, M.S., Weinberg, E.J. (1997): Phys. Rev. D **56**, 4604
11. Buonanno, A., Meissner, K.A., Ungarelli, C., Veneziano, G. (1998): Phys. Rev. D **57** 2543
12. Veneziano, G. (1997): Phys. Lett. B **406**, 297
13. Belinskii, V.A., Khalatnikov, I.M. (1970): Sov. Phys. JETP **30**, 1174;
    Belinskii, V.A., Lifshitz, E.M., Khalatnikov, I.M. (1971): Sov. Phys. Uspekhi **13**, 745
14. Damour, T., Henneaux, M. (2000): Phys. Rev. Lett. **85**, 920;
    (2000): Phys. Lett. B **488** 108;
    (2000): Erratum Phys. Lett. B **491**, 377; hep-th/0012172
15. Buonanno, A., Damour, T., Veneziano, G. (1999): Nucl. Phys. B **543**, 275
16. Linde, A.D., Kaloper, N., Bousso, R. (2000): Phys. Rev. D **59**, 043508
17. Buonanno, A., Damour, T. (2001): Phys. Rev. D**64**, 043501; gr-qc/0102102
18. Buonanno, A., Meissner, K., Ungarelli, C., Veneziano, G. (1998): JHEP **9801**, 004;
    Brustein, R., Hadad, M. (1998): Phys. Rev. D **57**, 725
19. Brustein, R., Gasperini, M., Giovannini, M., Mukhanov, S., Veneziano, G. (1995) Phys. Rev. D **51**, 6744;
    Brustein, R., Gasperini, M., Giovannini, M., Veneziano, G. (1995): Phys. Lett. B **361** 45;
    Buonanno, A., Maggiore, M., Ungarelli, C. (1997): Phys. Rev. D;
    Maggiore, M. (2000): Phys. Rep. **331** 283
20. Buonanno, A., Lemoine, M., Olive, K.A. (2000): Phys. Rev. D **62**, 083513
21. Gasperini, M., Giovannini, M., Veneziano, G. (1995): Phys. Rev. Lett. **75**, 3796;
    Lemoine, D., Lemoine, M. (1995): Phys. Rev. D **52**, 1955

22. Copeland, E., Easther, R., Wands, D. (1997): Phys. Rev. D **56**, 874
23. Abramovici, A., Althouse, W.E., Drever, R.W.P., Gursel, Y., Kawamura, S., Raab, F.J., Shoemaker, D., Sievers, L., Spero, R.E., Thorne, K.S., Vogt, R.E., Weiss, R., Whitcomb, S.E., Zucker, M.E. (1992): Science **256**, 325;
    Bradaschia, C. et al. (1990): Nucl. Instrum. Meth. A **289**, 518
24. Durrer, R., Gasperini, M., Sakellariadou, M., Veneziano, G. (1998): Phys. Lett. B **436**, 66;
    (1999): Phys. Rev. D **59**, 043511;
    Melchiorri, A., Vernizzi, F., Durrer, R., Veneziano, G. (1999): Phys. Rev. Lett. **83**, 4464
25. Veneziano, G. (1994): Strings, cosmology, ... and a particle. CERN-TH-7502-94
26. Gaillard, M., Murayama, H., Olive, K.A. (1995): Phys. Lett. B **355**, 71;
    Campbell, B.A., Gaillard, M., Murayama, H., Olive, K.A. (1999): Nucl. Phys. B **538**, 351
27. Affleck, I., Dine, M. (1985): Nucl. Phys. B **249**, 361
28. Damour, T. (1983): In *Gravitational radiation*, ed. by N. Deruelle, T. Piran, Amsterdam, pp. 59–144
29. Brady, P.R., Creighton, J.D.E., Thorne, K.S. (1998): Phys. Rev. D **58**, 061501
30. Clark, J.P.A., Eardley, D.M. (1977) Astrophys. J. **215** 311;
    Blackburn, J.K., Detweiler, S. (1992): Phys. Rev. D **46**, 2318;
    Kidder, L.E., Will, C.M., Wiseman, A.G. (1992): Class. Quantum Grav. **9**, L127;
    (1993): Phys. Rev. D **47**, 3281;
    Wex, N., Schäfer, G. (1993): Class. Quantum Grav. **10**, 2729;
    Baumgarte, T.W., Cook, G.B., Scheel, M.A., Shapiro, S.L., Teukolsky, S.A. (1998): Phys. Rev. D **57** 7299;
    Baumgarte, T.W. (2000): Phys. Rev. D **62**, 024018
31. Buonanno, A., Damour, T. (1999): Phys. Rev. D **59**, 084006
32. Chandrasekhar, S., Detweiler, S. (1975): Proc. R. Soc. Lond. A **344**, 441
33. Baker, J., Brugmann, B., Campanelli, M., Lousto, C.O., Takahashi, R.: gr-qc/0102037
34. Buonanno, A., Damour, T. (2000): Phys. Rev. D **62**, 064015
35. Brézin, E., Itzykson, C., Zinn-Justin, J. (1970): Phys. Rev. D **1**, 2349
36. Buonanno, A. (2000): Phys. Rev. D **62**, 104022
37. Damour, T., Iyer, B.R., Sathyaprakash, B.S. (1998): Phys. Rev. D **57**, 885;
    (2000): Phys. Rev. D **62**, 084036
38. Ori, A., Thorne, K.S. (2000): Phys. Rev. D **62**, 124022
39. Damour, T., Iyer, B.R., Sathyaprakash, B.S. (2001): Phys. Rev. D **63**, 044023
40. Damour, T., Jaranowski, P., Schäfer, G. (2000): Phys. Rev. D **62**, 084011

2D Dynamical Triangulations
and the Weil–Petersson Measure

M. Carfora, A. Marzuoli, P. Villani

Abstract. Our goal here is to present an approach connecting the anomalous scaling properties of 2D simplicial quantum gravity to the geometry of the moduli space $\overline{\mathfrak{M}}_{g,N_0}$ of genus g Riemann surfaces with N_0 punctures. In the case of pure gravity we prove that the scaling properties of the set of dynamical triangulations with N_0 vertices are directly provided by the large N_0 asymptotics of the Weil–Petersson volume of $\overline{\mathfrak{M}}_{g,N_0}$, recently discussed by Manin and Zograf.

1 Introduction

Dynamical triangulations (DT) provide one of the most powerful technique for analysing two-dimensional quantum gravity also in regimes which are not accessible to the standard field-theoretic formalism. This is basically due to the fact that in such a discretized setting the quantum measure of the theory, describing the gravitational dressing of conformal operators in the continuum theory, reduces to a suitably constrained enumeration of distinct triangulations admitted by a surface of given topology, (see e.g., [1] for a review). The geometrical origin of such a property is rather elusive and it is not clear how the counting for dynamical triangulations factorizes, so to speak, in terms of a discrete analogous of a moduli space measure and of a Liouville measure over the conformal degrees of freedom of the theory. Some results in such a direction have been recently discussed by S. Catterall and E. Mottola [2] with an emphasis on the numerical simulation (and indicating a possible strategy for further analytical work). P. Menotti and P. Peirano [3] discussed a similar issue in connection with the Regge measure in simplicial quantum gravity. However in such a case the problem takes on a rather different flavour, being more directly connected with the issue of the diffeomorphism invariance of the resulting measure.

In order to provide (partial) analytical answers to such basic questions, we discuss in this paper the connection between the Weil–Petersson measure of the (compactified) moduli space of genus g Riemann surfaces with N_0 punctures $\overline{\mathfrak{M}}_{g,N_0}$, and the entropy function associated with the counting of distinct triangulations of surfaces with N_0 vertices. The starting point of our analysis is the observation that to each dynamical triangulation we can associate an (open) generic region of the moduli space $\overline{\mathfrak{M}}_{g,N_0}$. This is just a restatement in the language of dynamical triangulations of the well-known combinatorial parametrization of $\overline{\mathfrak{M}}_{g,N_0}$ in terms of ribbon graph theory (i.e., of graphs which can be drawn on surfaces) [4]. Such a parametrization has been successfully exploited in Kontsevich–Witten theory, where ribbon graphs are used to construct the generating function of generalized volume forms on $\overline{\mathfrak{M}}_{g,N_0}$

connecting the rational cohomology of $\overline{\mathfrak{M}}_{g,N_0}$ to the KdV integrable hierarchy [5]. Recently, Y. Manin and P. Zograf [6], motivated by a conjecture due to C. Itzykson, have provided a remarkable connection between Kontsevich–Witten theory and the computation of the Weil–Petersson volume of $\overline{\mathfrak{M}}_{g,N_0}$. The large N_0 asymptotics of the resulting volume carries a striking resemblance with the (canonical) entropy function of DT theory, and one wonders if this represents an incidental fact or if it has some deeper origin. The explanation of such an origin, which to our knowledge has not appeared explicitly elsewhere, will be discussed here in detail. It has important implications for non-critical string theory (coupled to matter) that will be discussed in a forthcoming paper.

2   Triangulations as singular Euclidean surfaces

Let T denote a 2-dimensional simplicial complex with underlying polyhedron $|T|$ and f-vector $(N_0(T), N_1(T), N_2(T))$, where $N_i(T) \in N$ is the number of i-dimensional sub- simplices σ^i of T. If we consider the (first) barycentric subdivision of T, then the closed stars, in such a subdivision, of the vertices of the original triangulation T form a collection of 2-cells $\{\rho^2(i)\}_{i=1}^{N_0(T)}$ characterzing the polytope P barycentrically dual to T. A Regge triangulation of a 2-dimensional PL manifold M (without boundary), is a homeomorphism $|T_l| \to M$ where each face of T is geometrically realized by a rectilinear simplex of variable edge-lengths $l(\sigma^1(k))$ of the appropriate dimension. A dynamical triangulation $|T_{l=a}| \to M$ is a particolar case of a Regge PL-manifold realized by rectilinear and equilateral simplices of edge-length $l(\sigma^1(k)) = a$. Similarly, a rectilinear presentation $|P_{T_L}| \to M$ of the dual cell complex P (with edge-lengths $L = L(l)$) characterizes the Regge polytope (and its rigid equilateral specialization $|P_{T_a}| \to M$) baricentrically dual to $|T_l| \to M$. The metric structure of a Regge triangulation is locally Euclidean everywhere except at the vertices σ^0, (the *bones*), where the sum of the dihedral angles, $\theta(\sigma^2)$, of the incident σ^2's is in excess (negative curvature) or in defect (positive curvature) with respect to the 2π flatness constraint. The corresponding deficit angle r is defined by $r = 2\pi - \sum_{\sigma^2} \theta(\sigma^2)$, where the summation is extended to all 2-dimensional simplices incident on the given bone σ^0. For dynamical triangulations, the deficit angles are generated by the string of integers, the *curvature assignments*, $\{q(k)\}_{k=1}^{N_0} \in N^{N_0(T)}$, providing the numbers of top-dimensional simplices incident on the vertex $\sigma^0(i)$, and in terms of which we can write $r(i) = 2\pi - q(i) \arccos(1/2)$. As recalled, a Regge triangulation $|T_l| \to M$ defines on the PL manifold M a polyhedral metric with conical singularities associated with the bones $\{\sigma^0(i)\}_{i=1}^{N_0(T)}$ of the triangulation, but which is otherwise flat and smooth everywhere else. Such a metric has important special features, in particular it induces on the PL manifold M a geometrical structure which turns out to be a particular case of the theory of *Singular Euclidean Structure* (in the sense of M. Troyanov, and W. Thurston [7]). In particular, let us denote by $C|lk(\sigma^0(k))|$ the cone over the link of the vertex $\sigma^0(k)$. On any such a disc $C|lk(\sigma^0(k))|$ we can introduce a locally uniformizing complex coordinate $\zeta_k \in C$ in terms of which we can explicitly write down the singular (conformal) Euclidean metric locally characterizing the singular

Euclidean structure of $|T_l| \to M$, *viz.*,

$$ds^2_{(k)} \doteq e^{2u} \left| \zeta_k - \zeta_k(\sigma^0(k)) \right|^{-2\left(\frac{r(k)}{2\pi}\right)} |d\zeta_k|^2 , \tag{1}$$

where $r(k)$ is the corresponding deficit angle, and $u : B^2 \to R$ is a continuous function (C^2 on $B^2 - \{\sigma^0(k)\}$). Note that up to the presence of the conformal factor e^{2u}, we immediately recognize in such an expression the metric of a Euclidean cone of total angle $\theta(k) = r(k) - 2\pi$.

3 Regge surfaces and ribbon graphs

The geometrical realization of the 1-skeleton of the dual polytope $|P_{T_L}| \to M$ of a Regge surface is a 3-valent graph $\Gamma = (\{\varrho^o(k)\}, \{\rho^1(j)\})$. The vertex set $\{\varrho^o(k)\}_{k=1}^{N_2(T)}$ of such a graph is identified with the barycenters of the triangles $\{\sigma^2(k)\}_{k=1}^{N_2(T)} \in |T_l| \to M$, whereas each edge $\rho^1(j) \in \{\rho^1(j)\}_{j=1}^{N_1(T)}$ is generated by two half-edges $\rho^1(j)^+$ and $\rho^1(j)^-$ joined through the barycenters $\{W(h)\}_{h=1}^{N_1(T)}$ of the edges $\{\sigma^1(h)\}$ belonging to the original triangulation $|T_l| \to M$. Thus, if we formally introduce a degree-2 vertex at each middle point $\{W(h)\}_{h=1}^{N_1(T)}$, the actual graph naturally associated to the 1-skeleton of $|P_{T_L}| \to M$ is

$$\Gamma_{ref} = \left(\{\varrho^o(k)\} \bigsqcup \{W(h)\}, \left\{\rho^1(j)^+\right\} \bigsqcup \left\{\rho^1(j)^-\right\} \right), \tag{2}$$

the so called edge-refinement [4] of $\Gamma = (\{\varrho^o(k)\}, \{\rho^1(j)\})$. The relevance of such a notion stems from the observation that the natural automorphism group $\mathrm{Aut}(P_L)$ of $|P_{T_L}| \to M$, (i.e., the set of bijective maps $\Gamma \to \widetilde{\Gamma}$ preserving the incidence relations defining the graph structure), is not the automorphism group of Γ but rather the (larger) automorphism group of its edge refinement [4], i.e., $\mathrm{Aut}(P_L) \doteq \mathrm{Aut}(\Gamma_{ref})$. The locally uniformizing complex coordinate $\zeta_k \in C$ in terms of which we can explicitly write down the singular Euclidean metric (1) around each vertex $\sigma^0(k) \in |T_l| \to M$, provides a (counterclockwise) orientation in the 2-cells of $|P_{T_L}| \to M$. Such an orientation gives rise to a cyclic ordering on the set of half-edges $\{\rho^1(j)^{\pm}\}_{j=1}^{N_1(T)}$ incident on the vertices $\{\varrho^o(k)\}_{k=1}^{N_2(T)}$. According to such remarks, the 1-skeleton of $|P_{T_L}| \to M$ is a three-valent ribbon (or fat) graph [4], *viz.*, a graph Γ together with a cyclic ordering on the set of half-edges incident to each vertex of Γ. Since we want to keep track of the fact that Γ is associated with a polytope $|P_{T_L}| \to M$ dual to a Regge triangulation, it is natural to attach to the oriented boundaries $\partial M(\Gamma)$ the length structure naturally associated with $|P_{T_L}| \to M$, and glue to the boundary components of the ribbon graph Γ a corresponding set of punctured discs $D^2_{\mathrm{punct}}(j)$. The punctures can be profitably identified with the vertices $\{\sigma^0(k)\}$ of the Regge triangulation $|T_l| \to M$ which, upon barycentrical dualization, gives rise to $|P_{T_L}| \to M$. In this way (the edge-refinement of) the 1-skeleton of

$|P_{T_L}| \to M$ acquires from the underlying **Regge** triangulation the structure of a metric ribbon graph. In this connection, it is worth noticing that not all trivalent ribbon graphs are dual to regular triangulations (e.g., degenerate triangulations with *pockets*, where two triangles are incident on a vertex, give rise to trivalent dual ribbon graphs with loops, i.e. Regge polytopes containing 2-gons). Our analysis must be extended to such degenerate case as well, and in what follows we shall refer to such generalized triangulations and their associated baricentrically dual polytopes. In general, the set of all possible metrics on a (trivalent) ribbon graph Γ with given edge-set $e(\Gamma)$ can be characterized (see [6], Definition 3.1) as a topological space homeomorphic to $R_+^{|e(\Gamma)|}$, ($|e(\Gamma)|$ denoting the number of edges in $e(\Gamma)$), topologized by the standard ϵ-neighborhoods $U_\epsilon \subset R_+^{|e(\Gamma)|}$. On such a space there is a natural action of Aut(Γ), the automorphism group of Γ defined by the homomorphism Aut$(\Gamma) \to \mathfrak{G}_{e(\Gamma)}$ where $\mathfrak{G}_{e(\Gamma)}$ denotes the symmetric group over $|e(\Gamma)|$ elements. Thus, the resulting space $R_+^{|e(\Gamma)|}/$Aut(Γ) is a differentiable orbifold, (the quotient of a manifold by a finite group). Let Aut$_\partial(P_L) \subset$ Aut(P_L), denote the subgroup of ribbon graph automorphisms of the (trivalent) 1-skeleton Γ of $|P_{T_L}| \to M$ that preserve the (labeling of the) boundary components of Γ. Then, the space of 1-skeletons of generalized Regge polytopes $|P_{T_L}| \to M$, with $N_0(T)$ labelled boundary components, on a surface M of genus g can be defined by [4,8]

$$RGP_{g,N}^{\mathrm{met}} = \bigsqcup_{\Gamma \in RGB_{g,N}} \frac{R_+^{|e(\Gamma)|}}{\mathrm{Aut}_\partial(P_L)}, \tag{3}$$

where the disjoint union is over the subset of all trivalent ribbon graphs (with labelled boundaries) satisfying the stability topological condition $2 - 2g - N_0(T) < 0$, and which are dual to generalized triangulations. It follows, (see [4] theorems 3.3, 3.4, and 3.5), that $RGP_{g,N}^{\mathrm{met}}$ is locally modelled on a stratified space constructed from the components (rational orbicells) $R_+^{|e(\Gamma)|}/$Aut$_\partial(P_L)$ by means of a (Whitehead) expansion and collapse procedure for ribbon graphs, which basically amounts to collapsing edges and coalescing vertices. Explicitly, if $L(t) = tL$ is the length of an edge $\rho^1(j)$ of a ribbon graph $\Gamma_{L(t)} \in RGP_{g,N}^{\mathrm{met}}$, then, as $t \to 0$, we get the metric ribbon graph $\widehat{\Gamma}$ which is obtained from $\Gamma_{L(t)}$ by collapsing the edge $\rho^1(j)$. Note that the cells of top dimension $(6g - 6 + 2N_0(T))$ are labelled by trivalent graphs, and since the dual of any such a graph is a (generalized) triangulation $|T_L| \to M$ of the surface M, it follows that the cells $\frac{R_+^{|e(\Gamma)|}}{\mathrm{Aut}_\partial(P_L)}$ can be equivalently labelled by the triangulation associated with $|P_{T_L}| \to M$, (see also [8]). Also note that $RGP_{g,N}^{\mathrm{met}}$ can be extended to a suitable closure $\overline{RGP}_{g,N}^{\mathrm{met}}$ which is described in detail in [8].

3.1 Quadratic differentials and singular Euclidean structures

Recall [4] that a holomorphic quadratic differential, ψ, (a transverse traceless rank two tensor), on a Riemann surface M is defined, in a locally uniformizing complex coordinate chart (U, ζ), by a holomorphic function $\mu : U \to C$ such that

$\psi = \mu(\zeta)d\zeta \otimes d\zeta$. For genus $g > 0$ the complex vector space of quadratic differentials, $Q(M)$, is non-empty with complex dimension $\dim_C Q(M) = 3g - 3$, ($\dim_C Q(M) = 1$, for $g = 1$). The geometry of $Q(M)$ is directly related with the characterization of the Teichmüller space of M, $\mathcal{T}_g(M)$, the space of all conformal structures on M under the equivalence relation given by pullback by diffeomorphisms isotopic to the identity map $id : M \to M$. It is well known that $\mathcal{T}_g(M)$ is a smooth finite dimensional manifold that can be identified with a $6g - 6$ (R)-dimensional cell defined by the open unit ball (in a suitable norm) in the space of quadratic differentials. Conversely, the tangent space to $\mathcal{T}_g(M)$ at a reference quadratic differential ψ, is C-anti-linear isomorphic to $Q(M)$. In other words $Q(M)$ can be canonically identified with the cotangent space to $\mathcal{T}_g(M)$. If we assume that M has finite hyperbolic area, then the Weil–Petersson metric is defined on the cotangent space $Q(M) \simeq T^*\mathcal{T}_g(M)$ as the inner product between quadratic differentials corresponding to the L^2-norm defined by

$$\|\psi\|_{WP}^2 \doteq \int_M h^{-2}(\zeta)\,|\psi(\zeta)|^2\,|d\zeta|^2\,, \tag{4}$$

where $\psi \in Q(M)$ and $h(\zeta)\,|d\zeta|$ is the hyperbolic metric $(1 - |\zeta|^2)^{-2}\,|d\zeta|^2$ on M. Note that $\frac{\psi}{h}$ is a Beltrami differential on M, (i.e., locally we can write $\frac{\psi}{h}\frac{\partial}{\partial\zeta} \otimes \overline{d\zeta}$ as a harmonic representative of a Kodaira-Spencer class), thus if we introduce a basis $\{\mu_\alpha\}_{\alpha=1}^{3g-3}$ of the vector space of harmonic Beltrami differentials on M, we can write

$$G_{\alpha\bar\beta} = \int_M \mu_\alpha\overline{\mu_\beta}h(\zeta)\,|d\zeta|^2\,, \tag{5}$$

for the components of the Weil–Petersson metric on the tangent space to $\mathcal{T}_g(M)$. Since $G_{\alpha\bar\beta}$ is Kähler we can introduce the corresponding Weil–Petersson Kähler form according to

$$\omega_{WP} = \sqrt{-1}G_{\alpha\bar\beta}d\zeta^\alpha \wedge d\overline{\zeta^\beta}. \tag{6}$$

Let $\pi_1(M)$ denote the fundamental group of the surface M. The quotient of $\mathcal{T}_g(M)$ by the action of the outer automorphism group of $\pi_1(M)$, i.e., $\mathfrak{M}_g = \mathcal{T}_g(M)/Out(\pi_1(M))$, is the Riemann moduli space parametrizing conformal equivalence classes of Riemann surfaces of genus g. If we fix λ distinct points $x_1, \ldots, x_\lambda \in M$, corresponding to which M is punctured, then the corresponding moduli space acquires one extra (complex) dimension for each puncture, i.e.,

$$\dim_C \mathfrak{M}_{g,\lambda} = 3g - 3 + \lambda. \tag{7}$$

It is well-known that $\mathfrak{M}_{g,\lambda}$ is connected and that, although in general non complete, it admits a stable curve compactification (Deligne–Mumford) into an orbifold space $\overline{\mathfrak{M}}_{g,\lambda}$. Since the Kähler potential of ω_{WP} can be made invariant under the mapping class group $Out(\pi_1(M))$, the Weil–Petersson volume 2-form ω_{WP} on $\mathcal{T}_g(M)$ descends on $\mathfrak{M}_{g,\lambda}$, and it has a (differentiable) extension, in the sense of orbifold, to $\overline{\mathfrak{M}}_{g,\lambda}$.

The geometry of the moduli space is strictly connected to a combinatorial stratification of $\overline{\mathfrak{M}}_{g,\lambda}$ in terms of the graphical data describing inequivalent quadratic differentials. The rationale underlying such stratification is the observation that a quadratic differential ψ may be pictured by a transverse measured foliation, (i.e., a foliation generalizing the slicing of C by horizontal lines endowed with the transverse measure $|dy|$), which induces on M singular Euclidean structures whose moduli can be connected with $\overline{\mathfrak{M}}_{g,\lambda}$. If the measured foliation generated by a quadratic differential $\psi \in Q(M) - \{0\}$ has closed horizontal leaves (up to a set of measure zero on the surface), then such a ψ decomposes the surface M into the maximal ring domains foliated by the closed leaves, (typically annuli or punctured disks). Such a property characterizes the Jenkins-Strebel (JS) quadratic differentials, and allows for the parametrization of the holomorphic structure of a Riemann surface into the combinatorial data of metric ribbon graphs. Such a parametrization defines a bijective mapping (a homeomorphism of orbifolds) between the space of ribbon graphs $RGB_{g,N}^{\mathrm{met}}$ and the moduli space $\mathfrak{M}_{g,N}$ of Riemann surfaces M of genus g with N ordered marked points (punctures) [4,8],

$$h : \mathfrak{M}_{g,N} \times R_+^N \rightarrow RGB_{g,N}^{\mathrm{met}} \ (M, L_i) \longmapsto \Gamma, \tag{8}$$

where $(L_1, \ldots, L_N)$ is an ordered n-tuple of positive real numbers and Γ is a metric ribbon graphs with N labelled boundary lengths $\{L_i\}$ defined by the corresponding JS quadratic differential. Note that such a bijection extends suitably to $\overline{\mathfrak{M}}_{g,N} \times R_+^N \rightarrow \overline{RGB}_{g,N}^{\mathrm{met}}$. In the DT framework it can be explicitly described according to

Proposition 1. *Let*

$$\mathcal{DT}\left[\{q(k)\}_{k=1}^{N_0}\right]$$
$$\doteq \left\{|T_{l=a}| \rightarrow M \ : q(\sigma^0(k)) = q(k) \geq 1, \quad k = 1, \ldots, N_0(T)\right\}, \tag{9}$$

denote the set of distinct generalized dynamically triangulated surfaces of genus g, with a given set of ordered curvature assignments $\{q(i), \sigma^0(i)\}_{i=1}^{N_0(T)}$ over its $N_0(T)$ labelled vertices. Then, we can associate with each polygonal 2-cell $\{\rho^2(i)\}_{i=1}^{N_0(T)}$ of the polytope $|P_{T_a}| \rightarrow M$ dual to a $|T_{l=a}| \rightarrow M \in \mathcal{DT}[\{q(k)\}_{k=1}^{N_0}]$ the quadratic differential

$$\rho^2(k) \longmapsto \psi(k) \doteq -\frac{\left(\frac{\sqrt{3}}{3}aq(k)\right)}{4\pi^2\zeta^2(k)}d\zeta(k) \otimes d\zeta(k), \tag{10}$$

where $\zeta(k)$ is a locally uniformizing complex coordinate in the unit disk. The set of such $\{\psi(k)\}_{i=1}^{N_0(T)}$ naturally restricts to the edges $\{\rho^1(j)\}_{j=1}^{N_1(T)}$ of $|P_{T_a}| \rightarrow M$ and characterizes uniquely a meromorphic quadratic differential $\psi d\zeta \otimes d\zeta$ associated with $|T_{l=a}| \rightarrow M$. By associating to the JS quadratic differential ψ the corresponding punctured Riemann surface $(M/\{\sigma^0(i)\}_{i=1}^{N_0(T)}\}, \psi)$ it follows that, if the stability *condition $g \geq 0$, $2 - 2g - N_0(T) > 0$ holds, then there is an injective mapping*

$$h_{T_a} : \mathcal{DT}\left[\{q(k)\}_{k=1}^{N_0}\right] \longrightarrow \mathfrak{M}_{g,N_0} \times \left(\frac{\sqrt{3}}{3}a\right) N_+^{N_0} \tag{11}$$

$$\left(|T_{l=a}| \to M; \{q(k)\}_{k=1}^{N_0}\right) \longmapsto \left(M/\{\sigma^0(i)\}_{i=1}^{N_0(T)}\}, \psi\right),$$

defining the dynamical triangulations $|T_{l=a}| \to M \in \mathcal{DT}[\{q(k)\}_{k=1}^{N_0}]$ as distinguished elements in the orbicells $h\left(\dfrac{R_+^{|e(\Gamma)|}}{\mathrm{Aut}_\partial(P_L)}\right) \subset \mathfrak{M}_{g,N_0}$.

Proof. The expression (10) for the quadratic differential $\psi(k)$ associated with each polygonal cell $\{\rho^2(i)\}_{i=1}^{N_0(T)}$, of the polytope $|P_{T_a}| \to M$ dual to a $|T_{l=a}| \to M \in \mathcal{DT}[\{q(k)\}_{k=1}^{N_0}]$, follows immediately from the Schwartz-Christoffel transformation applied to the polygonal boundary of $\rho^2(i)$, and, in the generic case of a metric ribbon graph, has been worked out in detail in [4] (§§4 and 5) to which we refer for details. The injectivity of the map follows from the unicity of such JS differential. Surjectivity fails since the metric ribbon graphs associated with generalized triangulations in $\mathcal{DT}[\{q(k)\}_{k=1}^{N_0}]$ do not span the whole $RGB_{g,N_0}^{\mathrm{met}}$, but only a subset of $RGB_{g,N_0}^{\mathrm{met}}$ generated by ribbon graphs whose labelled boundaries have lengths which are provided by $\oint_{\partial(\rho^2(k))} \sqrt{\psi(k)} = \left(\frac{\sqrt{3}}{3}a\right) q(k)$.

More generally, the construction in [4] always gives rise to an explicit map from the whole space of metric ribbon graphs $RGB_{g,N_0}^{\mathrm{met}}$ to $\mathfrak{M}_{g,N} \times R_+^N$. Such a map is defined by a meromorphic quadratic differential $\tilde{\psi}$ which can be locally given by [4],

$$\left(|P_{T_L}| \to M\right) \to \tilde{\psi} \doteq \begin{cases} \psi(h)|_{\rho^1(h)} = dz(h) \otimes dz(h), \\[4pt] \psi(j)|_{\rho^0(j)} = \dfrac{9}{4}w(j)dw(j) \otimes dw(j), \\[4pt] \psi(k)|_{\rho^2(k)} = -\dfrac{\left[L(\partial(\rho^2(k)))\right]^2}{4\pi^2 \zeta^2(k)} d\zeta(k) \otimes d\zeta(k), \end{cases} \tag{12}$$

where $z(h)$ is a complex uniformizing coordinate associated with the generic edge $\rho^1(h)$ of the Regge polytope $|P_{T_L}| \to M$, and defined in the strip $U_{\rho^1(h)} \doteq \{z(h) \in C | 0 < \mathrm{Re}z(h) < L(\rho^1(h))\}$, $L(\rho^1(h))$ being the length of the edge; $w(j)$ is a complex coordinate associated to the generic 3-valent vertex $\rho^0(j)$, (with $w(\rho^0(j)) = 0$), and defined in an open neighborhood $U_{\rho^0(j)}$ of $w(j) = 0$. Finally $\zeta(k)$ is a locally uniformizing complex coordinate defined in the unit disc $U_{\rho^2(k)}$ and associated with the generic face $\rho^2(k)$ with boundary length $L(\partial(\rho^2(k)))$. Note that $L(\partial(\rho^2(k))) = \sum_{h=1}^{q(k)} L(\rho^1(h))$. The (punctured) Riemann surface associated, via

$\widetilde{\psi}$, to the generic trivalent ribbon graph is defined by

$$\left(M/\left\{\sigma^0(i)\right\}_{i=1}^{N_0(T)}, \widetilde{\psi}\right)$$
$$\doteq \bigcup_{\{\rho^0(j)\}} (U_{\rho^0(j)}, \psi(j)) \bigcup_{\{\rho^1(h)\}} (U_{\rho^1(h)}, \psi(h)) \bigcup_{\{\rho^2(k)\}} (U_{\rho^2(k)}, \psi(k)). \quad (13)$$

Let us denote by

$$\Omega_{T_a}(\{q(k)\}_{k=1}^{N_0}) \doteq \frac{R_+^{|e(\Gamma)|}}{\text{Aut}_\partial(P_{T_a})}, \quad (14)$$

the rational cell associated with the dynamical triangulation

$$|T_{l=a}| \to M \in \mathcal{DT}\left[\{q(k)\}_{k=1}^{N_0}\right].$$

Such a rational cell contains the ribbon graph associated with the polytope $|P_{T_a}| \to M$ dual to $|T_{l=a}| \to M$, and all (trivalent) metric ribbon graphs $|P_{T_L}| \to M$ with the same combinatorial structure of $|P_{T_a}| \to M$ but with all possible length assignments $\{L(\rho^1(h))\}_1^{N_1(T)}$ associated with the corresponding set of edges $\{\rho^1(h)\}_1^{N_1(T)}$. Note that $\text{Aut}_\partial(P_{T_a})$ is isomorphic to the isotropy subgroup, in the mapping class group $Out(\pi_1(M))$, of the generic Riemann surface $(M/\{\sigma^0(i)\}_{i=1}^{N_0(T)}, \widetilde{\psi})$ in the orbicell $\Omega_{T_a}(\{q(k)\}_{k=1}^{N_0})$. Let us define the forgetful mapping

$$\phi : \mathfrak{M}_{g,N_0} \times R_+^{N_0} \to \mathfrak{M}_{g,N_0}, \quad (15)$$

as the map which forgets the decorations on $\mathfrak{M}_{g,N_0} \times R_+^{N_0}$ provided by the perimeters $\left(\frac{\sqrt{3}}{3}a\right)\{q(k)\}_{i=1}^{N_0(T)}$, (the map ϕ associates to $(M/\{\sigma^0(i)\}_{i=1}^{N_0(T)}, \widetilde{\psi})$ the underlying Riemann surface $(M/\{\sigma^0(i)\}_{i=1}^{N_0(T)}, [\widetilde{\psi}])$ corresponding to the conformal class of the metric defined by the quadratic differential $\widetilde{\psi}$). Then the map

$$h_{T_a} : \Omega_{T_a}\left(\{q(k)\}_{k=1}^{N_0}\right) \longrightarrow \mathfrak{M}_{g,N_0} \times R_+^{N_0} \xrightarrow{\phi} \mathfrak{M}_{g,N_0} \quad (16)$$
$$(|P_{T_L}| \to M) \longmapsto \left(M/\{\sigma^0(i)\}_{i=1}^{N_0(T)}, \widetilde{\psi}\right) \longmapsto \left(M/\{\sigma^0(i)\}_{i=1}^{N_0(T)}, [\widetilde{\psi}]\right),$$

defines an open cell in the moduli space $\mathfrak{M}_{g,N_0}$, labelled by the given (generalized) dynamical triangulation $|T_{l=a}| \to M \in \mathcal{DT}[\{q(k)\}_{k=1}^{N_0}]$. As $|T_{l=a}| \to M$ varies in the discrete set $\mathcal{DT}[\{q(k)\}_{k=1}^{N_0}]$, we get in this way a cell decomposition of $\mathfrak{M}_{g,N_0}$ which is parametrized by $\mathcal{DT}[\{q(k)\}_{k=1}^{N_0}]$, i.e.

$$\mathfrak{M}_{g,N_0} = \bigsqcup_{T_a \in \mathcal{DT}[\{q(k)\}_{k=1}^{N_0}]} h_{T_a}\left(\Omega_{T_a}(\{q(k)\}_{k=1}^{N_0})\right). \quad (17)$$

Note that, as the notation suggests, such a cell decomposition depends on the set of curvature assignments considered $\{q(k)\}_{k=1}^{N_0}$. Distinct curvature assignments $\{\widehat{q(k)}\}_{k=1}^{N_0}$ give rise to possibly distinct cell decompositions of $\mathfrak{M}_{g,N_0}$ labelled by the corresponding $\mathcal{DT}[\{\widehat{q(k)}\}_{k=1}^{N_0}]$. We can integrate the Weil–Petersson volume form associated to (6) over $\overline{\mathfrak{M}}_{g,N_0}$ and obtain

$$\mathrm{VOL}\left(\overline{\mathfrak{M}}_{g,N_0}\right) = \frac{1}{N_0!} \int_{\overline{\mathfrak{M}}_{g,N_0}} \frac{\omega_{WP}^{3g-3+N_0(T)}}{(3g-3+N_0(T))!}, \tag{18}$$

where $\mathrm{VOL}\left(\overline{\mathfrak{M}}_{g,N_0}\right)$ denotes the Weil–Petersson volume of the (compactified) moduli space $\overline{\mathfrak{M}}_{g,N_0}$, and where we have divided by $N_0(T)!$ since we have to factor out the labelling of the $N_0(T)$ punctures. On the other hand, the integral appearing at the right member of (18) can be evaluated as an orbifold integration,(the orbifold integration over moduli space is defined in [9]) by exploiting the explicit cell-decomposition of $\overline{\mathfrak{M}}_{g,N_0}$ provided by (17), *viz.*,

$$\frac{1}{N_0!} \int_{\overline{\mathfrak{M}}_{g,N_0}} \frac{\omega_{WP}^{3g-3+N_0(T)}}{(3g-3+N_0(T))!}$$

$$= \sum_{T \in \mathcal{DT}[\{q(i)\}_{i=1}^{N_0}]} \frac{1}{|\mathrm{Aut}_\partial(P_{T_a})|} \int_{\Omega_{T_a}(\{q(k)\}_{k=1}^{N_0})} \frac{h_{T_a}^*(\omega_{WP})^{3g-3+N_0(T)}}{(3g-3+N_0(T))!}, \tag{19}$$

where $h_{T_a}^*(\omega_{WP})$ denotes the pull-back of the Weil–Petersson form to Ω_{T_a} under the map h_{T_a}, (thus $h_{T_a}^*(\omega_{WP})$ is the form associated with the quadratic differentials $\widetilde{\psi}(|P_{T_L}| \to M)$), and where the summation is over all distinct dynamical triangulations with given unlabelled curvature assignments weighted by the order $|\mathrm{Aut}_\partial(P_T)|$ of the automorphisms group of the corresponding dual polytope. Thus, we get the nice relation

$$\mathrm{VOL}\left(\overline{\mathfrak{M}}_{g,N_0}\right)$$

$$= \sum_{T \in \mathcal{DT}[\{q(i)\}_{i=1}^{N_0}]} \frac{1}{|\mathrm{Aut}_\partial(P_{T_a})|} \int_{\Omega_{T_a}(\{q(k)\}_{k=1}^{N_0})} \frac{h_{T_a}^*(\omega_{WP})^{3g-3+N_0(T)}}{(3g-3+N_0(T))!}. \tag{20}$$

If we introduce the average value of the Weil–Petersson volume of the cells $\Omega_{T_a}(\{q(k)\}_{k=1}^{N_0})$ according to

$$\left\langle \Omega_{T_a}(\{q(k)\}_{k=1}^{N_0}) \right\rangle$$

$$\doteq \frac{\sum_{T \in \mathcal{DT}[\{q(i)\}_{i=1}^{N_0}]} \frac{1}{|\mathrm{Aut}_\partial(P_{T_a})|} \int_{\Omega_{T_a}(\{q(k)\}_{k=1}^{N_0})} \frac{h_{T_a}^*(\omega_{WP})^{3g-3+N_0(T)}}{(3g-3+N_0(T))!}}{\sum_{T \in \mathcal{DT}[\{q(i)\}_{i=1}^{N_0}]} \frac{1}{|\mathrm{Aut}_\partial(P_{T_a})|}}, \tag{21}$$

then we can write

$$\sum_{T \in \mathcal{DT}[\{q(i)\}_{i=1}^{N_0}]} \frac{1}{|\mathrm{Aut}_\partial(P_{T_a})|} = \frac{\mathrm{VOL}\left(\overline{\mathfrak{M}}_{g,N_0}\right)}{\left\langle \Omega_{T_a}(\{q(k)\}_{k=1}^{N_0})\right\rangle}. \tag{22}$$

Recall that the relation between the counting of distinct (generalized) triangulations with given curvature assignments $\mathcal{DT}[\{q(i)\}_{i=1}^{N_0}]$, and the enumeration of all distinct (generalized) triangulations with a given number N_0 of unlabelled vertices is provided by

$$\mathrm{Card}\left[\mathcal{DT}(N_0)\right] = \sum_{\{q(i)\}_{i=1}^{N_0}} \sum_{T \in \mathcal{DT}[\{q(i)\}_{i=1}^{N_0}]} \frac{1}{|\mathrm{Aut}_\partial(P_{T_a})|}, \tag{23}$$

where the summation $\sum_{\{q(i)\}_{i=1}^{N_0}}$ is over all possible curvature assignments $\{q(i)\}_{i=1}^{N_0}$ on the N_0 vertices. Thus, according to (22) we get

$$\mathrm{Card}\left[\mathcal{DT}(N_0)\right] = \mathrm{VOL}\left(\overline{\mathfrak{M}}_{g,N_0}\right) \sum_{\{q(i)\}_{i=1}^{N_0}} \frac{1}{\left\langle \Omega_{T_a}(\{q(k)\}_{k=1}^{N_0})\right\rangle}, \tag{24}$$

which provides a non trivial connection between the canonical entropy function, $\mathrm{Card}\left[\mathcal{DT}(N_0)\right]$, for (generalized) dynamical triangulations and the Weil–Petersson volume $\mathrm{VOL}\left(\overline{\mathfrak{M}}_{g,N_0}\right)$ of the moduli space. In order to discuss some of the consequences of such a connection, let us recall that the large N_0 asymptotics of $\mathrm{Vol}_{W-P}(\overline{\mathfrak{M}}_{g,N_0})$ has been discussed by Manin and Zograf [6]. They obtained

$$\mathrm{Vol}_{W-P}(\overline{\mathfrak{M}}_{g,N_0}) = \pi^{2(3g-3+N_0)}$$

$$\times (N_0+1)^{\frac{5g-7}{2}} C^{-N_0} \left(B_g + \sum_{k=1}^{\infty} \frac{B_{g,k}}{(N_0+1)^k}\right),$$

where $C = -\frac{1}{2} j_0 J_0'(j_0)$, ($J_0(z)$ is the Bessel function, j_0 its first positive zero); (note that $C \simeq 0.625\ldots$). The genus dependent parameters B_g are explicitly given by

$$\begin{cases} B_0 \dfrac{1}{A^{1/2}\Gamma(-\frac{1}{2})x_0^{1/2}}, & B_1 = \dfrac{1}{48}, \\[2ex] B_g \dfrac{A^{\frac{g-1}{2}}}{2^{2g-2}(3g-3)!\,\Gamma(\frac{5g-5}{2})x_0^{\frac{5g-5}{2}}} \left\langle \tau_2^{3g-3}\right\rangle, & g \geq 2, \end{cases} \tag{25}$$

where $A \doteq -j_0^{-1} J_0'(j_0)$, and $\left\langle \tau_2^{3g-3}\right\rangle$ is a Kontsevich–Witten intersection number, (the coefficients $B_{g,k}$ can be computed similarly-see [6] for details). However, it is well known that, for large $N_0(T)$, we get

$$\mathrm{Card}\left[\mathcal{DT}[N_0]\right] \sim N_0(T)^{\gamma_g+N_0-3} e^{\mu_0 N_0}\left(1 + O\left(\frac{1}{N_0}\right)\right), \tag{26}$$

where

$$\gamma_g \doteq \frac{5g - 1}{2},\tag{27}$$

is the genus-g pure gravity critical exponent, and μ_0 is a (non-universal) parameter independent of g and N_0. Thus, from a moduli theory point of view, the above analysis explains the origin of the critical exponents in the canonical entropy function (26). According to the factorization (24) and to the Manin–Zograf asymptotics, the critical exponents for pure gravity provide the subleading polynomial asymptotics of the volume of the moduli space $\overline{\mathfrak{M}}_{g,N_0}$. The curvature assignments term

$$\sum_{\{q(i)\}_{i=1}^{N_0}} \frac{1}{\left\langle \Omega_{T_a}(\{q(k)\}_{k=1}^{N_0}) \right\rangle},\tag{28}$$

provides, in this setting, the discretized counterpart of the Liouville measure over the conformal mode (represented, according to $\oint_{\partial(\rho^2(k))} \sqrt{\psi(k)} = \left(\frac{\sqrt{3}}{3}a\right) q(k)$, by the curvature assignments $q(k)$). It is interesting to note that such a term can at most renormalize the non-universal parameter μ_0. In order to extend the factorization (24) to the case of gravity coupled to matter, we need to discuss how the discretization of the matter fields can be represented on the Riemann surfaces $(M/\{\sigma^0(i)\}_{i=1}^{N_0(T)}\}, \widetilde{\psi})$ associated with dynamical triangulations in $\mathcal{DT}[\{q(i)\}_{i=1}^{N_0}]$. Such a representation is strictly related to the study of certain matter-related determinant line bundles on $\overline{\mathfrak{M}}_{g,N_0}$ and we be explicitly dealt with in a forthcoming paper.

Acknowledgements. One of us (M.C.) would like to thank Jan Ambjørn for useful discussion. This work was supported in part by the Ministero dell'Universita' e della Ricerca Scientifica under the PRIN project *The geometry of integrable systems*.

References

1. Ambjørn, J., Durhuus, B., Jonsson, T. (1997): Quantum geometry. Cambridge monograph on mathematical physics, Cambridge University Press, Cambridge
2. Catterall, S., Mottola, E. (2000): Reconstructing the conformal mode in simplicial gravity. Nucl. Phys. B [Proc. Suppl.] **83, 84**, 748–750;
   Catterall, S., Mottola, E. (2000): The conformal mode in 2D simplicial gravity; hep-lat/9906032
3. Menotti, P., Peirano, P. (1996): Functional integration on two-dimensional Regge geometries. Nucl. Phys. B **473**, 426
   (1995): Phys. Lett. B **353**, 444
4. Mulase, M., Penkava, M. (1998): Ribbon graphs, quadratic differentials on Riemann surfaces, and algebraic curves defined over $\overline{Q}$. math-ph/9811024 v2;
   Strebel, K. (1984): Quadratic differentials. Springer-Verlag, Berlin Heidelberg New York
5. Kontsevich, M. (1992): Intersection theory on moduli space of curves. Commun. Math. Phys. **147**, 1;
   Witten, E. (1991): Two dimensional gravity and intersection theory on moduli space. Surveys in Diff. Geom. **1**, 243

6. Manin, Y., Zograf, P. (2000): Invertible cohomological field theories and Weil–Petersson volumes. Ann. Inst. Fourier. **50** 519–535.
   Zograf, P. (1992): Weil–Petersson volumes of moduli spaces and the genus expansion in two dimensional gravity. math.AG/9811026;
   Kaufmann, R., Manin, Yu., Zagier, D. (1996): Higher Weil–Petersson volumes of moduli spaces of stable n-pointed curves. Commun. Math. Phys. **181**, 763–787
7. Troyanov, M. (1991): Prescribing curvature on compact surfaces with conical singularities. Trans. Amer. Math. Soc. **324**, 793;
   Troyanov, M. (1986): Les surfaces euclidiennes a' singularites coniques. L'Enseignment Mathematique **32**, 79;
   Thurston, W.P. (1998): Shapes of polyhedra and triangulations of the sphere. Geometry and Topology Monographies, Vol. **1**, p. 511
8. Looijenga, E. (1992–93): Intersection theory on Deligne–Mumford compactifications. Seminaire BOURBAKI, N **768**
9. Penner, R.C. (1992): Weil–Petersson volumes. J. Diff. Geom. **35**, 559–608

Nöther Conserved Quantities and Entropy in General Relativity

G. Allemandi, L. Fatibene, M. Ferraris, M. Francaviglia, M. Raiteri

Abstract. In the framework of classical field theories, the notions of conserved quantities and entropy for stationary solutions of covariant theories of gravitation, e. g. of Einstein field equations of General Relativity, are discussed. Nöther theorem is used to provide the correct definition of (covariantly) conserved quantities such as mass and angular momentum. The variation of entropy is then defined as a macroscopical quantity which satisfies a Clausius-like first principle of thermodynamics. Finally, a proposal for the entropy of non-stationary solutions is discussed.

1 Introduction

In the geometric framework for classical field theories, the kinematics of a system is described by sections of a configuration bundle C over a spacetime manifold M, which is assumed to be connected, paracompact and of arbitrary dimension m. A variation principle based on a Lagrangian describes the dynamics by inducing field equations which in turn single out the *solutions* (also called *critical sections*). Several important physical theories can in fact be dealt with in this framework, e.g. the theories of the gravitational field, possibly in interaction with electromagnetic or gauge fields, as well as with other kinds of matter.

If the field theory under consideration is generally covariant, i. e. covariant with respect to all spacetime diffeomorphisms, one can use Nöther theorem to obtain conservation laws. These allow in turn to define physical quantities (e.g. mass, angular momentum and so on) which parametrize classes of solutions. In other words one can find families of solutions, which depend on some parameters which are called Nöther charges. Since Nöther charges are not local but rather quasilocal in their intrinsic geometric nature (see [1,2]), the geometrical framework (or some other equivalent way of controlling global properties of the relevant objects) plays an essential role.

One could consider these parameters as representing the macroscopical degrees of freedom of the system. In analogy with gas thermodynamics, it seems reasonable to look for dynamical laws which select, among all the macroscopic processes in General Relativity, the ones which are physically allowed. In fact, these processes have to obey at least some conservation law (e.g. conservation of the total energy of the system) as S. Hawking proposed in [3] for the particular case of black hole solutions of vacuum Einstein theory. Such conservation laws are encoded by the so-called *first principle of black hole thermodynamics* (see e. g. [4–7]) which constrains the infinitesimal variations of the relevant macroscopical quantities occuring in physical processes. A temperature parameter naturally arises in the first principle of thermodynamics as a Pfaffian integrating factor. This fact led to an initial *empasse* since, at least classically, a black hole is expected to be, by definition, perfectly *cold* (see, e.g. [1,8]).

However, Hawking himself discovered the so-called *evaporation* of black holes [3], according to which the black hole actually radiates because of quantum corrections. This radiation is due to the production of couples of particles near the horizon and the temperature appearing in the first principle can be eventually related to the temperature of such a radiation.

There are many other ways of defining black hole entropy (for a review see, e.g. [9,10]); the relation between different prescriptions deserves further investigation and the ultimate meaning of black hole entropy is still partially unknown. Under the most fundamental and general perspective, the entropy of a system is related to the lack of information in describing its degrees of freedom. Accordingly, it seems to be better defined and described from a microscopical point of view by microstate counting. Unfortunately, there is yet no general agreement on how to define microstates of a gravitational system. String theories seem today to be the most promising framework in this context although they seem to apply only to extreme and quasi-extreme black holes. String results, moreover, cannot be compared with the macroscopic results since, in the extreme case, the temperature defined through a semiclassical approximation *á la* Hawking identically vanishes. As a consequence, the entropy of extreme black holes is still out of the scope of the macroscopic framework we are going to present.

A different approach to the problem was proposed by J. D. Brown and J. W. York [6,11]. They described the gravitational field in a region of finite spatial extent as a microcanonical system. Accordingly, they defined an ad hoc action functional, *the microcanonical action functional*, by adding suitable boundary terms to the standard gravitational Lagrangian. The microcanonical action functional was then used to define the density of states of the microcanonical system and subsequently to give a definition of entropy based on a statistical approach. It has been recently proved that the results obtained in this way coincide with the ones obtained via Nöther theorem, provided the correct boundary conditions are imposed. Hence, quasilocal Brown–York quantities (entropy as well) turn out to be nothing but Nöther charges relative to well-defined spacetime symmetries [12,13].

Hereafter we propose a geometrical definition of entropy by requiring that it satisfies a form of the first principle of black hole thermodynamics in which mass and angular momentum (as well as all extensive variables) are calculated via Nöther theorem while the temperature and the angular velocity (as well as all intensive variables) are considered as given. It has been recognized [14] that the values of the temperature and of the angular velocity cannot be predicted within any macroscopic and geometric framework and they have to be provided by means of some other physical consideration; in other words they have to be introduced as "external data".

A similar to ours but insufficient prescription was first given by R. Wald et al. in [4]. This prescription was later shown to rely on unessential hypotheses (see [5]) and because of them was proven to fail in many cases of physical interest. Furthermore, as already noticed in [5], the original prescription of [4] cannot be generalized to non-stationary black holes. These drawbacks are completely overcome by our new geometrical framework.

The mathematical framework we are going to use hereafter is based on the variational formulation of field theories. The Lagrangian of the theory is geometrically interpreted as a bundle morphism, field equations are derived from its variation and from Hamilton's principle [15–18]. The general covariance principle, as well as the gauge covariance principle when gauge fields are also considered, provide us a group of symmetries for the theory. These symmetries can be in turn interpreted from a geometrical and physical viewpoint as the Nöther generators of mass, angular momentum and all other possible gauge charges [5,15,18,19]. Using the first order Lagrangian formalism developed in [20] we will show that is possible to define algorithmically the conserved quantities up to the choice of a reference background metric which, in turn, can be considered a definition of the "vacuum state" for the gravitational field. The first principle of black hole thermodynamics [4] is then used to define the entropy for natural and gauge natural theories. This scheme encompasses in a correct and easily manageable mathematical framework not only the case of General Relativity in interaction with the electromagnetic field, but also other gravitational theories as well as gauge field theories and various kinds of matter fields interacting with them.

2 Natural and gauge natural theories

In this section we shall shortly review the mathematical foundations of classical field theories and, in particular, we shall consider natural and gauge natural theories which are just enough to describe gauge fields theories, General Relativity and various kinds of matter (bosonic as well as fermionic matter fields). For further details see [21,22] and references therein.

Spacetime is a connected paracompact manifold M of dimension $\dim M = m$ with local coordinates $\{x^\mu\}$. We shall denote by $ds = dx^1 \wedge dx^2 \wedge \ldots \wedge dx^m$ the standard local volume form induced by the coordinates.

Let (C, M, π, F) be the *configuration bundle*. We denote by $\{x^\mu, y^i\}$ fibered coordinates on it. The local expression of a *configuration*, namely a section $\sigma :$ $M \to C$, reads as $\sigma : x^\mu \mapsto (x^\mu, y^i = \sigma^i(x))$ and describes the fields y^i in each point of spacetime.

A *Lagrangian of order k* is a bundle morphism between the k-jet bundle $J^k C$ of the configuration bundle C and the bundle $A_m(M)$ of m-forms on spacetime:

$$L : J^k C \to A_m(M), \tag{1}$$

and in local coordinates

$$L = \mathcal{L}(x^\mu, y^i, y^i{}_{\mu_1}, \ldots, y^i{}_{\mu_1 .. \mu_k}) ds, \tag{2}$$

where $(x^\mu, y^i, y^i{}_{\mu_1}, \ldots, y^i{}_{\mu_1 .. \mu_k})$ are fibered coordinates over $J^k C$ induced by the fibered coordinates (x^μ, y^i) on C. Throughout the paper, all morphisms with values in a bundle of forms will be identified if necessary with the corresponding differential forms (see [23]).

Vector fields on the configuration bundle which are tangent to fibers are called *vertical vector fields* and they are sections of the so-called *vertical bundle* $V(C)$; in local coordinates a vertical vector field reads as $X = X^i(x, y)\partial_i$.

The *action functional*

$$A_D(\sigma) = \int_D (j^k\sigma)^*L, \tag{3}$$

is obtained by integrating over a compact region $D \subset M$ the pull–back of the Lagrangian L via (the prolongation of) a configuration σ. According to the Hamilton principle, field equations are obtained imposing the action $A_D(\sigma)$ to be stationary along any compactly supported vertical field X. In this way one obtains the standard Euler–Lagrange field equations. The variation of the Lagrangian with respect to the vertical vector field X can be defined as a global bundle morphism $\delta L : J^kC \rightarrow V^*(J^kC) \otimes A_m(M)$ by:

$$\langle \delta L \circ j^k\sigma \mid j^k X \rangle = \frac{d}{ds}\left(L \circ J^k\Psi_s \circ j^k\sigma\right)\Big|_{s=0}, \tag{4}$$

where Ψ_s is the (vertical) flow of the vector field X on C. For the moment, we shall consider only natural theories, namely those field theories for which the following two axioms hold:

(a) The configuration bundle C is natural, i. e. each spacetime vector field ξ naturally (i. e. preserving the commutators) lifts to a vector field $\hat{\xi}$ over the configuration bundle and consequently up to any J^kC. This means that for each $\xi = \xi^\mu\partial_\mu \in \mathfrak{X}(M)$ we have a naturally lifted vector field $\hat{\xi} \in \mathfrak{X}(C)$ defined by:

$$\hat{\xi} = \xi^\mu(x^\mu)\partial_\mu + \xi^i(x^\mu, y^j)\partial_i. \tag{5}$$

The components ξ^i are understood to depend algorithmically on the components ξ^μ together with their derivatives up to some finite order s, which is called the order of the natural bundle C (for example, $s = 1$ when dealing with tensors or tensor densities, while $s = 2$ for linear connections).

(b) The Lagrangian is *natural* (or *covariant*), i. e. each spacetime vector field (which because of axiom (a) induces an infinitesimal transformation in C) is an *infinitesimal Lagrangian symmetry*. This means that $(J^k\Phi_s)^*L = L$, where Φ_s is the flow of $\hat{\xi}$. This covariance condition is equivalent to the following infinitesimal condition:

$$\langle \delta L \mid j^k \pounds_\xi \sigma \rangle = \pounds_\xi L, \tag{6}$$

where $\pounds_\xi \sigma$ is the *Lie derivative of a section* σ defined as

$$\pounds_\xi \sigma = T\sigma(\xi) - \hat{\xi} \circ \sigma \equiv (\pounds_\xi y^i)\partial_i. \tag{7}$$

In local coordinates it is possible to express the variation (4) of the Lagrangian as:

$$\langle \delta L \mid j^k X \rangle = (p_i X^i + p_i{}^\mu X^i{}_\mu + \cdots) \otimes ds, \tag{8}$$

where p^i, $p_i{}^\mu$, are the so-called *naive momenta* and they are defined as the partial derivatives of the Lagrangian density $\mathcal{L}$ with respect to the fields and their derivatives. Integrating (covariantly) by parts this formula we obtain the *first variation formula*

$$\langle \delta L \mid j^k X \rangle = \langle \mathbb{E}(L) \mid X \rangle + \mathrm{Div}\langle \mathbb{F}(L) \mid j^{k-1} X \rangle, \tag{9}$$

where

$$\begin{aligned}
&\mathbb{E}(L) : J^{2k} \to V^*(C) \otimes A_m(M), \\
&\mathbb{F}(L) : J^{2k-1} \to V^*(J^{k-1}C) \otimes A_{m-1}(M),
\end{aligned} \tag{10}$$

are two global bundle morphisms. $\mathbb{E}(L)$ is called the *Euler–Lagrange morphism* and $\mathbb{F}(L)$ is called the *Poincaré–Cartan morphism*; generally the latter depends on a background connection γ (see [23]). One can now easily apply Hamilton's principle to obtain field equations, which are thence expressed as the Euler–Lagrange equations $\mathbb{E}(L) \circ j^{2k}\sigma = 0$. In this formalism a generalized form of the Nöther theorem arises if we consider that (6) can be recast as:

$$\langle \delta L \mid j^k \pounds_\xi \sigma \rangle = \mathrm{Div}(i_\xi L), \tag{11}$$

where i_ξ denotes the inner product. By considering that the Lie derivative generates a vertical vector field, see (7), and inserting the first variation formula (9) into (11) we obtain that for each vector field ξ on spacetime (thence a symmetry for the covariant Lagrangian) it is possible to define a covariantly conserved current, called the *Nöther current*, by setting:

$$\mathcal{E}(L, \xi) = \langle \mathbb{F}(L) \mid j^{k-1}\pounds_\xi \sigma \rangle - i_\xi L. \tag{12}$$

This is a *weakly conserved current*, i.e. we can write:

$$\begin{aligned}
\mathrm{Div}\, \mathcal{E}(L, \xi) &= \mathcal{W}(L, \xi), \\
\mathcal{W}(L, \xi) &= -\langle \mathbb{E}(L) \mid \pounds_\xi \sigma \rangle.
\end{aligned} \tag{13}$$

Here $\mathcal{W}$ is the so-called *work-form* and it vanishes on shell, i.e. along solutions.

Integrating covariantly by parts these quantities, it is possible to decompose both $\mathcal{E}$ and $\mathcal{W}$ as sums of two (essentially unique and global) morphisms:

$$\mathcal{E}(L, \xi) = \tilde{\mathcal{E}}(L, \xi) + \mathrm{Div}\, \mathcal{U}(L, \xi), \tag{14}$$

$$\mathcal{W}(L, \xi) = \mathcal{B}(L, \xi) + \mathrm{Div}\, \tilde{\mathcal{E}}(L, \xi), \tag{15}$$

where $\mathcal{B}(L, \xi)$ identically vanishes and it generates the so-called *Bianchi identities*, $\tilde{\mathcal{E}}(L, \xi)$ is the *reduced current* and it vanishes on-shell; finally $\mathcal{U}(L, \xi)$ is the *superpotential* of the theory. We stress that all of these quantities are global and algorithmically defined out of the Lagrangian by covariant integration by parts.

We also stress that all of the morphisms above are defined at the bundle level (while the corresponding spacetime differential forms which they originate are obtained, *a posteriori*, by evaluation on a specific configuration σ). The difference $\mathcal{E} - \tilde{\mathcal{E}}$ between the Nöther current and the reduced current is *strongly conserved*; by this we mean that it is closed also off-shell, i.e. out of the set of solutions. This holds because of $\mathrm{Div}^2 = 0$.

Finally, since $\tilde{\mathcal{E}}$ vanishes on-shell, we can define the Nöther charges $\hat{Q}(L, \xi)$ for a solution σ of the field equations as:

$$\hat{Q}(L, \xi) = \int_B \mathcal{U}(L, \xi, \sigma) = \int_B \mathcal{U}(L, \xi) \circ j^{2k-2}\sigma, \tag{16}$$

where B is any $(m-2)$-region of spacetime.

In general, once a Lagrangian L is given, this formalism allows one to calculate a charge $Q(L, \xi)$ for the symmetries generated by ξ [15]. For example, it would be natural to expect that the conserved quantity associated with the timelike vector field $\xi = \partial_t$ relative to an ADM foliation of spacetime in $t = $ constant spacelike hypersurfaces Σ_t, could be identified with the mass of the system. However, we shall see that this is not true in general and some correction is still needed.

Let us first specialize the framework so far developed to the case of vacuum General Relativity described by the Hilbert Lagrangian $L_H = \sqrt{g} R > ds$. The superpotential $\mathcal{U}(L_H, \xi)$ associated to the spacetime vector field ξ on M is the so-called *Komar superpotential* [15]:

$$\mathcal{U}_K(L_H, \xi) = \sqrt{g} g^{\nu\rho} \nabla_\rho \xi^\mu ds_{\mu\nu}. \tag{17}$$

This superpotential is uniquely and geometrically defined, but it is affected by the *anomalus factor problem* ([24]). In fact, when we are interested in defining the mass and the angular momentum of a solution of Einstein field equations, the quantity (16) computed with the Komar potential (17) does not in general coincide with the expected value. For example, the result is just one-half of the expected mass for the Kerr solution when one lets B tend to spatial infinity ∞ and ξ is chosen to be ∂_t. A solution to this problem is obtained by a suitable redefinition of the conserved quantities as it is suggested by a *covariant ADM formalism* [20]. Let us come back to a general natural Lagrangian L; the variation $\delta_X Q(L, \xi)$ of the *corrected* conserved charge $Q(L, \xi)$ is given by:

$$\delta_X Q(L, \xi) = \int_B \delta_X \mathcal{U}(L, \xi) - i_\xi \langle \mathbb{F}(L) \mid j^{k-1} X \rangle, \tag{18}$$

where $X \in V(C)$ is a vertical vector field which drags solutions into solutions (i.e. it is a solution of linearized field equations) and where evaluation on a solution σ has been implicitly assumed in the right hand side of (18). As first pointed out in [20,25], if there exists an $(m-2)$-form $\Delta(L, \xi)$ such that

$$\delta_X \Delta(L, \xi)\Big|_B = i_\xi \langle \mathbb{F}(L) \mid j^{k-1} X \rangle\Big|_B, \tag{19}$$

from (18) we obtain that the conserved quantity Q is given by

$$Q(L, \xi) = \int_B [\mathcal{U}(L, \xi) - \Delta(L, \xi)]. \tag{20}$$

Here $\Delta(L, \xi)$ is the *correction term*. By its own definition the prescription (19) takes into account the behaviour of the fields on the "boundary" B and for this reason it allows to handle a wide class of solutions with different boundary behaviours. It reproduces, in particular, the Regge–Teitelboim correction term for asymptotically flat metric solutions and, in more general cases, it allows to avoid anomalous factor problems as well as divergence problems; see [20,26]. In fact when dealing with the Hilbert Lagrangian L_H such a *covariant* form $\Delta(L_H, \xi)$ does not exist in general. However Δ can be always defined if we introduce a background connection $\bar{\Gamma}^\alpha_{\beta\mu}$ (which, even if not strictly necessary, may be considered as the Levi–Civita connection of a background metric $\bar{g}$). In this case we have

$$\Delta(L_H, \xi) = -\xi^{[\lambda} w^{\mu]}{}_{\alpha\beta} g^{\alpha\beta} \sqrt{g}\, ds_{\lambda\mu}, \tag{21}$$

where we set

$$\begin{cases} w^\mu{}_{\alpha\beta} = u^\mu{}_{\alpha\beta} - \bar{u}^\mu{}_{\alpha\beta}, \\ u^\mu{}_{\alpha\beta} = \Gamma^\mu{}_{\alpha\beta} - \delta^\mu{}_{(\alpha} \Gamma^\lambda{}_{\beta)\lambda}, \\ \bar{u}^\mu{}_{\alpha\beta} = \bar{\Gamma}^\mu{}_{\alpha\beta} - \delta^\mu{}_{(\alpha} \bar{\Gamma}^\lambda{}_{\beta)\lambda}. \end{cases} \tag{22}$$

Moreover, the relevant form of the identity (19), namely

$$\delta_X \Delta(L_H, \xi)\Big|_B = i_\xi \langle \mathbb{F}(L_H) \mid j^1 X \rangle\Big|_B, \tag{23}$$

holds true provided we assume the boundary condition

$$\delta_X g|_B = 0. \tag{24}$$

Equation (20) can be also regarded as a definition of conserved quantities *relative* to a fixed background. Let us in fact choose the *first order Lagrangian*

$$L_1 = \left[\sqrt{g} R - d_\lambda(\sqrt{g} w^\lambda_{\alpha\beta} g^{\alpha\beta}) - \sqrt{\bar{g}} \bar{R} \right] ds. \tag{25}$$

For this Lagrangian and the boundary condition (24), one has $\Delta = 0$. Because of the divergence term, the Lagragian L_1 is first order in g but it is second order in $\bar{g}$; the two metrics do not really interact and they are both subjected to Einstein vacuum field equations. From a variational viewpoint both g and $\bar{g}$ have to be regarded as dynamical fields. From a physical viewpoint g is related to the gravitational field, while $\bar{g}$ cannot be directly observed but it selects the *ground state* for the gravitational field. This interpretation is related to the fact that metrics are sections of a bundle which is not a vector bundle (as it often happens for matter fields): thence there is no canonical choice of a preferred ground state. We stress however that also with vector matter fields the canonical choice (the zero section) may be not the most convenient choice (think about the Higgs field in a low energy regime!). All extensive physical quantities have to be regarded as *relative* in nature and thence they have in fact to be referred to an arbitrary ground state.

The superpotential for the Lagrangian L_1 is:

$$\mathcal{U}(L_1, \xi) = [\sqrt{g}\, g^{\nu\rho} \nabla_\rho \xi^\mu + \xi^{[\mu} w^{\nu]}{}_{\alpha\beta} g^{\alpha\beta} \sqrt{g} - \sqrt{\bar{g}}\, \bar{g}^{\nu\rho} \bar{\nabla}_\rho \xi^\mu]\, ds_{\mu\nu}. \tag{26}$$

When integrated at space infinity ∞, according to definition (20), this superpotential provides the numerical values for the physically expected quantities. The variation of these conserved quantities is given in general by

$$\begin{aligned}
\delta_X Q(L_1, \xi) &= \int_\infty \delta_X \mathcal{U}(L_1, \xi), \\
&= \int_\infty \delta_X \mathcal{U}_K(L_H, \xi) - i_\xi \langle \mathbb{F}(L_H) | j^1 X \rangle \\
&\quad - \int_\infty \delta_X \mathcal{U}_K(\bar{L}_H) - i_\xi \langle \mathbb{F}(\bar{L}_H) | j^1 X \rangle,
\end{aligned} \tag{27}$$

where $\bar{L}_H = \sqrt{\bar{g}}\,\bar{R}$ denotes the Hilbert Lagrangian for the background metric $\bar{g}$.

It is also possible to define the conserved quantities within a finite closed $(m-2)$-surface B homologous to spatial infinity (i. e. when $B - \infty$ is the boundary of a $(m-1)$ dimensional region). The integrated form in (18) is in fact closed whenever ξ is a Killing vector for the solution σ, see [5].

The framework described above can be generalized to the case of gauge natural theories, i. e. to Lagrangians which beside differential covariance admit a gauge group invariance. This generalization is necessary since Yang-Mills theories are not correctly described in a natural framework. In the particular case of electromagnetism there is a global gauge invariance with respect to the group $U(1)$ in addition to the general covariance, i.e. covariance with respect to general coordinate transformations. The local potential A_μ of the electromagnetic field is not a tensor and the Lagrangian of the theory is invariant with respect to the following local gauge trasformations:

$$A'_\mu(x') = J_\mu{}^\nu \left(A_\nu(x) + d_\nu \alpha(x) \right), \tag{28}$$

where $J_\mu{}^\nu$ is the Jacobian of the coordinate trasformations and $\alpha(x)$ is a local function on spacetime. Gauge trasformations (28) can be recast as

$$A'_\mu = J_\mu{}^\nu e^{-i\alpha}(A_\nu e^{i\alpha} + d_\nu e^{i\alpha}), \tag{29}$$

from which they appear as pointwise changes of the global phase. The absolute phase is not a physical quantity, every choice describing the same physical situation. This is due to the so-called *hole argument* and it is related to the existence of compactly supported gauge transformations [27]. By means of these compact supported transformations one can easily produce two solutions of field equations which differ on a compact support. If determinism has to be saved, then the only way out is to assume that any pair of such solutions actually represents the same physical situation.

Mathematically, this can be interpreted as a Lie group G acting on the configuration bundle; on the other hand we would like the dynamical field A_μ to be a global section of a suitable configuration bundle. This can be done since there exists a *structure bundle*, namely a principal bundle $(P, M, \pi; G)$, such that the dynamical field A_μ is a connection over P. Let $\mathrm{Con}(P)$ be the bundle of principal connections on P, which is an affine bundle "associated" to P. A principal automorphism $\Phi \in \mathrm{Aut}(P)$ acts pointwise on the configuration fields by means of elements of the gauge group G. The choice of a particular structure bundle selects the class of dynamical fields on a (possibly) topologically non-trivial manifold M.

A *gauge natural bundle* is a fiber bundle $C = (C, M, \pi, F)$, with a rule which naturally (i. e. preserving compositions) associates to each automorphism of P an automorphism on C. One can construct the so-called *gauge natural prolongation bundle of order (r,s)*, with $s \geq r$, denoted by $W^{(r,s)}P = J^r P \times_M L^s(M)$, which is a principal bundle and where $L^s(M)$ is the frame bundle of order s, see [22]. A *gauge natural bundle* is a bundle associated to some gauge natural prolongation of P. For example $\mathrm{Con}(P)$ is a gauge natural bundle of order $(1, 1)$ associated to the structure bundle P, i. e. it refers to $J^1 P \times_M L(M)$. An automorphism $\Phi \in \mathrm{Aut}(P)$ naturally induces an automorphism $\Phi^{(r,s)}$ on the gauge natural prolongation $W^{(r,s)}P$ of P and thence a gauge transformation Φ_C on the gauge natural bundle C associated to P. Let $\Xi = \xi^\mu(x)\partial_\mu + \xi^A(x)\rho_A$ be the infinitesimal generator of a 1-parameter family of principal automorphisms on P (ρ_A is a local basis for vertical right invariant vector fields over P). It projects onto the vector field $\xi = \xi^\mu \partial_\mu$ on M. This vector field Ξ over P induces in a canonical way a vector field $\tilde{\Xi}$ over C which still projects onto ξ. The following diagram thence commutes

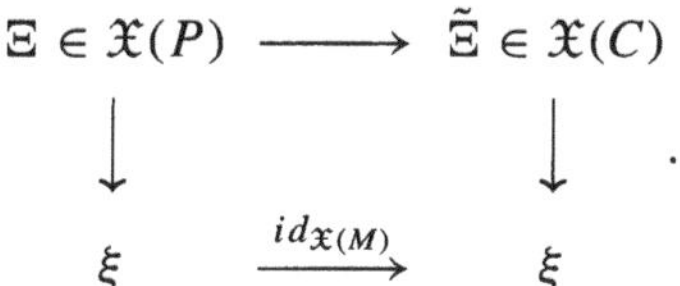

$$\Xi \in \mathfrak{X}(P) \longrightarrow \tilde{\Xi} \in \mathfrak{X}(C)$$

$$\downarrow \qquad\qquad\qquad \downarrow$$

$$\xi \xrightarrow{\; id_{\mathfrak{X}(M)} \;} \xi$$

As in the natural case, in local fiber coordinates (x^λ, y^i) of C we have $\tilde{\Xi} = (\xi^\mu \partial_\mu + \xi^i \partial_i)$ where ξ^i depends on (ξ^μ, ξ^A) and their derivatives up to some finite order (r, s), respectively. The pair (s, r) is called the order of the gauge natural bundle C so that, e.g., connections are gauge natural objects of order $(1, 1)$.

In this gauge natural framework we can define the Lie derivative of a section σ with respect to Ξ, as:

$$\pounds_\Xi \sigma = T\sigma(\xi) - \tilde{\Xi} \circ \sigma. \tag{30}$$

With this formalism we can reproduce step by step the same costruction considered above for natural objects. By means of the Nöther theorem we can obtain the conserved quantities; see [5,28]. Gauge natural theories are the most satisfactory framework to treat symmetries in a field theory, once the transformation laws (or equivalently the Lie derivatives of the dynamical fields) are specified. We can easily calculate the conserved charges, corrected by the anomalous factor problem, with a formula analogous to equation (18), see [5].

We stress that in some particular topological situation (e. g. when M is Minkowski space), it is also possible to regard electromagnetism as a natural theory, since is possible to define $U(1)$-bundles P which carry also a natural structure [29]. The natural structure is thence inherited by the bundle of principal connections where A_μ lives. However the natural structure bundles so constructed are necessarily trivial so that the natural framework for electromagnetism can be defined for trivial structure bundles only.

3   Boundary terms and conserved quantities

A number of problems about uniqueness arises and they have to be solved before the conservation laws (20) or (18) are used to define the physical conserved quantities. Are these quantities unique? Do they depend on the boundary terms in the Lagrangian? How do they depend on the reference background $\bar{g}$?

Just to quote an example, the vector field $\xi = \partial_t$ commonly used to define the mass of a gravitating system is coordinate-depending. It is not intrinsically nor globally related to the solution g, in the sense that there is no way of algorithmically defining it out of the metric alone. The problem about globality is not a serious one; e.g., since ∂_t is the generator of an ADM foliation in t=constant hypersurfaces, this foliation may be required to be a global foliation for a wide class of reasonable spacetimes (e.g. all globally hyperbolic spacetimes) [30]. Of course the foliation itself is not unique. This non-uniqueness can be interpreted as the freedom in choosing the evolution of the observers or, in other words, the vector field $\xi = \partial_t$ which describes the spacetime velocity of such observers. Hence, different foliations (i.e., different evolutions of the same initial configuration of the observer) give rise to different conserved quantities which are eventually related by "boost" transformations; see [31,32]. The situation is analogous to the case of Mechanics where an observer is identified with a connection on the configuration bundle $(C, \mathbb{R}, p; Q)$ which, being a first rank distribution (thence involutive), induces a horizontal foliation $C = \mathbb{R} \times Q$ of the space of events [33]. Different connections, i. e. different ways to lift the standard vector field ∂_t, correspond to different choices for the velocity of the observer.

Secondly, we stress that the integrand terms of equation (27) are invariant with respect to the addition of a pure divergence to the Lagrangian. Thence, even if the Lagrangian itself is not uniquely defined, the conserved quantities are never affected by pure divergence terms possibly added to L. Indeed, as better clarified below, different divergences added to the Lagrangian merely reflect on different boundary conditions the solutions have to satisfy. But, as we already remarked, the second term in the right hand side of (18) takes somehow into account the boundary conditions, so that additional terms in the superpotential (due to additional divergences of the Lagrangian) are exactly balanced by additional terms in the term $i_\xi \langle \mathbb{F} \mid J^{k-1} X \rangle$.

Finally, the definition of conserved quantities we give here using Nöther theorem depends on the choice of the background field $\bar{g}$. Thanks to the structure of the first order Lagrangian L_1, see (25), the dependence on the background can be interpreted as fixing the zero energy level. This is reasonable first of all since there is no interaction between g and $\bar{g}$. If there were interactions one could in principle observe the field $\bar{g}$ by means of its effects on g and thence $\bar{g}$ would be endowed with a direct physical meaning.

We thence believe that one can accept those ambiguities on a physical ground. We stress, however, that these ambiguities are not unrelated to each other. First of all there is a well known relation between pure divergence terms in the Lagrangian and boundary conditions of the problem. In fact for a generic k-order Lagrangian L the corresponding Poincaré–Cartan morphism $\mathbb{F}(L)$ is of order $k - 1$ (see definition (10)) so that when the $(k - 1)$-jet prolongation of the deformation X is required to vanish at the boundary, field equations can be obtained. Of course, in some particular cases, e.g. for the Lagrangian L_1, the Poincaré–Cartan morphism may happen to be "degenerate" so that weaker boundary conditions are enough. Thence, in the family of boundary terms which can be added to the Lagrangian there can be some representative which simplifies boundary conditions. The addition of a boundary term in the action certainly results in different minimal boundary conditions (namely, different boundary behaviour for the variation X) to be imposed to obtain field equations.

The above relation between boundary conditions and divergence terms is further revealed by the following fact: if one decides to rely on the Lagrangian which defines suitable boundary conditions and defines variations which preserve these boundary conditions, then the ambiguity in defining conserved quantities disappears. In fact, to any Lagrangian with its boundary term, we associate different conserved quantities. However, the differences in the *variation* of these different conserved quantities vanishes because of the constraint that variations of fields have to satisfy to respect boundary conditions themselves. Consequently, different representatives of the Lagrangian give rise to the same numerical values for the conserved quantities just by implementing variations in the appropriate fundamental class.

Another relation may be found between the choice of a background and boundary conditions. This relation is very closely related to what happens with the first order Lagrangian L_1: once a background $\bar{g}$ has been fixed, one is implicitly restricting from the set of all the metrics g to the ones which agree with $\bar{g}$ on the boundary ∂D of the region chosen. Namely, the fixing of the background identifies the boundary behaviour of the solution, i.e. its boundary conditions. The reason to do that is a topological one: the set of (pseudo-Riemannian) metrics over a manifold M may be disconnected into homotopical classes depending on the topology of M and the signature chosen. In other words, there may exist two metrics g_1 and g_2 which cannot be continuously deformed one onto the other. In that case, especially when dealing with conserved quantities in a classical (i. e. non quantum) context, trying to define the relative energy between g_1 and g_2 is quite meaningless, since one cannot smoothly pass from one configuration to the other. In those cases, it is physically reasonable to limit the definition of conserved quantities to deal with solutions which lie in the same given homotopy class. Of course, fixing a background is equivalent to selecting a homotopy class of metrics. If now one considers a generic, possibly not metric, field theory it can happen that all the sections are homotopic with each other (e. g. this happens for sections of a vector bundle). In that case, of course, one can choose the background arbitrarily. However the set of sections is in general disconnected into homotopy classes and in that case one is forced to choose a background to select the homotopy class one is interested in. Accordingly, we argue that *in principle* one should choose a background in all field theories, even if in some particular cases this choice may turn out to be trivial. This is in agreement with the fact that *in principle* the energy of any system has always a relative meaning. We stress, however, that studying in its full generality the space of all homotopy classes of sections (in particular of pseudo-Riemannian metrics) is by no means easy. Consequently, studying just the variation of conserved quantities, which is done infinitesimally, is often easier than dealing with finite conserved quantities since in that case one can avoid to choose explicitly a background with given boundary conditions.

Since the integrand is a closed form under some quite general hypotheses [5], when defining the corrected conserved quantities as in (18), it is possible to choose the closed surface in the homology class of space infinity, i. e. there is a homological invariance for the definition of the conserved charges. It is important to stress that this property holds true only if we assume explicitly that ξ is a symmetry vector for the solution σ of the Euler–Lagrange equations (namely we must have $\pounds_\xi \sigma = 0$). In the case of gravitational theories, then, ξ has to be Killing.

We also stress that the conserved quantities corrected in this way are at the basis of a *covariant* Hamiltonian formulation of field theories. When dealing with General Relativity they can in fact be considered as a suitable generalization of the ADM conserved quantities [4,20,34]. When integrated over a surface of *finite* spatial extent which is homologically equivalent to ∞ they again give the same correct total conserved quantities. Integrating the appropriate superpotential on a surface enclosing the metric singularities, we have obtained in [5] the expected conserved quantities for the Kerr and the Kerr–Newman solutions. It is also possible to treat correctly some non-asymptotically flat solutions like BTZ and Taub–Bolt solutions, see [35,36]. Finally this formalism is applicable to calculate energy and angular momentum for a system of N relativistic particles with a dilatonic Lagrangian in two dimensional gravity; see [37]. Remarkably enough, in the latter case it was also shown how the energy of the system, calculated via the Nöther theorem, correctly reproduces the Hamiltonian dynamics for the motion of the particles.

Within this formalism it is also possible to calculate the energy for a spatially finite gravitational system. In this situation it is even possible to compare the results obtained starting from the Lagrangian (27) with the results previously obtained by Brown and York in [6,11]; these two approaches give the same result for the quasilocal conserved quantities. The covariantly conserved quantity calculated via the Nöther theorem is analogous with the quasilocal energy calculated by Brown and York, provided we choose as a generating vector field ξ the normal vector u^μ to the spacelike slices of spacetime rather than the time traslation generator ∂_t, see [12].

4   Entropy

The first principle of thermodynamics in a classical statistical theory was first introduced by Clausius in the form:

$$\delta U = T\delta S - \delta W, \tag{31}$$

where U denotes the internal energy, T the temperature, W the mechanical work and S the entropy [38]. A formally analogous principle was proven to exist for stationary black holes. In this case, the variational equation for the macroscopical quantities of the system is [9,10]:

$$\delta M = \frac{\mathrm{K}}{8\pi}\delta A + \Omega\delta J + b^A\delta Q_A, \tag{32}$$

where K is the surface gravity of the black hole, A is the area of the horizon and $T = \frac{\mathrm{K}}{2\pi}$ is the temperature of the black hole horizon (in geometric units) so that $S = \frac{A}{4}$ is the entropy of black hole [3,8]; Ω is the angular velocity of the horizon. The last term on the right hand side of the equation (32) is directly related to the gauge fields; Q_A is the gauge charge and b^A a suitable constant which makes the first principle integrable. If we are considering a natural theory, no gauge symmetry is present so that this extra term is not present. In a general gauge natural theory, one has pure gauge symmetries, generated by vertical right invariant vector fields $\Xi = b^A\rho_A$ on the structure bundle P (then inducing infinitesimal gauge transformations on the configuration bundle C). These are the symmetries ρ_A inducing the Nöther charges Q_A. To define entropy in General Relativity one has essentially to assume that a timelike Killing vector exists and to have null quadrupole momentum. These latter requirements are equivalent to claim that the system under consideration is non-radiating (thence defining closed systems); they are satisfied in particular by stationary black holes [9].

Under these hypotheses it is possible to prove that, given the vector field $\Xi = \partial_t + \Omega\partial_\varphi - b^A\rho_A$ over the structure bundle P, projecting onto the vector field $\xi = \partial_t + \Omega\partial_\varphi$ on M, the definition for the variation of entropy naturally arises from (32), under the explicit form:

$$\delta_X S = \frac{1}{T}\delta_X Q(L, \Xi) = \frac{1}{T}\int_\infty \delta_X \mathcal{U}(L, \Xi) - i_\xi \langle \mathbb{F}(L) \mid j^{k-1}X\rangle, \tag{33}$$

namely, it is nothing but the variation of a (corrected) Nöther charge. Let us consider as an example the first order Lagrangian L_1. This is a suitable choice to deal with solutions of vacuum Eistein equations. Because of the form of L_1 and of the invariance of the integrand of Eq. (33) with respect to boundary terms, the variation $\delta_X S$ splits into a contribution coming from the dynamical metric g, namely $\delta_X S_g$, plus an analogous contribution $\delta_X S_{\bar{g}}$ coming from the reference background metric. The relative entropy is thence $\delta_X S = \delta_X S_g - \delta_X S_{\bar{g}}$ (see [35]).

If the solution σ for a gauge natural field theory is stationary, Ξ is a simmetry vector for σ and the vertical vector field X is a solution of the linearized field equations, then it follows that:

$$\mathrm{Div}(\delta_X \mathcal{U}(L, \Xi) - i_\xi \langle \mathbb{F}(L) \mid j^{k-1} X \rangle) = 0, \tag{34}$$

so that the $(m - 2)$-form under parentheses is closed [5]. Under these hypotheses it is then possible to define entropy integrating on a surface B which wraps around the singularities (if any) and which belongs to the same homotopy class of spatial infinity ∞. This means that we can compute the variation of the entropy (33) on a surface B by means of:

$$\delta_X S = \frac{1}{T} \int_B \delta_X \mathcal{U}(L, \Xi) - i_\xi \langle \mathbb{F}(L) \mid j^{k-1} X \rangle. \tag{35}$$

We stress that no requirements on the existence of horizons nor on their properties is required here. Moreover, no coordinate conditions have to be imposed. In fact, the surface B where the integral is evaluated needs only to enclose the singularities (if any) and to belong to the same homotopy class of the space infinity surface. Obviously, when a bifurcate Killing horizon exists and we integrate (35) on the particular surface $\bar{\Sigma}$ which is called the *bifurcation surface*, our definition (35) reproduces the same formula given by Wald [4]:

$$\delta_X S = \int_{\bar{\Sigma}} \delta_X \mathcal{U}(L, \Xi) \tag{36}$$

In fact, the term $i_\xi \langle \mathbb{F}(L) \mid j^{k-1} X \rangle$ vanishes on this surface as, by definition of $\bar{\Sigma}$, ξ is equal to zero on it. The existence of such a surface, however, is guaranteed only in the maximal analytic extension of the solution and does not at all constitute a simplification of the problem. This fact prevents, e.g. , to apply Wald formula to the Taub–Bolt solution which does not admit such a bifurcate horizon. Furthermore, a direct application of Wald's prescription in more trivial topologies (e.g. Schwarzschild) is also hindered by the quite essential use of Kruskal-like coordinates around the horizon (see [35]), which are not easy to find by explicit integration.

The prescription for entropy given in (35) has been applied to to case of Kerr–Newman solutions and the result obtained, after choosing a suitable lift of the vector field ξ to the configuration bundle, is in agreement with the area theorem that states entropy to be one quarter of the area of the horizon [5]. The same definition (35) allows us to compute entropy for BTZ solutions for standard $(2+1)$ General Relativity with a negative cosmological constant [35]. In this case the results can be once again obtained also by means of Wald prescription, i. e. integrating on the bifurcate Killing horizon (which exists). Of course the BTZ solution is not asymptotically flat so that ADM mass is undefined and one has to use our generalized notion of mass. The results obtained in the two different ways are of course the same, but the calculation based on the general formula (35) are by far much simpler with respect to the calculation based on (36).

The last and the most relevant example we treated is the application of our formalism to Taub–Bolt solutions, which do not satisfy Wald regularity hypotheses. In this case it has been shown in [36] that, although the Misner string is not enclosed by a Killing horizon and the solution is not asymptotically flat, the geometrical approach we propose can be succesfully applied. Using as background metric the Taub-NUT solution with the suitable matching conditions at infinity, the results obtained agree perfectly with the already known results obtained from a statistical point of view [39,40]. The deviation from the one-quarter area paradigm is thence reproduced as it was predicted by different techniques by Hawking, Hunter and Page.

Thence this example shows that our prescription is a genuine and much more powerfull generalization of the prescription proposed by Wald in his paper [4]), since it can be applied in a much larger class of solutions [5,35,36], which moreover do not need asymptotically flat conditions at infinity.

Our prescription seems to apply also to the case of non-stationary solutions of Einstein equations, provided suitable changes are made. We suppose that the same first principle of thermodynamics still holds for these solutions in the case of non-stationary black holes without a quadrupole momentum. From a physical point of view this is a model for an oscillating black hole which does not emit radiation. In this case it is no longer true that ξ is a Killing vector for the solution; this implies that (34) is no longer a closed differential form and we have to add to (35) a volume term integrated on the volume Π enclosed between the surface B and the spatial infinity ∞. The proposal for a definition of entropy in this case is the following formula:

$$T\delta_X S_{dyn} = \int_B (\delta_X \mathcal{U} - i_\xi \langle \mathbb{F}(L) | j^{k-1} X \rangle) + \int_\Pi \omega(L, \sigma, \xi, X), \tag{37}$$

where ω is a suitable $(m-1)$ form which can be explicitly built out of the Lagrangian. This formula is a corrected version of the one (which is wrong, as he also realized and stated) given by R. Wald in [4]. Our proposal [41] satisfies in particular the reliability conditions imposed in [4]. Explicit examples are still to be worked out due to the lack of exact *global* solutions which correspond either to the vacuum case or to the case with matter described by global Lagrangians.

5   Conclusions and perspectives

A review of the results obtained in the last years, appling Nöther theorem to the calculation of entropy for singular solutions of Einstein field equations, has been given in this paper. The fundamental role played by the parameters of the theory as the background metric and the boundary conditions for the definition of the corrected conserved quantities has been stressed. In the last section, some concrete applications and examples of the theory were briefly discussed.

There are still same topics which deserve investigation. We have still not found a global non-stationary solution to test our last proposal (37). All of the solution we have found in literature, in fact, are either not well defined from a variational point of view or they contain fields (for example pressure or velocities of some dust-like matter) which are not well defined in a geometrical framework.

References

1. Misner, W., Thorne, S., Wheeler, J.A. (1970): Gravitation. W.H. Freeman and Co, San Francisco
2. Penrose, R. (1982): Proc. R. Soc. London A, **381**, 53
3. Hawking, S.W. (1975): Commun. Math. Phys. **43**, 199
4. Iyer, V., Wald, R. (1994): Phys. Rev. D **50**, 846
5. Fatibene, L., Ferraris, M., Francaviglia, M., Raiteri, M. (1999): Annals of Phys. **275**, 27; hep-th/9810039
6. Brown, J.D., York, J.W. (1993): Phys. Rev. D **47**, 1420
7. Carter, B. (1979): In *Black Holes*, ed. by S.W. Hawking, W. Israel, Cambridge University Press, Cambridge
8. Hawking, S.V., Ellis, G.F.R. (1973): The large scale structure of spacetime. Cambridge University Press, Cambridge
9. Frolov, V.P., Fursaev, D.V.: Thermal Fields, Entropy, and Black Holes. hep-th/9802010
10. Frolov, D.V.: Black holes entropy, and physics at Planckian scales. hep-th/9510156v3
11. Brown, J.D., York, J.W. (1993): Phys. Rev. D **47**, 1407
12. Fatibene, L., Ferraris, M., Francaviglia, M., Raiteri, M.: Noether charges, Brown York quasi local energy and related topics. gr-qc/0003019, J. Math. Phys., in press
13. Fatibene, L., Ferraris, M., Francaviglia, M., Raiteri, M.: Entropy in General Relativity, in *Proceedings for the conference in honour of De Ritis*, in press
14. Brown, J.D., York, J.W.: gr-qc9506085
15. Ferraris, M., Francaviglia, M. (1991): In *Mechanics, Analysis and Geometry: 200 Years after Lagrange*, ed. by M. Francaviglia, Elsevier
16. Gotay, M.J., Isenber, J., Marsden, J.E., Montgomery, R.: Momentum maps and classical relativistc fields. The Lagrangian and Hamiltonian structure of classical field theories with constrains; hep-th/9801019
17. Giachetta, G., Sardanashvily, G.: Stress energy momentum tensor in Lagrangian field theory. Part I: Superpotentials. E-print: gr-qc@xxx. lanl. gov. gr-qc/9510061
18. Trautman, A. (1962): Gravitation: an introduction to current research. Ed. by L. Witten Whiley, New York, pp. 168–198
19. Trautman, A. (1967): Commun. Math. Phys. **6**, 248
20. Ferraris, M., Francaviglia, M., Sinicco, I.: Il Nuovo Cimento, Vol. **107B**, N. 11
21. Fatibene, L., Ferraris, M., Francaviglia, M. (1997): J. Math. Phys. **38**, (8), 3953–3967
22. Kolar, I., Michor, P.W., Slovak, J. (1993): Natural operations in differential geometry. Springer-Verlag, New York
23. Ferraris, M.: In *proceedings of the conference on Differential Geometry and its applications, Part 2, Geometrical methods in Physics*, ed. by D. Krupka, Brno, pp. 61–91
24. Katz, J. (1985): Class. Quantum Grav. **2**, 423
25. Iyer, V., Wald, R.: gr-qc 9503052
26. Regge, T., Teitelboim, C. (1974): Ann. Phys. **88**, 286
27. Fatibene, L., Francaviglia, M. (2002): Natural and Gauge Natural Formalism for Classical Field Theories. A Geometric Perspective including Spinors and Gauge Theories, in press

28. Fatibene, L. (1999): Gauge-natural formalism for classical field theories. Ph.D. Thesis, University of Turin
29. Francaviglia, M., Ferraris, M., Reina, C. (1983): Ann. Inst. Henri Poincarè, **XXXVIII**, N. 4, 371–383
30. Geroch, R.P. (1970): J. Math. Phys. **11**, 437–449
31. Brown, J.D., Lau, S.R., York, J.W.: gr-qc/0010024
32. Booth, I.: gr-qc/0008030
33. Mangiarotti, L., Sardanashvily, G. (2000) Connections in classical and quantum field theory. World Scientific, Singapore
34. Ferraris, M., Francaviglia, M., Robutti, O. (1984): In *Atti del VI convegno nazionale di Relatività Generale e Fisica della Gravitazione*, ed. by M. Modugno, Pitagora, Bologna, pp. 137–150
35. Fatibene, L., Ferraris, M., Francaviglia, M., Raiteri, M. (1999): Phys. Rev. D **60**, 124012, 124013; gr-qc/9902063, gr-qc/9902065
36. Fatibene, L., Ferraris, M., Francaviglia, M., Raiteri, M. (2000): Annals of Phys. **197**, 2; gr-qc/9906114
37. Mann, R.B., Potvin, G., Raiteri, M. (2000): Class. Quantum Grav. **17**, 23
38. Landau, L., Lifchitz, L. (1970): Statistical mechanics. 3rd edition, MIR, Moscow
39. Hawking, S.W., Hunter, C.J., Page, D.N.: hep-th/9809035
40. Hawking, S.W., Hunter, C.J.: hep-th/9808085 and hep-th/9807010
41. Allemandi, G., Fatibene, L., Francaviglia, M.: Remarks on the entropy of non stationary black holes. J. Math. Phys.
42. Gibbons, G.W., Hawking, S.W. (1977): Phys. Rev. D**15**, 2752

Spinors, Supergravity
and the Signature of Space-Time

S. Ferrara

Abstract. Supersymmetry algebras embedding space-time in any dimension and signature are considered. Different real forms of the R-symmetries arise both for usual space-time signature (one time) and for Euclidean or exotic signatures (more than one times). Application of these superalgebras are found in the context of supergravities with 32 supersymmetries, in any dimension $D \leq 11$. These theories are related to $D = 11$, M, M^* and M' theories or $D = 10$, IIB, IIB* theories when compactified on Lorentzian tori. All dimensionally reduced theories fall in three distinct phases specified by the number of (128 bosonic) positive and negative norm states: $(n^+, n^-) = (128, 0), (64, 64), (72, 56)$. 10D super Yang-Mills theories in (9,1) and (5,5) space-times are also considered, yielding $N = 4$, $D = 4$ gauge theories with (3,1), (4,0) and (2,2) signature.

1 Introduction

We describe superconformal algebras embedding space-time $V_{s,t}$ with arbitrary dimension $D = s + t$ and signature $\rho = s - t$ [1]. The relation between R-symmetries and space-time signatures is elucidated [2]. For supergravities with 32 supersymmetries, there exist theories in three distinct phases specified by the number of (128 bosonic) positive and negative metric "states": $(n^+, n^-) = (128, 0), (64, 64), (72, 56)$. This fact is closely related to the compactification of M, M^* and M' theories on Lorentzian tori [3].

When dimensionally reduced to $D = 3$, two of these phases are Minkowskian with σ-model $E_{8(8)}/SO(16)$ and $E_{8(8)}/SO(8, 8)$; one is Euclidean with σ-model $E_{8(8)}/SO^*(16)$. The three real forms of D_8 correspond to the superconformal algebras OSP(16/4, R), OSP(8, 8/4, R) and OSP(16*/2, 2) associated to real and quaternionic spinors respectively.

The paper is organized as follows: in Sects. 2 and 3, we describe symmetry and reality properties of spinors in arbitrary space-time signatures and dimensions. In Sects. 4 and 5, we consider real forms of classical Lie algebras and superalgebras. In Sect. 6, we study extended superalgebras. Supersymmetry algebras with non-compact R-symmetries arise in the context of Euclidean theories as well as in theories with exotic (more than one time) signature. In Sect. 7, we show that non-compact R-symmetries arise when $D = 11$ and M, M^*, M' theories introduced by C. Hull [3] are compactified to lower dimensions on Lorentzian tori.

Euclidean supergravity from time-reduction [4] of M-theory was considered by Hull and Julia [5] and BPS branes in theories with more than one time were studied by Hull and Khuri [6]. Hull's observation is that theories on exotic space-time signatures arise by T-duality on time-direction from conventional M theory on string theory so this can perhaps be regarded as different phases of the same non-perturbative M-theory [3]. Irrespectively of whether it is sensible to use time-like T-duality in string theory [7], supergravity theories in exotic space-time dimensions certainly exist, on the basis of invariance principles due to the existence of appropriate superalgebras [1,2]. U-duality also implies the existence of two new kinds of type II string theories, IIA*, IIB* in $V_{9,1}$ space-time where all bosonic RR fields have reversed sign for the metric [8]. A strong evidence of this reasoning is that the U-duality groups in all lower dimensions are the same, independently of the space-time signature. BPS states are classified by orbits of the U-duality group. These orbits were given in Refs. [9,10] and [11]. Since they depend only on the U-duality group and not on the R-symmetry, it follows that the orbit classification is insensitive to the signature of space-time and is still valid for M^*, M' theories.

Finally, in Sect. 8 we describe state counting of supergravities in exotic space-time. This is given in $D = 4$ and $D = 3$ dimensions, as well as in the $D = 11, 10$ original dimensions. This counting shows that there are essentially three distinct phases, in any dimension and signature where the (128) bosonic degrees of freedom fall in positive and negative norm classes $(n^+, n^-) = (128, 0), (64, 64), (72, 56)$. The first two are Minkowskian phases (one time or more than one times), the latter is the Euclidean phase (no time). State counting in theories with 16 supersymmetries is also given, and again three different phases emerge.

2 Properties of spinors of SO(V)

Let V be a real vector space of dimension $D = s + t$ and $\{v_\mu\}$ a basis of it. On V there is a non degenerate symmetric bilinear form, which in the basis is given by the matrix

$$\eta_{\mu\nu} = \mathrm{diag}(+, \dots (s \text{ times}) \dots, +, -, \dots (t \text{ times}) \dots, -).$$

We consider the group $\mathrm{Spin}(V)$, the unique double covering of the connected component of $\mathrm{SO}(s, t)$ and its spinor representations. A spinor representation of $\mathrm{Spin}(V)^{\mathbb{C}}$ is an irreducible complex representation whose highest weights are the fundamental weights corresponding to the right extreme nodes in the Dynkin diagram. These do not descend to representations of $\mathrm{SO}(V)$. A spinor-type representation is any irreducible representation that does not descend to $\mathrm{SO}(V)$. A spinor representation of $\mathrm{Spin}(V)$ over the reals is an irreducible representation over the reals whose complexification is a direct sum of spin representations [12–14,18].

Two parameters, the signature $\rho \bmod(8)$ and the dimension $D \bmod(8)$ classify the properties of the spinor representation. Through this paper we will use the following notation:

$$\rho = s - t = \rho_0 + 8n, \qquad D = s + t = D_0 + 8p,$$

where $\rho_0, D_0 = 0, \ldots, 7$. We set $m = p - n$, so

$$s = \frac{1}{2}(D + \rho) = \frac{1}{2}(\rho_0 + D_0) + 8n + 4m,$$

$$t = \frac{1}{2}(D - \rho) = \frac{1}{2}(D_0 - \rho_0) + 4m.$$

The signature $\rho \bmod(8)$ determines if the spinor representations are of the real($\mathbb{R}$), quaternionic ($\mathbb{H}$) or complex ($\mathbb{C}$) type. Also note that reality properties depend only on $|\rho|$ since $\mathrm{Spin}(s, t) = \mathrm{Spin}(t, s)$.

Table 1. Reality properties of spinors

ρ_0(odd)	Real dim(S)	Reality	ρ_0(even)	Real dim($S^{\pm}$)	Reality
1	$2^{(D-1)/2}$	$\mathbb{R}$	0	$2^{D/2-1}$	$\mathbb{R}$
3	$2^{(D+1)/2}$	$\mathbb{H}$	2	$2^{D/2}$	$\mathbb{C}$
5	$2^{(D+1)/2}$	$\mathbb{H}$	4	$2^{D/2}$	$\mathbb{H}$
7	$2^{(D-1)/2}$	$\mathbb{R}$	6	$2^{D/2}$	$\mathbb{C}$

Table 2. Properties of morphisms

D	k even		k odd	
	Morphism	Symmetry	Morphism	Symmetry
0	$S^{\pm} \otimes S^{\pm} \to \Lambda^k$	$(-1)^{k(k-1)/2}$	$S^{\pm} \otimes S^{\mp} \to \Lambda^k$	
1	$S \otimes S \to \Lambda^k$	$(-1)^{k(k-1)/2}$	$S \otimes S \to \Lambda^k$	$(-1)^{k(k-1)/2}$
2	$S^{\pm} \otimes S^{\mp} \to \Lambda^k$		$S^{\pm} \otimes S^{\pm} \to \Lambda^k$	$(-1)^{k(k-1)/2}$
3	$S \otimes S \to \Lambda^k$	$-(-1)^{k(k-1)/2}$	$S \otimes S \to \Lambda^k$	$(-1)^{k(k-1)/2}$
4	$S^{\pm} \otimes S^{\pm} \to \Lambda^k$	$-(-1)^{k(k-1)/2}$	$S^{\pm} \otimes S^{\mp} \to \Lambda^k$	
5	$S \otimes S \to \Lambda^k$	$-(-1)^{k(k-1)/2}$	$S \otimes S \to \Lambda^k$	$-(-1)^{k(k-1)/2}$
6	$S^{\pm} \otimes S^{\mp} \to \Lambda^k$		$S^{\pm} \otimes S^{\pm} \to \Lambda^k$	$-(-1)^{k(k-1)/2}$
7	$S \otimes S \to \Lambda^k$	$(-1)^{k(k-1)/2}$	$S \otimes S \to \Lambda^k$	$-(-1)^{k(k-1)/2}$

The dimension-$D \bmod(8)$ determines the nature of the (Spin V)-morphisms of the spinor representation S. Let $g \in \mathrm{Spin}(V)$ and let $\Sigma(g) : S \longrightarrow S$ and $L(g) : V \longrightarrow V$ the spinor and vector representations of $l \in \mathrm{Spin}(V)$ respectively. Then a map A

$$A : S \otimes S \longrightarrow \Lambda^k,$$

where $\Lambda^k = \Lambda^k(V)$ are the k-forms on V, is a $\mathrm{Spin}(V)$-morphism if

$$A(\Sigma(g)s_1 \otimes \Sigma(g)s_2) = L^k(g)A(s_1 \otimes s_2).$$

In Tables 1 and 2, reality and symmetry properties of spinors are reported.

3 Orthogonal, symplectic and linear spinors

We now consider morphisms

$$S \otimes S \longrightarrow \Lambda^0 \simeq \mathbb{C}.$$

If a morphism of this kind exists, it is unique up to a multiplicative factor. The vector space of the spinor representation then has a bilinear form invariant under $\mathrm{Spin}(V)$. Looking at Table 2, one can see that this morphism exists except for $D_0 = 2, 6$, where instead a morphism

$$S^{\pm} \otimes S^{\mp} \longrightarrow \mathbb{C},$$

occurs.

We shall call a spinor representation orthogonal if it has a symmetric, invariant bilinear form. This happens for $D_0 = 0, 1, 7$ and $\mathrm{Spin}(V)^{\mathbb{C}}$ (complexification of $\mathrm{Spin}(V)$) is then a subgroup of the complex orthogonal group $\mathrm{SO}(n, \mathbb{C})$, where n is the dimension of the spinor representation (Weyl spinors for D even). The generators of $\mathrm{SO}(n, \mathbb{C})$ are $n \times n$ antisymmetric matrices. These are obtained in terms of the morphisms

$$S \otimes S \longrightarrow \Lambda^k,$$

which are antisymmetric. This gives the decomposition of the adjoint representation of $\mathrm{SO}(n, \mathbb{C})$ under the subgroup $\mathrm{Spin}(V)^{\mathbb{C}}$. In particular, for $k = 2$ one obtains the generators of $\mathrm{Spin}(V)^{\mathbb{C}}$.

A spinor representation is called symplectic if it has an antisymmetric, invariant bilinear form. This is the case for $D_0 = 3, 4, 5$. $\mathrm{Spin}(V)^{\mathbb{C}}$ is a subgroup of the symplectic group $\mathrm{Sp}(2p, \mathbb{C})$, where $2p$ is the dimension of the spinor representation. The Lie algebra $\mathrm{sp}(2p, \mathbb{C})$ is formed by all the symmetric matrices, so it is given in terms of the morphisms $S \otimes S \to \Lambda^k$, which are symmetric. The generators of $\mathrm{Spin}(V)^{\mathbb{C}}$ correspond to $k = 2$ and are symmetric matrices.

For $D_0 = 2, 6$ one has an invariant morphism

$$B : S^+ \otimes S^- \longrightarrow \mathbb{C}.$$

One of the representations S^+ and S^- is one the contragradient (or dual) of the other. The spin representations extend to representations of the linear group $\mathrm{GL}(n, \mathbb{C})$, which leaves the pairing B invariant. These spinors are called linear. $\mathrm{Spin}(V)^{\mathbb{C}}$ is a subgroup of the simple factor $\mathrm{SL}(n, \mathbb{C})$.

These properties depend exclusively on the dimension [18]. When combined with the reality properties, which depend on ρ, one obtains real groups embedded in $\mathrm{SO}(n, \mathbb{C})$, $\mathrm{Sp}(2p, \mathbb{C})$ and $\mathrm{GL}(n, \mathbb{C})$, which have an action on the space of the real spinor representation S^{σ}. The real groups contain $\mathrm{Spin}(V)$ as a subgroup.

We first need some general facts about real forms of simple Lie algebras [18]. Let S be a complex vector space of dimension n, which carries an irreducible representation of a complex Lie algebra $\mathcal{G}$. Let G be the complex Lie group associated to $\mathfrak{g}$. Let σ be a conjugation or a pseudoconjugation on S such that $\sigma X \sigma^{-1} \in \mathfrak{g}$ for all $X \in \mathfrak{g}$. Then the map

$$X \mapsto X^\sigma = \sigma X \sigma^{-1},$$

is a conjugation of $\mathfrak{g}$. We shall write

$$\mathfrak{g}^\sigma = \{X \in \mathfrak{g} \mid X^\sigma = X\}.$$

$\mathfrak{g}^\sigma$ is a real form of $\mathfrak{g}$; if $\tau = h\sigma h^{-1}$, with $h \in \mathfrak{g}$, $\mathfrak{g}^\tau = h\mathfrak{g}^\sigma h^{-1}$. We have $\mathfrak{g}^\sigma = \mathfrak{g}^{\sigma'}$ if and only if $\sigma' = \epsilon\sigma$, for ϵ a scalar with $|\epsilon| = 1$; in particular, if $\mathfrak{g}^\sigma$ and $\mathfrak{g}^\tau$ are conjugate by G, σ and τ are both conjugations or both pseudoconjugations. The conjugation can also be defined on the group G, $g \mapsto \sigma g \sigma^{-1}$.

4 Real forms of the classical Lie algebras

We describe the real forms of the classical Lie algebras from this point of view [1]. (See also [16].)

Linear algebra, (sl S).

(a) If σ is a conjugation of S, then there is an isomorphism $S \to \mathbb{C}^n$ such that σ goes over to the standard conjugation of $\mathbb{C}^n$. Then $\mathfrak{g}^\sigma \simeq \mathrm{sl}(n, \mathbb{R})$. (The conjugation acting on $\mathrm{gl}(n, \mathbb{C})$ gives the real form $\mathrm{gl}(n, \mathbb{R})$).

(b) If σ is a pseudoconjugation and $\mathfrak{g}$ does not leave invariant a non-degenerate bilinear form, then there is an isomorphism of S with $\mathbb{C}^n$, $n = 2p$ such that σ goes over to

$$(z_1, \ldots, z_p, z_{p+1}, \ldots, z_{2p}) \mapsto (z^*_{p+1}, \ldots, z^*_{2p}, -z^*_1, \ldots, -z^*_p).$$

Then $\mathfrak{g}^\sigma \simeq \mathrm{su}^*(2p)$. (The pseudoconjugation acting in on $\mathrm{gl}(2p, \mathbb{C})$ gives the real form $\mathrm{su}^*(2p) \oplus \mathrm{so}(1, 1)$.)

To see this, it is sufficient to prove that $\mathfrak{g}^\sigma$ does not leave invariant any non-degenerate hermitian form, so it cannot be of the type $\mathrm{su}(p, q)$. Suppose that $\langle \cdot, \cdot \rangle$ is a $\mathfrak{g}^\sigma$-invariant, non-degenerate hermitian form. Define $(s_1, s_2) := \langle \sigma(s_1), s_2 \rangle$. Then $(\cdot, \cdot)$ is bilinear and $\mathfrak{g}^\sigma$-invariant, so it is also $\mathfrak{g}$-invariant.

(c) The remaining cases, following E. Cartan's classification of real forms of simple Lie algebras, are $\mathrm{su}(p, q)$, where a non-degenerate hermitian bilinear form is left invariant. They do not correspond to a conjugation or pseudoconjugation on S, the space of the fundamental representation. (The real form of $\mathrm{gl}(n, \mathbb{C})$ is in this case $\mathrm{u}(p, q)$).

Orthogonal algebra, so(S). $\eth$ leaves invariant a non-degenerate, symmetric bilinear form. We will denote it by $(\cdot, \cdot)$.

(a) If σ is a conjugation preserving $\eth$, one can prove that there is an isomorphism of S with $\mathbb{C}^n$ such that $(\cdot, \cdot)$ goes to the standard form and $\eth^\sigma$ to $so(p, q)$, $p+q = n$. Moreover, all $so(p, q)$ are obtained in this form.

(b) If σ is a pseudoconjugation preserving $\eth$, $\eth^\sigma$ cannot be any of the $so(p, q)$. By Cartan's classification, the only other possibility is that $\eth^\sigma \simeq so^*(2p)$. There is an isomorphism of S with $\mathbb{C}^{2p}$ such that σ goes to

$$(z_1, \ldots, z_p, z_{p+1}, \ldots, z_{2p}) \mapsto (z^*_{p+1}, \ldots, z^*_{2p}, -z^*_1, \ldots, -z^*_p).$$

Symplectic algebra, sp(S). We denote by $(\cdot, \cdot)$ the symplectic form on S.

(a) If σ is a conjugation preserving $\eth$, it is clear that there is an isomorphism of S with $\mathbb{C}^{2p}$, such that $\eth^\sigma \simeq sp(2p, \mathbb{R})$.

(b) If σ is a pseudoconjugation preserving $\eth$, then $\eth^\sigma \simeq usp(p, q)$, $p + q = n = 2m$, $p = 2p'$, $q = 2q'$. All the real forms $usp(p, q)$ arise in this way. There is an isomorphism of S with $\mathbb{C}^{2p}$ such that σ goes to

$$(z_1, \ldots z_m, z_{m+1}, \ldots z_n) \mapsto J_m K_{p',q'}(z^*_1, \ldots z^*_m, z^*_{m+1}, \ldots z^*_n),$$

where

$$J_m = \begin{pmatrix} 0 & I_{m \times m} \\ -I_{m \times m} & 0 \end{pmatrix}, \quad K_{p',q'} = \begin{pmatrix} -I_{p' \times p'} & 0 & 0 & 0 \\ 0 & I_{q' \times q'} & 0 & 0 \\ 0 & 0 & -I_{p' \times p'} & 0 \\ 0 & 0 & 0 & I_{q' \times q'} \end{pmatrix}.$$

In Sect. 2 we saw that there is a conjugation on S when the spinors are real and a pseudoconjugation when they are quaternionic [1] (both denoted by σ). We have a group, $SO(n, \mathbb{C})$, $sp(2p, \mathbb{C})$ or $GL(n, \mathbb{C})$ acting on S and containing $Spin(V)^\mathbb{C}$. We note that this group is minimal in the classical group series. If the Lie algebra $\eth$ of this group is stable under the conjugation

$$X \mapsto \sigma X \sigma^{-1},$$

then the real Lie algebra $\eth^\sigma$ acts on S^σ and contains the Lie algebra of $Spin(V)$. We shall call it the $Spin(V)$ algebra.

Let B be the space of $Spin(V)^\mathbb{C}$-invariant bilinear forms on S. Since the representation on S is irreducible, this space is at most one-dimensional. Let it be one-dimensional, let σ be a conjugation or a pseudoconjugation, and let $\psi \in B$. We define a conjugation in the space B as

$$B \longrightarrow B,$$
$$\psi \mapsto \psi^\sigma,$$
$$\psi^\sigma(v, u) = \psi(\sigma(v), \sigma(u))^*.$$

It is then immediate that we can choose $\psi \in B$ such that $\psi^\sigma = \psi$. Then if X belongs to the Lie algebra preserving ψ, so does $\sigma X \sigma^{-1}$.

One can determine the real Lie algebras in each case [1]. All of the possible cases must be studied separately. All dimension and signature relations are mod(8). In the following, a relation like $\mathrm{Spin}(V) \subseteq G$ for a group G will mean that the image of $\mathrm{Spin}(V)$ under the spinor representation is in the connected component of G. The same applies for the relation $\mathrm{Spin}(V) \simeq G$. For $\rho = 0, 1, 7$ spin algebras commute with a conjugation, for $\rho = 3, 4, 5$ they commute with a pseudoconjugation. For $\rho = 2, 6$ they are complex. The complete classification is reported in Table 3.

5   Spin (V) superalgebras

We now consider the embedding of $\mathrm{Spin}(V)$ in simple real superalgebras. We require in general that the odd generators be in a real spinor representation of $\mathrm{Spin}(V)$. In the cases $D_0 = 2, 6$, $\rho_0 = 0, 4$ we have to allow the two independent irreducible representations S^+ and S^- in the superalgebra, since the relevant morphism is

$$S^+ \otimes S^- \longrightarrow \Lambda^2.$$

The algebra is then non-chiral.

We first consider minimal superalgebras [17,18], i.e. those with the minimal even subalgebra. From the classification of simple superalgebras [19–21] one obtains the results listed in Table 4.

We note that the even part of the minimal superalgebra contains the $(\mathrm{Spin}\,V)$ algebra obtained in Section 4 as a simple factor. For all quaternionic cases, $\rho_0 = 3, 4, 5$, a second factor $\mathrm{SU}(2)$ or $\mathrm{SO}^*(2)$ is present. For the linear cases there is an additional non-simple factor, $\mathrm{SO}(1,1)$ or $\mathrm{U}(1)$, as discussed in Sect. 4.

For $D = 7$ and $\rho = 3$ there is actually a smaller superalgebra, the exceptional superalgebra $f(4)$ with bosonic part $\mathrm{Spin}(5, 2) \times \mathrm{su}(2)$. The superalgebra appearing in Table 4 belongs to the classical series and its even part is $\mathrm{so}^*(8) \times \mathrm{su}(2)$, $\mathrm{so}^*(8)$ being the $\mathrm{Spin}(5, 2)$-algebra.

Keeping the same number of odd generators, the maximal simple superalgebra containing $\mathrm{Spin}(V)$ is an orthosymplectic algebra with $\mathrm{Spin}(V) \subset \mathrm{Sp}(2n, \mathbb{R})$ [22–26], $2n$ being the real dimension of S. The various cases are displayed in Table 5. We note that the minimal superalgebra is not a subalgebra of the maximal one, although it is so for the bosonic parts.

6   Extended superalgebras

The present analysis can be generalized to the case of N copies of the spinor representation of $\mathrm{spin}(s, t)$ algebras [2]. By looking at the classification of classical simple

Table 3. Spin(s, t) algebras

Orthogonal $D_0 = 1, 7$	Real, $\rho_0 = 1, 7$	$so\left(2^{\frac{(D-1)}{2}}, \mathbb{R}\right)$ if $D = \rho$
		$so\left(2^{\frac{(D-1)}{2}-1}, 2^{\frac{(D-1)}{2}-1}\right)$ if $D \neq \rho$
	Quaternionic, $\rho_0 = 3, 5$	$so^*\left(2^{\frac{(D-1)}{2}}\right)$
Symplectic $D_0 = 3, 5$	Real, $\rho_0 = 1, 7$	$sp\left(2^{\frac{(D-1)}{2}}, \mathbb{R}\right)$
	Quaternionic, $\rho_0 = 3, 5$	$usp\left(2^{\frac{(D-1)}{2}}, \mathbb{R}\right)$ if $D = \rho$
		$usp\left(2^{\frac{(D-1)}{2}-1}, 2^{\frac{(D-1)}{2}-1}\right)$ if $D \neq \rho$
Orthogonal $D_0 = 0$	Real, $\rho_0 = 0$	$so\left(2^{\frac{D}{2}-1}, \mathbb{R}\right)$ if $D = \rho$
		$so\left(2^{\frac{D}{2}-2}, 2^{\frac{D}{2}-2}\right)$ if $D \neq \rho$
	Quaternionic, $\rho_0 = 4$	$so^*\left(2^{\frac{D}{2}-1}\right)$
	Complex, $\rho_0 = 2, 6$	$so\left(2^{\frac{D}{2}-1}, \mathbb{C}\right)_{\mathbb{R}}$
Symplectic $D_0 = 4$	Real, $\rho_0 = 0$	$sp\left(2^{\frac{D}{2}-1}, \mathbb{R}\right)$
	Quaternionic, $\rho_0 = 4$	$usp\left(2^{\frac{D}{2}-1}, \mathbb{R}\right)$ if $D = \rho$
		$usp\left(2^{\frac{D}{2}-2}, 2^{\frac{D}{2}-2}\right)$ if $D \neq \rho$
	Complex, $\rho_0 = 2, 6$	$sp\left(2^{\frac{D}{2}-1}, \mathbb{C}\right)_{\mathbb{R}}$
Linear $D_0 = 2, 6$	Real, $\rho_0 = 0$	$sl\left(2^{\frac{D}{2}-1}, \mathbb{R}\right)$
	Quaternionic, $\rho_0 = 4$	$su^*\left(2^{\frac{D}{2}-1}\right)$
	Complex, $\rho_0 = 2, 6$	$su\left(2^{\frac{D}{2}-1}\right)$ if $D = \rho$
		$su\left(2^{\frac{D}{2}-2}, 2^{\frac{D}{2}-2}\right)$ if $D \neq \rho$

superalgebras [17–21], we find extensions for all N, where the number of supersymmetries is always even if spinors are quaternionic (because of reality properties) or orthogonal (because of symmetry properties).

In Table 6 the classification is analogous to the one given in Table 4. Super-Poincaré algebras can be obtained from the simple superalgebras either by contraction $\mathrm{Spin}(s, t) \to \mathrm{InSpin}(s, t - 1)$ or as subalgebras $\mathrm{Spin}(s, t) \to \mathrm{InSpin}(s - 1, t - 1)$.

Table 4. Minimal (Spin V) superalgebras

D_0	ρ_0	(Spin V) algebra	(Spin V) superalgebra
1,7	1,7	$\mathrm{so}\left(2^{(D-3)/2}, 2^{(D-3)/2}\right)$	
1,7	3,5	$\mathrm{sop}^*\left(2^{(D-1)/2}\right)$	$\mathrm{osp}\left(2^{(D-1)/2})^*|2\right)$
3,5	1,7	$\mathrm{sp}\left(2^{(D-1)/2}, \mathbb{R}\right)$	$\mathrm{osp}\left(1|2^{(D-1)/2}, \mathbb{R}\right)$
3,5	3,5	$\mathrm{usp}\left(2^{(D-3)/2}, 2^{(D-3)/2}\right)$	$\mathrm{osp}\left(2^*|2^{(D-3)/2}, 2^{(D-3)/2}\right)$
0	0	$\mathrm{so}\left(2^{(D-4)/2}, 2^{(D-4)/2}\right)$	
0	2,6	$\mathrm{so}\left(2^{(D-2)/2}, \mathbb{C}\right)^{\mathbb{R}}$	
0	4	$\mathrm{so}^*\left(2^{(D-2)/2}\right)$	$\mathrm{osp}\left(\left(2^{(D-2)/2}\right)^*|2\right)$
2,6	0	$\mathrm{sl}\left(2^{(D-2)/2}, \mathbb{R}\right)$	$\mathrm{sl}\left(2^{(D-2)/2}|1\right)$
2,6	2,6	$\mathrm{su}\left(2^{(D-4)/2}, 2^{(D-4)/2}\right)$	$\mathrm{su}\left(2^{(D-4)/2}, 2^{(D-4)/2}|1\right)$
2,6	4	$\mathrm{su}^*\left(2^{(D-2)/2}\right)$	$\mathrm{su}^*\left(2^{(D-2)/2}\right)|2)$
4	0	$\mathrm{sp}\left(2^{(D-2)/2}, \mathbb{R}\right)$	$\mathrm{osp}\left(1|2^{(D-2)/2}, \mathbb{R}\right)$
4	2,6	$\mathrm{sp}\left(2^{(D-2)/2}, \mathbb{C}\right)^{\mathbb{R}}$	$\mathrm{osp}\left(1|2^{(D-2)/2}, \mathbb{C}\right)$
4	4	$\mathrm{usp}\left(2^{(D-4)/2}, 2^{(D-4)/2}\right)$	$\mathrm{osp}\left(2^*|2^{(D-4)/2}, 2^{(D-4)/2}\right)$

It is important to observe that the R-symmetry may be non-compact for different signatures of space-time.

All these superalgebras have a space-time symmetry that commutes with the R-symmetry group. We can further extend these superalgebras to orthosymplectic superalgebras where these symmetries no longer commute. The number of fermionic generators remains unchanged. The bosonic part is simply a real symplectic algebra, which contains as a subalgebra the direct sum of the space-time and R-symmetry algebra [24–27]. These extensions are reported in Table 7. These superalgebras contain by contraction or as subalgebras, all Poincaré superalgebras in any dimension with all possible "central charges".

7 11D and 10D supergravities on $\mathbf{V}_{s,t}$ revisited

Theories with N super-Poincaré supersymmetries can be obtained as subalgebras of superconformal algebras with $2N$ spinorial generators. This is so because a conformal spinor $S_{s,t}$ of the conformal group $SO(s, t)$ always decomposes into two Poincaré

Table 5. Orthosymplectic Spin(V) superalgebras

D_0	ρ_0	Orthosymplectic
3,5,	1,7	$\mathrm{osp}\left(1\vert 2^{(D-1)/2},\mathbb{R}\right)$
1,7	3,5	$\mathrm{osp}\left(1\vert 2^{(D+1)/2},\mathbb{R}\right)$
3,5	3,5	$\mathrm{osp}\left(1\vert 2^{(D+1)/2},\mathbb{R}\right)$
0	4	$\mathrm{osp}\left(1\vert 2^{D/2},\mathbb{R}\right)$
4	0	$\mathrm{osp}\left(1\vert 2^{(D-2)/2},\mathbb{R}\right)$
4	4	$\mathrm{osp}\left(1\vert 2^{D/2},\mathbb{R}\right)$
4	2,6	$\mathrm{osp}\left(1\vert 2^{D/2},\mathbb{R}\right)$
2,6	0	$\mathrm{osp}\left(1\vert 2^{D/2},\mathbb{R}\right)$
2,6	4	$\mathrm{osp}\left(1\vert 2^{(D+2)/2},\mathbb{R}\right)$
2,6	2,6	$\mathrm{osp}\left(1\vert 2^{D/2},\mathbb{R}\right)$

spinors of opposite dimensions:

$$S_{s,t} \to s^{1/2}_{s-1,t-1} + s^{-1/2}_{s-1,t-1}.$$

The R-symmetry is inherited from the simple superalgebra whose odd generators contain the spin group of the conformal group. Poincaré superalgebras with at most 32 supersymmetries are obtained as subalgebras of superconformal algebras with at most 64 charges. Then the R-symmetries of these theories are read from their superconformal extension. In particular, although a super-Poincaré graviton exists with at most 32 supersymmetries, a conformal graviton multiplet exists with at most 64 supercharges [26–31].

From Table 6, we can read the R-symmetries of super-Poincaré algebras by replacing (D, ρ) with $(D - 2, \rho)$.

Let us now consider maximal supergravity theories in $D = 11$ ($N = 1$) and $D = 10$, IIA, IIB. For $D = 11$, we know that a real spinor exists for $\rho = \pm 1$ mod (8). This implies the existence of two more theories other than M theory with signature (9,2) and (6,5).

These theories were introduced by C. Hull and called M^* and M' theories [3]. They were discovered on the basis of time-like T duality of type IIA and IIB string theory, which requires the existence of new kinds of type II strings called IIA* and IIB* string theories [8]. The strong coupling limit of IIA* theory is M^* theory.

By dimensional reduction of M, M^* and M' theories to $D = 10$, new type IIA theories with signatures (10,0), (9,1), (8,2), (6,4) and (5,5) are found [3]. Type IIA

Table 6. N-extended Spin(s, t) superalgebras

D_0	ρ_0	R-symmetry	Spin (s, t) superalgebra
1,7	1,7	sp $(2N, \mathbb{R})$	osp $\left(2^{\frac{D-3}{2}}, 2^{\frac{D-3}{2}} \| 2N, \mathbb{R}\right)$
1,7	3,5	usp $(2N - 2q, 2q)$	osp $\left(2^{\frac{D-1}{2}} * \| 2N - 2q, 2q\right)$
3,5	1,7	so $(N - q, q)$	osp $\left(N - q, q \| 2^{\frac{D-1}{2}}\right)$
3,5	3,5	so* $(2N)$	osp $\left(2N^* \| 2^{\frac{D-3}{2}}, 2^{\frac{D-3}{2}}\right)$
0	0	sp $(2N, \mathbb{R})$	osp $\left(2^{\frac{D-4}{2}}, 2^{\frac{D-4}{2}} \| 2N\right)$
0	2,6	sp $(2N, \mathbb{C})_{\mathbb{R}}$	osp $\left(2^{\frac{D-2}{2}} \| 2N, \mathbb{C}\right)_{\mathbb{R}}$
0	4	usp $(2N - 2q, 2q)$	osp $\left(2^{\frac{D-2}{2}} * \| 2N - 2q, 2q\right)$
2,6	0	sl $(N, \mathbb{R})$	sl $\left(2^{\frac{D-2}{2}} \| N, \mathbb{R}\right)$
2,6	2,6	su $(N - q, q)$	su $\left(2^{\frac{D-4}{2}}, 2^{\frac{D-4}{2}} \| N - q, q\right)$
2,6	4	su* $(2N)$	su* $\left(2^{\frac{D-2}{2}} \| 2N\right)$
4	0	so $(N - q, q)$	osp $\left(N - q, q \| 2^{\frac{D-2}{2}}\right)$
4	2,6	so $(N, \mathbb{C})_{\mathbb{R}}$	osp $\left(N \| 2^{\frac{D-2}{2}}, \mathbb{C}\right)_{\mathbb{R}}$
4	4	so* $(2N)$	osp $\left(2N^* \| 2^{\frac{D-4}{2}}, 2^{\frac{D-4}{2}}\right)$

(10,0) is IIA$_E$ Euclidean supergravity, $(9, 1)^*$ supergravity is type IIA*. In the (8,2) (6,4) supergravities, there is a single (16) dimensional complex Weyl spinor such that $\psi_L^* = \psi_R$.

In type IIB, the story is rather different: (2,0) supergravity has an R-symmetry that can be either SO(2) or SO(1,1) for $\rho = 0$ or SO*(2) = SO(2) for $\rho = 4$ (see Table 6 for $d = 4$). This leads to the existence of two types of IIB theories for $\rho = 0$, (9,1) and (5,5) signature called type IIB (SO(2)) and IIB* (SO(1,1)) by C. Hull [8] and one theory for (7,3) signature with SO(2) R-symmetry. Their respective $10D$ σ-model is $\frac{SL(2, R)}{SO(2)}$ for type IIB and $\frac{SL(2, R)}{SO(1,1)}$ for IIB* theories [8,4].

Most interestingly, in type II* theories, the sign of the kinetic term in the RR bosonic fields is reversed with respect to the NS fields case. Lifting string theory to $D = 11$ M theory, the above implies that in M^* theory the three-form part of the action has a reversed sign with respect to the M and M' theories case [3].

We have just seen that a non-compact R-symmetry arises in IIB* theory in $D = 10$. To find other non-compact R-symmetries, it is sufficient to compactify M, M^* and

Table 7. Orthosymplectic superalgebras

D_0	ρ_0	$\mathrm{osp}\,(1/2n,\,R) \supset \mathrm{sp}\,(2n,\,R)$
1,7	1,7	$\mathrm{sp}\left(2N \times 2^{\frac{D-1}{2}}\right)$
1,7	3,5	$\mathrm{sp}\left(2N \times 2^{\frac{D-1}{2}}\right)$
3,5	1,7	$\mathrm{sp}\left(N \times 2^{\frac{D-1}{2}}\right)$
3,5	3,5	$\mathrm{sp}\left(2N \times 2^{\frac{D-1}{2}}\right)$
0	0	$\mathrm{sp}\left(2N \times 2^{\frac{D-2}{2}}\right)$
0	2,6	$\mathrm{sp}\left(2N \times 2^{\frac{D}{2}}\right)$
0	4	$\mathrm{sp}\left(2N \times 2^{\frac{D-2}{2}}\right)$
2,6	0	$\mathrm{sp}\left(N \times 2^{\frac{D}{2}}\right)$
2,6	2,6	$\mathrm{sp}\left(N \times 2^{\frac{D}{2}}\right)$
2,6	4	$\mathrm{sp}\left(2N \times 2^{\frac{D}{2}}\right)$
4	0	$\mathrm{sp}\left(N \times 2^{\frac{D-2}{2}}\right)$
4	2,6	$\mathrm{sp}\left(N \times 2^{\frac{D}{2}}\right)$
4	4	$\mathrm{sp}\left(2N \times 2^{\frac{D-2}{2}}\right)$

M' theories on Lorentzian tori $T^{(p,q)}$. A property of R-symmetries is that they must contain, as a subgroup, the orthogonal group $SO(p,\,q)$ related to the classical moduli space of a Lorentzian torus

$$\mathcal{M}_{T^{(p,q)}} = \frac{GL(p+q,\,R)}{SO(p,\,q)}.$$

Moreover, the R-symmetries must also be a subgroup of the U-duality group $E_{11-D(11-D)}$ and be related to reality properties and dimensions of space-time spinors from Table 6. This uniquely fixes the non-compact form of the R-symmetry H_D.

As illustrative examples, let us consider the $D = 5,\,4$ and 3 cases for space-time of all possible signatures. For $D = 5,\,4,\,3$, M, M^* and M' theories we get

$$V_{(s,11-s)} \to V_{(s',t')} \times T^{(p,q)} \quad (s = 10, 9, 6 \quad (s' + p = s,\ t' + q = 11 - s).$$

For $D = 5,\,4,\,3$, the R-symmetries must be different real forms of the C_4, A_7 and D_8 Lie algebras, appropriate to spinors of $V_{s',t'}$, and they must contain $SO(p,\,q)$ as subgroup. The appropriate real forms can be read from Table 8 and are a consequence of Table 6.

Note that all real forms of C_4, A_7 and D_8 are a maximal subgroup of the U-duality groups $E_{6(6)}$, $E_{7(7)}$, $E_{8(8)}$. The associated moduli spaces of these theories are therefore $E_{11-D(11-D)}/H_D$, when H_D are the appropriate real forms of the R-symmetries in Table 8.

8 State counting and ghosts in M, M^*, M' theories and IIB* theories

In this section, we count the degrees of freedom and show that in any theory, with any signature, supersymmetry implies that there are at most three distinct phases, two Minkowskian (at least one time), where all 128 bosons either are no-ghost or they split in 64^+ positive norm and 64^- negative norm states. The other phase is Euclidean (no time) and the states arrange in a 72^+ positive norm and 56^- negative norm sector. For the Euclidean theories, care is needed to give a meaning to a state and its norm since there are no truly massless particles in this case. Our analysis is made as counting scalar degrees of freedom with their kinetic term factor after all these theories are dimensionally reduced to three dimensions. In this extreme situation, there are only three possibilities, as shown in Table 8, since the 128 bosons are coordinates of the σ-models $E_{8(8)}/SO(16)$ or $E_{8(8)}/SO(8,8)$ for Minkowskian phases or $E_{8(8)}/SO^*(16)$ for the Euclidean phase. Note that no other possibility is allowed since there are no other real forms of D_8 contained in $E_{8(8)}$ [32].

The first case has no negative norm states since $SO(16)$ is the maximal compact subgroup of $E_{8(8)}$. For the other two cases, we precisely get $n^- = 64, 56$.

The rest of this section is devoted to showing how these states are arranged with the spin degrees of freedom when these theories are lifted to higher dimensions.

To do state counting in Minkowskian spaces with arbitrary signatures, one must use some rules, which are dictated by the underlying gauge invariance of the higher-dimensional supergravity with respect to coordinate transformations and gauge transformations of the p-form gauge potentials. For the metric part, one uses the fact that a massless graviton, in a Minkowskian space $V_{(s,t)}$, is associated to the coset $SL(s+t-2)/SO(s-1,t-1)$. The number of positive and negative norm states is

$$n^+ = \frac{(s+t-2)(s+t-1)}{2} - (s-1)(t-1) - 1,$$

$$n^- = \frac{(s+t-2)(s+t-3)}{2} - \frac{(s-1)(s-2)+(t-1)(t-2)}{2}.$$

Similar formulae exist for p-forms gauge fields. The above analysis does not apply for the Euclidean case ($t = 0$). However, in this case, one can use the rule that a positive norm $3d$ Euclidean vector is dual to a negative norm Euclidean scalar. By lifting this rule, using the KK ansatz, one is led to conclude that a Euclidean graviton in D dimensions parametrizes the coset $SL(D-2, R)/SO(D-3, 1)$. Curiously, this

Table 8. R-symmetries of $11D$ supergravity compactified on Lorentzian tori

(s, t)	(s', t')	(p, q)	R-symmetry
$D = 5$			
(10,1)	(4,1)	(6,0)	usp(8)
	(5,0)	(5,1)	usp(4, 4)
(9,2)	(4,1)	(5,1)	usp(4, 4)
	(3,2)	(6,0)	sp(8, R)
	(5,0)	(4,2)	usp(4, 4)
(6,5)	(4,1)	(2,4)	usp(4, 4)
	(1,4)	(5,1)	usp(4, 4)
	(5,0)	(1,5)	usp(4, 4)
	(0,5)	(6,0)	usp(8)
	(3,2)	(3,3)	sp(8, R)
	(2,3)	(4,2)	sp(8, R)
$D = 4$			
(10,1)	(3,1)	(7,0)	su(8)
	(4,0)	(6,1)	su*(8)
(9,2)	(3,1)	(6,1)	su(4, 4)
	(4,0)	(5,2)	su*(8)
	(2,2)	(7,0)	sl(8, R)
(6,5)	(3,1)	(3,4)	su(4, 4)
	(1,3)	(5,2)	su(4, 4)
	(4,0)	(2,5)	su*(8)
	(0,4)	(6,1)	su*(8)
	(2,2)	(4,3)	sl(8, R)
$D = 3$			
(10,1)	(2,1)	(8,0)	so(16)
	(3,0)	(7,1)	so*(16)
(9,2)	(2,1)	(7,1)	so(8, 8)
	(1,2)	(8,0)	so(8, 8)
	(3,0)	(6,2)	so*(16)
(6,5)	(2,1)	(4,4)	so(8, 8)
	(1,2)	(5,3)	so(8, 8)
	(3,0)	(3,5)	so*(16)
	(0,3)	(6,2)	so*(16)

is identical to the coset of a graviton in $V_{D-2,2}$ Minkowski space. The reason for this is that, after compactification on T_{D-4}, one gets $4d$ gravity in (4,0) and (2,2) space-time and they both give, upon S_1 reduction, the σ-model $SL(2, R)/SO(1, 1)$, where the role of the two scalars has just been interchanged. Note that for (3,1) gravity, the S^1 reduction gives the standard $SL(2, R)/SO(2)$ coset. Another explanation of this rule, related to M^* theory, is given later. The above rules are also consistent with the fact that upon reduction on a time-like S^1, gravity gives a KK vector with a reversed sign of the metric (and a scalar with the correct sign), while a p-form gives a $(p-1)$ form with a wrong sign of the metric (in addition to a p-form with the original sign) [5].

Let us now consider M, M^* and M' theories. On an $11D$ Minkowskian background, the state counting goes as follows:

$$
\begin{aligned}
M \text{ theory} \quad & 128 = 44^+ + 84^+, \\
M^* \text{ theory} \quad & 128 = 64^+ + 64^-, \\
& 44 = 36^+ + 8^-, \\
& 84 = 28^+ + 56^-, \\
M' \text{ theory} \quad & 128 = 64^+ + 64^-, \\
& 44 = 24^+ + 20^-, \\
& 84 = 40^+ + 44^-.
\end{aligned}
$$

If we reduce M^* theory on S^1, we get IIA* theory and the negative norm states correspond to the RR vector and three forms. In a similar way, if we consider IIB* theory, the NS and RR states have reversed signs of the metric and since they are equal in number, they give $128 = 64^+_{NS} + 64^-_{RR}$ on a (9,1) background. For the Euclidean gravity, IIA$_E$, the counting can be obtained by time-reduction on S^1 from M theory and, using the previous rules, the ghost counting goes as follows:

$$
\begin{aligned}
128 &= 72^+ + 56^-, \\
44 &= 30^+ + 14^-, \\
84 &= 42^+ + 42^-.
\end{aligned}
$$

Note that in type IIA$_E$, the RR vector and the NS antisymmetric tensor have reversed sign, not the RR three-form. If we instead consider M^* theory as a space-like S^1, we get an (8,2) theory where the RR vector and the RR three-forms have reversed sign with respect to the IIA$_E$ theory. In the (8,2) theory, these states contribute as $(36^-, 28^+)$ to the total $(64^+, 64^-)$ states. If we flip the signs of these states, we get $(36^+, 28^-)$ added to $(36^+, 28^-)$, explaining the $(72^+, 56^-)$ signs in the Euclidean case. In three dimensions, each phase corresponds to a different σ-model and there is a one-to-one correspondence. For higher dimensions, many different theories can be in the same phase. For instance in $D = 4$, $(64^+, 64^-)$ correspond to two $N = 8$ supergravities with σ-models $E_{7(7)}/SU(4, 4)$ and $E_{7(7)}/SL(8, R)$, corresponding to (3,1) signature and (2,2) signature, respectively. The $(72^+, 56^-)$ phase corresponds to $N = 8$ Euclidean supergravity with σ-model $E_{7(7)}/SU^*(8)$ [5]. The higher the dimension, the more theories correspond to the same phase.

8.1   Theories with 16 supersymmetries

The previous analysis can be extended to any theory with a lower number of super-symmetries. Here we just consider theories with 16 supersymmetries. In $D = 10$ the $(1,0)$ algebra can be constructed for signatures $(9,1)$ and $(5,5)$. This algebra admits both a supergravity and a Yang–Mills theory whose Lagrangians are identical to the standard $(9,1)$ Lagrangians since their space-time is related by mod (8) periodicity. No other signatures are possible for real Weyl fermions.

By compactification on Lorentzian tori, one can get several theories in lower dimensions. In particular, descending to $D = 4$, one gets $N = 4$ supergravities and $N = 4$ superconformal Yang–Mills theories with space-time, with signature $(3,1)$ $(2,2)$ and $(4,0)$ [33–35]. The relevant superalgebras are $SU(2, 2/4)$, $SU(2, 2/2, 2)$ for a $(3,1)$ signature (corresponding to super Yang–Mills on T_6 and $T_{2,4}$ respectively), $SL(4/4)$ for a $(2,2)$ signature (corresponding to super-Yang–Mills on $T_{(3,3)}$) and $SU^*(4/4)$ for a $(4,0)$ signature (corresponding to super-Yang–Mills on $T_{5,1}$). In $D = 4$ the moduli space of $(1,0)$ supergravity on the corresponding tori are:

$$
(3, 1) \times (6, 0), \qquad \frac{SO(6, 6)}{SU(4) \times SU(4)} \times \frac{SU(1, 1)}{U(1)},
$$

$$
(3, 1) \times (2, 4), \qquad \frac{SO(6, 6)}{SU(2, 2) \times SU(2, 2)} \times \frac{SU(1, 1)}{U(1)},
$$

$$
(4, 0) \times (5, 1), \qquad \frac{SO(6, 6)}{SU^*(4) \times SU^*(4)} \times \frac{SU(1, 1)}{SO(1, 1)},
$$

$$
(2, 2) \times (3, 3), \qquad \frac{SO(6, 6)}{SL(4, R) \times SL(4, R)} \times \frac{SU(1, 1)}{SO(1, 1)}.
$$

Further reducing to $D = 3$, we get three σ-models:

$$
\frac{SO(8, 8)}{SO(8) \times SO(8)}, \qquad \frac{SO(8, 8)}{SO(4, 4) \times SO(4, 4)}, \qquad \frac{SO(8, 8)}{SO^*(8) \times SO^*(8)}.
$$

This again leads to three phases for the 64 (bosonic) degrees of freedom. In the Minkowskian phases, either $n^- = 0$ ($n^+ = 64$) or $n^- = 32$ ($n^+ = 32$). In the Euclidean phase, $n^- = 24$ ($n^+ = 40$). Lifting the theory to $D = 10$, the graviton has $35 = 23^+ + 12^-$ degrees of freedom, the antisymmetric tensor $28 = 16^+ + 12^-$, and the dilaton 1^+. The graviton states span the σ-model $SL(8, R)/SO(6, 2)$. The positive and negative norm states fall, as usual, in the representation of the maximal compact subgroup $U(4)$.

Acknowledgements. Part of this report is based on collaborations with R. D'Auria, M.A. Lledó and R. Varadarajan. Enlightening conversations with Y. Oz, A. van Proeyen and R. Stora are also acknowledged. This work has been supported in part by the European Commission RTN network HPRN-CT-2000-00131 (Laboratori Nazionali di Frascati, INFN) and by the D.O.E. grant DE-FG03-91ER40662, Task C.

References

1. D'Auria, R., Ferrara, S., Lledó, M.A., Varadarajan, V.S.: hep-th/0010124
2. D'Auria, R., Ferrara, S., Lledó, M.A.: Embedding of space-time groups in simple super-algebras. CERN preprint CERN-TH/2000-380;
   Ferrara, S.: Letter in Math. Phys.; hep-th/0012186
3. Hull, C. (1998): JHEP **9807**, 021; hep-th/9807127
4. Cremmer, E., Lavrinenko, I.V., Lu, H., Pope, C.N., Stelle, K.S., Tran, T.A. (1998): Nucl. Phys. B **534**, 40; hep-th/9803259
5. Hull, C., Julia, B. (1998): Nucl. Phys. **534**, 250
6. Hull, C., Khuri, R. (1998): Nucl Phys. B **536**, 219; hep-th/9808069;
   (2000): Nucl. Phys. B **575**, 251; hep-th/9911082
7. Moore, G.: hep-th/9305139
8. Hull, C. (1998): Nucl. Phys. B **536**, 219; hep-th/9806146
9. Ferrara, S., Günaydin, M. (1998): Int. J. Mod. Phys. A **13**, 2075
10. Ferrara, S., Maldacena, J. (1998): Class. Quant. Grav. **15**, 749
11. Lu, H., Pope, C., Stelle, K.S. (1998): Class. Quant. Grav. **15**, 537
12. Chevalley, C. (1954): The algebraic theory of spinors. Columbia University Press
13. Bourbaki, N. (1958): Algèbre. Ch. 9, Hermann, Paris
14. Choquet-Bruhat, Y., DeWitt-Morette, C. (1989): Analysis, manifols and physics. Part II, Applications. North-Holland, Amsterdam
15. Deligne, P. (1999): Notes on Spinors, in *Quantum Fields and Strings: A Course for Mathematicians*, American Mathematical Society
16. Helgason, S. (1978): Differential geometry, Lie groups and symmetric spaces. Academic Press, New York
17. Nahm, W. (1978): Nucl. Phys. B **135**, 149
18. Van Proeyen, A.: hep-th/9910030
19. Nahm, W., Rittenberg, V., Scheunert, M. (1976): J. Math. Phys. **21**, 1626
20. Kac, V.G. (1977): Commun. Math. Phys. **53**, 31;
    Kac, V.G. (1977): Adv. Math. **26**, 8;
    Kac, V.G. (1980): J. Math. Phys. **21**, 689
21. Parker, M. (1980): J. Math. Phys. **21**, 689
22. D'Auria, R., Fré, P. (1982): Nucl. Phys. B **201**, 101
23. van Holten, J.W., van Proeyen, A. (1982): J. Phys. A **15**, 3763
24. Bergshoeff, E., van Proeyen, A.: hep-th/0003261, hep-th/0010194, hep-th/0010195
25. Bars, I. (2000): Phys. Rev. D **62**, 046007;
    (1997): Phys. Lett. B **403**, 257–264;
    (1997): Phys. Rev. D **55** 2373–2381
26. Günaydin, M. (1998): Nucl. Phys. B **528**, 432
27. West, P. (2000): JHEP **0008**, 007
28. Günaydin, M., Marcus, N. (1985): Class. Qaunt. Grav. **2**, 219
29. Hull, C.: JHEP **0012.007.2000**, hep-th/0011215
30. Schwarz, J.H.: hep-th/0009009

31. Ferrara, S., Sokatchev, E.: hep-th/0005151
32. Gilmore, L. (1974): In *Lie group, Lie algebras, and some of their applications*. John Wiley & Sons, New York, Table 9.7
33. Siegel, W. (1992): Phys. Rev. D **46** 3235; hep-th/9205075;
    (1993):Phys. Rev. D **47**, 2504; hep-th/9207043;
    (1995):Phys. Rev. D **521** 1042
34. McKeon, D.G.C. (2000): Can. J. Phys. **78**, 261;
    McKeon, D.G.C., Sherry, T.M. (2000): Ann. Phys. **285**, 221; hep-th/9810118;
    McKeon, D.G.C. (2000): Nucl. Phys. B **591**, 591;
    Brandt, F.J., McKeon, D.G.C. (2000): Mod. Phys. Lett. A **15**, 1349
35. Belitsky, A.V., Vandoren, S., van Nieuwenhuizen, P. (2000): Phys. Lett. B **477**, 335; hep-th/0001010

VIRGO: An Interferometric Detector of Gravitational Waves

A. Giazotto, S. Braccini

Abstract. The VIRGO project refers to the construction of a 3 km-long arms interferometric gravitational wave detector, close to Pisa (Italy). The working principle of the detector is outlined in Sect. 1. Then the nature of the gravitational waves (Sect. 2) and the amplitude of the signals expected by astrophysical sources (Sect. 3) is discussed. The design and the techniques used to increase the sensitivity of VIRGO are illustrated in Sect. 4 while the status of construction can be found in Sect. 5.

1 VIRGO working principle

VIRGO aims to detect gravitational waves by measuring the small effects induced on a Michelson interferometer with optical components suspended like a pendulum. In the next section we will show that a gravitational wave, impinging on the plane of the suspended interferometer, "stretches" one of its arm and "compresses" the other one, inducing the opposite effect half cycle later. The amplitude of the mirror displacement is given (in the limit of GW wavelength $< L$) by the product between the arms length L and the dimensionless amplitude of the gravitational wave usually indicated with $h(t)$ and defined in the next section. A detection of the gravitational wave can be provided by a standard interferometric measurement of the phase shift between the interfering beams. This means to detect the variation of the interference pattern at the output photodetector arising from the change of the optical path difference between the two arms. The phase shift to be detected is

$$\Delta\varphi(t) = \frac{4\pi h(t) L}{\lambda}, \tag{1}$$

where λ is the laser wavelength. Since the small phase shift is proportional to the length of the arms, an interferometer with very long arms is desired for the purpose.

In order to evaluate the sensitivity that VIRGO has to have, let us consider the typical signal emitted by a supernova in the Virgo cluster. This signal, at the Earth, is a pulse of a few milliseconds with dimensionless amplitude h ranging from 10^{-22} to 10^{-20}. In a 3 km-long arms interferometer (as VIRGO) the passage of such a signal should induce a pulse displacement of the mirrors of about 10^{-19}–10^{-17} m and therefore a phase shift of 10^{-12}–10^{-10} radians (where a laser wavelength of $1\,\mu$m has been considered). It is clear that the interferometer has to be characterized by an extremely high sensitivity.

A few interferometric gravitational wave antennas are under construction around the world with the aim to reach a sufficient sensitivity to detect gravitational waves emitted by astrophysical sources in the frequency range from a few tens of Hz to a few kHz. At lower frequencies the detection is hindered by seismic noise, namely by continuous vibrations of the ground which induce spurious motions of the optical components and mask the small displacements induced by gravitational waves. Since many astrophysical sources, such as pulsars and coalescing binaries, are expected to emit mainly low frequency gravitational waves (from a fraction of a Hz to a few tens of Hz), it is of great importance to lower the frequency detection threshold as much as possible. VIRGO aims to lower the detection band down to a few Hz by suspending each optical component of the interferometer from a special attenuation chain (superattenuator) which reduces the transmission of seismic noise.

2   Gravitational waves and their effects

In order to understand the effects induced by gravitational waves on VIRGO, it is necessary to review some results of the theory of general relativity. In this theory, a gravitational wave can be viewed as a perturbation of the space-time metric traveling at the speed of light. The evolution of this perturbation is governed by Einstein's field equations determining the form of the metric tensor $\mathbf{g}_{ik}$ [1]. In general, since the Einstein's equations are not linear, the evaluation of the metric tensor is very difficult, at least analytically. Nevertheless, the gravitational field on Earth is weak enough to allow us to use a linear approximation. In the weak field approximation, the perturbation of the metric can be expressed as a small deviation (described by the tensor h_{ik}) from a flat space-time metric which is given by the pseudo-Euclidean tensor $\eta_{ik} = \mathrm{Diag}\{1, 1, 1, -1\}$:

$$g_{ik} \approx \eta_{ik} + \mathbf{h}_{ik} \quad |\mathbf{h}_{ik}| \ll 1. \tag{2}$$

In this hypothesis, with a suitable choice of gauge (called "harmonic gauge"), the Einstein's equations in vacuum take the form of a wave equation that determines the propagation of the perturbation $\mathbf{h}_{ik}$:

$$\frac{\partial^2 \mathbf{h}_{ik}}{\partial x^2} - \frac{1}{c^2}\frac{\partial^2 \mathbf{h}_{ik}}{\partial t^2} = 0. \tag{3}$$

As in electromagnetism, any solution of this equation can be written as a superposition of traveling plane waves. With another suitable choice of gauge, called "Transverse Traceless gauge" or simply "TT gauge", these tensorial plane waves assume the following form:

$$\mathbf{h}_{ik}^{\mathrm{TT}}(t) = \mathbf{h}_{ik}^0 \mathbf{e}^{-i(\omega t - kz)} = \left(\mathbf{h}_+ e_{ik}^+ + \mathbf{h}_{\times} e_{ik}^{\times}\right) \mathbf{e}^{-i(\omega t - kz)}, \tag{4}$$

where z is the propagation axis, while h_+ and h^X are the amplitudes of two independent polarization states described by the tensors $\mathbf{e}_{ik}^+$ and $\mathbf{e}_{ik}^X$:

$$
\mathbf{e}_{ik}^+ = \begin{pmatrix} 0 & 0 & 0 & 0 \\ 0 & 1 & 0 & 0 \\ 0 & 0 & -1 & 0 \\ 0 & 0 & 0 & 0 \end{pmatrix}, \quad
\mathbf{e}_{ik}^X = \begin{pmatrix} 0 & 0 & 0 & 0 \\ 0 & 0 & 1 & 0 \\ 0 & 1 & 0 & 0 \\ 0 & 0 & 0 & 0 \end{pmatrix}.
\tag{5}
$$

In general, the polarization tensor of the wave, $\mathbf{h}_{ik}^0$, reduces to two scalar amplitudes which describe the superposition of a couple of independent polarization states. In this particular case, the scalar amplitudes h_+ and h_X provide the superposition of the two polarization states $\mathbf{e}_{ik}^+$ and $\mathbf{e}_{ik}^X$:

The meaning of the two polarization states $\mathbf{e}_{ik}^+$ and $\mathbf{e}_{ik}^X$ can be understood by reviewing the equation determining the change in the spatial distance ($\vec{\xi}$) between two freely falling masses. In the weak field approximation, this equation assumes the following form:[1]

$$
\ddot{\xi}_i = \frac{1}{2}\ddot{\mathbf{h}}_{ik}^{TT}\xi^k.
\tag{6}
$$

When the change in the spatial distance ($\delta\xi_i(t)$) is small with respect to the original distance, the solution of the last equation is

$$
\delta\xi_i(t) = \frac{1}{2}\mathbf{h}_{ik}^{TT}(t)\xi^k.
\tag{7}
$$

By replacing in Eq. (3) the polarization tensor $\mathbf{h}_{ik}^0$ with the tensors $\mathbf{e}_{ik}^+$ and $\mathbf{e}_{ik}^X$, respectively, one can evaluate from the last equation the effects that two waves, "+" and "X" polarized, traveling in the z direction, induce on a ring of freely falling masses placed in the x-y plane (Fig. 1). In an experimental approach, the perturbation induced by a gravitational wave is evaluated by its "dimensionless amplitude" (h), defined as twice the relative variation of the distance (L) between two freely falling masses, namely

$$
\Delta L(t) = \frac{1}{2}h(t)L.
\tag{8}
$$

The last equation justifies the statement at the beginning of the previous section; the relative displacement of any pair of masses, such as the mirrors of the VIRGO arms (which can be considered as freely-falling masses above the pendulum resonant frequency), is approximately given by the product between their initial separation and the dimensionless amplitude of the wave.

---

[1] This equation is valid only if the wavelength of the gravitational radiation is large with respect to the distance between the masses. This condition is respected in VIRGO for usual signals.

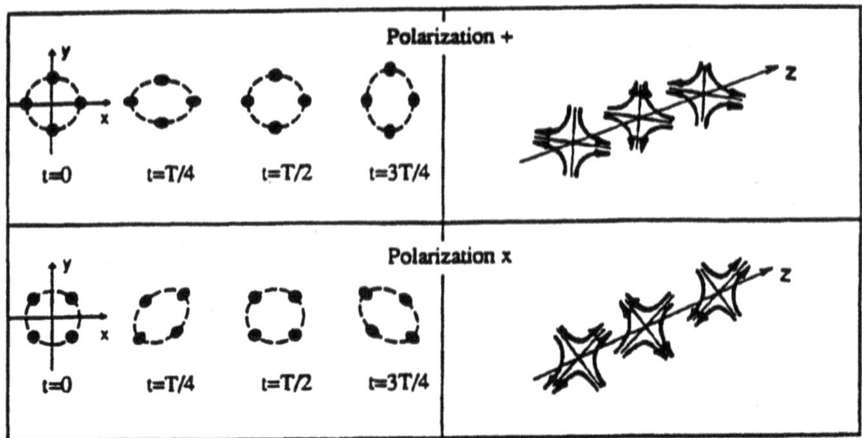

Fig. 1. Response of a ring of freely falling masses placed on the x-y plane of a gravitational wave traveling in the direction z. The strain induced by the wave, for each of the two polarization states ("+" and "X"), is illustrated at each quarter of the period (T). The force field engendered by the two polarization states is represented on the right side

3   Sources of gravitational waves

General relativity predicts that an accelerated system of masses could emit gravitational waves. The main difference with respect to the case of electromagnetism is that the lowest order radiation is now quadrupolar and not dipolar. The strongest gravitational signals that can reach an observer are expected to be emitted by astrophysical processes in which.highly accelerated coherent motions of large masses take place [2,3]. These astrophysical sources of gravitational waves can be distinguished as belonging to three main categories: burst sources, periodic sources, and stochastic background. The following predictions depend crucially on the adopted theoretical models and they are affected by rather large uncertainties (a few orders of magnitude).

Burst sources produce gravitational signals for a short time only (lasting a few cycles). This category includes coalescing compact binaries (of black holes and neutron stars), infalls of massive objects into black holes, supernovae, and so on. As described above, a typical supernova in the Virgo cluster is expected to radiate a gravitational pulse of a few milliseconds, having a dimensionless amplitude h, at the Earth, of 10^{-22}–10^{-20}.

Periodic sources, such as pulsars and binary neutron stars, are expected to emit gravitational waves continuously. The periodicity of these signals could favor their detection despite the very low amplitudes which are involved. Some thousands of pulsars in our galaxy are expected to produce signals in the frequency range from a fraction of a Hz to a few hundreds of Hz, with dimensionless amplitudes h, at the Earth, ranging from 10^{-24} to 10^{-28}.

Stochastic background is a superposition of random signals, coming from random directions at all frequencies. This background is produced by the radiation originated in the early evolution of the universe, and by all the gravitational signals which come from unknown sources. Its amplitude is expected to be well below the detection threshold of single ground-based detectors. In order to detect this small background, coincidence measurements between more detectors are necessary.

4 Design and performances of the VIRGO interferometer

In VIRGO the technique employed to enhance the phase shift induced by the gravitational signal is to increase the optical path of the light in the two arms of the interferometer. This result can be achieved by allowing the light to bounce back and forth several times along the two arms before the recombination on the beam splitter. In this way, the phase shift to be detected ($\Delta\varphi$) is magnified by the number of reflections made by the light in each arm. As shown in Fig. 2, this increase of the optical path can be obtained by replacing each arm of the interferometer with a Fabry–Perot cavity, resonating at the laser frequency. In this way each photon goes back and forth in each arm an average number of times related to the finesse F of the cavity. Typical values of the finesse of the Fabry–Perot range, in interferometric antennae, from a few tens to about 100 (in VIRGO a finesse of 50 has been chosen). By the use of this technique the relevant phase shift can be amplified by a factor F/π. Several noise processes can generate spurious signals in the antenna, masking the effect induced by the gravitational wave. Some of these processes, such as seismic noise and thermal noise, induce displacements of the mirrors ("displacement noise"). Others, such as the noise induced by the frequency fluctuations of the laser, affect the phase of the optical rays even if a real movement of the mirrors is not present ("phase noise"). In order to define the sensitivity of the antenna it is necessary to compare the real signal $h_{gw}(t)$ to each relevant fake signal $h_{\text{noise}}(t)$. This comparison is usually expressed in terms of the "linear spectral density" which is defined as the square root of the power spectrum of the signal. In the case of dimensionless amplitudes, the linear spectral density can be therefore expressed in units of $\text{Hz}^{-1/2}$. The sensitivity curve of VIRGO, described in the following, is written by adding uncoherently the linear spectral density of all the fake signals associated to spurious processes.

Poisson statistical fluctuations of the number of photons in the light beam generate a shot-noise on the interference signal; the linear spectral density of this noise decreases like the square root or the beam power. The beam power can be increased by "recycling" the light that is back-reflected from the interferometer [4]. This means to use a semi-transparent mirror located in the position indicated in Fig. 2 in order to build a resonating cavity between this mirror and the whole interferometer. With this technique, and with a very high power laser source (in VIRGO a 20 W Nd:Yag laser), an impinging power of 1 kW on the beam splitter can be obtained. In this condition shot-noise is a white noise on the interference phase corresponding to a linear spectral density of the wave amplitude close to $10^{-23}\ \text{Hz}^{-1/2}$. Any other noise source

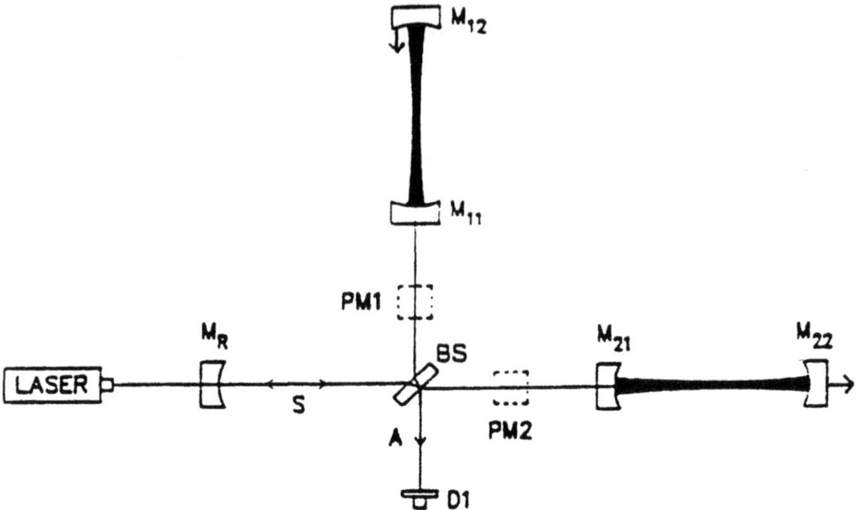

Fig. 2. Effect of the gravitational wave propagating in the positive z-direction and "+" polarized on an interferometer whose suspended arms are built along the x and y axis. Alternately one arm is shortened while the other is lengthened. Each arm is replaced by a Fabry–Perot cavity resonating at the laser frequency so that the phase shift engendered by this effect is enhanced

affecting the antenna is usually compared to this level of noise, called "shot-noise limit".

The static path difference between the two arms of the interferometer generates noise through frequency fluctuations of the laser source [5]. In fact, because of an unavoidable asymmetry between the two arms, the fields which interfere at the beam splitter are originated from the laser at different times, and a fluctuation of the laser frequency is converted into a fluctuation of the phase difference $\Delta\varphi(t)$. In order to keep this noise below the shot-noise limit, it is adequate to develop a very efficient system for the stabilization of the laser frequency. The goal is to achieve a stability of the frequency v of $\delta\tilde{v}(f) < 10^{-5}\,\mathrm{Hz/Hz}^{1/2}$.

Fluctuations of the gas density in the two arms generate a variation in the refractive index, and therefore in the speed of light in the two cavities. Since these fluctuations are not correlated, they will produce a fluctuating phase difference and thus a spurious signal. For this reason the 3 km-long arms of the interferometer must be kept under vacuum. As described in [6], a pressure of 10^{-8} mbar in the vacuum chambers and pipes containing the whole interferometer is required.

By the improvement of the techniques described above it is possible to develop an antenna of very high sensitivity able to detect several gravitational signals at high frequencies (above a few tens of Hz). At lower frequencies the detection is hindered by seismic noise which can cause spurious vibrations of the optical components of the interferometer. The linear spectral density of the seismic displacement of a point on the ground is found to be of the same order of magnitude in all the directions and to vary, to a good approximation, as A/f^2 (where A ranges from 10^{-6} to $10^{-9}\,\mathrm{m\,Hz^{3/2}}$, depending on the location of the measurement). In order to keep the mirrors spurious motion well below the displacements generated by gravitational signals, it is necessary to design a suspension system able to attenuate the propagation of the seismic vibrations. As mentioned above, VIRGO is the unique interferometric gravitational wave detector which plans to incorporate sophisticated suspension systems (superattenuators) for the optical components that will be able to isolate the antenna from seismic noise. The VIRGO superattenuator is essentially a six meters high five stages pendulum where each mass is replaced by a special mechanical filter designed to reduce the transmission of seismic vibrations in all the degrees of freedom. The goal of the VIRGO superattenuators is to attenuate the seismic noise transmitted to the optical components by a factor larger than 10^{10} starting from a few Hz. The technical design of the apparatus is illustrated in [7]. Recently the performances of the apparatus have been tested, obtaining an attenuation larger than 10^{10} in all degrees of freedom starting from about 4 Hz. With this solution and starting from this frequency, one can obtain a mirror displacement sensitivity of about $3 \times 10^{-18}\,\mathrm{m\,Hz^{-1/2}}$ limited by thermal noise acting on the optical components. This corresponds to a gravitational wave sensitivity of about $10^{-21}\,\mathrm{Hz^{-1/2}}$ in the low frequency range.

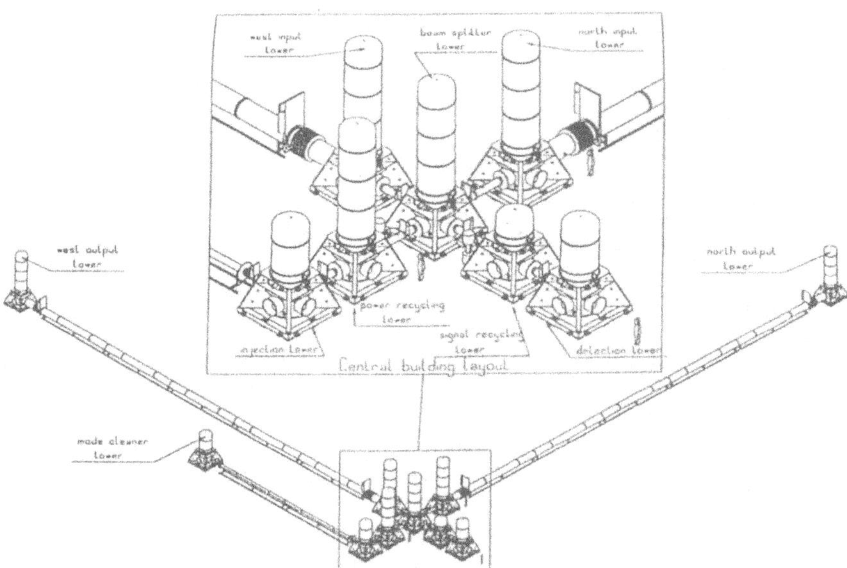

Fig. 3. A schematic design of the external structure of the antenna

In Fig. 3 the external structure of the VIRGO antenna is shown. One can note the 3 km-long vacuum pipes in which the two Fabry-Perot cavities are contained. In each

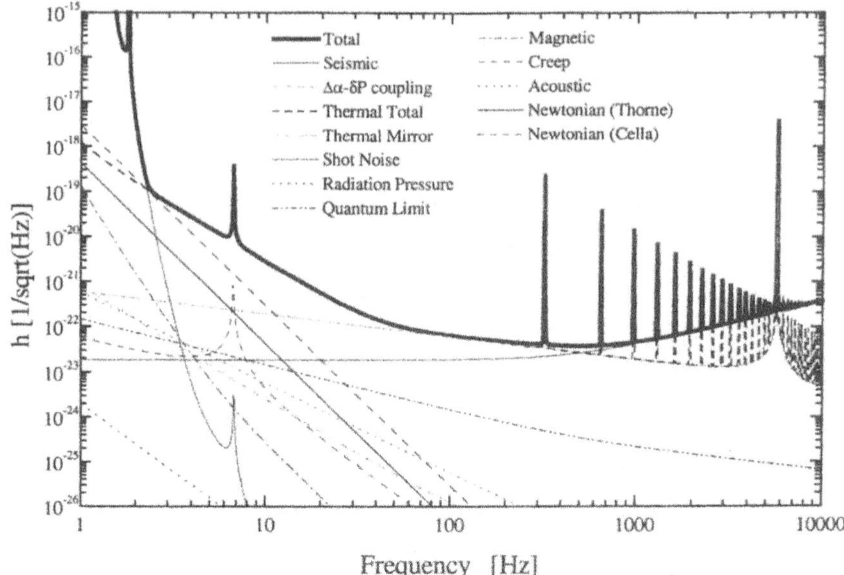

Total
Seismic
Δα-δP coupling
Thermal Total
Thermal Mirror
Shot Noise
Radiation Pressure
Quantum Limit
Magnetic
Creep
Acoustic
Newtonian (Thorne)
Newtonian (Cella)
h [1/sqrt(Hz)]
Frequency   [Hz]

mirrors are provisionally used as end mirrors of a 6 m-long arm Michelson interferometer (with the addition of the Recycling Mirror, already assembled between the input bench and the beam splitter). This small scale interferometer, named "central interferometer", will be used to test the locking strategies and to characterize the main parts of the apparatus (injection system, suspensions, vacuum system and so on). The "commissioning" of the central interferometer will end as soon the terminal mirrors are made available, namely at the end of 2002. The experience accumulated during this phase should make the extension to the full-scale interferometer less arduous.

References

1. Misner, C.W., Thorne, K.S., Wheeler, J. (1973): Gravitation. W.H. Freeman, San Francisco
2. Thorne, K.S. (1987): Gravitational radiation, in *300 years of gravitation*, ed. by S.W. Hawking, W. Israel, Cambridge University Press, New York, Vol. 9, pp. 330–458
3. Shapiro, H.V., Teukoisky, S.A. (1984): The physics of compact objects. J. Wiley & Sons, New York
4. Vinet, J.Y., Meers, B., Man, C.N., Brillet, A. (1988): Phys. Rev. D **38**, 433
5. Hough, J. (1991): The stabilisation of lasers for interferometric gravitational wave detectors, in *The detection of gravitational waves*, ed. by D. G. Blair, Cambridge University Press, New York, Vol. 13, pp. 329–352
6. Brillet, A. (1985): Ann. de Phys. **10**, 219
7. Braccini, S. (1997): Design of the superattenuators for the VIRGO construction. VIRGO Internal Report (VIR-TRE-PIS-4600-134)

Oscillation and Instabilities of Relativistic Stars

K.D. Kokkotas, N. Andersson

Abstract. In this short review we discuss the relevance of ongoing research into stellar oscillations and associated instabilities for the detection of gravitational waves and the future field of "gravitational-wave astronomy".

1 Introduction

As we enter the new millennium there is a focussed worldwide effort to construct devices that will enable the first undisputed detection of gravitational waves. A network of large-scale ground-based laser-interferometer detectors (LIGO, VIRGO, GEO600, TAMA300) is due to come on-line soon, and the sensitivity of the several resonant mass detectors that are already in operation continues to be improved. An integral part in this effort is played by theoretical modeling of the expected sources. Theorists are presently racking their brains to think of various sources of gravitational waves that may be observable once the new ultra-sensitive detectors operate at their optimum level, and of any piece of information one may be able to extract from such observations.

The theory of stellar pulsation is richly endowed with interesting phenomena, and ever improving observations suggest that most stars exhibit complicated modes of oscillation. Thus it is natural to try to match theoretical models to observed data in order to extract information about the dynamics of distant stars. This interplay between observations and stellar pulsation theory is known as asteroseismology.

One of the most challenging goals that can (at least, in principle) be achieved via gravitational-wave detection is the determination of the equation of state of matter at supranuclear densities. We have recently argued that observed gravitational waves from the various nonradial pulsation modes of a neutron star can be used to infer both the mass and the radius of the star with surprisingly good accuracy, and thus put useful constraints on the equation of state [1–5]. But it is not clear that astrophysical mechanisms can excite the various oscillation modes to a detectable level. It seems likely that only the most violent processes, such as the actual formation of a neutron star following a supernova or a dramatic starquake following, for example, an internal phase-transition, will be of relevance. The strengthening evidence for magnetars [6], in which a starquake could release large amounts of energy, is also very interesting in this respect. One can estimate that these events must take place in our immediate neighborhood (the Milky Way or the Local Group) in order to be observable.

The spectrum of a pulsating relativistic star is known to be tremendously rich, but most of the associated pulsation modes are of little relevance for gravitational-wave detection. From the gravitational-wave point of view one would expect the most important modes (in addition to the unstable r-modes [7]) to be the fundamental (f) mode of fluid oscillation, the first few pressure (p) modes and the first gravitational-wave (w) modes [8]. For more details on the theory of relativistic stellar pulsation we refer the reader to two recent review articles [9,10]. That the bulk of the energy from an oscillating neutron star is, indeed, radiated through these modes has been demonstrated by numerical experiments [11–17].

Consequently it is relevant to try to device a strategy for detecting gravitational waves from pulsating neutron stars, and see what information such a detection could provide regarding the stars parameters. Such a strategy can potentially be of great importance to gravitational-wave astronomy, since most stars are expected to oscillate nonradially. In principle, one would expect the modes of a star to be excited in any dynamical scenario that leads to significant asymmetries. Of course, one can only hope to observe gravitational waves from the most compact stars (neutron stars and possibly strange stars) and only the most violent processes are of interest. Still, there are several scenarios in which the various pulsation modes may be excited to an interesting level:

(1) A supernova explosion is expected to form a wildly pulsating neutron star that emits gravitational waves. The current estimates for the energy radiated as gravitational waves from supernovae is rather pessimistic, suggesting a total release of the equivalent to $10^{-6} M_\odot c^2$, or so. However, this may be a serious underestimate if the gravitational collapse in which the neutron star is formed is strongly non-spherical. Optimistic estimates suggest that as much as $10^{-2} M_\odot c^2$ may be released in extreme events.

(2) Another potential excitation mechanism for stellar pulsation is a starquake, e.g. associated with a pulsar glitch. The typical energy released in this process may be of the order of the maximum mechanical energy that can be stored in the crust, estimated to be at the level of $10^{-9} - 10^{-7} M_\odot c^2$ [18,19]. This is also an interesting possibility considering the recent suggestion that the soft-gamma repeaters are magnetars, neutron stars with extreme magnetic fields [20], that undergo frequent starquakes. It seems very likely that some pulsation modes are excited by the rather dramatic events that lead to the most energetic bursts seen from these systems. Indeed, Duncan [6] has recently argued that toroidal modes in the crust should be excited. If modes are excited in these systems, an indication of the energy released in the most powerful bursts is the $10^{-9} M_\odot c^2$ estimated for the March 5 1979 burst in SGR 0526-66. The maximum energy should certainly not exceed the total supply in the magnetic field $\sim 10^{-6}(B/10^{15}G)^2 M_\odot c^2$ [20]. The possibility that a burst from a soft gamma-ray repeater may have a gravitational-wave analogue is very exciting.

(3) The coalescence of two neutron stars at the end of binary inspiral may form a pulsating remnant. It is, of course, most likely, that a black hole is formed when two neutron stars coalesce, but even in that case the eventual collapse may be halted long enough (many dynamical timescales) that several oscillation modes could potentially be identified [21]. Also, stellar oscillations can be excited by the tidal fields of the two stars during the inspiral phase that precedes the merger [22].

(4) The star may undergo a dramatic phase-transition that leads to a mini-collapse. This would be the result of a sudden softening of the equation of state (for example, associated with the formation of a condensate consisting of pions or kaons). A phase-transition could lead to a sudden contraction during which a considerable part of the stars gravitational binding energy would be released, and it seems inevitable that part of this energy would be channeled into pulsations of the remnant. Large amounts of energy that could be released in the most extreme of these scenarios: a contraction of (say) 10% can easily lead to the release of $10^{-2} M_\odot c^2$. Transformation of a neutron star into a strange star is likely to induce pulsations in a similar fashion. It is reasonable to assume that the bulk of the total energy of the oscillation is released through a few of the stars quadrupole pulsation modes in all of these scenarios. We will assume that this is the case and assess the likelihood that the associated gravitational waves will be detected. Having done this we discuss the inverse problem, and investigate how accurately the neutron star parameters can be inferred from the gravitational wave data. Finally, we briefly discuss the gravitational-wave driven instability of the so-called r-modes.

2 Nonradial stellar oscillations: Theoretical minimum

A neutron star has a large number of families of pulsation modes with more or less distinct character. For the simplest stellar models, the relevant modes are high frequency pressure p-modes and the low frequency gravity g-modes [23]. For a typical nonrotating neutron star model the fundamental p-mode (usually referred to as the f-mode), whose eigenfunction has no nodes in the star, has frequency in the range 2–4 kHz, while the first overtone lies above 4 kHz. The g-modes depend sensitively on the internal composition and temperature distribution, but they typically have frequencies of a few hundred Hz. The standard mode-classification dates back to the seminal work of Cowling [24], and is based on identifying the main restoring force that influences the fluid motion. As the stellar model is made more detailed and further restoring forces are included new families of modes come into play. For example, a neutron star model with a sizeable solid crust separating a thin ocean from a central fluid region will have g-modes associated with both the core and the ocean as well as modes associated with shearing motion in the crust [25,26]. Of particular interest to relativists is the existence of a class of modes uniquely associated with the spacetime itself [8,27–29]; the so-called w-modes (for gravitational wave). These modes essentially arise because the curvature of spacetime that is generated by the background density distribution can temporarily trap impinging gravitational waves. The w-modes typically have high frequencies (above 6 kHz) and damp out in a fraction of a millisecond. It is not yet clear whether one should expect these modes to be excited to an appreciable level during (say) a gravitational collapse following a supernova. One might argue that they provide a natural channel for the release of any initial deformation of the spacetime, but there are as yet no solid evidence indicating a significant level of w-mode excitation in a realistic scenario [11,12,10,14–17].

3   Gravitational wave asteroseismology

Once gravitational waves are detected, the first task will be to identify the source. This should be possible from the general character of the waveform and may not require very accurate theoretical models, but such models will be of crucial importance for a deduction of the parameters of the source. That is, for gravitational-wave "astronomy".

The idea behind this presentation is a familiar one in astronomy: For many years, studies of the light variation of variable stars have been used to deduce their internal structure [23]. The Newtonian theory of stellar pulsation was to a large extent developed in order to explain the pulsations of Cepheids and RR Lyrae. This approach, known as Asteroseismology (Helioseismology in the specific case of the Sun), has been remarkably successful in recent years. In comparison, the relativistic theory of stellar pulsation, which has now been developed for thirty years, has not yet been applied in a similar way. So far, the relativistic theory has no immediate connections to observations (that are not already provided by the Newtonian theory). We believe that this situation will change once the gravitational-wave window to the universe is opened, and in this review we discuss how the information carried by the gravitational-wave signal can be inverted to estimate the parameters of pulsating stars. That is, we take a first step towards gravitational-wave asteroseismology.

3.1   What can we learn from observations?

Our present understanding of neutron stars comes mainly from X-ray and radio-timing observations. These observations provide some insight into the structure of these objects and the properties of supranuclear matter. The most commonly and accurately observed parameter is the rotation period, and we know that radio pulsars can spin very fast (the shortest observed period being the $1.56\,\mathrm{ms}$ of PSR $1937+21$). Another basic observable, that can be obtained (in a few cases) with some accuracy from todays observations, is the mass of the neutron star. As Finn [30] has shown, the observations of radio pulsars indicate that the mass lies in the range $1.01 < M/M_\odot < 1.64$. Similarly, van der Kerkwijk et al. [31] find that data for X-ray pulsars indicate that $1.04 < M/M_\odot < 1.88$. The data used in these two studies is actually consistent with (if one includes error bars) $M < 1.44 M_\odot$. We now recall that the various EOS that have been proposed by theoretical physicists can be divided into two major categories: (i) the "soft" EOS which typically lead to neutron star models with maximum masses around $1.4 M_\odot$ and radii usually smaller than 10 km, and (ii) the "stiff" EOS with maximum values $M \sim 1.8 M_\odot$ and $R \sim 15$ km [32]. We thus see that, even though the constraint put on the neutron star mass by present observations seems strong, it actually does not rule out many of the proposed EOS. In order to arrive at a more useful result we are likely to need detailed observations also of the stellar radius. Unfortunately, available data provide little information about the radius. The recent observations of quasiperiodic oscillations in low mass X-ray binaries indicate that $R < 6M$, but again this is not a severe constraint. Although a number of attempts have been made, using either X-ray observations [33] or the limiting spin period of neutron stars [34], to put constraints on the mass-radius relation, we do not yet have a method which can provide the desired answer. In view of this situation, any method that can be used to infer neutron star parameters is a welcome addition. Of specific interest may be the new possibilities offered once gravitational-wave observations become reality.

Let us suppose that a nearby supernova explodes, say in the Local Group of galaxies, and is followed by a core collapse that leads to the formation of a compact object. As the dust from the collapse settles the compact object pulsates wildly in its various oscillation modes, generating a gravitational-wave signal which is composed of an overlapping of different frequencies. We will assume that the results of Allen et al. [11] can be brought to bear on this situation, i.e. that most of the energy is radiated through the f-mode, a few p-modes and the first w-mode. Our detector picks up this signal, and a subsequent Fourier analysis of the data stream yields the frequencies and the energy carried by each mode.

The first question to be answered by the gravitational-wave astronomer concerns what kind of compact object could produce the detected signal. Is it a black hole or a neutron star? The pulsation of these objects lead to qualitatively similar gravitational waves, eg. exponentially damped oscillations, but the question should nevertheless be relatively easy to answer. If more than one of the stellar pulsation modes is observed the answer is clear, but even if we only observe only one single mode the two cases should be easy to distinguish. The fundamental (quadrupole) quasinormal mode frequency of a Schwarzschild black hole follows from

$$f \approx 12 \, \text{kHz} \left(\frac{M_\odot}{M} \right), \tag{1}$$

while the associated e-folding time is

$$\tau \approx 0.05 \, \text{ms} \left(\frac{M}{M_\odot} \right). \tag{2}$$

That is, the oscillations of a 10 $M_\odot$ black hole lie in the frequency range of the f-mode for a typical neutron star. But the two signals will differ greatly in the damping time, the e-folding time of the black hole being nearly three orders of magnitude shorter than that of the neutron star f-mode.

Having excluded the possibility that our signal came from a black hole, we want to know the mass and the radius of the newly born neutron star. We also want to decide which of the proposed EOS that best represents this star. To address these questions we can use a set of empirical relations that can be used to estimate the mass, the radius and the EOS of the neutron star with good precision.

3.2   Addressing the inverse problem

Considering the possibility of a future detection it is relevant to pose the "inverse problem" for gravitational waves from pulsating stars. Once we have observed the waves, can we deduce the details of the star from which they originated? To answer this question we have calculated the frequencies and damping times of the modes that we expect to lead to the strongest gravitational waves for a selection of EOS. Nearly twenty years ago Lindblom and Detweiler [35] tabulated the frequencies and damping times of the f-mode for a number of EOS. Recently we [2] extended their calculation by adding more recent EOS and providing data also for the w- and p-modes. These numerical data were then used to to create useful "empirical" relations between the "observables" (frequencies and damping times) and the parameters of the star (mass, radius and possibly the EOS). We will now outline how these relations can be used to infer the stellar parameters from detected mode data.

Let us first consider the frequency of the f-mode. It is well known that the characteristic time-scale of any dynamical process is related to the mean density ($\bar{\rho}$) of the mass involved. This notion should be relevant for the fluid oscillation modes of a star, and we consequently expect that $\omega_f \sim \bar{\rho}^{1/2}$. That is, we should normalize the f-mode frequency with the average density of the star. As shown in [2] the relation between the f-mode frequencies and the mean density is almost linear, and a linear fitting leads to the following simple relation:

$$\omega_f(\mathrm{kHz}) \approx 0.78 + 1.635 \left(\frac{\bar{M}}{\bar{R}^3}\right)^{1/2}, \tag{3}$$

where we have introduced the dimensionless variables

$$\bar{M} = \frac{M}{1.4 M_\odot} \quad \text{and} \quad \bar{R} = \frac{R}{10\,\mathrm{km}}. \tag{4}$$

From Eq. (3) follows that the typical f-mode frequency is around 2.4 kHz.

To deduce a similar relation for the damping rate of the f-mode, we can use the rough estimate given by the quadrupole formula. That is, the damping time should follow from

$$\tau_f \sim \frac{\text{oscillation energy}}{\text{power emitted in GWs}} \sim R \left(\frac{R}{M}\right)^3. \tag{5}$$

Using this relation and the numerical results for the damping time of the f-mode for various stellar parameters M and R, we find that [2]:

$$\frac{1}{\tau_f(\mathrm{s})} \approx \frac{\bar{M}^3}{\bar{R}^4} \left[22.85 - 14.65 \left(\frac{\bar{M}}{\bar{R}}\right)\right], \tag{6}$$

and it follows that the typical damping time of the f-mode is around 0.15 sec.

Analogous empirical relations for the p-mode data are much less robust and useful. This is natural since the p-modes are sensitive to changes in the matter distribution inside the star (as manifested through changes in the sound speed). In contrast, the gravitational-wave w-modes lead to very robust results. It is well known [8,29] that the w-modes do not excite a significant fluid motion. Thus, they are more or less independent of the characteristics of the fluid: the frequencies do not depend on the sound speed and the damping times cannot be modeled by the quadrupole formula. Analytic results for model problems for the w-modes [27,36], show that the frequency of the w-mode is inversely proportional to the size of the star. Meanwhile, the damping time is related to the compactness of the star, i.e. the more relativistic the star is the longer the w-mode oscillation lasts. These properties have already been discussed in some detail in [29] for uniform density stars. One can find the following relations for the frequency and damping of the first w-mode:

$$\omega_w \text{ (kHz)} \approx \frac{1}{\bar{R}}\left[20.92 - 9.14\left(\frac{\bar{M}}{\bar{R}}\right)\right],\tag{7}$$

and

$$\frac{1}{\tau_w \text{(ms)}} \approx \frac{1}{\bar{M}}\left[5.74 + 103\left(\frac{\bar{M}}{\bar{R}}\right) - 67.45\left(\frac{\bar{M}}{\bar{R}}\right)^2\right].\tag{8}$$

We see that a typical value for the w-mode frequency is 11–12 kHz, but since the frequency depends strongly on the radius of the star it varies greatly for different EOS. For example, for a very stiff EOS (L) the w-mode frequency is around 6 kHz while for the softest EOS in our set (G) the typical frequency is around 14 kHz. The w-mode damping time is comparable to that of an oscillating black hole with the same mass, i.e. it is typically less than a tenth of a millisecond.

3.3   Noisy signals

Suppose that one tries to detect the gravitational waves associated with the stellar pulsation modes that are excited when (say) a neutron star forms after a supernova explosion. Since all modes are relatively short lived, the detection situation is similar to that for a perturbed rotating black hole [37,38]. For each individual mode the signal is expected to have the following form:

$$h(t) = \begin{cases} 0 & \text{for } t < T, \\ \mathcal{A}e^{-(t-T)/\tau}\sin[2\pi f(t - T)] & \text{for } t \geq T. \end{cases}\tag{9}$$

Here, $\mathcal{A}$ is the initial amplitude of the signal, T is its arrival time, and f and τ are the frequency and damping time of the oscillation, respectively. Since the violent formation of a neutron star is a very complicated event, the above form of the waves becomes realistic only at the late stages when the remnant is settling down and its pulsations can be accurately described as a superposition of the various modes, either fluid or spacetime ones, that have been excited. At earlier times ($t < T$) the waves are expected to have a random character that is completely uncorrelated with the intrinsic noise of an earth-bound detector. This partly justifies our simplification of setting the waveform equal to zero for $t < T$.

The energy flux F carried by any weak gravitational wave h is given by

$$F = \frac{c^3}{16\pi G}|\dot{h}|^2,\tag{10}$$

where c is the speed of light and G is Newton's gravitational constant. Thus, when gravitational waves emitted from a pulsating neutron star hit such a detector on Earth, their initial amplitude will be [39]

$$
\mathcal{A} \sim 2.4 \times 10^{-20} \left(\frac{E_{\mathrm{gw}}}{10^{-6} M_\odot c^2} \right)^{1/2} \left(\frac{10\mathrm{kpc}}{r} \right) \left(\frac{1\mathrm{kHz}}{f} \right) \left(\frac{1\mathrm{ms}}{\tau} \right)^{1/2}, \qquad (11)
$$

where E_{gw} is the energy released through the mode and r is the distance between detector and source. In order to dig out this kind of signal from the noisy output of a detector one could use templates of the same form as the expected signal (so called matched filtering). Following the analysis of Echeverria [37] the signal-to-noise ratio is found to be

$$
\left(\frac{S}{N} \right)^2 = \rho^2 \equiv 2\langle h \mid h \rangle = \frac{4Q^2}{1 + 4Q^2} \frac{\mathcal{A}^2 \tau}{2 S_n}, \qquad (12)
$$

with

$$
Q \equiv \pi f \tau, \qquad (13)
$$

being the quality factor of the oscillation, and S_n the spectral density of the detector (assumed to be constant over the bandwidth of the signal).

3.4   Are the modes detectable?

Two separate questions must be addressed in any discussion of gravitational-wave detection. The first one concerns identifying a weak signal in a noisy detector, thus establishing the presence of a gravitational wave in the data. The second question regards extracting the detailed parameters of the signal, e.g., the frequency and e-folding time of a pulsation mode. To address either of these issues we need an estimate of the spectral noise density S_n of the detector.

The pulsation modes of a (non-rotating) neutron star that may be detectable through the associated gravitational waves all have rather high frequencies; typically of the order of several kHz. To illustrate this we show the mode-frequencies for all models considered in [2] as a function of the stellar mass in Fig. 1. From this figure we immediately see that a detector must be sensitive to frequencies of the order of 8–12 kHz and above to observe most w-modes. Of course, it is also clear that some equations of state yield w-modes with lower frequencies. For example, for massive neutron stars with $M \approx 1.8 - 2.3 M_\odot$ (stiff EOS), which may be needed to account for data for low-mass X-ray binaries, the w-mode frequency could be as low as 6 kHz (see also recent results for the axial w-modes [4]). The p-modes lie mainly in the range 4–8 kHz, while all f-modes have frequencies lower than 4 kHz. This means that the mode-signals we consider lie in the regime where an interferometric detector is severely limited by the photon shot noise. For this reason a detection strategy based on resonant detectors (bars, spheres or even networks of small resonant detectors [40]) or laser interferometers operating in dual recycling mode, in order to

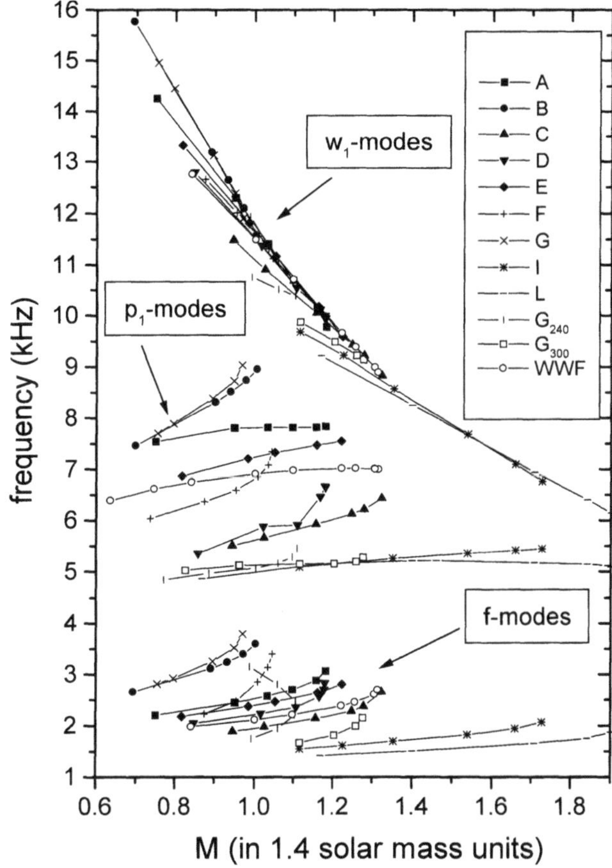

Fig. 1. This diagram shows the range of mode-frequencies for twelve different equations of state as functions of the normalized mass ($\bar{M} = M/1.4 M_\odot$) of the star

overcome the overwhelming photon shot noise at high frequencies [41], seems the most promising. In fact, the range of mode-frequencies in Fig. 2 provides strong motivation for detailed studies into the prospects for construction of dedicated ultrahigh frequency detectors.

In the following we will compare three different detectors: the initial and advanced LIGO interferometers, for which

$$S_n^{1/2} \approx h_m \left(\frac{f}{\alpha f_m} \right)^{3/2} \frac{1}{\sqrt{f}} \, \mathrm{Hz}^{-1/2}, \tag{14}$$

with $h_m = 3.1 \times 10^{-22}$, $\alpha = 1.4$ and $f_m = 160\,\mathrm{Hz}$ for the initial configuration, and $h_m = 1.4 \times 10^{-23}$, $\alpha = 1.6$ and $f_m = 68\,\mathrm{Hz}$ for the advanced configuration [42]. We also consider an "ideal" detector that is tuned to the frequency of the mode and has sensitivity of the order of $S_n^{1/2} \approx 10^{-24}\,\mathrm{Hz}^{-1/2}$. This is the sensitivity goal of the new generation of detectors under construction, see Fig. 3. As an example of a suitably advanced instrument we will take the so called EURO detector, for which the noise-level curves have been estimated by Sathyaprakash and Schutz (private communication). It should be noted that the Advanced LIGO estimates are roughly valid also for spherical detectors such as TIGA, cf. Harry, Stevenson and Paik [43].

The detectability of the f, p and w-modes for different detectors can be assessed from (12). The main problem in doing this is the lack of realistic simulations providing information about the level of excitation of various modes in an astrophysical situation. Still, given the frequency and damping rate of a specific mode we can ask what amount of energy must be channeled through the mode in order for it to be detectable by a given detector. We immediately find that detection of pulsating neutron stars from outside our own galaxy is very unlikely. Let us consider a "typical" stellar model for which the f-mode has parameters $f_f = 2.2\,\text{kHz}$ and $\tau_f = 0.15\,\text{s}$, this corresponds to a $1.4 M_\odot$ neutron star according to the Bethe–Johnson equation of state [2]. For this example we find that the f-mode in the Virgo cluster (at 15 Mpc) must carry an energy equivalent to more than $0.3 M_\odot c^2$ to lead to a signal-to-noise ratio of 10 in our ideal detector. Given that the total energy estimated to be radiated as gravitational waves in a supernova is at the level of $10^{-5} - 10^{-6} M_\odot c^2$, we cannot realistically expect to observe mode-signals from far beyond our own galaxy.

This means that the number of detectable events may be rather low. Certainly, one would not expect to see a supernova in our galaxy more often than once every thirty years or so. Still, there are a large number of neutron stars in our galaxy, all of which may be be involved in dramatic events (see the introduction for some possible scenarios) that lead to the excitation of pulsation modes. The energies required to make each mode detectable (with a signal-to-noise ratio of 10) from a source at the center of our galaxy (at 10 kpc) are listed in Table 1. In the table we have used the data for the "typical" stellar model, for which the characteristics of the f-mode were given above, $f_p = 6\,\text{kHz}$ and $\tau_p = 2\,\text{s}$, and $f_w = 11\,\text{kHz}$ and $\tau_w = 0.02\,\text{ms}$. This data indicates that, even though the event that excites the modes must be violent, the energy required to make each mode detectable is not at all unrealistic. In fact, the energy levels required for both the f- and p-modes are such that detection of violent events in the life of a neutron star should be possible, given the Advanced LIGO detectors (or alternatively spheres with the sensitivity proposed for TIGA). On the other hand, detection of w-modes with the broad band configuration of LIGO seems unlikely. Detection of these modes, which would correspond to observing a purely relativistic phenomenon, requires dedicated high frequency detectors operating in the frequency range above 6 kHz. We believe that the data in Table 1 illustrates that neutron star pulsation modes may well be detectable from within our galaxy. The first detection may come as soon as the first generation of LIGO detectors come on line, but it may be more realistic to expect that we need a third generation detector (such as EURO) to truly probe the pulsation modes of neutron stars.

Table 1. The estimated energy (in units of $M_\odot c^2$) required in each mode in order to lead to a detection with signal-to-noise ratio of 10 from a pulsating neutron star at the center of our galaxy (10 kpc), cf. Eqs. (3) and (4). The given data correspond to a $1.4 M_\odot$ star with the Bethe–Johnson equation of state

Detector	f-mode	p-mode	w-mode
LIGO I	4.9×10^{-5}	4.0×10^{-3}	6.8×10^{-2}
LIGO II	8.7×10^{-7}	7.0×10^{-5}	1.2×10^{-3}
Ideal	1.4×10^{-8}	1.3×10^{-7}	6.4×10^{-7}

3.5   How well can we determine the mode parameters?

Let us now discuss the precision with which we can hope to infer the details of each pulsation mode. We can compute the relative measurement error in the frequency and the damping time of the waves by some appropriately designed detector [3]. After introducing a convenient parameter $\mathcal{P}$, defined by

$$\mathcal{P}^{-1} = \left(\frac{S_n^{1/2}}{10^{-24}\,\mathrm{Hz}^{-1/2}} \right) \left(\frac{r}{10\,\mathrm{kpc}} \right) \left(\frac{E_{\mathrm{gw}}}{10^{-6} M_\odot c^2} \right)^{-1/2}, \tag{15}$$

we find that the error estimates take the following form

$$\frac{\sigma_f}{f} \simeq 0.0042\, \mathcal{P}^{-1} \sqrt{\frac{1 - 2Q^2 + 8Q^4}{4Q^4}} \left(\frac{\tau}{1\,\mathrm{ms}} \right)^{-1}, \tag{16}$$

and

$$\frac{\sigma_\tau}{\tau} \simeq 0.013\, \mathcal{P}^{-1} \sqrt{\frac{10 + 8Q^2}{Q^2}} \left(\frac{f}{1\,\mathrm{kHz}} \right). \tag{17}$$

Also, for the time of arrival of the gravitational wave signal we get from

$$\sigma_T \simeq 0.0042\, \mathcal{P}^{-1}\,\mathrm{ms}. \tag{18}$$

To illustrate these results we list in Table 2 the relative errors associated with the parameter extraction for the "typical" $1.4 M_\odot$ stellar model we used in the previous section. We assume that each mode carries the energy required for it to be observed with signal-to-noise ratio of 10, cf. Table 1. (This is a convenient measure since it is independent of the particulars of the detector.)

From the sample data in Table 2 one sees clearly that, while an extremely accurate determination of the frequencies of both the f- and the p-mode is possible, it would be much harder to infer their respective damping rates. It is also clear that an accurate determination of both the w-mode frequency and damping will be difficult. To illustrate this result in a different way, we can ask how much energy must be channeled through each mode in order to lead to a 1% relative error in the frequency

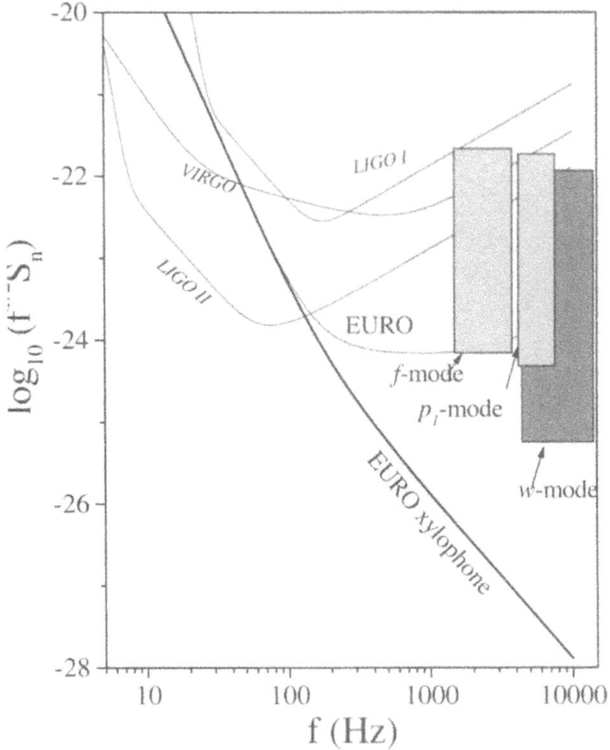

Fig. 2. The spectral noise density for the new generation of laser interferometric gravitational wave detectors. EURO is a third generation detector with ideal characteristics for fgravitational wave asteroseismology. The boxes represent the sensitivity needed to detect a mode into which an energy of $10^{-6} M \odot c^2$ is deposited. The upper limit of each box corresponds to an event in our galaxy and the lower in the Virgo cluster

Table 2. The relative errors in the extraction of the mode parameters assuming a signal-to-noise ratio of 10. The relatively large error in estimating f_w is due to the rapid damping of these modes (small Q factor), cf. Eq. (12). In other words, we will not be able to detect more than a couple of cycles of a w-mode whereas many hundred cycles of both the f- and the p-modes could be observed

Mode	σ_f/f	σ_τ/τ	σ_T (10^{-6} s)7	
f-mode	8×10^{-6}	0.02	1.1	7
p-mode	3×10^{-7}	0.02	0.38	7
w-mode	0.01	0.03	0.177	7

or the damping rate, respectively. Let us call the corresponding energies E_f and E_τ. This measure will then be detector dependent, so we list the relevant estimates for the three detector configurations used in Table 1. When the data is viewed in this way, cf. Table 3, we see that an accurate extraction of w-mode data will not be possible unless a large amount of energy is released through these modes. Furthermore, one would clearly need a detector that is sensitive at ultrahigh frequencies.

Table 3. The estimated energy (in units of $M_\odot c^2$) required in each mode in order to lead to a relative error of 1% in the inferred mode-frequency (E_f) and damping rate (E_τ). In cases where no entry is given, the required energy is unrealistically high (typically larger than $M_\odot c^2$). The distance to the source is assumed to be 10 kpc

Detector	f-mode		p-mode		w-mode	
	E_f	E_τ	E_f	E_τ	E_f	E_τ
LIGO I	3×10^{-9}	2×10^{-2}	3×10^{-10}	–	–	–
LIGO II	5×10^{-11}	3×10^{-4}	5×10^{-12}	3×10^{-2}	0.2	–
Ideal	9×10^{-13}	6×10^{-6}	9×10^{-15}	5×10^{-5}	9×10^{-5}	6×10^{-4}

3.6 Revealing the position of the source

In the previous section we discussed issues regarding the detectability of a mode-signal, and the accuracy with which the parameters of the mode could be inferred from noisy gravitational wave data. Let us now assume that we have detected the mode and extracted the relevant parameters. We then naturally want to constrain the supranuclear equation of state by deducing the mass and the radius of the star. In principle, the mass and the radius can be deduced from any two observables, [2]. In the absence of detector noise, several combinations look promising, but in reality only few combinations are likely to be useful.

As with other kinds of gravitational-wave sources, a network of at least three detectors is needed to pinpoint the location of the source in the sky. The difference in arrival time for the three detectors could be used to determine the position of the source. The higher the accuracy in measuring the time of arrival at each detector, the more precise will be the positioning of the source. Two remote detectors, at a distance d apart from each other will receive the signal with a temporal difference of

$$\Delta T = \frac{d}{c} \cos \theta, \tag{19}$$

where c is the speed of light, and θ is the angle between the line joining the two detectors and the line of sight of the source. Therefore, the accuracy by which this angle can be measured is

$$\Delta \theta = \frac{\sqrt{2} \sigma_T c}{d \sin \theta}. \tag{20}$$

The $\sqrt{2}$ arises from the measurement errors of the two times of arrival. If one assumes an 'L' shaped network of 3 detectors with arm length of $d = 10,000$ km, Eqs. (18) and (20) lead to an error box on the sky with angular sides of 1°, at most (for specific areas of the sky, and large signal-to-noise ratios the angular sizes could be much smaller). This is quite interesting since one could then correlate the detection of gravitational waves with radio, X-ray or gamma-ray observations directed towards that specific corner of the sky.

4 The unstable r-modes

So far we have discussed the modes of (essentially) non-rotating stars. A more optimistic scenario is based on the notion that various modes of oscillation may be unstable in a rotating star. Following the serendipitous discovery of such an instability in the so-called r-modes [44,45], this area of research has attracted considerable attention. Should such an instability operate in a young neutron star it may lead to the emission of copious amounts of gravitational waves [46]. Such gravitational waves have been estimated to be detectable for sources in the Virgo cluster (at 15–20 Mpc). If we suppose that most newly born neutron stars pass through a phase where this kind of instability is active, several such events should be observed per year once the advanced interferometers come into operation. This is a very exciting prospect, indeed.

In this review article we give a brief introduction to these recent ideas and suggestions. For an exhaustive discussion we refer the reader to [7].

The r-modes are unstable to the emission of gravitational waves via a mechanism that was first suggested by Chandrasekhar [47]. Subsequent work by Friedman and Schutz [48] showed that this instability is a generic feature of all rotating fluids, and in the case of r-modes Friedman and Morsink [45] showed that that the new instability is generic for toroidal perturbations of relativistic stars.

In a simple description, the instability works as follows: In a rapidly rotating star a backward moving mode (as measured by a co-rotating observer) can be dragged forward according to an inertial observer. This means that the mode radiates positive angular momentum, even though the angular momentum of the mode remains negative because the perturbed star has lower net angular momentum than the unperturbed star. As positive angular momentum is removed from the mode, its angular momentum becomes increasingly negative, implying that its amplitude increases. The mode grows due to the emission of gravitation radiation. In other words, a mode is unstable if it is prograde relative to infinity and retrograde relative to the star. For many years the investigations into the relevance of the CFS instability was focussed on the spheroidal f-mode (for a recent review, see [49]). The reason for this is that this mode was considered the most important one as far as gravitational waves are concerned. Hence, it came as some surprise that the instability associated with the toroidal r-modes is considerably stronger than the f-mode one.

In fact, toroidal modes of relativistic stars attracted little attention until recently. Such perturbations of non-rotating stars lead to a set of zero frequency modes in

Newtonian theory, complemented by gravitational-wave modes (w-modes) in the relativistic description [50,51]. Toroidal modes of a Newtonian star describe stationary horizontal fluid currents, and do not induce variations in the density and pressure. When the star is set into rotation the Coriolis force provides a weak restoring force that gives the toroidal modes true dynamics, and the fluid undergoes oscillations with a frequency (measured by an inertial observer at infinity)

$$\sigma = -m\Omega + \frac{2m\Omega}{\ell(\ell+1)},\qquad(21)$$

where Ω is the rotational frequency of the star and ℓ and m are the spherical harmonic indices. This result, that identifies the r-modes, follows from a Newtonian treatment of oscillations of slowly rotating stars. As one can easily deduce from their frequency the r-modes are always retrograde in the corotating frame and their phase velocity is always smaller than that of the rotation of the star; thus they are generically unstable independently of the rotation rate of the star.

That a mode is formally unstable in a perfect fluid star does not in itself mean that it will be allowed to grow and affect, for example, the star's spin evolution. In any "realistic" star there are dissipation mechanisms that may halt growth of an instability. In the simplest model, one must account for the effects of bulk and shear viscosity. Comparison between the damping times due to viscosity and the growth time due to gravitational radiation provides a criterion for the significance of the instability. The damping/growth time associated with each mechanism can be estimated from the ratio of the energy loss to the available mode-energy (measured in the rotating frame) i.e.

$$\frac{1}{\tau_{\text{diss}}} = -\frac{\dot{E}_{\text{diss}}}{2E_{\text{mode}}}.\qquad(22)$$

In our case the onset of the instability is signalled by

$$\frac{1}{\tau} = -\frac{1}{\tau_{gw}} + \frac{1}{\tau_{sv}} + \frac{1}{\tau_{bv}} = 0.\qquad(23)$$

In this relation, the growth time due to gravitational waves (gw) is temperature independent while both bulk (bv) and shear viscosity (sv) are strongly dependent on the internal temperature of the star. We can now find the critical rotation frequency at which the r-mode becomes unstable for the relevant values of the core temperature of the star. Detailed calculations have shown that bulk viscosity damps any fluid motion for temperatures higher than 10^{10} K, while shear viscosity dominates for temperatures below 10^6 K. This means that there is a "window of opportunity" between $10^6 - 10^{10}$ K where the r-mode instability is active and may play an astrophysical role [6,8,52,53]. This instability window is shown in Fig. 3.

Finally, it is worth pointing out that the gravitational radiation emitted from the r-modes comes primarily from the time-dependent *mass-currents*. This is the gravitational analogue of magnetic multipole radiation and the r-mode instability is unique among expected astrophysical sources of gravitational radiation in radiating primarily by gravitomagnetic effects. The detectability of these gravitational waves is discussed in detail in [7].

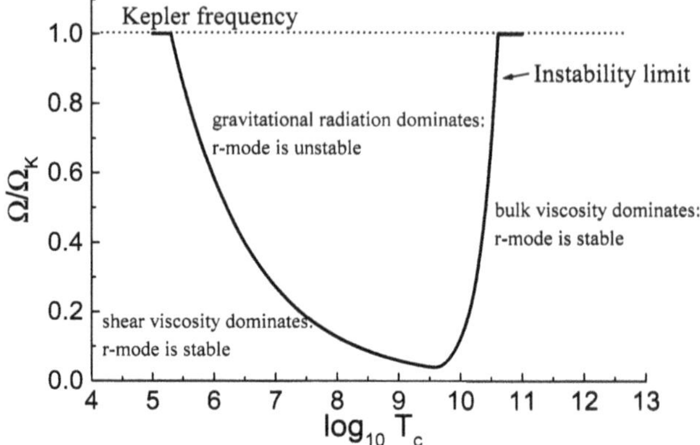

Fig. 3. The critical rotation rate at which shear viscosity (at low temperatures) and bulk viscosity (at high temperatures) balance the r-mode instability

4.1 A novel scenario

Before we conclude this review, it is appropriate to comment on the fact that we neglected the effects of both rotation and magnetic fields on the pulsation modes in our discussion of asteroseismology. This was (obviously!) not because rotation and magnetic fields play an insignificant role. On the contrary, we expect rotation to be highly relevant in many cases (in particular since various instabilities may set in above a critical spin rate), and the recent evidence for magnetars make it clear that one cannot neglect the role of the magnetic field. Still, as far as most pulsation modes are concerned, one would expect rotation to have a significant effect only for neutron stars with very short period, and the present study may well be reasonable for stars with periods longer than, say, 20 ms. Furthermore, it has been argued that neutron stars are typically born slowly rotating [54]. If that is the typical case, then our results could be relevant also for most newly born neutron stars. A more pragmatic reason for not including rotation in the present study is that detailed data for modes of rotating neutron stars is not yet available. Once such data has been computed the present study can be extended to incorporate rotational effects. There are currently efforts by various groups around the world to estimate the effects of rotation on f-, p- and w-modes (and also calculate the r-modes) in the framework of general relativity. There are already preliminary results for slowly-rotating stars [55,44] but a complete study of both slowly and fast rotating stars is outstanding.

One of the main rotational effects is the splitting of the various frequencies, i.e. the frequency is no longer degenerate with respect to the harmonic index m. For example, the quadrupole f-mode oscillations ($l = 2$) will emit gravitational waves with all five possible values of m $(-2, -1, 0, 1, 2)$. This splitting of the spectrum will create significant problems in using the empirical relations such as (3). One has to derive a similar formula with an additional term of the form $m\Omega(M/R^3)^{1/2}$ in order to include in an appropriate way the rotational corrections. In this case the prior knowledge of the rotation rate of the star may be needed. Potentially, this information can be deduced from gravitational waves due to the r-mode instability. In this case the frequency is directly proportional to the rotation rate of the star, cf. Eq. (21), and the r-mode instability signal is expected to be much stronger than those of the other modes. Hence, one can immediately infer the rotation of the star. The rotation rate will obviously change due to the r-mode, but for time intervals of the order of 1–2 secs (the typical damping time of other modes) it can be considered as constant. The information about the rotation rate can then be combined with empirical formulas that can be derived for the modes of rotating stars in order to extract all of the parameters of the star.

Strong magnetic fields can also affect the mode frequencies of the neutron stars. Preliminary results for Newtonian stars have shown that typical magnetic fields of the order of 10^{11}–10^{14} Gauss, do not produce significant shifts in the mode frequencies. One must to consider extremely strong magnetic fields, like in magnetars $(10^{15}$–10^{16} G) in order to get a significant frequency shift. In fact, this highlights an interesting issue for the magnetars: if one can detect gravitational waves from the starquakes induced by their strong magnetic fields, a possible shift in the spectrum may provide a measure of the magnetic field strength.

Finally, it is worth mentioning that a detection mode pulsation modes from old, cold neutron stars could also provide unique insights into the superfluid nature of neutron star cores. It is generally believed that once a neutron star cools below a few times 10^9 K (a few months after its birth) the bulk of its core will become superfluid. Thus, the more than 1000 observed pulsars provide useful laboratories for studying large scale superfluidity. In the simplest description, a superfluid neutron star core can be discussed in terms of two distinct fluids. One of these fluids represents the superfluid neutrons and the other fluid represents all charged components (which are expected to be coupled on a relatively short timescale). The fact that these two fluids – the "neutrons" and the "protons" – can flow more or less independently provides one of the main distinguishing dynamical features of a superfluid neutron star. In particular, one can show that two sets of pulsation modes are interlaced in the spectrum of a superfluid neutron star core [56,57]. One set of modes are the familiar p-modes, for which the two fluids tend to move together. The other set of modes are distinguished by the fact that the protons and neutrons are largely "countermoving". This class of modes is unique to the two-fluid system. Of particular interest for our current discussion is the fact that the "superfluid mode" frequencies are (locally) approximated by

$$\omega_s^2 \approx \frac{m_p}{m_p^*}\frac{l(l+1)}{r^2}c_p^2, \tag{24}$$

where c_p^2 is (roughly) the sound speed in the proton fluid, r is the radial coordinate, and l is the index of the relevant spherical harmonic $Y_{lm}(\theta, \varphi)$ used to describe the angular dependency of the mode. From this relation it is clear that an observation of these modes would provide potentially unique information regarding the nature of large scale superfluidity, and could put useful constraints on crucial parameters such as the ratio between the "bare" and "effective" proton masses m_p and m_p^* (estimates to lie in the range $0.3 \leq m_p^*/m_p \leq 0.8$). We think this is a very exciting prospect that should motivate future efforts in this field.

Acknowledgements. We appreciated useful comments by Th. Apostolatos, J. Ruoff and N. Stergioulas. This work has been supported by the EU Network Contract No. HPRN-CT-2000-00137.

References

1. Andersson, N., Kokkotas, K.D. (1996): Phys. Rev. Lett. **77**, 4134
2. Andersson, N., Kokkotas, K.D. (1998): MNRAS, **299**, 1059
3. Kokkotas, K.D., Apostholatos, Th., Andersson, N. (2001): MNRAS **100**, 100
4. Benhar, O., Berti, E., Ferrari, V. (1999): MNRAS **310**, 797
5. Yip, C.W., Chu M.-C., Leung, P.T. (1999): ApJ, **513**, 849
6. Duncan, R.C. (1998): ApJ, **498**, L45
7. Andersson, N., Kokkotas. K.D. (2001): Int. J. Mod. Phys. D **10**, 381
8. Kokkotas, K.D., Schutz, B.F. (1992): MNRAS **255**, 119
9. Kokkotas, K.D. (1997): Pulsating relativistic stars, in *Relativistic gravitation and gravitational radiation*, ed. by J.-A. Marck, J.-P. Lasota, Cambridge University Press, pp. 89
10. Kokkotas, K.D., Schmidt, B.G. (1999): Living Reviews in Relativity, 1999-2; http://www.livingreviews.org/Articles/Volume2/1999-2kokkotas
11. Allen, G., Andersson, N., Kokkotas, K.D., Schutz, B.F. (1998): Phys. Rev. D **58**, 124012
12. Allen, G., Andersson, N, Kokkotas, K.D., Laguna, P., Pullin, J.A., Ruoff J. (1999): Phys. Rev. D **60**, 104021
13. Ruoff, J. (2001): Phys. Rev. D**63**, 064018
14. Tominaga, K., Saijo, M., Maeda, K. (1999): Phys. Rev. D **60**, 24004
15. Andrade, Z., Price, R.H. (1999): Phys. Rev. D **60**, 104037
16. Ferrari, V., Kokkotas, K.D. (2000): Phys. Rev. D **62**, 107504
17. Ruoff, J., Laguna, P., Pullin, J. (2001): Phys. Rev. D **63**, 064019
18. Blaes, O., Blandford, R., Goldreich, P., Madau, P. (1989): ApJ **343**, 839
19. Mock, P.C., Joss, P.C. (1998): ApJ **500**, 374
20. Duncan, R.C., Thompson, C. (1992): ApJ **392**, L9
21. Baumgarte, T.W., Janka, H-T., Keil, W., Shapiro, S.L., Teukolsky, S.A. (1996): ApJ **468**, 823
22. Kokkotas, K.D., Schäfer, G. (1995): MNRAS **275**, 301
23. Unno, W., Osaki, Y., Ando, H., Shibahashi, H. (1989): Nonradial oscillations of stars. University of Tokyo Press, Cambridge
24. Cowling, T.G. (1941): MNRAS **101**, 367
25. McDermott, P.N., van Horn, H.M., Hansen, C.J. (1988): ApJ **325**, 725

26. Strohmayer, T.E. (1991): ApJ **372**,573
27. Kokkotas, K.D, Schutz, B.F. (1986): Gen. Rel. Grav. **18**, 913
28. Kojima, Y. (1988): Prog. Theor. Phys. **79**, 665
29. Andersson, N., Kojima, Y., Kokkotas, K.D. (1996): **462**, 855
30. Finn, L.S. (1994): Phys. Rev. Lett. **73**, 1878
31. van Kerkwijk, M.H., van Paradijs, J., Zuiderwijk, E.J. (1995): A & A **303**, 497
32. Arnett, W.D., Bowers, R.L. (1977): ApJS **33**, 415
33. Lewin, W.H.G., van Paradijs, J., Taam, R.E. (1993): Space Sci. Rev. **62**, 223
34. Friedman, J.L., Ipser, J.R., Parker, L. (1986): ApJ **305**, 115
35. Lindblom, L., Detweiler, S. (1983): Ap. J. Suppl. **53**, 73
36. Andersson, N. (1996): Gen. Relativ. Gravitation **28**, 1433
37. Echeverria, F. (1989): Phys. Rev. D. **40**, 3194
38. Finn, L.S. (1992): Phys. Rev. D **46** 5236
39. Schutz, B.F. (1997:) The detection of gravitational waves, in *Relativistic gravitation and gravitational radiation*, ed. by J.-A. Marck, J.-P. Lasota, Cambridge University Press, Cambridge, pp. 447
40. Frasca, S., Papa, M.A. (1995): Int. J. Mod. Phys. **4**, 1
41. Meers, B.J. (1988): Phys. Rev. D **38**, 2317
42. Flanagan, E.E., Hudges, S. (1998): Phys. Rev. D **57**, 4235
43. Harry, G.M., Stevenson, T.R., Paik, H.J. (1996): Phys. Rev. D **54**, 2409
44. Andersson, N. (1998): ApJ **502**, 708
45. Friedman, J.L., Morsink, S. (1998): ApJ **502**, 714
46. Owen, B.J., Lindblom, L., Cutler, C., Schutz, B.F., Vecchio, A., Andersson, N. (1998): Phys. Rev. D **58**, 084020
47. Chandrasekhar, S. (1970): Phys. Rev. Lett. **24**, 611
48. Friedman, J.F., Schutz B.F. (1978): Ap. J. **222**, 281
49. Stergioulas, N. (1998): Living Reviews in Relativity. 1998-8; http://www.livingreviews.org/Articles/Volume1/1998-8stergio/
50. Chandrasekhar, S., Ferrari, V. (1991): Proc. R. Soc. London Ser. A **433**, 423
51. Kokkotas, K.D. (1994): M.N.R.A.S. **268**, 1015
52. Lindblom, L., Owen, B.J., Morsink, S.M. (1998): Phys. Rev. Lett. **80** 4843
53. Andersson, N., Kokkotas, K.D., Schutz, B.F. (1999): ApJ **510**, 846
54. Spruit, H., Phinney, E.S. (1998): Nature **393**, 139
55. Kojima, Y. (1993): ApJ **414**, 247
56. Comer, G.L., Langlois, D., Lin, L.M. (1999): Phys. Rev. D **60**, 104025
57. Andersson, N., Comer, G.L. (2001): MNRAS, in press, preprint astro-ph/0101193

Active Galactic Nuclei and the Properties of Supermassive Black Holes

L. Maraschi, F. Tavecchio

Abstract. Advances in the capabilities of optical and X-ray telescopes allow to derive increasingly precise information on the presence of supermassive black holes at the centres of galaxies. Here we focus on a special class of Active Galactic Nuclei (AGN) called blazars. For these objects the broad band sensitivity of the *Beppo*SAX satellite extending up to 100 keV has allowed unprecedented studies of their X-ray emission, believed to derive from relativistic jets. For a group of blazars with emission lines, it is possible to estimate both the luminosity in the jet and the luminosity of the accretion disk. Implications for the origin of the power carried by relativistic jets, possibly involving rapidly spinning supermassive black holes are discussed.

1 Introduction

The scope of this paper is to bring to the attention of an audience of relativists some important advances of astrophysical research about AGN and its relation with the theory of Relativity. While Relativity is essential to this field in many respects we are not yet at the stage where we can *test* Relativity with observations of AGN. This is because the "ambient" around the supermassive black holes (SMBH) at the cores of AGN is not clean. Most of the relevant physics at the basis of AGN theory involves indeed strong gravitational fields and relativistic bulk motion. However, an understanding of the observed phenomena requires hydro- and magnetohydrodynamics, introducing complex additional physical uncertainties.

The original argument which led to hypothesise SMBH as basic engines for the AGN phenomenon [1,2] was extremely "simple". A source of high luminosity which varies rapidly must be very "compact". Since its size R must be less than cT_{var} the photon density $L/(4\pi R^2 c)$ is extremely high. A source of high compactness (defined as $l = L\sigma_T/Rmc^3$) must be very efficient as shown most clearly by Fabian [3]. Accretion onto black holes can be orders of magnitude more efficient than the ordinary nuclear reactions powering stars.

Even assuming 100% efficiency, there is a maximum compactness that any source powered by accretion cannot exceed. However, observationally some AGN violate even this limit. Early results concerned the excessive brightness temperatures inferred from the variability of compact radio sources. More recently the observation of gamma-rays from the same type of sources exhibiting the fast radio variability provided new independent evidence of violation of the fundamental compactness limit. For this relatively small fraction of AGN, called blazars, relativistic motion of the emitting plasma was invoked in order to reconcile the observed properties with basic physics [4].

It is now generally accepted that the blazar "phenomenon" (highly polarised and rapidly variable radio/optical continuum) is due to a relativistic jet pointing close to the line of sight. We further propose that all of the objects exhibiting the blazar "phenomenon" belong to a single population, despite diversities in other properties, most notably the presence or absence in their optical spectra of emission lines (flat spectrum quasars (FSQ) vs. BL Lacs). An analogous approach was suggested by Maraschi and Rovetti [5] on the basis of the "continuity" of the radio and X-ray luminosity functions of the two classes of blazars.

Here and in the following we will therefore assume that Quasars with Flat Radio Spectrum (FSQs), which include Optical Violently Variable (OVVs) and Highly Polarized Quasars (HPQs) and BL Lac objects are essentially "similar" objects in the sense that the nature of the central engine is similar apart from some basic scales. Our aim is to start from a (common) physical comprehension of the phenomenology with the goal of understanding the role of more fundamental parameters, like the central black hole mass, angular momentum and accretion rate, in determining the properties of the jets and of the associated accretion disks.

2   The unified framework for the SEDs of blazars

It was noted early on that the Spectral Energy Distribution (SEDs) of blazars exhibited remarkable systematic properties [6,7]. The subsequent discovery by the Compton Gamma Ray Observatory of gamma-ray emission from blazars (a summary can be found in [8]) was a major step forward, showing that in many cases the bulk of the luminosity was emitted in this band and questioning the importance of previous studies of the SEDs at lower frequencies.

A systematic investigation on the SEDs of the main complete samples of blazars (X-ray selected, radio-selected and Quasar-like, [9]) including gamma-ray data showed that the systematic trends found previously indeed persisted, suggesting a continuity of spectral properties (spectral sequence). All the SEDs show two broad components with peaks in the $10^{13} - 10^{18}$ Hz and $10^{21} - 10^{25}$ Hz ranges respectively. Both peaks appear to shift to higher frequencies with decreasing luminosity. We will call red and blue the objects at the different extremes of the sequence.

Beamed synchrotron and inverse Compton emission from a single population of relativistic electrons accounts very well for the observed SEDs except in the radio to mm range where effects of selfabsorption and inhomogeneity are important (see also [10]). We recall that the relativistic particle spectrum must be "curved" in order to explain the peaks observed in the SEDs. The curvature is often modelled with a broken power law. The change in spectral index must be quite large to explain the emission peaks and the energy γ_b at which the change (or break) occurs is the energy of electrons which radiate at the peak.

The model predicts that the synchrotron and IC emissions should vary in a correlated fashion since they derive from the same electron population. In particular, the emission at frequencies near the peaks derives form electrons in the same energy interval (in the absence of Klein–Nishina effects). Despite the difficulty of getting adequate data, this has been verified at least in some well studied objects. In the following we will assume that this emission model holds in general.

Ghisellini et al. [11] derived the physical parameters of jets of different luminosities along the sequence applying the above model with seed photons of internal (SSC) as well as external (EC) origin. The results suggest that (i) the importance of external seed photons increases with increasing jet luminosity; (ii) the "critical" energy of the radiating electrons decreases with increasing (total) radiation energy density. The latter dependence is physically plausible since the radiation energy density determines the energy losses of relativistic particles. If the critical electron energies were determined by a balance between injection/acceleration and cooling processes the latter dependence could be understood.

This "unified" theoretical scheme, needs to be tested in many respects. One can think of at least two ways of doing so: i) determine the physical parameters in individual objects with improved data; ii) understand the mechanisms of particle acceleration and injection from detailed variability studies.

In a broader perspective, if FSQs and BL Lacs contain "similar" jets (at least close to the nucleus) as suggested by the continuity of the SEDs, we still need to understand the differences in emission line properties. Also in this respect continuity could hold in the sense that the accretion rate may decrease continuously along the sequence but the emission properties of the disk may not simply scale with the accretion rate.

3   Studies of individual objects

We discuss below the case of one of the best studied sources, 3C 279, a prototype of red blazars and the first one to be discovered as a powerful gamma-ray emitter. Its spectral energy distribution illustrates well the presence of two main continuum components, attributed to the Synchrotron and inverse Compton mechanisms. Two SEDs are shown in Fig. 1 obtained respectively in 1996 and 1997 with simultaneous optical X-ray and gamma-ray data. The 1997 X-ray data derive from observations with *Beppo*SAX [12,13]. The two SEDs differ largely in brightness: that of 1996 is close to a historical maximum, while that of 1997 is rather faint.

Both states have been reproduced using the same theoretical model (e.g., [14]) with different parameters [13]. The contributions of synchrotron photons and of external photons to the Inverse Compton emission are shown separately (dot dashed and dashed lines respectively). Clearly, the gamma-ray variability amplitude is much larger than that in any other band. We can model the two states by varying mainly the bulk Lorentz factor Γ of the emitting plasma in the jet. While this picture is probably oversimplified, we wish to mention an interesting application to blazars of the "internal shock" model developed for Gamma-Ray Bursts. In the latter [15], most of the variability is attributed to the collisions of plasma sheaths moving along the jet with different Γs. This scenario is very promising for explaining the full range of variability of 3C 279.

For red blazars the study of the synchrotron component is difficult, because the Synchrotron peak, falls in the poorly covered IR–FIR range. Furthermore, the study

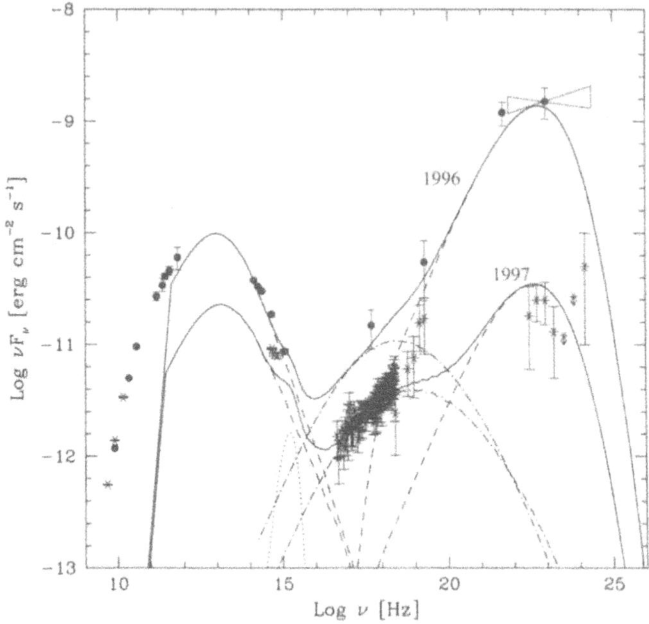

Fig. 1. Quasi-simultaneous SEDs of the quasar 3C279 obtained at two epochs 1996 and 1997. The 1996 SED represents a historical maximum. The X-ray data of 1997 were obtained with *Beppo*SAX. The continuous lines represent synchrotron plus Inverse Compton models computed to reproduce the observations at the two epochs (see text). For both epochs subcomponents of the Inverse Compton emission are shown as dot-dashed (SSC) and dashed (EC) lines respectively. The dotted component is a blackbody approximating the estimated emission from an accretion disk, assumed to be constant. The models for the two epochs differ mainly in the value of Γ

of the gamma-ray component in the MeV-GeV region of the spectrum has been difficult in the last few years due to the loss of efficiency of EGRET and is now impossible after reentry of CGRO. Substantial progress will have to await the launch of new gamma-ray satellites, like AGILE planned by the Italian Space Agency (ASI) and GLAST by NASA.

In the last years high energy observations have concentrated on blue blazars. For several sources of this class the Synchrotron component peaks in the X-ray band, where numerous satellites can provide good data. In few bright extreme BLLac objects the high energy γ-ray component is observable from ground with TeV telescopes (for a general account see [16]). In these particular cases the contemporaneous X-ray/TeV monotoring demonstrated well the correlation between the Synchrotron and the IC components. A very good example is Mkn 421 for which a rapid flare was observed simultaneously in the TeV and X-ray bands by the Whipple observatory

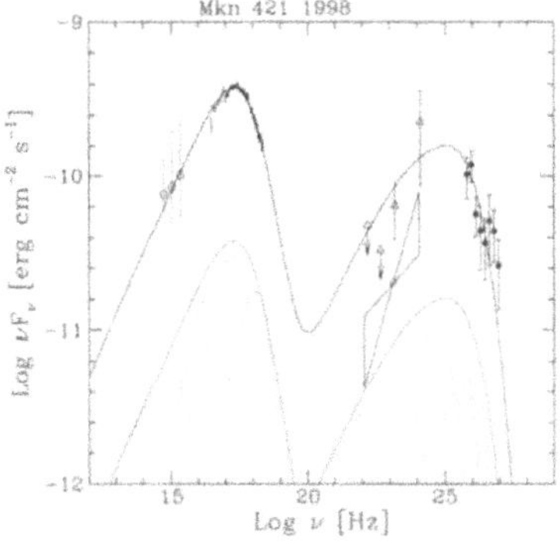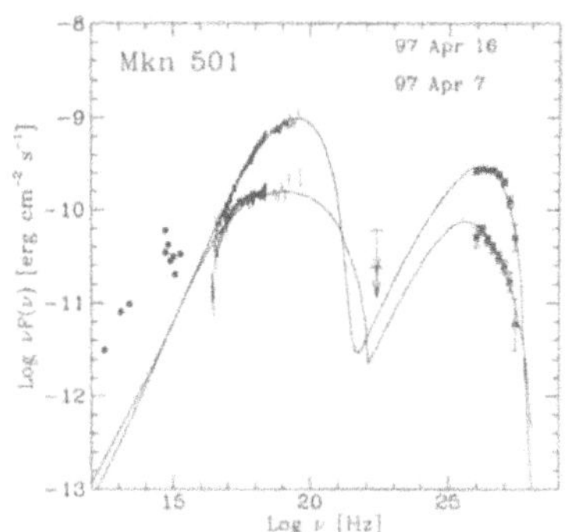

Fig. 2. *Left:* Overall SED of Mkn 421 obtained from observations taken in 1998 April (from [17]). The X-ray and TeV data are exactly simultaneous. The solid line is the spectrum computed with the SSC model. Below the actual model, subcomponents due to electrons in 4 fixed energy intervals are shown both for the synchrotron and SC processes.
Right: Overall SED of Mkn 501 observed simultaneously by *Beppo*SAX and CAT during the major flare in April 1997 [14]. The solid lines are the spectra computed with the SSC model. The models for the high and low state differ only in the value of $\gamma_b mc^2$, the energy of the electrons radiating at the peak of the synchrotron component. For both sources the observational constraints on the two peaks allow to obtain robust estimates of the physical parameters of the jet

and by *Beppo*SAX satellite in 1998. This observation probed for the first time the existence of correlation on short (hour) time scales [17–19].

When the position of the synchrotron and SSC peaks can be well determined observationally, as is possible in this type of sources, robust estimates of the physical parameters of the jet can be obtained (e.g. [20]). This was done for both Mkn 421 and Mkn 501 as illustrated in Fig. 2 [17,21,22]. In particular for Mkn 421 subcomponents due to electrons in 4 fixed energy intervals are shown both for the synchrotron and SSC process in order to give an intuitive view of the correlation between different energy ranges.

4   Jet power vs. accretion power

We now turn to discuss luminous blazars with emission lines. These fall at the high-luminosity end of the sequence, with the Synchrotron peak in the FIR region. In these sources the beamed X-ray emission is believed to be produced through the IC scattering between soft photons external to the jet (produced and/or scattered by the Broad Line Region) and *electrons at the low energy end of their energy distribution.* It is important to stress that the broad band sensitivity of *Beppo*SAX allowed to measure the X-ray spectra from 0.3 up to 100 KeV for a number of these objects. A further interesting fact is that for this type of source the hard X-ray emission has luminosity comparable to that measured in gamma-rays, due to the fact that the EC peak falls in between the two ranges.

Measuring the X-ray spectra and adapting a broad band model to their SEDs yields reliable estimates of the total number of relativistic particles involved, which is dominated by those at the lowest energies. This is interesting in view of a determination of the total energy flux along the jet (e.g. [23,24]). The "kinetic" luminosity of the jet can be written as:

$$L_{\mathrm{j}} = \pi R^2 \beta c\, U \Gamma^2, \tag{1}$$

where R is the jet radius, Γ is the bulk Lorentz factor and U is the total energy density in the jet, including radiation, magnetic field, relativistic particles and eventually protons. If one assumes that there is 1 (cold) proton per relativistic electron, the proton contribution is usually dominant.

In high luminosity blazars the UV bump is often directly observed and/or can be estimated from the measurable emission lines, yielding direct information on the accretion process in the hypothesis that the UV emission derives from an accretion disk. Thus the relation between accretion power and jet power can be explored.

This approach was started by Celotti et al. [23] but their estimates of L_{jet} were obtained applying the SSC theory to VLBI radio data which refer to larger scales. Our study involves the analysis of the *Beppo*SAX data, modeling the overall SED and deriving the physical parameters for the jet. For three sources the results have been already published [14]. Their SEDs are shown in Fig. 3 together with the emission models computed to reproduce the data. The low energy peak is due to synchrotron emission, the high energy peak is dominated by IC on the ambient photons. Note that for these objects the hard X-ray luminosity is comparable to that emitted in gamma-rays (non simultaneous data). We have preliminary results for 6 other sources with similar characteristics, all observed with *Beppo*SAX.

We further included in the analysis blazars with less prominent emission lines, hazinger3C 279 and four BL Lac objects for which we had previuos good quality *Beppo*SAX data (namely BL Lac, ON231, Mkn 501 and Mkn 421) [14,17,25,26].

In all cases we estimated physical parameters by means of a homogeneous SSC+EC model and derived accordingly the kinetic luminosity of the jet including 1 cold proton per electron, L_{P}, as well as the total luminosity radiated by the jet in the observer frame (L_{rad}). The luminosity of the disk could be estimated for all objects except the latter three BL Lac, for which we could set only upper limits on the luminosity of their putative accretion disks. For 3C 279 and BL Lac, the presence of broad Ly_α and H_α respectively allowed to estimate the ionizing continuum (e.g. [27]).

The results are shown in Fig. 4a, where the total radiative luminosity L_{rad} and the kinetic luminosity of the jet L_{P} are compared. The ratio between these two quantities gives directly the "radiative efficiency" of the jet, which turns out to be $\eta \simeq 0.1$, though with large scatter. The line traces the result of a least-squares fit: we found a slope ~ 1, indicating a rather constant radiative efficiency along the Blazar sequence (note that the data cover a wide range of about 5 orders of magnitude).

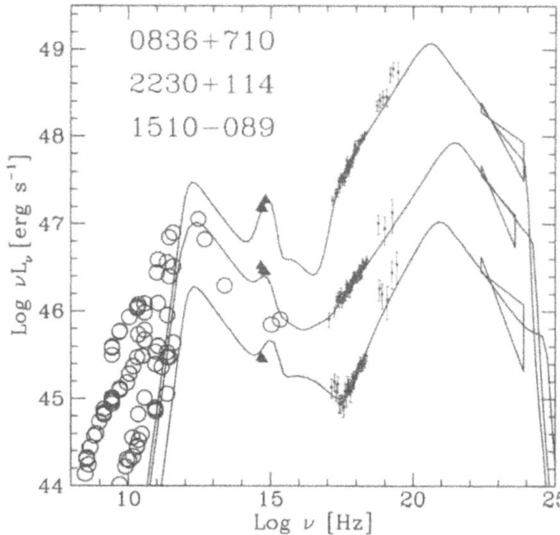

Fig. 3. Overall SEDs of three powerful emission-lines Blazars (from [14]). The objects are characterized by the presence of a strong UV-bump, allowing the determination of the luminosity of the accretion disk

In Fig. 4b we compare the luminosity of the jet, L_{rad}, which is a *lower limit* to L_{jet}, with the luminosity of the disk, L_{disk}.

A first important result is that on average the minimal power transported by the jet is *of the same order* as the luminosity released in the accretion disk. This result poses an important constraint for models elaborated to explain the formation of jets.

Two main classes of models consider either extraction of rotational energy from the black hole itself or magnetohydrodynamic winds associated with the inner regions of accretion disks. Let us parametrize the two possibilities as follows. Blandford & Znajek [28] summarize the result of their complex analysis of extraction of rotational energy from a black hole in the well known expression:

$$P_{BZ} \simeq B_0^2 r_g^2 a^2 c. \tag{2}$$

Assuming maximal rotation for the black hole ($a = 1$), the critical problem is the estimate of the intensity reached by the magnetic field threading the event horizon, which must be provided by the accreting matter. Using a spherical free fall approximation with $B_0^2/8\pi \simeq \rho c^2$ one can write:

$$P_{BZ} \simeq g\dot{M}c^2, \tag{3}$$

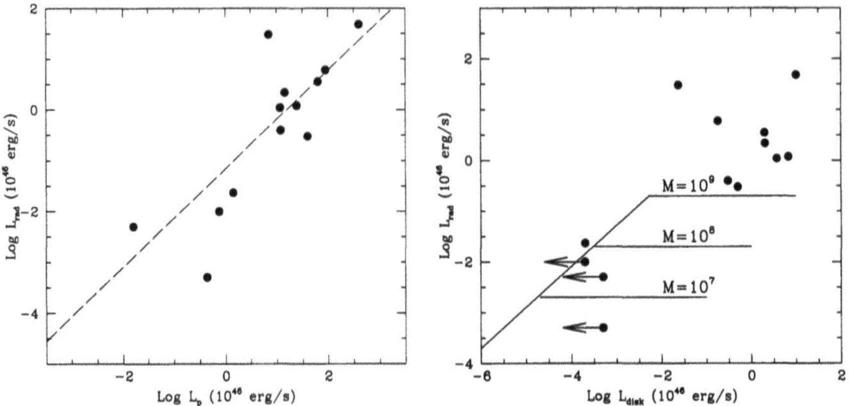

Fig. 4. a Radiative luminosity vs. jet power for the sample of Blazars discussed in the text (open circles ·represent BL Lac objects). The dashed line indicates the least-squares fit to the data. **b** Radiative luminosity of jets vs disk luminosity. The solid lines represent the *maximum* jet power estimated for the Blandford & Znajek model for black holes with different masses (in Solar units)

where $P_{\text{acc}} = \dot{M}c^2$ is the accretion power and g is of order 1 in the spherical case but in fact it is a highly uncertain number since it also depends on the field configuration. Several authors have recently discussed this difficult issue in the case of an accretion disk: the arguments discussed by Ghosh & Abramovicz [29] and Livio, Ogilvie & Pringle [30] plus equipartition within an accretion disk described by the Shakura and Sunyaev model [31] lead to $g \simeq 1$ when gas pressure dominates. Unfortunately at high accretion rates, when radiation pressure dominates, the pressure and consequently the estimated magnetic field do not increase further with $\dot{M}$ but saturate at the transition value (see Fig. 4b). Frame dragging by the rotating hole may however increase g to values even larger than 1 [32].

As argued strongly by Livio et al. [30] the accretion flow itself may power jets through a hydromagnetic wind. However for consistency only some fraction $f\dot{M}c^2$ can be used to power the jet. Further recall that the luminosities observed from the jet and disk are related to their respective powers by efficiency factors $L_{\text{rad}} = \eta P_{\text{jet}}$; $L_{\text{disk}} = \epsilon P_{\text{acc}}$.

Using the condition that $P_{\text{jet}} \leq (P_{BZ} + f P_{\text{acc}})$ together with the previous relations we finally find

$$L_{\text{rad}} \leq \frac{\eta(g + f)}{\epsilon} L_{\text{disk}}. \tag{4}$$

One can account for the observed relation between L_{rad} and L_{disk} if for instance $\eta \simeq \epsilon \simeq 0.1$ and f or g or both are close to 1. It is also possible that $\epsilon << 0.1$ if the optically thick emission derives from radii larger than $3r_G$ or if the disk is optically thin as may happen at low accretion rates [33]. This may be necessary at low luminosities to explain why disks are not seen (upper limits). However at high luminosities there is an additional condition deriving from the total luminosities observed. At 10^{47} erg s^{-1} the Eddington luminosity implies a mass of 10^9 $M_\odot$ and an accretion rate of 10 $M_\odot$ y^{-1} for $\epsilon = 10^{-1}$. It seems then difficult to invoke lower efficiencies.

For comparison we show in Fig. 4b the estimates for the rotational power derived by Ghosh and Abramovicz [29] for various values of the mass of the central black hole as a function of the luminosity observed from the disk. The latter is related to the accretion rate which appears in the formulae of Ghosh and Abramovicz [29] adopting $\epsilon \simeq 0.1$. Clearly the model fails to explain the large power observed in the jets of bright quasars, even for BH masses ($M \sim 10^9 M_\odot$). This is because in the radiation pressure dominated region of the disk the pressure used to estimate the magnetic field via equipartition does not increase with the accretion rate.

5   Conclusions

The study of blazars yields unique information on the physical conditions and emission processes in relativistic jets. A unified approach is possible whereby the jets in all blazars are similar and their power sets the basic scale. While the phenomenological framework is suggested to be "simple" (e.g. "red" blazars are highly luminous, have low average electron energies and emit GeV gamma-rays while "blue" blazars have low luminosity, high average electron energies and emit TeV γ-rays) we do not yet know what determines the emission properties of jets of different power nor what determines the jet power in a given AGN. There is however the exciting prospect that such problems can be tackled with data that can be gathered in the near future.

References

1. Zeldovich Y.B., Novikov, I.D. (1964): Usp. Fiz. Nauk **84**, 377
2. Salpeter, E. E. (1964): ApJ **140**, 796
3. Fabian, A.C. (1979): Proc. Royal. Soc. **366**, 449
4. Blandford, R.D., Rees, M.J. (1978): Pittsburgh Conference on BL Lac Objects, Proceedings (A79-3002611-90) Pittsburgh, PA, University of Pittsburgh, pp. 328–341
5. Maraschi, L., Rovetti, F. (1994): ApJ **436**, 79
6. Landau, R., Gohsch, B., Jones, T.J. et al. (1986): ApJ **308**, 78
7. Sambruna, R.,M., Maraschi, L., Urry, C.M. (1996): ApJ **463**, 444
8. Mukherjee, R., Bertsch, D.L., Bloom, S.D. et al. (1997): ApJ **490**, 116
9. Fossati, G., Maraschi, L., Celotti, A. et al. (1998): MNRAS **299**, 433
10. Kubo, H., Takahashi, T., Madejski, G. et al. (1998): ApJ **504**, 693
11. Ghisellini, G., Celotti, A., Fossati, G. et al. (1998): MNRAS **301**, 451
12. Hartman, R.C., Buttcher, M., Bertsch, D. et al. (2001): ApJ **553**, 683
13. Ballo et al. (2001): APJ, submitted
14. Tavecchio, F., Maraschi, L., Ghisellini, G. et al. (2000): ApJ **543**, 535
15. Spada, M. et al. (2001): in press
16. Catanese, M., Weekes, T.C. (1999); PASP **111**, 1193
17. Maraschi, L. (1999): ApJ **526**, L81

18. Takahashi, T., Madejski, G., Kubo, H. (1999): Astroparticle Physics **11**, 177
19. Catanese, M., Sambruna, R.M. (2000) ApJ **534**, L39
20. Tavecchio, F., Maraschi, L., Ghisellini, G. (1998): ApJ **509**, 608
21. Pian, E., Vacant, G., Tagliaferri, G. et al. (1998): ApJ **492**, L17
22. Tavecchio, F. (2001): APJ **554**, 125
23. Celotti, A., Padovani, P., Ghisellini, G. (1997): MNRAS **286**, 415
24. Sikora, M. (1997): ApJ **484**, 108
25. Tagliaferri, G., Ghisellini, G., Giommi, P. et al. (2001): Proceedings of the Conference "X-ray Astronomy'99: Stellar Endpoints, AGN and Diffuse Background", 2000, Astrophysical Letters and Communications, AIP, Conf. series, in press
26. Tagliaferri, G., Ghisellini, G., Giommi, P. (2000): A & A **354**, 431
27. Corbett, E.A., Robinson, A., Axon, D. (2000): MNRAS **311**, 485
28. Blandford, R.D., Znajek, R.L. (1977): MNRAS **179**, 433
29. Ghosh, P., Abramowicz, M.A. (1997): MNRAS **292**, 887
30. Livio, M., Ogilvie, G. I., Pringle, J. E. (1999): ApJ**512**, 100
31. Shakura, N. I., Sunyaev, R. A. (1973): A & A **24**, 337
32. Meier, D. L. (1999): ApJ **522**, 753
33. Blandford, R.D. (1990): Lecture Notes, Saas Fee, SSAA, Springer-Verlag, pp. 161–269

Huygens' Principle and MAPLE's NPspinor Package

K.C. Chu, S.R. Czapor, R.G. McLenaghan

Abstract. With the assistance of the computer algebra system MAPLE's NPspinor package, two propositions are proved regarding the validity of Huygens' principle for the non-self-adjoint scalar wave equation on a Petrov type D spacetime. A decom-position of the problem is given according to the alignment of the principal spinors of the Maxwell and Weyl spinors. The first proposition states that the validity of Huygens' principle implies a certain product involving four of the spin coefficients is real. The second proposition states that if the associated Maxwell spinor of a non-self-adjoint scalar wave operator is algebraically degenerate and its principal spinor is aligned with one of the doubly degenerate Weyl principal spinors, then that wave operator cannot be Huygens'.

1 Introduction

The scalar wave equation on a spacetime $(\mathcal{M}, g_{\alpha\beta})$ has the form:

$$Pu = (\Box + A^\alpha \nabla_\alpha \nabla_\beta + B)u = f, \tag{1}$$

where ∇_α is the Levi–Civita connection on $(\mathcal{M}, g_{\alpha\beta})$, A^α is any smooth vector field, B and f are smooth scalar fields, and

$$\Box := g_{\alpha,\beta} \nabla_\alpha \nabla_\beta = \frac{1}{|g|} \frac{\partial}{\partial x^\alpha} \left(\sqrt{|g|} g^{\alpha\beta} \frac{\partial}{\partial x^\beta} \right)$$

is the d'Alembertian operator on $(\mathcal{M}, g_{\alpha\beta})$. If A^α is zero, the equation is said to be self-adjoint; otherwise, it is said to be non-self-adjoint.

Equation (1), or equivalently the operator P, is said to satisfy Huygens' principle if, roughly speaking, its solutions do not exhibit "ripples" behind the wave front. The precise formulation of Huygens' principle is given as a property of the support of the solution to the Cauchy problem for the partial differential equation (1) and can be found in [1].

One of the current working conjectures (proposed by Carminati and McLenaghan) is that Huygens' principle is valid only on an underlying spacetime which is either conformally at or conformally equivalent to an exact plane wave spacetime.

In an attempt to establish the Carminati–McLenaghan conjecture, researchers have exploited (and extended) a series of conditions necessary for the validity of Huygens' principle. For computational reasons, disjoint classes of spacetimes have been studied separately according to Petrov type. Numer- ous results have been obtained for spacetimes of Petrov types 0, N, III, D and II; for an extensive summary, the reader is referred to [2]. The present paper gives two partial results for the case of the non-self-adjoint wave equation on a type D background spacetime.

The proofs are highly computational, and make extensive use of the MAPLE system. In particular the NPspinor package, which we describe brieyly in the next section, is central to our approach. Details of the proofs may be found in [3].

2   The necessary conditions and the MAPLE Package NPspinor

The proofs of Propositions 2 and 3 (see below) are based on the first six necessary conditions for Huygens' principle derived from Hadamard's criterion, which is a necessary and sufficient condition for Huygens' principle. A proof of Hadamard's criterion can be found in [1]. As is customary, the spinor form of these necessary conditions will be used. They are referred to as the 0-index, 1-index, . . . , and 5-index conditions, which are listed below in that order:

$$0 = b - \frac{1}{2}\Lambda^{\alpha}{}_{;\alpha} - \frac{1}{4}A^{\alpha} - \frac{1}{6}R, \tag{2}$$

$$0 = \varphi_{AK;}{}^{K}{}_{\dot{A}}, \tag{3}$$

$$\begin{aligned}
0 = {}&\Psi_{ABKL}{}^{K}{}_{\dot{A}}{}^{L}{}_{\dot{B}} + \bar{\Psi}_{\dot{A}\dot{B}\dot{K}\dot{L}}{}^{\dot{K}}{}_{A}{}^{\dot{L}}{}_{B} \\
&+ \Psi_{AB}{}^{KL}\varphi_{KL\dot{A}\dot{B}} + \bar{\Psi}^{\dot{K}\dot{L}}_{\dot{A}\dot{B}}\Phi_{AB\dot{K}\dot{L}} + 10\varphi_{AB}\bar{\varphi}_{\dot{A}\dot{B}}, \tag{4}
\end{aligned}$$

$$\begin{aligned}
0 = {}&3\Psi_{ABCK;}{}^{K}{}_{(\dot{A}\varphi\dot{B}\dot{C})} + 3\Psi_{\dot{A}\dot{B}\dot{C}\dot{K};}{}^{\dot{K}}{}_{(A\varphi BC)} \\
&- \Psi_{ABC}{}^{K}\bar{\varphi}_{(\dot{A}\dot{B};\dot{C})K} - \Psi_{\dot{A}\dot{B}\dot{C}}{}^{\dot{K}}\bar{\varphi}_{(ABC)\dot{K}}, \tag{5}
\end{aligned}$$

$$\begin{aligned}
0 = {}&3\Psi_{ABCD;K\dot{K}}\bar{\Psi}^{K\dot{K}}_{\dot{A}\dot{B}\dot{C}\dot{D};} - 40\Psi_{ABC|K|;}{}^{K}{}_{(\dot{A}}\bar{\Psi}^{\dot{K}}_{\dot{B}\dot{C}\dot{D})\dot{K}D)} \\
&+ 4\Psi^{K}{}_{(ABC;D)(A}\bar{\Psi}^{\dot{\textit{J}}}_{\dot{B}\dot{C}\dot{D})\dot{L};K} + 4\bar{\Psi}^{\dot{K}}_{\dot{A}\dot{B}\dot{C};\dot{D}(A}\Psi_{BCD)L;}{}^{L}_{\dot{K}} \\
&- 4\Psi^{K}{}_{(ABC}\bar{\Psi}^{\dot{K}}_{(\dot{A}\dot{B}\dot{C}|\dot{K};K|\dot{D})D)} - 4\bar{\Psi}^{\dot{K}}{}_{\dot{A}\dot{B}\dot{C}}\Psi_{(ABC|K;\dot{K}|D)\dot{D}} \\
&+ 12\Psi^{K}{}_{(ABC}\bar{\Psi}^{\dot{K}}_{\dot{A}\dot{B}\dot{C}|\dot{K}|;\dot{D}K\dot{D})} + 12\bar{\Psi}^{\dot{K}}{}_{(\dot{A}\dot{B}\dot{C}}\Psi_{ABC|K|;\dot{D})\dot{K}D)} \\
&- 16\Psi^{K}{}_{(ABC}\Phi_{D)K\dot{K}(\dot{A}}\bar{\Psi}^{\dot{K}}_{\dot{B}\dot{C}\dot{D})} - 32\Psi_{ABCD}\bar{\Psi}_{\dot{A}\dot{B}\dot{C}\dot{D}} \\
&- 6\varphi_{(AB;CD)(\dot{C}\dot{D}}\bar{\varphi}_{\dot{A}\dot{B})} - 6\bar{\varphi}_{(\dot{A}\dot{B};\dot{C}\dot{D})(CD}\varphi_{AB)} \\
&- 42\varphi_{(AB}\varphi_{CD)}\bar{\Psi}_{\dot{A}\dot{B}\dot{C}\dot{D}} - 42\bar{\varphi}_{(\dot{A}\dot{B}}\bar{\varphi}_{\dot{C}\dot{D})}\Psi_{ABCD} \\
&+ 16\varphi_{(AB:C(\dot{C}}\bar{\varphi}_{\dot{A}\dot{B};\dot{D})D)} + 36\varphi_{(AB}\Psi_{CD)(\dot{C}\dot{D})}\bar{\varphi}_{\dot{A}\dot{B}}, \tag{6}
\end{aligned}$$

$$\begin{aligned}
0 = S\Big[&-6\Psi^{K}_{ABC;K\dot{A}}\dot{A}_{\dot{B}\dot{C};D\dot{D}E\dot{E}} + 6\Psi^{K}_{ABC\;;D\mathcal{D}}\bar{\varphi}_{\dot{A}\dot{B}:K\dot{C}E\dot{E}} \\
&- 24\Psi^{K}_{ABC;K\dot{A}DD\dot{D}}\varphi_{\dot{B}\dot{C};E\dot{E}} + 24\Psi^{K}{}_{ABC}\Phi_{KD\dot{A}\dot{D}}\bar{\varphi}_{\dot{B}\dot{C};E\dot{E}} \\
&- 18\Psi^{K}_{ABC;E\dot{E}}\Phi_{KD\dot{A}\dot{D}}\bar{\varphi}_{\dot{B}\dot{C}} + 18\Psi^{k}_{ABC;K\dot{A}}\Phi_{DE\dot{D}\dot{E}}\bar{\varphi}_{\dot{B}\dot{C}} \\
&- 36\Psi^{K}_{ABC}\bar{\Psi}_{\dot{B}\dot{C}\dot{D}\dot{E};K\dot{A}}\varphi_{DE} - 138\Psi^{K}_{ABC K\dot{A}}\bar{\Psi}_{\dot{B}\dot{C}\dot{D}\dot{E}}\varphi_{DE} \\
&+ 6\Psi^{K}_{ABC;D\dot{D}}\Psi_{\dot{B}\dot{C}\dot{E}\dot{A}}\varphi_{KE} + 6\Psi_{ABCD;E\dot{E}}\bar{\Psi}^{K}_{\dot{A}\dot{B}\dot{C}}\bar{\varphi}_{\dot{K}\dot{D}} \\
&+ c.c.\Big], \tag{7}
\end{aligned}$$

where "S" refers to the symmetric part, and "c.c." refers to the complex conjugate of the entire preceding expression. We refer the reader to [4–6], and [7] for further discussion of the notation used above.

The necessary conditions were originally derived in tensor form by a number of researchers including Günther, McLenaghan, and Anderson. Their derivations can be found in [5, 8, 9]. The conversion from tensor form to spinor form is a straightforward (but extremely lengthy) process. A descrip- tion of this process can be found, for example, in [4].

Next, in order to exploit our assumption of Petrov type D, we must convert the spinor quantities (2)–(7) to their component equations with respect to a *canonical* spinor dyad $\{o_A, \ell_B\}$ in which the only nonvanishing component of the Weyl spinor is Ψ_2. Note that a symmetric spinor with n dotted and n undotted indices has $(n+1)^2$ (symmetrized) components; hence (2)–(7) yield $2 + \cdots + 6^2 = 91$ component equations. Moreover, when expressed in terms of the Newman–Penrose (NP) formalism, many of these equations will involve hundreds of terms. It is this considerable loss of compactness that makes the task of computing and manipulating such quantities so well-suited to a symbolic algebra system such as MAPLE.

The NPspinor package, developed by Czapor et al. [10, 11], addresses the above problem by providing representations for the NP scalars and derivatives, and such spinors as typically appear in the Huygens' conditions; it also provides sufficient functionality to convert these spinor quantities into their dyad components and manipulate them in NP form.

In NPspinor, the NP quantities are represented as simply as possible, using ASCII/English equivalents for Greek letters. For example, the NP quan tity $\Delta\Psi_0 - D\Phi_{02} + 2(\bar{\pi} + \bar{\beta})$ would become the the MAPLE expression

$$V(W0) - D(R02) + 2 * (pc + bc) * L.$$

Note that, in general, the complex conjugate of a quantity is denoted by appending/removing "C" to/from its name. The procedure `conj()` perfroms this function, whili noting quantities that are real/Hermitian (e.g Δ, Φ_{00}). The NP derivatives themselves are simply MAPLE procedures, known to the `conj()` function, which behave as general differential operators. These, in turn, are the basis for the NP commutators; for example, $[\Delta, D]$ becomes the procedure $V_D()$. Note that the NP Ricci and Bianchi identities are also provided; the function `eqns()` loads them into the (global) variables `eq1,...,eq29`. Typically, this list is extended by the user as new equations are derived.

In order to represent spinor quantities as simply as possible, NPspinor requires that they be written solely in terms of covariant (i.e. lower index) spinors with the exception of the spin metric ϵ^{AB}. Hence, we may write

$$o_A \sim \text{o[A]}, \qquad \iota_A \sim \text{i[A]}, \quad \epsilon^{AB} \sim \text{eps[a,B]}$$
$$\Psi_{ABCD} \sim \text{psi[A,B,C,D]}, \quad \Phi_{A,B,\dot{A}\dot{B}} \sim \text{phi[A,B,Ac,Bc]},$$

without ambiguity, and can still represent contractions. For example, the object F $\Phi_{AB\dot{A}\dot{B}}\bar{o}^{\dot{B}}$ may be written as $\epsilon^{\dot{B}\dot{E}}\Phi_{AB\dot{A}\bar{o}_{\dot{E}}}$, and thus translates to the MAPLE expression

$$\text{eps[Bc,Ec]} * \text{phi[A,B,Ac,Bc]} * \text{oc[Ec]}.$$

Note that while this expression contains a contraction, NPspinor cannot *simplify* it until after the Ricci spinor is expanded in terms of the dyad. This is done with the function dyad(), which contains templates for expanding the Weyl and Ricci spinors (in terms of the spinor dyad and the NP curvature scalars); it may also be given user-defined templates for other objects. We may then force simplification of any explicit contractions of *dyad* spinors using the contract() function.

Covariant derivatives are represented using the del() function; for example, $\Psi_{ABCD;E\dot{E}}$ becomes

del(psi[A,B,C,D], E,Ec).

As with the NP Pfaffians, del() simply behaves as a general differential operator modulo the rules $\epsilon_{AB;C\dot{C}} = 0$,

$$\varphi_{;A\dot{A}} = \Delta\varphi\, o_A\bar{o}_{\dot{A}} - \delta\varphi\iota_A\bar{o}_{\dot{A}} + D\varphi\iota_A\bar{\iota}_{\dot{A}}$$

(where φ is a scalar), and the expansions of $o_{A;B\dot{B}}$ and $\iota_{A;B\dot{B}}$ in terms of the dyad and the NP spin coefficients. Hence, derivatives of Ψ_{ABCD} or $\Phi_{AB\dot{A}\dot{B}}$ are also inert until they are expanded using dyad().

Finally, there are facilities for symmetrization; the most basic is symm(), which symmetrizes an expression over a list of indices.

We conclude this section with a small example. Consider the second term of (5), namely $3\bar{\Psi}_{\dot{A}\dot{B}\dot{C}\dot{K}}{}^{K}{}_{(A}\varphi_{BC)}$. When $\varphi_{AB} = 2\varphi_1 o_{(A}\iota_{B)}$ (corresponding to Case 2 in Sect. 4) we may compute this as follows:

```
# Assume a space of Petrov type D, and canonical dyad.
W0 :=0: WO := 0: Wic :=0:
# Simplify the contracted derivative first.
term2 := 3*eps[Kc,Lc]*del( psic[Ac,Bc,Cc,Kc], Lc,A ):
term2 := contract( dyad(term2) ):
# Define the Maxwell spinor.
template[F] := F[A,B]=-2*f1*symm( o[A]*i[B], [A,B] ):
# Extend the product, and symmetrize over undotted indices.
term2 := dyad( term2*F[B,C] ):
term2 := symm( term2, [A,B,C] ):
term2 := findsymm( term2, [A,B,C] ):
```

The final line in the above isolates subexpressions which are symmetric products of dyad spinors (e.g. $o\,0_{(AB)}\iota^C\bar{o}_{\dot{A}\dot{B}\dot{C}}$; then the component equations may be easily separated and manipulated in NP form.

We refer the reader to [11] for a complete description (and examples) of all package components.

It bears mention that NPspinor is not the *only* MAPLE package on which we rely. Parts of the proof of Proposition 2 require the use of the theory of Gröbner bases to simplify certain multivariate polynomial systems of equations. The reader is referred to [12] and [13] for an account of the theory of Gröbner bases. In particular, the function gsolve() within MAPLE's Groebner package, developed by Czapor, is used to perform the pertinent Gröbner basis computations. Regarding the algorithm and implementation of gsolve(), the reader is referred to [14] and [15].

3   The first proposition

Proposition 2. *The validity of Huygens' principle for any non-self-adjoint scalar wave equation on a Petrov type D spacetime implies that with respect to any canonical spinor dyad (one in which the only non-vanishing component of the Weyl spinor is Ψ_2), the following equation holds:*

$$\kappa\bar{\sigma}\bar{\lambda}\nu - \bar{\kappa}\sigma\lambda\bar{\nu} = 0. \tag{8}$$

Proof. Let $\mathcal{M}$ be a Petrov type D spacetime on which there exists a non-self-adjoint scalar wave equation that satis
es Huygens' principle. Suppose on the contrary that (8) does not hold with respect to some canonical spinor dyad on $\mathcal{M}$. Note, first of all, that this implies none of κ, σ, λ and ν can vanish with respect to this dyad.

Since the dyad is canonical, all components of the Weyl spinor vanish except Ψ_2. The $o_{(AB}\iota_{C)}$, $\iota_{ABC}\bar{\iota}_{\dot{A}\dot{B}\dot{C}}$, $\iota_{ABC}\bar{\iota}_{\dot{A}\dot{B}\dot{C}}$, and $o_{ABC}\bar{o}_{\dot{A}\dot{B}\dot{C}}$ component equations of the 3-index condition then lead to the following homo- geneous linear system of equations in $\varphi_{\bar{0}}$, φ_2 and their conjugates:

$$\begin{pmatrix} \bar{\Psi}_2\bar{\lambda} & 0 & & \Psi_2\sigma \\ 0 & \Psi_2\lambda & \bar{\Psi}_2\bar{\sigma} & 0 \\ \bar{\Psi}_2\bar{\kappa} & \Psi_2\kappa & 0 & 0 \\ 0 & 0 & \bar{\psi}_2\bar{\nu} & \Psi_2\nu \end{pmatrix} \begin{pmatrix} \varphi_0 \\ \bar{\varphi}_0 \\ \varphi_2 \\ \bar{\varphi}_2 \end{pmatrix} = \begin{pmatrix} 0 \\ 0 \\ 0 \\ 0 \end{pmatrix}. \tag{9}$$

The determinant of the coefficient matrix in (9) is

$$-\Psi_2^2\bar{\Psi}_2^2(\kappa\bar{\sigma}\bar{\lambda}\nu - \bar{\kappa}\sigma\lambda\bar{\nu}).$$

Since Ψ_2 does not vanish on $\mathcal{M}$, if (8) does not hold, then both φ_0, and φ_2 must vanish (and so must their conjugates). This in turn implies that $\varphi_1 \neq 0$ and $\bar{\varphi}_1 \neq 0$, since by hypothesis the scalar wave equation is non-self-adjoint.

With the conditions that $\varphi_0 = \bar{\varphi}_0 = \varphi_2 = \bar{\varphi}_2 = 0$ the $o_{(A}\iota_{BC)}\bar{\iota}_{\dot{A}\dot{B}\dot{C}}$ and $\iota_{ABC}\bar{o}_{(\dot{A}}\bar{\iota}_{\dot{B}\dot{C})}$ component equations of the 3-index condition lead to the following homogeneous linear system of equations in φ_1 and $\bar{\varphi}_1$:

$$\begin{pmatrix} 2\bar{\Psi}_2\kappa & 6\Psi_2\kappa \\ -6\bar{\Psi}_2\bar{\kappa} & -2\Psi_2\bar{\kappa} \end{pmatrix} \begin{pmatrix} \varphi_1 \\ \bar{\varphi}_1 \end{pmatrix} = \begin{pmatrix} 0 \\ 0 \end{pmatrix}. \tag{10}$$

Since $\varphi_1 \neq 0$ and $\bar{\varphi}_1 \neq 0$, the detreminant of the coefficient matrix in (10) must vanish. This determinant is $32\Psi_2\bar{\Psi}_2\kappa\bar{\kappa}$, implying that κ vanishes; this contradicts our assumption that (8) does not hold.

4 Alignment of principal null directions between the Maxwell spinor and the Weyl spinor

We decompose the question of the validity of Huygens' principle on a Petrov type D spacetime into sub-cases according to the alignment of the two principal null directions of the Maxwell spinor with the two doubly degenerate principal null directions of the Weyl spinor.

We fix a canonical spinor dyad, $\{o_A, \iota_B\}$, for the underlying Petrov type D spacetime, i.e. a spinor dyad with respect to which the only non-vanishing component of the Weyl spinor is Ψ_2. Since it is totally symmetric, the Maxwell spinor φ_{AB} takes the form:

$$\varphi_{AB} = \xi_{(A}\zeta_{b)}, \tag{11}$$

where ξ_A and ζ_A are the principal spinors of the Maxwell spinor.

The alignment of the Maxwell principal spinors with the Weyl principal spinors determines which of the Maxwell spinor components vanish with respect to the chosen canonical spinor dyad. This is shown in Table 1, where 0 and N represent the vanishing and non-vanishing of the corresponding Maxwell spinor component respectively. Note that

1. Case 0 is the self-adjoint case.
2. Case 1 and Case 4 are equivalent; so are Case 3 and Case 6. We can see this equivalence by interchanging o_A and ι_B. So, there are in fact only five geometrically distinct sub-cases that are non-self-adjoint.

We now give the geometric meaning of these sub-cases. First we express φ_{AB} with respect to the chosen canonical spinor dyad $\{o_A, \iota_B\}$ as follows:

$$\varphi_{AB} = \varphi_0\,\iota_A \otimes \iota_B - 2\varphi_1 o_{(A} \otimes \iota_{B)} + \varphi_2\,o_A \otimes o_B \tag{12}$$

If both Maxwell principal spinors are aligned with, say ι_A, then we see from (12) that $\varphi_1 = \varphi_2 = 0$. This corresponds to Case 4 in Table 1. If both Maxwell principal

Table 1. Possible alignments between the Maxwell and Weyl principal spinors

Case	φ_{AB}	φ_0	φ_1	φ_2
0	0	0	0	0
1	$o_A o_B$	0	0	N
2	$o_{(A}\iota_{B)}$	0	N	0
3		0	N	N
4	$\iota_A \iota_B$	N	0	0
5		N	0	N
6		N	N	0
7		N	N	N

spinors are aligned with o_A instead, then again by (12), we must have $\varphi_0 = \varphi_1 = 0$. This corresponds to Case 1 in Table 1. Also, if one of the Maxwell principal spinors is aligned with o_A and the other with ι_A, then clearly $\varphi_0 = \varphi_2 = 0$. This corresponds to Case 2. The geometric interpretations of the other cases are similar.

5   The second proposition

Proposition 3. *Let $P := \Box + A^\alpha \nabla_\alpha + B$ be a non-self-adjoint scalar wave operator on a Petrov type D spacetime. If the Maxwell spinor (or tensor) associated to A^α is algebraically special and its degenerate principal null di- rection coincides with one of the doubly degenerate principal null directions of the Weyl spinor (or tensor) of the underlying spacetime, then P is not a Huygens' operator.*

We shall include only an outline of the proof. The complete details of the proof can be found in [3].

Outline of proof. Suppose Proposition 3 is false. Then there exists a Petrov type D spacetime $(\mathcal{M}, \langle_{\alpha\beta})$ and a Huygens' scalar wave operator on $\mathcal{M}$ such that either the Case 1 or Case 4 conditions are satisfied. It suffices to establish only the inadmissibility of Case 4, since Case 1 is geometrically equivalent to it.

Under the Case 4 assumptions, the Maxwell spinor φ_{AB} associated to A^α has the form $\varphi_{AB} = \varphi_0 \iota_A \iota_B$, with $\varphi_0 \neq 0$, where $\{0_A, \iota_A\}$ is a spinor dyad canonical to the Weyl spinor of $(\mathcal{M}, \}_{\alpha\beta})$. Let w be the smooth function defined on $\mathcal{M}$ by

$$e^w = \frac{\Psi_2}{\varphi_0}.$$

Then under the dyad transformation

$$o'_A = e^{W/2} o_A, \quad \iota'_A = e^{-w/2} \iota_A,$$

the Maxwell spinor component φ_0 transforms as follows:

$$\varphi'_0 = e^w \varphi_0 = \frac{\Psi_2}{\varphi_0} = \Psi_2 = \Psi'_2,$$

while φ'_1 and φ'_2 remain zero. The above assertions follow from the transformation laws of the respective spinor components which can be found in [3].

Next, let φ be the smooth function on $\mathcal{M}$ defined by $e^{4\varphi} = \Psi'_2 \bar{\Psi}'_2$. Under a conformal transformation to $(\mathcal{M}, \rceil^{\in\varphi}\}_{\alpha\beta})$, and the following choices for the van der Waerden correspondence and spinor dyad:

$$\epsilon''AA = \epsilon'^{AB}, \qquad \epsilon''_{AB} = \epsilon'_{AB},$$

$$\sigma''^{A\dot{A}}_\alpha = \epsilon^\varphi \sigma'^{A\dot{A}}_\alpha, \quad \sigma''^\alpha = e^{-\varphi}\sigma'^\alpha{}_{A\dot{A}},$$

$$o''_A = e^{\frac{r-1}{2}} o'_A, \qquad \iota''_A = e^{\frac{1-r}{2}} \iota'_A,$$

the quantity Ψ_2' transforms as follows:

$$\Psi_2'' = e^{-2\varphi}\Psi_2'.$$

As a result

$$\Psi_2''\bar{\Psi}_2'' \equiv 1,$$

i.e. Ψ_2'' is identicall unimodular. If we choose $r = 1$, the φ_0' transforms as follows:

$$\varphi_0'' = e^{(r-3)\varphi}\varphi_0' = e^{(r-3)\varphi}\Psi_2' = e^{-2\varphi}\Psi_2' = \Psi_2''.$$

Therefore, under the above transformations, we have set

$$\Psi_2''\bar{\Psi}_2'' \equiv 1 \quad \text{and} \quad \varphi_0'' \equiv \Psi_2''. \tag{13}$$

Since the Huygens' nature of P is preserved under trivial transformations, P'', the transformed operator of P under the above transformations, is also Huygens'. In what follows, we assume that these transformations have been made, and hence that (13) holds; we then drop the double prime.

Step 1 We will make use of the necessary conditions to solve for a large number of the Pfaffian derivatives and some of the spin coefficients as algebraic expressions of the remaining spin and curvature coeffients, thereby simplifying our system of necessary conditions from an algebraic-differential system to a largely algebraic system.

The manipulations involved in this step are purely algebraic and straightforward (though computationally intensive due to the large size of the system). We will therefore omit the details here but refer the reader to [3]. We will however list all of the quantities being solved for in this step and give explicitly the substitutions that we will use later:

- The following quantities are shown to vanish:

$$0 = \lambda = \bar{\lambda} = \nu = \bar{\nu} = \mu = \bar{\mu} = \gamma = \bar{\gamma} = \Delta\bar{\alpha} = \Delta\tau, \tag{14}$$
$$0 = \Psi_{21} = \Psi_{12} = \Psi_{22} = \Delta\Psi_{11} = \Delta\Psi_{20} = \Delta\Lambda. \tag{15}$$

- The following are shown to hold:

$$\bar{\kappa} = -\kappa, \tag{16}$$
$$\delta\Psi_2 = -\Psi_2\pi + 2\Psi_2\alpha, \tag{17}$$
$$\delta\Psi_2 = \Psi_2\bar{\pi} - 2\Psi_2\bar{\alpha}, \tag{18}$$
$$\beta = \frac{7}{2}\bar{\pi} - 2\bar{\alpha}. \tag{19}$$
$$\Delta\sigma = 4\bar{\pi}\bar{\alpha} + 6\bar{\alpha}^2 - \frac{1}{2} + 3\tau^2 + 7\bar{\pi}\tau + 10\tau\bar{\alpha}. \tag{20}$$

- The following quantities are determined (i.e. solved for) as algebraic expressions of the remaining spin and curvature coefficients:
$\Psi_{12}, \Lambda, \Delta\pi, \bar{\delta}\mu, \delta\mu, \Delta\Psi_2, D\Psi_2, \Delta\epsilon, \delta\pi, \bar{\delta}\alpha, \delta\alpha, \delta\tau, \delta\rho, \delta\bar{\rho}, \bar{\delta}\tau,$
$D\kappa, D\pi, D\bar{\pi}, D\tau, D\alpha, D\bar{\alpha}, \delta\kappa, \delta\epsilon, \delta\bar{\epsilon}, \delta\rho, \delta\bar{\rho}, \delta\bar{\sigma}, \bar{\delta}\rho, \bar{\delta}\bar{\rho}, \bar{\delta}\sigma.$

Step 2 In this step we make use of the substizutions we have obtained in Step 1 to further somplify the remaining necessary conditions. Our analysis will split into a number of sub-cases arising from the factorization of the simplified equations. In each sub-case, on one hand, Ψ_2 will be solved for as a homogeneous multivariate polynomiaö expressin involving only α, π, τ and their conjugates; on the other hand, a system of polynomial equations involving the same spin coefficients will be derived by simplifying the remaining necessary conditions. Each such system is then shown, using Gröbner basis methods, to admit only the trivial (zero) solution, i.e. all of α, π, τ and their conjugates must vanish if Huygens' principle is to hold. Since Ψ_2 is a homogeneous multivariate polynomial of these spin coefficients, this in turn implies Ψ_2 vanishes, which contradicts the assumption that the underlying spacetime is of Petrov type D.

Making all the substitutions from Step 1 in the $o_{(AB}\iota_{ABCDE)}\bar{o}_{\dot{A}\dot{B}\dot{C}\dot{D}\dot{E}}$ component equation of the 5-index condition we obtain

$$- \Psi_2\bar{\Psi}_2\bar{\sigma}\big(12\alpha^2 + 10\pi\alpha - 21\bar{\tau}\pi - 18\alpha\bar{\tau} + 6\pi^2$$
$$- 44\bar{\Psi}_2^2 - 12\alpha\bar{\alpha} - 21\tau\pi - 18\alpha\tau - 12\pi\bar{\alpha}\big) = 0. \tag{21}$$

Equation (21) implies we have either:

Scenario 1 $\sigma = \bar{\sigma} = 0,$ or

Scenario 2 $12\alpha^2 + 10\pi\alpha - 21\bar{\tau}\pi - 18\alpha\bar{\tau} + 6\pi^2 + 44\bar{\Psi}_2$
$\qquad\qquad\quad - 12\alpha\bar{\alpha} - 21\tau\pi - 18\alpha\tau - 14\pi\bar{\alpha} = 0.$

Scenario 1 Substitution of $\sigma = \bar{\sigma} = 0$ into the NP Ricci identities gives

$$\delta(\kappa) = \kappa\tau + \frac{19}{2}\kappa\bar{\pi} + 7\kappa\bar{\alpha}. \tag{22}$$

Making the Step 1 substitutions and using the conjugate of (22) in the Ricci identities we find

$$D(\rho) = \bar{\tau}\kappa - 7\kappa\pi + 2\kappa\alpha + \rho^2 + \rho\epsilon + \rho\bar{\epsilon} - \bar{\kappa}\tau + \Psi_{00}. \tag{23}$$

Similar substitutions into the $\iota_{ABCD}\bar{o}_{(\dot{A}\dot{B}}\bar{\iota}_{\dot{C}\dot{D})}$ component of the 4-index condition give:

$$\Psi_2 = -\frac{7}{2}\bar{\pi}^2 - \frac{8}{7}\bar{\alpha}^2 - \frac{8}{7}\bar{\pi}\bar{\alpha} + \frac{26}{7}\kappa\bar{\pi} + \frac{16}{7}\kappa\tau + \frac{20}{7}\kappa\bar{\alpha}. \tag{24}$$

Using the commutator relation $[\bar{\delta}, \delta]\Psi_2$ and the Ricci identities with the Step 1 substitutions, we may solve for Φ_{11} and Λ as multivariate polynomials in α, π, τ and their conjugates:

$$\left\{\begin{array}{l}
\Phi_{11} = \frac{10}{21}\bar{\pi}\bar{\alpha} + \frac{21}{4}\bar{\tau}^2 + \frac{7}{2}\bar{\alpha}\bar{\tau} + \frac{21}{2}\tau\pi + \frac{1}{2}\pi\bar{\pi} - \frac{155}{168}\pi^2 + \frac{49}{4}\bar{\tau}\pi \\[4pt]
\quad + 7\pi\bar{\alpha} + \frac{21}{4}\tau\bar{\tau} + \frac{10}{21}\bar{\alpha}^2 + \frac{433}{42}\alpha^2 + \frac{5}{42}\bar{\pi}^2 - \frac{20}{21}\kappa\tau - \frac{10}{21}\kappa\alpha \\[4pt]
\quad - \frac{13}{21}\kappa\pi - \frac{65}{42}\kappa\bar{\pi} - \frac{25}{21}\kappa\bar{\alpha} + \frac{21}{2}\alpha\tau + \frac{143}{21}\pi\alpha - \frac{8}{21}\bar{\tau}\kappa + \frac{35}{2}\alpha\bar{\tau}, \\[6pt]
\Lambda = \frac{10}{21}\bar{\pi}\bar{\alpha} + \frac{3}{4}\bar{\tau}^2 + \frac{1}{2}\bar{\alpha}\bar{\tau} + \frac{3}{2}\tau\pi - \frac{1}{2}\pi\bar{\pi} - \frac{5}{168}\pi^2 + \alpha\bar{\alpha} + \frac{7}{4}\bar{\tau}\pi \\[4pt]
\quad + \pi\bar{\alpha} + \frac{3}{4}\tau\bar{\tau} + \frac{10}{21}\bar{\alpha}^2 + \frac{79}{42}\alpha^2 - \frac{5}{42}\bar{\pi}^2 - \frac{20}{21}\kappa\tau + \frac{20}{21}\kappa\alpha + \frac{26}{21}\kappa\pi \\[4pt]
\quad - \frac{65}{42}\kappa\bar{\pi} - \frac{25}{21}\kappa\bar{\alpha} + \frac{3}{2}\alpha\tau + \frac{29}{21}\pi\alpha + \frac{16}{21}\bar{\tau}\kappa - \frac{5}{2}\alpha\bar{\tau}
\end{array}\right. \tag{25}$$

We have proved that the Scenario 1 assumption implies both (24) and (25). We next list six other multivariate polynomial equations in α, τ, π, $\bar{\alpha}$, $\bar{\tau}$, $\bar{\pi}$ and κ which hold in Scenario 1 and which will be used repeatedly to establish the inadmissibility of Scenario 1:

- the $o_{(A}{}^l{}_{BC)}\bar{o}_{(\dot{A}}{}^l{}_{\dot{B}\dot{C}}$ component of the 3-index condition:

$$2\alpha + 3\bar{\tau} + 2\bar{\alpha} + 3\tau = 0. \tag{26}$$

We have solved for $\delta\sigma$ in Step 1. Hence, the Scenario 1 assumption that $\sigma = \bar{\sigma} = 0$ gives us the following equation and its conjugate:

$$-8\bar{\pi}\bar{\alpha} - 12\bar{\alpha}^2 + \bar{\pi}^2 - 6\tau^2 - 14\bar{\pi}\tau - 20\tau\bar{\alpha} = 0, \tag{27}$$

$$-8\pi\alpha - 12\alpha^2 + \pi^2 - 6\bar{\tau}^2 - 14\bar{\tau}\pi - 20\alpha\bar{\tau} = 0, \tag{28}$$

- The real and imaginary parts of the $o_{(AB}{}^l{}_{CD)}\bar{o}_{(\dot{A}\dot{B}}{}^l{}_{\dot{C}\dot{D})}$ component equation of the 4-index condition, under the Step 1 substitutions, simplify respectively to

$$\begin{array}{l}
16\bar{\pi}\bar{\alpha} + 12\bar{\tau}^2 + 2\bar{\alpha}\bar{\tau} + 15\bar{\pi}\bar{\tau} + 15\tau\pi - 43\pi\bar{\pi} - 2\pi^2 \\[4pt]
+ 20\alpha\bar{\alpha} + 28\bar{\tau}\pi - 8\pi\bar{\alpha} - 30\tau\bar{\tau} + 12\tau^2 + 40\tau\bar{\alpha} + 24\bar{\alpha}^2 \\[4pt]
+ 24\alpha^2 - 2\bar{\pi}^2 + 2\alpha\tau - 8\alpha\bar{\pi} + 16\pi\alpha + 28\bar{\pi}\tau + 40\alpha\bar{\tau} = 0,
\end{array} \tag{29}$$

$$\begin{array}{l}
-152\bar{\pi}\bar{\alpha} + 126\bar{\tau}^2 - 168\bar{\alpha}\bar{\tau} - 252\bar{\pi}\bar{\tau} + 252\tau\pi - 25\pi^2 \\[4pt]
+ 294\bar{\tau}\bar{\alpha} + 168\pi\bar{\alpha} - 126\tau^2 - 420\tau\bar{\alpha} - 236\bar{\alpha}^2 + 236\alpha^2 \\[4pt]
+ 25\bar{\pi}^2 - 32\kappa\tau - 40\kappa\alpha - 52\kappa\pi - 52\kappa\bar{\pi} - 40\kappa\bar{\alpha} \\[4pt]
+ 168\alpha\tau - 168\alpha\bar{\pi} + 152\pi\alpha - 32\bar{\tau}\kappa - 294\bar{\pi}\tau + 420\alpha\bar{\tau} = 0.
\end{array} \tag{30}$$

- Recall that we obtained in Step 1 that $\delta\Psi_2 = \psi_2\bar{\pi} - 2\Psi_2\bar{\alpha}$, which when substituted into (24) gives

$$\kappa\left(129\bar{\pi}^2 + 170\tau\bar{\alpha} + 242\bar{\pi}\bar{\alpha} + 40\tau^2 + 165\bar{\pi}\tau + 128\bar{\alpha}^2\right) = 0. \tag{31}$$

We remind the reader here that all of the following equations hold in Scenario 1:
(24)–(31). Equation (31) shows that Scenario 1 splits into further two sub-cases:

Scenario 1A $\kappa = 0,$ or $\qquad\qquad\qquad\qquad\qquad\qquad\qquad$ (32)

Scenario 2B $129\bar\pi^2 + 170\tau\bar\alpha + 242\bar\pi\bar\alpha + 40\tau^2 + 165\bar\pi\tau + 127\bar\alpha^2 = 0.$

$\qquad\qquad\qquad\qquad\qquad\qquad\qquad\qquad\qquad\qquad\qquad\qquad\qquad\qquad$ (33)

We first treat Scenario 1A. Recall again from Step 1 that $\bar\delta\Psi_2 = -\pi\Psi_2 + 2\Psi_2\alpha.$
Substituting into (32), (24) and (25) with the use of the Step 1 substitutions we obtain

$$
\begin{aligned}
(\bar\pi + 2\alpha)\big(4\bar\pi^2 &+ 16\bar\pi\bar\alpha + 126\pi\bar\pi + 84\alpha\bar\pi \\
&+ 126\bar\tau^2 + 200\pi\alpha + 252\alpha\tau + 126\tau\bar\tau + 84\bar\alpha\bar\tau \\
&+ 420\alpha\bar\tau + 420\pi\bar\alpha + 336\alpha\bar\alpha + 284\alpha^2 \\
&+ 294\bar\tau\pi - 13\pi^2 + 16\bar\alpha^2 + 252\tau\pi\big) = 0.
\end{aligned}
$$

$\qquad\qquad\qquad\qquad\qquad\qquad\qquad\qquad\qquad\qquad\qquad\qquad\qquad\qquad$ (34)

Equation (34) thus implies that Scenario 1A splits into two further sub- cases:

Scenario 1A.1 $\alpha = -\tfrac{1}{2}\pi, \quad \bar\alpha = -\tfrac{1}{2}\bar\pi,$ or $\qquad\qquad\qquad$ (35)

Scenario 1A.2
$$
\begin{aligned}
4\bar\pi^2 &+ 16\bar\pi\bar\alpha + 126\pi\bar\pi + 126\bar\tau^2 \\
&+ 200\pi\alpha + 252\alpha\tau + 126\tau\bar\tau + 84\bar\alpha\bar\tau \\
&+ 420\alpha\bar\tau + 420\pi\bar\alpha + 336\alpha\bar\alpha + 284\alpha^2 \\
&+ 294\bar\tau\pi - 13\pi^2 + 16\bar\alpha^2 + 252\tau\pi = 0 :
\end{aligned}
$$

$\qquad\qquad\qquad\qquad\qquad\qquad\qquad\qquad\qquad\qquad\qquad\qquad\qquad\qquad$ (36)

Scenario 1A.1, or equivalently (35), leads to a contradiction, as we presently show.
Under the Scenario 1A.1 assumption (i.e. (35)), Eq. (29) becomes:

$$-2\bar\pi + 6\bar\tau^2 + 7\bar\tau\bar\pi + 7\tau\pi - 15\pi\bar\pi - 2\pi^2 + 4\bar\tau\pi - 15\tau\bar\tau + 6\tau^2 + 4\bar\pi\tau = 0. \tag{37}$$

Recall that(27) holds in Scenario 1. It thus follows that the Scenario 1A.1 assumption
gives

$$-(\tau + \bar\pi)(3\tau - \bar\pi) = 0 \implies \begin{cases} \pi = -\bar\tau \\ \bar\pi = -\tau \end{cases} \text{ or } \begin{cases} \pi = 3\bar\tau \\ \bar\pi = 3\tau \end{cases}. \tag{38}$$

These two possibilities together with (37) yield respectively:

$$\text{either} \quad 44\tau\bar\tau = 0, \quad \text{or} \quad 108\tau\bar\tau = 0. \tag{39}$$

Equation (39) implies that $\tau = \bar\tau = 0$. Equations (38) and (35) then imply $\pi = \bar\pi = 0$
and $\alpha = \bar\alpha = 0$ respectively. Therefore, (24) implies that $\Psi_2 = \bar\Psi_2 = 0$, which
contradicts the assumption that Ψ_2 is non-vanishing. We have thus proved that (35)
leads to a contradiction, and hence Scenario 1A.1 is inadmissible.

We next consider Scenario 1A.2. In this scenario, (36) holds and the real and imaginary parts of (36) respectively are:

$$-72\bar{\pi}\bar{\alpha} - 42\bar{\tau}^2 - 112\bar{\alpha}\bar{\tau} - 84\bar{\pi}\bar{\tau} - 84\tau\pi - 84\pi\bar{\pi} + 3\pi^2$$
$$-224\alpha\bar{\alpha} - 98\bar{\tau}\pi - 168\pi\bar{\alpha} - 84\tau\bar{\tau} - 42\tau^2 - 140\tau\bar{\alpha} - 100\bar{\alpha}^2 \tag{40}$$
$$-100\alpha^2 + 3\bar{\pi}^2 - 112\alpha\tau - 168\alpha\bar{\pi} - 72\pi\alpha - 98\bar{\pi}\tau - 140\alpha\bar{\tau} = 0.$$

$$184\bar{\pi}\bar{\alpha} - 126\bar{\tau}^2 + 168\bar{\alpha}\bar{\tau} + 252\bar{\pi}\bar{\tau} - 252\tau\pi + 17\pi^2$$
$$-294\bar{\tau}\pi - 336\pi\bar{\alpha} + 126\tau^2 + 420\tau\bar{\alpha} + 268\bar{\alpha}^2 + 268\alpha^2 \tag{41}$$
$$-17\bar{\pi}^2 - 168\alpha\tau + 336\alpha\bar{\pi} - 184\pi\alpha + 294\bar{\pi}\tau - 420\alpha\bar{\tau} = 0.$$

Also, the Scenario 1A assumption that $\kappa = 0$ and (30) yield:

$$-152\bar{\pi}\bar{\alpha} + 126\bar{\tau}^2 - 168\bar{\alpha}\bar{\tau} - 252\bar{\pi}\bar{\tau} + 252\tau\pi - 25\pi^2 + 294\bar{\tau}\pi$$
$$+168\pi\bar{\alpha} - 126\tau^2 - 420\tau\bar{\alpha} - 236\bar{\alpha}^2 + 236\alpha^2 + 25\bar{\pi}^2 + 168\alpha\tau \tag{42}$$
$$-168\alpha\bar{\pi} + 152\pi\alpha - 294\bar{\pi}\tau + 420\alpha\bar{\tau} = 0.$$

Therefore, the equations (26), (27), (28), (29), (40), (41) and (42) form a system of seven multivariate polynomial equations in the six indeterminates $\alpha, \bar{\alpha}, \tau, \bar{\tau}, i, \bar{\pi}$. We will call this system System 1A.2, since it arises in Scenario 1A.2. We claim that this system has only the trivial solution, i.e. System 1A.2 implies that all of $\alpha, \bar{\alpha}, \tau,$ $\bar{\tau}, i, \bar{\pi}$ must vanish. This can be proved using the following lemma:

Lemma 1. *The Gröbner basis for the ideal (in $\mathbb{C}[\tau, \alpha, \pi, \bar{\pi}, \bar{\alpha}, \bar{\tau}]$) generated by the left-hand-sides[1] of the equations in System 1A.2 with respect to the pure lexicographical ordering with $\tau \succ \alpha \succ \pi \succ \bar{\pi} \succ \bar{\alpha} \succ \bar{\tau}$ decomposes into the components:*

$$\{\alpha, \bar{\alpha}, \tau, \bar{\tau}, \pi, \bar{\pi}\}, \{\bar{\tau}, \bar{\pi} + 2\bar{\alpha}, \pi, \alpha, 3\tau + 2\bar{\alpha}\}, \{\bar{\alpha}, \bar{\pi}, -3\bar{\tau} + \pi, 2\alpha + 3\bar{\tau}, \tau\}. \tag{43}$$

The decomposition asserted in Lemma 1 can be obtained using the MAPLE function `gsolve()` in MAPLE 's `Groebner` package. It is clear that each component in (43) leads to the trivial solution when we take into account that $\{\alpha, \bar{\alpha}\}$, $\{\tau, \bar{\tau}\}$, $\{\pi, \bar{\pi}\}$ are conjugate pairs. This in turn imply that Scenario 1A.2, or (36), imply that all of $\alpha, \bar{\alpha}, \tau, \bar{\tau}, \pi, \bar{\pi}$ must vanish. Hence, by (24) the quantity Ψ_2 vanishes, contradicting the assumption that Ψ_2 is non-vanishing. We have therefore proved that Scenario 1A.2 is also inadmissible.

Since both Scenario 1A.1 and Scenario 1A.2 are inadmissible, Scenario 1A is inadmissible.

The remaining Scenario 1B and Scenario 2 do not split into further sub-cases. The proofs that each leads to contradiction are similar to the one given above for

---

[1] Considered, in the present context, as multivariate polynomials in the indeterminates $\alpha, \bar{\alpha},$ $\tau, \bar{\tau}, \pi, \bar{\pi}$

Scenario 1A.2. Complete details of the proof can be found in [3] and will be omitted here. The proof of Proposition 3 is now complete.   □

Acknowledgements. This research was supported in part by Research Grants (S.R.C. and R.G.McL.) from the Natural Sciences and Engineering Research Council of Canada.

References

1. Friedlander, F.G. (1975): The Wave Equation on a Curved Space-time. Cambridge University Press, Cambridge
2. Czapor, S.R., McLenaghan, R.G., Sasse, F.D. (1999): Ann. Inst. Henri Poincaré **71**, 595
3. Chu, K.C. (2000): Contributions to the Study of the Validity of Huygens' Principle for the Non-self-adjoint Scalar Wave Equation on Petrov Type D Spacetimes. M. Math. Thesis, University of Waterloo, Waterloo, Ontario, Canada
4. Anderson, W.G. (1991): Contributions to the Study of Huygens' Principle for Non-self-adjoint Scalar Wave Equations on Curved Space-times. M. Math. Thesis, University of Waterloo, Waterloo, Ontario, Canada
5. Anderson, W.G., McLenaghan, R.G. (1994): Ann. Inst. Henri Poincaré **60**, 373
6. Carminati, J., McLenaghan, R.G. (1986): Ann. Inst. Henri Poincaré **44**, 115
7. McLenaghan, R.G., Walton, T.F. (1988): Ann. Inst. Henri Poincaré **48**, 267
8. Günther, P. (1952): Ber. Verh. Sachs. Akad. Wiss. Leipzig **100**, 1
9. McLenaghan, R.G. (1974): Ann. Inst. Henri Poincaré **20**, 153
10. Czapor, S.R., McLenaghan, R.G. (1987): General Relativity and Gravitation **19**, 623
11. Czapor, S.R., McLenaghan, R.G., Carminati, J. (1991): General Relativity and Gravitation **24**, 911
12. Adams, W.W., Loustaunau, P. (1994): Introduction to Gröbner Bases. American Mathematical Society, Providence
13. Geddes, K.O., Czapor, S.R., Labahn, G. (1992): Algorithms for Computer Algebra. Kluwer Academic Publishers, Boston
14. Czapor, S.R. (1988): Gröbner Basis Methods for Solving Algebraic Equations. Doctoral Thesis, University of Waterloo, Waterloo, Ontario, Canada
15. Czapor, S.R. (1989): J. Symbolic Computation **7**, 49

Hamiltonian Structure of 2+1 Dimensional Gravity

P. Menotti

Abstract. A summary is given of some results and perspectives of the hamiltonian ADM approach to $2+1$ dimensional gravity. After recalling the classical results for closed universes in the absence of matter, we go over the the case in which matter is present in the form of point spinless particles. Here the maximally slicing gauge proves most effective by relating $2+1$ dimensional gravity to the Riemann-Hilbert problem. It is possible to solve the gravitational field in terms of the particle degrees of freedom thus reaching a reduced dynamics which involves only the particle positions and momenta. Such a dynamics is proven to be hamiltonian and the hamiltonian is given by the boundary term in the gravitational action. As an illustration, the two body hamiltonian is used to provide the canonical quantization of the two particle system.

1 Introduction

The past fifteen years have witnessed a remarkable interest in gravity in 2+1 dimensions [1] both at the classical and quantum level.

In this paper we shall summarize some old results and some recent developments with regard to the hamiltonian formulation of the theory. We shall insist on the conceptual side; the technical details can be found in the published papers and reports.

2 2+1 dimensional gravity in absence of matter

In absence of matter the only degrees of freedom are given by the moduli of the space sections; thus we can deal only with genus $g \geq 1$.

In absence of boundaries the action of the gravitational field reduces to the Einstein–Hilbert term which can be put in hamiltonian form as

$$S_H = \int dt \int_{\Sigma_t} d^D x \left[\pi^{ij} \dot{g}_{ij} - N^i H_i - N H \right] \tag{1}$$

where we used the standard ADM metric [2]

$$ds^2 = -N^2 dt^2 + 2g_{z\bar{z}}(dz + N^z dt)(d\bar{z} + N^{\bar{z}} dt). \tag{2}$$

The choice of the gauge is of crucial importance in dealing with the problem. The well known York gauge in which the time slices are provided by the D (in our case $D = 2$) dimensional surfaces with $K = $ const., being K the trace of the intrinsic curvature tensor is particularly powerful as this gauge decouples the solution of the diffeomorphism constraint from that of the hamiltonian constraint. Exploiting this feature it is possible to solve the diffeomorphism constraint [3,4] and to provide the hamiltonian on the reduced phase space given by the moduli and their conjugate momenta. The number of moduli are $6g - 6$ for genus g larger than 1 and 2 for the torus topology. The explicit computation of the hamiltonian can be performed only in the simplest case of torus topology. It is given by

$$H = \sqrt{\tau_2^2(p_1^2 + p_2^2)}. \tag{3}$$

Quantization proceeds by replacing the canonical variables with operators [5,6] according to the correspondence principle. The ordering problem always subsists; the most natural ordering translates the classical hamiltonian into the square root of the Maass laplacian [7] giving rise to the Schroedinger equation

$$i\frac{\partial \psi(x, y, t)}{\partial t} = \sqrt{-y^2(\partial_x^2 + \partial_y^2)}\, \psi(x, y, t). \tag{4}$$

The Maass laplacian has been widely investigated by mathematicians [8–10]. The classical as well the quantum hamiltonians are invariant under modular transformations which in the case of the torus is given by the group $SL(2, Z)$ and thus the solutions should be invariant under such modular transformations. The eigenvalue problem under this condition is not trivial and the spectrum well studied [8–10]. Such an approach gives a complete quantum treatment of universes without matter with torus topology in the York gauge.

3 2+1 dimensional gravity coupled to particles

We come now to the more difficult and realistic case of gravity coupled to particles. It describes also a situation of 3+1 dimensional gravity, i.e. the interaction of parallel cosmic strings.

One starts from the action of the gravitational field coupled to a finite number of spinless point particles. The inclusion of spin to point particles is not a trivial issue as it gives rise to closed time like curves.

We must add to the action (1) the particle action

$$S_m = \int dt \sum_n \left(P_{ni}\, \dot{q}_n^i + N^i(q_n)P_{ni} - N(q_n)\sqrt{P_{ni}P_{nj}g^{ij}(q_n) + m_n^2} \right), \tag{5}$$

and for open universes we must also add boundary terms; we shall adopt here the so called Trace-K form completed of the terms which lye on two sub-manifolds of dimension $D - 1$, being $D + 1$ the dimension of space-time [11,12]. In hamiltonian form it reduces to

$$S_B = -\int dt\, H_B, \tag{6}$$

with

$$-H_B = 2 \int_{B_t} d^{(D-1)}x \; \sqrt{\sigma_{Bt}} N \left(K_{B_t} + \frac{\eta}{\cosh \eta} \mathcal{D}_\alpha v^\alpha \right)$$
$$2 \int_{B_t} d^{(D-1)}x \, r_\alpha \pi^{\alpha\beta}_{(\sigma_{Bt})} N_\beta. \tag{7}$$

The really important term turns out to be the first one i.e. $\sqrt{\sigma_{Bt}} N K_{B_t}$ where K_{B_t} is the extrinsic curvature of the $D-1$ dimensional boundary (in our case one-dimensional) of the time slices as a sub-manifold embedded in the D dimensional time slices and $d^{(D-1)}x\sqrt{\sigma_{Bt}}$ is the volume form induced by the space metric on the $D-1$ dimensional boundary.

We recall that the action $S_H + S_B + S_m$ is so constructed as to provide the correct equations of motion (i.e. Einstein's equations) when one computes a stationary point of the action by keeping the values of the metric fixed on the boundary. Such a procedure is equivalent [13] to the weaker requirement of keeping fixed the intrinsic metric of the boundary.

One could in principle adopt also here the York slicing. However the equations for the diffeomorphism and hamiltonian constraints are not at all trivial. In particular the hamiltonian constraint gives rise to an equation more complex than the inhomogeneous sine-Gordon equation. Progress has been achieved by the introduction of the instantaneous York gauge [14–18], or maximally slicing gauge. This is defined by all time slices having $K = 0$. Simple application of the Gauss- Bonnet theorem shows that such a gauge can be applied only to open universes, or universes with the topology of the sphere [15,16]. In addition, a closer inspection shows that for the sphere topology it can describe only the simple stationary case [17]. Thus application of the $K = 0$ gauge is practically restricted to open universes, but here it proves very powerful. The technical reason is the immediate solution of the diffeomorphism constraint given by

$$\pi^{\bar z}_z = -\frac{1}{2\pi} \sum_n \frac{P_n}{z - z_n} \equiv -2 \frac{\Pi_B(z - z_B)}{\mathcal{P}(z)}, \tag{8}$$

where z_n and P_n are the complex positions and canonical momenta of the particles, and the reduction of the hamiltonian constraint to an inhomogeneous Liouville equation, given by

$$2\Delta\tilde\sigma = -e^{-2\tilde\sigma} - 4\pi \sum_n \delta^2(z - z_n)(\mu_n - 1) - 4\pi \sum_B \delta^2(z - z_B), \tag{9}$$

to which powerful mathematical methods apply [19]. Here $\tilde\sigma$ is defined by

$$e^{2\sigma} = 2\pi^{\bar z}_z \pi^z_{\bar z} e^{2\tilde\sigma}, \tag{10}$$

and is related to the space metric by $g_{ij} = \delta_{ij} e^{2\sigma}$. In the above equation the sources are given by the particle and in addition by the zeros of Eq. (8) here denoted by z_B.

These are the points where the extrinsic curvature tensor vanishes, and through the Gauss–Codazzi relation also the intrinsic curvature scalar of the time slice vanishes. It is interesting that such z_B coincide with the positions of the accessory singularities which will appear in the uniformization problem. These accessory parameters summarize the gravitational interaction. An interesting restriction occurs on the conjugate momenta i.e. $\sum_n P_n = 0$ [20]. In the absence of such a restriction no solution exists to Eq. (9) with $\exp(2\sigma)$ behaving at infinity as $\exp(2\sigma) \approx s^2(z\bar{z})^{-\mu}$ which describes a space geometry asymptotic to a cone. This can be easily seen by applying the divergence theorem to the analogous of Eq. (9) for σ. The same reasoning excludes the addition of an holomorphic term to the $\pi^{\bar{z}}_z$ given by Eq. (8).

The solution of the inhomogeneous Liouville equation (9) can be understood as a variant of the Riemann- Hilbert problem and is reduced to a linear fuchsian differential equation [21]. By taking the ratio of two solutions of such equation one can build the function $f(z)$ which maps the metric of the Poincaré pseudo-sphere into the conformal factor which solves (9) according to the formula [19]

$$e^{-2\tilde{\sigma}} = \frac{8f'(z)\bar{f}'(\bar{z})}{(1 - f(z)\bar{f}(\bar{z}))^2} = \frac{\text{const}}{(\bar{y}_2(\bar{z})y_2(z) - \bar{y}_1(\bar{z})y_1(z))^2}; \quad f(z) = \frac{y_1(z)}{y_2(z)}. \tag{11}$$

It is a variant of the Riemann–Hilbert problem as we are not given directly with the monodromies but with the following information on them: all monodromies belong to the $SU(1, 1)$ group, otherwise the conformal factor (11) is not single valued and as such cannot solve the Liouville equation Eq. (9); in addition we are given, together with the conjugation classes of the monodromies around the particles (the particle masses), the conjugation class of the monodromy at infinity (the total energy) and the positions of the auxiliary singularities z_B, which are equivalent to the knowledge of the particle momenta P_n up to a multiplicative factor. Due to the relation $\sum_n P_n = 0$, the number of accessory singularities is $\mathcal{N} - 2$.

In order to count the physical degrees of freedom in addition to the particle positions, we must keep in mind that for $\mathcal{N}$ particles we have $\mathcal{N}$ $SU(1, 1)$ monodromies to which we have to subtract a $SU(1, 1)$ conjugation thus reaching $3\mathcal{N} - 3$ real degrees of freedom. These are equivalent to giving the $\mathcal{N}$ particle masses μ_i, the total energy μ and the complex positions z_B of the $\mathcal{N} - 2$ accessory singularities i.e. $3\mathcal{N} - 3 = \mathcal{N} + 1 + 2(\mathcal{N} - 2)$. The linear residues β_B of the auxiliary singularities are fixed by the solution of the Riemann- Hilbert problem while the linear residues β_i at the particle singularities are computed from the $\mathcal{N} - 2$ no logarithm condition at the auxiliary singularities and the first and second Fuchs conditions on the residues of the fuchsian differential equation.

The conformal factor $\tilde{\sigma}$ is the key quantity in all of the subsequent developments. In fact, a secondary constraint following from the primary gauge constraint $K = 0$ is

$$\Delta N = e^{-2\tilde{\sigma}} N, \tag{12}$$

and such N can be computed from the knowledge of $\tilde{\sigma}$. The reason is that the solution of the inhomogeneous Liouville equation (9) contains a free parameter μ which is related to the behavior at infinity of the conformal factor i.e. $\exp(2\sigma) \approx s^2(z\bar{z})^{-\mu}$. As the sources do not depend on μ a solution of Eq. (12) is given by

$$N = \frac{\partial(-2\tilde{\sigma})}{\partial M},$$

(13)

and one easily proves such a solution to be unique. The other secondary constraint on N^z is

$$\partial_{\bar{z}} N^z = -\pi^z_{\ \bar{z}}\, e^{-2\sigma}\, N,$$

(14)

solved by

$$N^z = -\frac{2}{\pi^{\bar{z}}_{\ z}(z)}\, \partial_z N + g(z).$$

(15)

Here $g(z)$ is a meromorphic function whose role is to cancel the poles occurring in the first member on the r.h.s. due to the zeros of $\pi^{\bar{z}}_{\ z}$. The expression of $g(z)$ in N^z is [16]

$$g(z) = \sum_B \frac{\partial \beta_B}{\partial M} \frac{1}{z - z_B} \frac{\mathcal{P}(z_B)}{\prod_{C \neq B}(z_B - z_C)} + p_1(z).$$

(16)

$p_1(z)$ is a first order polynomial related to the motion of the frame at infinity, more properly [22,23] to the gauge transformation of translations, rotations and dilatations which leave invariant the conformal structure of the space metric and leave fixed the point at infinity.

In conclusion the metric is obtained in a straightforward way from $\tilde{\sigma}$. The particle equations of motion are extracted from the variation of the action with respect to particle momenta and coordinates and take the form [16]

$$\dot{z}_n = -N^z(z_n) = -g(z_n)$$
$$-\sum_B \frac{\partial \beta_B}{\partial M} \frac{1}{z_n - z_B} \frac{\mathcal{P}(z_B)}{\prod_{C \neq B}(z_B - z_C)} - p_1(z_n),$$

(17)

$$\dot{P}_{nz} = 4\pi \frac{\partial \beta_n}{\partial M} + P_{nz},$$
$$g'(z_n) = 4\pi \frac{\partial \beta_n}{\partial M} - P_{nz} \sum_B \frac{\partial \beta_B}{\partial M} \frac{\mathcal{P}(z_B)}{(z_n - z_B)^2 \prod_{C \neq B}(z_B - z_C)} + P_{nz}\, p_1'(z_n).$$

(18)

As already mentioned, the linear polynomial $p_1(z)$ is related to the choice of the frame at infinity. The choice of the frame which does not rotate and dilatate at infinity imposes that N^z contains no linear term and thus fixes

$$p_1(z) = c_0 - \frac{1}{\sum_n P_n z_n}\, z.$$

(19)

An interesting generalized conservation law which holds for any number of particles can be derived from Eqs. (17,18) [16]

$$\sum_n P_n z_n = (1 - \mu)(t - t_0) - iL. \tag{20}$$

For the two body problem we have no auxiliary singularity and the fuchsian differential equation underlying the solution of Eq. (9) is the hypergeometric equation. The conformal factor can be easily expressed in term of hypergeometric functions, thus giving a complete information on the metric. Equations (17), (18) take a very simple form in the two body case due to the absence of apparent singularities. It is interesting that the equations of motion can be explicitly written just by looking at the local properties of the fuchsian differential equation which underlies the solution of Eq. (9). We obtain them going over to the relative coordinates $z_2' = z_2 - z_1$ and $P' = P_2 = -P_1$

$$\dot{z}_2' = \frac{1}{P_z'}; \quad \dot{P}_z' = -\frac{\mu}{z_2'}. \tag{21}$$

being $4\pi\mu = M$ the total mass (energy) of the system. One immediately realizes that such equations of motion (and their complex conjugate) are generated by the hamiltonian $H = h + \bar{h}$ with $h = \ln(P'z_2'^\mu)$ where h and $\bar{h}$ are constants of motion. Taking the ratio of $P'z_2'^\mu = \exp(h) = \text{const.}$ with Eq. (20) we obtain the solution

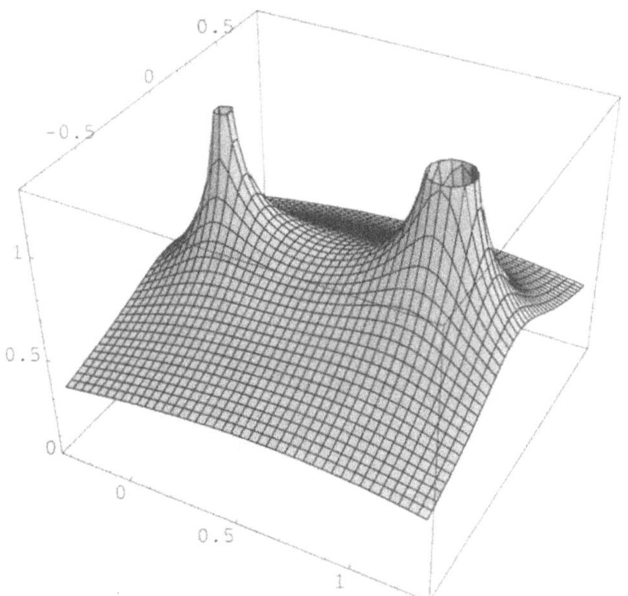

Fig. 1. Conformal factor for the classical two body problem

for the particle motion

$$z = \text{const}[(1 - \mu)(t - t_0) - iL]^{\frac{1}{1-\mu}}. \tag{22}$$

without the need to solve the system (21). Thus the two body problem is integrable. Solution (22) was first found in [14]. The fact that only the total energy $M = 4\pi\mu$ intervenes in Eq. (22) constitutes a proof of a conjecture by 't Hooft [24] about the independence of the solution of the two body problem of the masses of the constituent particles.

In Fig. 1 we report the conformal factor for the two body problem for $\mu_1 = 0.1$, $\mu_2 = 0.2$ and $\mu = 0.5$.

4   The N body problem

In the coordinate frame in which the polynomial p_1 vanishes the equations of motion in the relative coordinates $z'_n = z_n - z_1$, $P'_n = P_n$ for $n = 2\ldots\mathcal{N}$ take the form [18]

$$\dot{z}'_n = -\sum_B \frac{\partial\beta_B}{\partial\mu} \frac{\partial z'_B}{\partial P'_n}, \quad n = 2,\ldots\mathcal{N}, \tag{23}$$

and

$$\dot{P}'_n = \frac{\partial\beta_n}{\partial\mu} + \sum_B \frac{\partial\beta_B}{\partial\mu} \frac{\partial z'_B}{\partial z'_n}, \quad n = 2,\ldots\mathcal{N}. \tag{24}$$

These equations are canonically related to the ones with $p_1 \neq 0$ and thus is sufficient to examine them. The problem is now to show that such system is of hamiltonian nature; such a result is expected as we obtained the above equations by reduction of a hamiltonian system. Despite that, it is of interest to have a direct proof of it and an expression of the hamiltonian.

In the simpler case of three body, where we have a single auxiliary singularity, one immediately sees that the hamiltonian has to be of the form

$$H(z'_2, z'_3, P'_2, P'_3) = -\int_{z_0}^{z'_A} \frac{\partial\beta_A}{\partial\mu}(z'_2, z'_3, z''_A)\, dz''_A + f(z'_2, z'_3) + \text{c.c.} \tag{25}$$

H generates the equations for $\dot{z}'_n$ for any $f(z'_2, z'_3)$; the function f has to be determined by requiring that the same H generates also the equations for $\dot{P}'_n$. This imposes some integrability conditions on the function β_A; such conditions are satisfied due to the validity of the Garnier equations which give a constraint on the evolution of the auxiliary parameters under isomonodromic deformations [26,26]. The fact that our deformations are isomonodromic is a consequence of the constancy in $2+1$ dimensional gravity of the monodromies around the particle world lines. It is of interest that such equations can be derived directly form the ADM formalism as equations of motion [16].

The problem with four or more particles, when we are in the presence of two or more auxiliary singularities is more difficult and it is related to an interesting conjecture due to Polyakov [27] about the auxiliary parameters of the $SU(1, 1)$

172   P. Menotti

Riemann–Hilbert problem. Such a conjecture states that the regularized Liouville action [28]

$$
\begin{aligned}
S_\epsilon[\varphi] = {} & \frac{i}{2} \int_{X_\epsilon} \left(\partial_z\varphi\partial_{\bar{z}}\varphi + \frac{e^\varphi}{2} \right) dz \wedge d\bar{z} \\
& - \frac{i}{2} \sum_n (1 - \mu_n) \oint_{J_n} \varphi \left(\frac{d\bar{z}}{\bar{z} - \bar{z}_n} - \frac{dz}{z - z_n} \right) \\
& + \frac{i}{2} \sum_B \oint_{J_B} \varphi \left(\frac{d\bar{z}}{\bar{z} - \bar{z}_B} - \frac{dz}{z - z_B} \right) - \frac{i}{2}(\mu - 2) \oint_\infty \varphi \left(\frac{d\bar{z}}{\bar{z}} - \frac{dz}{z} \right)
\end{aligned}
$$

$$
- \pi \sum_n (1 - \mu_n)^2 \ln \epsilon^2 - \pi \sum_B \ln \epsilon^2 - \pi(\mu - 2)^2 \ln \epsilon^2, \tag{26}
$$

is the generator of the linear residues β_i, β_B in the fuchsian differential equation whose solutions provide the mapping function which solve the inhomogeneous Liouville equation, i.e.

$$
-\frac{1}{2\pi} dS_\epsilon = \sum_n \beta_n dz_n + \sum_B \beta_B dz_B + \text{c.c.} \tag{27}
$$

Such a conjecture has been proven by Zograf and Takhtajan [27] for the case of parabolic singularities and elliptic singularities of finite order. In the context of $2+1$-dimensional gravity with generic masses one should need the validity of such a conjecture for generic elliptic singularities. The relevance of such a conjecture is that it gives a new meaning to the auxiliary parameters; moreover it is a straightforward consequence of Eq. (27) that, apart from a constant

$$
H = \frac{1}{2\pi} \frac{\partial S_\epsilon}{\partial \mu}, \tag{28}
$$

as can be seen by computing $\frac{\partial H}{\partial P'_n}$, $\frac{\partial H}{\partial z'_n}$ and comparing with Eqs. (23), (24). We recall that in the nonrotating frame the hamiltonian contains an additional contribution, as already observed in Sect. 3. Its complete form in that frame is indeed given by [18]

$$
H = \ln \left[\left(\sum_n P_n z_n \right) \left(\sum_n \bar{P}_n \bar{z}_n \right) \right] + \frac{1}{2\pi} \frac{\partial S_\epsilon}{\partial \mu}. \tag{29}
$$

Note that this hamiltonian, being time-independent, provides a further conservation law in the $\mathcal{N}$-particle problem.

The hamiltonian H has been constructed from the equations of motion. On the other hand it should be possible to derive directly the reduced hamiltonian starting from the action after replacing into it the solution of the constraints. As in our gauge $\pi^{ij}\dot{g}_{ij} = 0$, the action reduces to

$$
S = \int dt \left(\sum_n P_{ni} \dot{q}_n^i - H_B \right), \tag{30}
$$

with H_B given by Eq. (7). The last term in H_B contributes to a constant while the term proportional to $\eta/\cosh\eta$ goes to zero at large distances. Thus we are left only with the term proportional to K_{Bt}, which for large r_0, can be computed to give [18]

$$H_B = -r_0 \ln r_0^2 \left(\frac{1}{r_0} + \partial_r \sigma \right) = (\mu - 1) \ln r_0^2. \tag{31}$$

We recall now that the equations of motion are obtained from the action by keeping the values of the fields fixed at the boundary, or equivalently [13] by keeping fixed the intrinsic metric of the boundary. In our case the variations should be performed keeping fixed the fields N, N^a, and σ at the boundary. We shall perform the computation for the boundary given by a circle of radius r_0 for a very large value of r_0. The asymptotic form of the conformal factor 2σ is

$$2\sigma = \ln s^2 - \mu \ln r_0^2. \tag{32}$$

If we change the particle positions and momenta, s^2 varies and in order to keep the value of σ fixed at the boundary we must vary μ as follows

$$0 = -\delta\mu \ln r_0^2 + \delta \ln s^2, \tag{33}$$

i.e. for large r_0

$$\delta\mu \approx \frac{1}{\ln r_0^2} \delta \ln s^2. \tag{34}$$

Substituting into Eq. (31) we have

$$H_B = \ln s^2 + \text{const.} \tag{35}$$

i.e. the hamiltonian is given by the logarithm of the coefficient of the asymptotic expansion of the conformal factor at large distances.

We want now to relate the result Eq. (35) to the results obtained directly from the equations of motion. Let us consider the value of the action S_ϵ on the solution of the Liouville equation and let us compute its derivative with respect to μ. As we are varying around a stationary point, the only contribution is provided by the terms in Eq. (26) which depend explicitly on μ, i.e.

$$\frac{\partial S_\epsilon}{\partial \mu} = -\frac{i}{2} \oint_\infty \varphi \left(\frac{d\bar{z}}{\bar{z}} - \frac{dz}{z} \right) - 2\pi(\mu - 2) \ln \epsilon^2, \tag{36}$$

and as $\varphi \equiv -2\tilde{\sigma}$ at infinity behaves like

$$\varphi \approx (\mu - 2) \ln z\bar{z} + \ln \left[\left(\sum_n P_n z_n \right) \left(\sum_n \bar{P}_n \bar{z}_n \right) \right] - \ln s^2 + \text{const.}, \tag{37}$$

we have

$$\frac{1}{2\pi} \frac{\partial S_\epsilon}{\partial \mu} = -\ln \left[\left(\sum_n P_n z_n \right) \left(\sum_n \bar{P}_n \bar{z}_n \right) \right] + \ln s^2 + \text{const.}, \tag{38}$$

which substituted in Eq. (35) gives

$$H = \ln\left[\left(\sum_n P_n z_n\right)\left(\sum_n \bar{P}_n \bar{z}_n\right)\right] + \frac{1}{2\pi}\frac{\partial S_\epsilon}{\partial \mu} + \text{const.,} \tag{39}$$

in agreement with the result (29) obtained through Polyakov's conjecture.

It is remarkable that the same hamiltonian is obtained independently from the boundary term of the $2+1$ gravitational action and from the equations of motion in conjunction with Polyakov conjecture. This lends support to the validity of Polyakov's conjecture in the general elliptic case or at least to a weak form of it, obtained by taking the derivative of Eq. (27) with respect to μ.

5   Quantization: the two particle case

Several quantization scheme have been put forward in the context of $2+1$ dimensional gravity (see e.g. [5,6,29–32]). Here we shall examine the canonical quantization which flows directly from the hamiltonian ADM formalism [2]. We recall that the classical two particle hamiltonian in the reference system which does not rotate at infinity is given by

$$H = \ln(Pz\bar{P}\bar{z}) + (\mu - 1)\ln(z\bar{z}) = \ln(Pz^\mu) + \ln(\bar{P}\bar{z}^\mu) = h + \bar{h}, \tag{40}$$

with $P = P_2'$ and $z = z_2'$. H can be rewritten in cartesian coordinates as

$$H = \ln\left((x^2 + y^2)^\mu((P_x)^2 + (P_y)^2)\right). \tag{41}$$

Keeping in mind that with our definitions P is the momentum multiplied by $16\pi G_N/c^3$ and applying the correspondence principle we have

$$[x, P_x] = [y, P_y] = il_P, \tag{42}$$

where $l_P = 16\pi G_n\hbar/c^3$, all other commutators equal to zero. H is converted into the operator

$$\ln[-(x^2 + y^2)^\mu \Delta] + \text{const.} \tag{43}$$

The argument of the logarithm is the Laplace–Beltrami Δ_{LB} operator on the metric $ds^2 = (x^2 + y^2)^{-\mu}(dx^2 + dy^2)$. We note that in the quantum problem only the simple metric of a cone (Fig. 2) intervenes and not the classical metric of Fig. 1. The angular deficit of the cone again in given by the total energy thus proving 't Hooft conjecture [24] at the quantum level. Following an argument similar to the one presented in [33] one easily proves that if we start from the domain of Δ_{LB} given by the infinite differentiable functions of compact support C_0^∞ which can also include the origin, then Δ_{LB} has a unique self-adjoint extension in the Hilbert space of functions square integrable on the metric $ds^2 = (x^2 + y^2)^{-\mu}(dx^2 + dy^2)$ [18] and as a result since Δ_{LB} is a positive operator, $\ln(\Delta_{LB})$ is also self-adjoint.

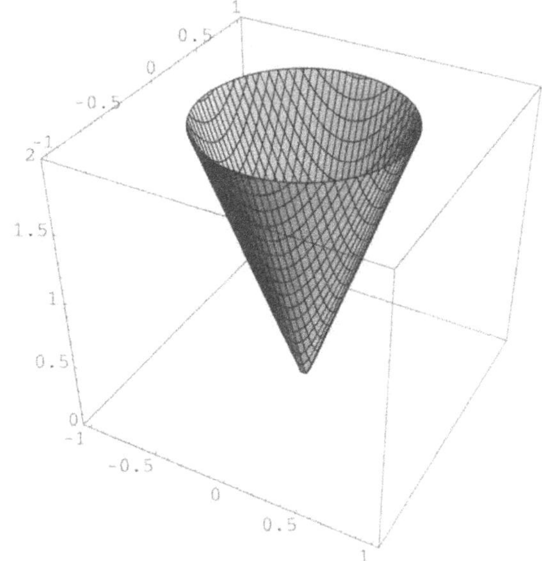

Fig. 2. Cone metric equivalent to the classical metric of Fig. 1, intervening in the quantum two body problem

Deser and Jackiw [34] considered the quantum scattering of a test particle on a cone both at the relativistic and non relativistic level. Most of the techniques developed there can be transferred here. The main difference is the following: instead of the hamiltonian $(x^2+y^2)^\mu(p_x^2+p_x^2)$ which appears in their non relativistic treatment, we have now the hamiltonian $\ln[(x^2+y^2)^\mu(p_x^2+p_y^2)]$. The partial wave eigenvalue differential equation

$$(r^2)^\mu \left[-\frac{1}{r}\frac{\partial}{\partial r} r \frac{\partial}{\partial r} + \frac{n^2}{r^2} \right] \varphi_n(r) = k^2 \varphi_n(r), \tag{44}$$

with $\mu = 1 - \alpha$ is solved by

$$\varphi_n(r) = J_{\frac{|n|}{\alpha}}\left(\frac{k}{\alpha} r^\alpha \right), \tag{45}$$

and out of them one can compute the quantum Green function which in our case can be expressed in terms of hypergeometric functions [18]. The ordering problem always subsists; the one adopted in (43) is the most natural. In the $\mathcal{N}$-body problem due to the complexity of the hamiltonian we expect the ordering problem to be more acute. Power expansions in some small parameter like the total kinetic energy or the particle masses may give useful indications.

Acknowledgements. This talk is based on papers written by the author in collaboration with Domenico Seminara and Luigi Cantini to whom the author is very grateful. The author is also grateful to Marcello Ciafaloni, Stanley Deser and Roman Jackiw for useful discussions.

References

1. Staruszkiewicz, A. (1963): Acta Phys. Polonica **24**, 734;
   Deser, S., Jackiw, R., 't Hooft, G. (1984): Ann. Phys. (NY) **152**, 220;
   Deser, S., Jackiw, R. (1984): Ann. Phys. **153**, 405
2. Arnowitt, R., Deser, S., Misner, C.W. (1962): The dynamics of general relativity, in *Gravitation: An introduction to current research*, ed. by L. Witten, John Wiley & Sons, New York, London, pp. 227–265
3. Moncrief, V. (1989): J. Math. Phys. **30**, 2907
4. Hosoya, A., Nakao, K. (1990): Class. Quantum Grav. **7**, 163
5. Hosoya, A., Nakao, K. (1990): Progr. Theor. Phys. **84**, 739
6. Carlip, S. (1990): Phys. Rev. D **42**, 2647;
   (1992): Phys. Rev. D **45**, 3584
7. Maass, H. (1964): Lectures on modular functions of one complex variable. Tata Institute, Bombay
8. Fay, J.D. (1977): J. Reine Angew. Math. **293**, 143
9. Terras, A. (1985): Harmonic analysis on symmetric spaces and applications. Springer-Verlag, Berlin Heidelberg New York
10. Puzio, R. (1994): Classical and Quantum Gravity **11**, 609
11. Hawking, S.W., Hunter, C.J. (1996): Class. Quantum Grav. **13**, 2735;
    Hayward, G. (1993): Phys. Rev. D **47**, 3275
12. Brown, J.D., Lau, S.R., York, J.W.: gr-qc/ 0010024
13. Wald, R.M. (1984): General relativity. The University of Chicago Press, Chicago and London
14. Bellini, A., Ciafaloni, M., Valtancoli, P. (1995): Physics Lett. B **357**, 532;
    (1996): Nucl. Phys. B **462**, 453
15. Welling, M. (1996): Class. Quantum Grav. **13**, 653;
    (1998): Nucl. Phys. B **515**, 436
16. Menotti, P., Seminara, D. (2000): Ann. Phys. **279**, 282
17. Menotti, P., Seminara, D. (2000): Nucl. Phys. [Proc. Suppl.] **88**, 132
18. Cantini, L., Menotti, P., Seminara, D. (2001): Classical and quantum gravity, **18**, 2253–2275
19. Liouville, J. (1853): J. Math. Pures Appl. **18**, 71;
    Poincaré, H. (1898): J. Math. Pures Appl. **5**, ser.2, 157;
    Ginsparg, P., Moore, G.: hep-th/9304011;
    Takhtajan, L. (1994): Topics in quantum geometry of Riemann surfaces: Two dimensional quantum gravity, in *Quantum groups and their applications in physics: Proceedings of the International School of Physics "Enrico Fermi"*, Course 127, Varenna on Lake Como, Villa Monastero 28 June–8 July 1994, ed. by L. Castellani, J. Wess, pp. 541–579; hep-th/9409088;
    Kra, I. (1972): Automorphic forms and Kleinian groups. Benjamin, Reading, MA
20. Deser, S. (1985): Class. Quantum Grav. **2** 489
21. Bolibrukh, A.A. (1990): Uspekhi Mat. Nauk **45**, 2, 3
22. Henneaux, M. (1984): Phys. Rev. D **29**, 2767

23. Ashtekar, A., Varadarajan, M. (1994): Phys. Rev. D **50**, 4944
24. 't Hooft, G. (1988): Commun. Math. Phys. **117**, 685
25. Yoshida, M. (1987): Fuchsian differential equations. Fried. Vieweg & Sohn, Braunschweig
26. Okamoto, K. (1986): J. Fac. Sci. Tokio Univ. **33**
27. Zograf, P.G., Tahktajan, L.A. (1988): Math. USSR Sbornik **60**, 143
28. Takhtajan, L.A. (1996): Mod. Phys. Lett. A **11** 93
29. 't Hooft, G. (1992): Class. Quantum Grav. **9**, 1335;
    (1993): Class. Quantum Grav. **10**, 1023;
    (1993): Class. Quantum Grav. **10**, 1653;
    (1996): Class. Quantum Grav. **13**, 1023
30. Waelbroeck, H. (1990): Class. Quantum Grav. **7**, 751;
    (1994): Phys. Rev. D **50**, 4982
31. Nelson, J., Regge, T. (1991): Phys. Lett. B **273**, 213
32. Welling, M. (1997): Class. Quant. Grav. **14**, 3313;
    Matschull, H.-J. and Welling, M. (1998): Class. Quantum Grav. **15**, 2981
33. Bourdeau, M., Sorkin, S.D. (1992): Phys. Rev. D **45**, 687
34. Deser, S., Jackiw, R. (1988): Commun. Math. Phys. **118**, 495

Covariant Quantum Mechanics and Quantum Symmetries

J. Janyška, M. Modugno, D. Saller

Abstract. We sketch the basic ideas and results on the covariant formulation of quantum mechanics on a curved spacetime with absolute time equipped with given gravitational and electromagnetic fields. Moreover, we analyse the classical and quantum symmetries and show their relations.

1   Introduction

Our fundamental picture of the physical world is due to the theory of general relativity and to the quantum field theory, which have found great theoretical and experimental success.

The well established historical steps in classical theory have been: non relativistic theory, special relativity, general relativity. Analogously, the well established steps in quantum theory have been: non relativistic quantum mechanics, special relativistic quantum field theory.

Unfortunately, these theories deal with different objects, use partially incompatible mathematical methods and fulfill different requirements of covariance. In particular, the standard formulation of quantum theories is highly based on concepts and methods strictly related to a flat spacetime and inertial observers, which conflict with general covariance on a curved spacetime.

So, a consistent formulation of quantum field theories and general relativity is a still open problem. The problem has at least two faces:

- General relativistic covariant formulation of quantum theories in a curved spacetime,
- quantum theory of gravitational field.

The model of *covariant quantum mechanics* discussed in this paper is aimed at contributing to the first face of the problem, by means of new ideas and methods [1–3,48,5–16]. Namely, we study a general relativistic covariant formulation of quantum mechanics on a classical background constituted by a curved spacetime fibred over absolute time and equipped with given spacelike Riemannian metric, and gravitational and electromagnetic fields. Thus, we restrict our investigation just to fundamental fields of classical and quantum mechanics, because we believe that this arena could possibly suggest us good ideas for unifying deeper fundamental theories of physics.

The framework of our model is allowed by the possible general relativistic formulation of classical physics in a curved spacetime with absolute time. This theory is well established in the literature [17–38], even if it is much less popular than the Einstein theory of relativity. This theory is rigorous and self-consistent from a mathematical viewpoint and describes the phenomena of classical physics by an approximation which is intermediate between the classical theory and the Einstein theory of relativity.

The standard term "relativistic theory" links the special or general covariance with the Minkowski or Lorentz metric. This usage is clearly motivated by the historical developments of the Einstein theory. However, it would be more appropriate to refer the word "relativistic theory" only to its semantic meaning related to covariance. Indeed, the standard usage would be highly misleading in our context. In fact, our model is general relativistic, in the sense of covariance, but it is not Minkowskian or Lorentzian.

Clearly, the Minkowski or Lorentz metric is physically related to the distinguished constant c. Actually, in our model this constant does not occur. The classical limit of Einstein general relativity for $c \to \infty$ [25] is quite delicate, if we wish to understand the limit of the geometric structures of the model and not only the limit of some measurements. In a sense, our model could be regarded as the "true" classical limit of Einstein general relativity.

Our model can be regarded as an intermediate step between the standard non relativistic quantum mechanics and a possible fully general relativistic quantum theory. This framework allows us to focus our attention on the general relativistic covariance and the curved spacetime, detaching them from the difficulties due to the Lorentz metric. Actually, our choice seems to be quite fruitful.

The main new methods and achievements can be summarised as follows.

First of all, our basic guide is the *covariance* (even more, the manifest covariance) of the theory as heuristic requirement. Nowadays, the concept of "covariance" has been formulated in a rigorous mathematical way through the geometric concept of "naturality" [39]. According to the covariance of the theory, *time* is not just a parameter, but a fundamental object of the theory; moreover, the main objects of the theory are not assumed to be split into time and space components. As classical *phase space* we take the first jet space of spacetime and not its tangent space; indeed, this minimal choice allows us to skip anholonomic constraints. Another consequence of our choices is that classical mechanics is ruled not by a symplectic structure, but by a *cosymplectic structure* [40]; actually, we do get a symplectic structure, but this describes only the spacelike aspects of classical theory and is insufficient to account for classical dynamics. An achievement of our theory is the Lie algebra of "*special quadratic functions*" (different from the Poisson algebra), which allows us to treat energy, momentum and spacetime functions on the same footing. We emphasize the fact that classical mechanics can be formulated in a covariant way by a *Lagrangian approach*, but not by a Hamiltonian approach, because the Hamiltonian function depends essentially on an observer.

As far as quantum mechanics is concerned, all objects are derived, in a covariant way, from three minimal objects. Here, we have some novelties. The quantum bundle lives *on spacetime* and not on the phase space and the quantum connection is "*uni-*

versal". These assumptions allow us to skip all problems of polarisations [41]. In a sense, we obtain naturally a covariant polarisation and this is sufficient for our purposes. Indeed, we replace the problematic search for such inclusions with a *method of projectability*, which turns out to be our implementation of covariance in the quantum theory. Another new assumption concerns the *Hermitian metric* of the quantum bundle, which takes its values in the space of spacelike volume forms. This assumption allows us to skip the problems related to half-densities. The Schroedinger equation is obtained, in a covariant way, through a *Lagrangian approach* and not through the standard non covariant Hamiltonian approach. Indeed, we exhibit an *explicit expression* of the Schroedinger equation for any quantum system. The quantum operators arise automatically, in a covariant way, from the *classification* of distinguished first and second order differential operators of the quantum framework and not from a quantisation requirement of a classical system [41]. The seat for the covariant probabilistic interpretation of quantum mechanics is a *Hilbert bundle*, naturally yielded by the quantum bundle, and not just a Hilbert space. Our theory provides explicit expressions of all objects for any *accelerated observer* and yields, at the same time, an interpretation in terms of gravitational field, according to the principle of equivalence.

In a few words, we start with really minimal geometric structures representing physical fields and proceed along a thread naturally imposed by the only requirement of general covariance. We take the well established *results* of classical and quantum mechanics as touchstone of our model. On the other hand, according to the aims of our theory, we disregard those standard *methods* for deriving quantum objects, which are incompatible with general covariance. Indeed, in the flat case, the results of our model reduce to the results of the standard classical and quantum mechanics.

In this paper, we deal just with a given gravitational and electromagnetic field; this is sufficient as classical background for our covariant model of quantum mechanics. On the other hand, our classical model can be completed by adding, in a covariant way, the equations linking the gravitational and electromagnetic fields to their mass and charge sources [3]. These equations are a covariant reduced version of the Einstein and Maxwell equations. In fact, due to the spacelike nature of the metric, there is no way to couple fully the gravitational and electromagnetic fields with the energy-momentum tensor and the charge current, respectively. Just this is the main point which makes the Einstein model physically much more complete than ours. On the other hand, in our quantum model, the gravitational and electromagnetic fields are "external", hence the relation of these fields with their sources does not play an effective role in this context.

The reader might be puzzled by the fact that we do not mention explicitly the representations of the (finite dimensional and infinite dimensional) groups involved in our theory. In fact, our natural geometric constructions provide these representations automatically. This is a byproduct of our manifestly covariant approach.

In our model we never make an essential use of the fact that the dimension of spacetime is $n = 1+3$. We just need $n \geq 1+2$. In fact we have applied our machinery to the quantisation of a rigid body, whose configuration space has dimension $n = 1 + 3 + 3$ [12].

Even more, in our model we never make an essential use of the fact that the spacelike metric of spacetime is Riemannian; we just need that it is non degenerate on each fibre. So, we could, for instance, apply our machinery to a model of dimension 5, with a fibring on an extra parameter, whose fibres are four dimensional Lorentzian manifolds. Such a model would work pretty well mathematically, but we do not know any interesting physical interpretation.

The scheme developed for covariant quantum mechanics of a scalar particle can be easily and nicely extended to the case of a spin particle [42].

In spite of the differences of the starting scheme of spacetime, several steps of the above methodology appeared to be usefully translable to the *Einstein case*. In particular, so far, we have been able to apply to the Einstein case the methods concerning the classical phase space, the algebra of quantisable functions and the algebra of pre-quantum operators [5,8,42–45].

We hope that the new methods arising in our model could yield fruitful hints for a possible generally covariant formulation of quantum field theory in an Einstein framework.

2   Covariant quantum mechanics

2.1   Classical background

We start by sketching our covariant model of classical curved spacetime fibred over absolute time, and the related formulation of classical mechanics. We recall the basic elements of the model and present new results, as well.

Classical spacetime.   According to [3,46], we postulate:

(C.1) A *classical spacetime E*, which is an oriented four dimensional manifold;

(C.2) the *absolute time T*, which is an oriented one dimensional affine space, associated with the vector space $\bar{\mathbb{T}}$;

(C.3) a *time fibring $t : E \to fT$*, which is a surjective map of rank 1;

(C.4) a "scaled" *spacelike metric g*, which is a "scaled" Riemannian metric of the fibres of spacetime;

(C.5) a *gravitational field $K^\natural$*, which is a linear connection of spacetime, which preserves the time fibring and the spacelike metric and whose curvature fulfills the typical symmetry of Riemannian connections;

(C.6) a "scaled" *electromagnetic field f*, which is a "scaled" closed 2–form of spacetime.

Here, the word "scaled" used for the spacelike metric and the electromagnetic field means that these objects are tensorialised by a suitable scale factor which accounts for the appropriate units of measurement.

A *time unit of measurement* will be denoted by $u^0 \in \bar{\mathbb{T}}$ and its dual by $u_0 \in \bar{\mathbb{T}}^*$. We refer to charts of spacetime $(x^\lambda) = (x^0, x^i)$ adapted to the time fibring, to the affine structure of time and to a time unit of measurement $u_0 \in \bar{\mathbb{T}}$.

With reference to a given particle of mass m and charge q, in order to get rid of any choice of length and mass units of measurement, it is convenient to "normalise" the spacelike metric and the electromagnetic field, by considering the Planck constant $\hbar$.

Thus, we consider the "re–scaled" spacelike metric $G := \frac{m}{\hbar} g$, which takes its values in $\bar{\mathbb{T}}$. Its coordinate expression is

$$G = G^0_{ij} u_0 \otimes \check{d}^i \otimes \check{d}^j,$$

where $\check{d}^i$ is the spacelike differential of the coordinate x^i. Analogously, we consider the "re–scaled" electromagnetic field $F := \frac{q}{\hbar} f$, which is a true form. Accordingly, all objects derived from G and F will be re-scaled and will include the mass and the charge of the particle, and the Planck constant as well.

As *phase space* for the classical particle we take the first order jet space $J_1 E$ of the spacetime fibring [39]. We recall that $J_1 E$ can be naturally identified with the affine subspace of $\bar{\mathbb{T}}^* \otimes T E$, whose elements v are normalised according to the condition $v^0_0 = 1$ (which is independent from the choice of a unit of measurement of time). The chart naturally induced on the phase space by a spacetime chart is denoted by (x^0, x^i, x^i_0).

We have assumed a projection of spacetime over time, but, according to the principle of general relativity, not a distinguished splitting of spacetime into space and time. In other words, for each spacetime vector X, we obtain, in a covariant way, its projection on time $X^0 u_0$, but not a timelike and a spacelike component.

On the other hand, an *observer* is defined to be a section $o : E \to J_1 E$. The coordinate expression of an observer o is of the type $o = u^0 \otimes (\partial_0 + o^i_0 \partial_i)$. An observer o yields a splitting of each spacetime vector X into its *observed timelike and spacelike components* $v = v^0 (\partial_0 + o^i_0 \partial_i) + (v^i - v^0 o^i_0)\partial_i$. A spacetime chart is said to be adapted to an observer if $o^i_0 = 0$. Conversely, each spacetime chart determines the observer, whose coordinate expression is $o = u^0 \otimes \partial_0$.

According to the principle of general relativity, we do not assume distinguished observers.

The above objects C.1,..., C.6 yield in a covariant way [3,46]:

- The scaled *time form dt* : $E \to \bar{\mathbb{T}} \otimes T^* E$ of spacetime;
- a *spacelike volume form η* and a *spacetime volume form υ* of spacetime;
- a *2–form* $\Omega^\natural$: $J_1 E \to \Lambda^2 T^* J_1 E$ of the phase space;
- a *dt-vertical 2-vector* $\Lambda^\natural$: $J_1 E \to \Lambda^2 V J_1 E$ of the phase space,
- a *second order connection* $\gamma^\natural$: $J_1 E \to \bar{\mathbb{T}}^* \otimes T J_1 E$ of spacetime,

Here, we have used the symbol $^\natural$ to label objects derived from the gravitational field.

We obtain the following identities

$$i(\gamma^\natural) dt = 1, \quad i(\gamma^\natural) \Omega^\natural = 0, \quad dt \wedge \Omega^\natural \wedge \Omega^\natural \wedge \Omega^\natural \neq 0,$$
$$d\Omega^\natural = 0, \qquad L[\gamma^\natural] \Lambda^\natural = 0, \quad [\Lambda^\natural, \Lambda^\natural] = 0.$$

Hence, the pair $(dt, \Omega^\natural)$ turns out to be a *cosymplectic structure* of the phase space [40]. Moreover, $\Lambda^\natural$ and $\Omega^\natural$ yield inverse linear isomorphisms between the vector spaces of dt-vertical vectors and $\gamma^\natural$-horizontal forms of the phase space.

The Lie derivative of the spacelike metric G and of the spacelike volume form η with respect to a vector field of E is well defined provided that the vector field is projectable on T.

If X is a vector field of E projectable on T, then we define its *spacelike divergence* by means of the equality $\mathrm{div}_\eta X = L[X]\eta$. We have the coordinate expression

$$\mathrm{div}_\eta X = X^0 \frac{\partial_0 \sqrt{|g|}}{\sqrt{|g|}} + \frac{\partial_i (X^i \sqrt{|g|})}{\sqrt{|g|}}.$$

It is convenient to add an electromagnetic term to the gravitational field, in a covariant way [3,46], according to the formula

$$K = K^\natural + K^e := K^\natural + \tfrac{1}{2}(dt \otimes \hat{F} + \hat{F} \otimes dt),$$

i.e., in coordinates,

$$K_h{}^i{}_k = K^\natural{}_h{}^i{}_k, \quad K_0{}^i{}_k = K^\natural{}_0{}^i{}_k + \tfrac{1}{2} G_0^{ij} F_{jk},$$
$$K_0{}^i{}_0 = K^\natural{}_0{}^i{}_0 + G_0^{ij} F_{j0},$$

where $\hat{F} := G_0^{ij} F_{j\lambda} d^0 \otimes \partial_i \otimes d^\lambda$.

Then, the "total" object K turns out to be a connection of spacetime, which fulfills the same properties postulated in (C.5). Moreover, all main formulas in classical and quantum mechanics concerning the given particle and involving the gravitational and electromagnetic fields can be expressed through the "total" K and its derived objects, without the need of splitting it into its gravitational and electromagnetic components.

Proceeding with the total spacetime connection K as before, we obtain the "total" second order connection, 2-form and 2-vector

$$\gamma = \gamma^\natural + \gamma^e, \quad \Omega = \Omega^\natural + \Omega^e, \quad \Lambda = \Lambda^\natural + \Lambda^e,$$

where the electromagnetic terms γ^e, Ω^e and Λ^e turn out to be, respectively, the (re-scaled) Lorentz force, $\tfrac{1}{2}$ the (re-scaled) electromagnetic field and $\tfrac{1}{2}$ the (re-scaled) contravariant spacelike electromagnetic field, i.e. the (re-scaled) magnetic field. These total objects fulfill all properties fulfilled by the gravitational objects as above.

The total *cosymplectic 2-form* Ω encodes the full structure of spacetime (metric, gravitational field and electromagnetic field), hence it plays a central role in the theory.

We obtain the following coordinate expressions

$$\gamma_0{}^i_0 = K_h{}^i{}_k x_0^h x_0^k + 2K_0{}^i{}_k x_0^k + K_0{}^i{}_0,$$
$$\Omega = G_{ij}^0 (d_0^i - \gamma_0{}^i_0 d^0 - (K_h{}^i{}_0 + K_h{}^i{}_k x_0^k)(d^h - x_0^h d^0)) \wedge (d^j - x_0^j d^0),$$

$$\Lambda = G_0^{ij} \left(\partial_i + (K_i{}^h{}_0 + K_i{}^h{}_k x_0^k) \partial_h^0 \right) \wedge \partial_j^0 .$$

Classical mechanics. The classical mechanics can be achieved as follows.

The second order connection γ yields, in a covariant way, the *generalised Newton law* $\nabla j_1 s = 0$, for a motion $s : \boldsymbol{T} \to \boldsymbol{E}$. Clearly, this equation splits into its gravitational and electromagnetic components as $\nabla^\natural j_1 s = \gamma^\mathfrak{e} \circ j_1 s$.

Moreover, the classical dynamics can be derived from Ω, by a Lagrangian formalism, in the following covariant way [7,46].

Proposition 2.1 *The closed 2-form Ω admits locally* horizontal potentials $\Theta : J_1 \boldsymbol{E} \to T^* \boldsymbol{E}$, *which are defined up to closed 1-forms of spacetime.*

The horizontal potentials Θ have coordinate expression of the type

$$\Theta = -\left(\tfrac{1}{2} G_{ij}^0 x_0^i x_0^j - A_0 \right) d^0 + \left(G_{ij}^0 x_0^j + A_i \right) d^i ,$$

with $A \in \mathrm{Sec}(\boldsymbol{E}, T^* \boldsymbol{E})$.

A horizontal potential Θ and an observer o yield the *classical potential $A := o^* \Theta$* : $\boldsymbol{E} \to T^* \boldsymbol{E}$, which is defined locally up to a closed form and depends on the observer.

Proposition 2.2 *Let us consider a given horizontal potential Θ; if o and $\bar{o} = o + v$ are two observers, then the associated potentials A and $\bar{A}$ are related, in a chart adapted to o, by the formula*

$$\bar{A} = A - \tfrac{1}{2} G_{ij}^0 v_0^i v_0^j d^0 + G_{ij}^0 v_0^j d^i .$$

Therefore, each horizontal potential Θ determines a distinguished observer; in fact, there is a unique observer o, such that the spacelike component of the associated potential A vanishes.

An observer o yields the *observed 2-form $\Phi := 2 o^* \Omega : \boldsymbol{E} \to \Lambda^2 T^* \boldsymbol{E}$.*

Proposition 2.3 *We have $\Phi_{\lambda\mu} = \partial_\lambda A_\mu - \partial_\mu A_\lambda$.*
We obtain also $\Phi_{0k} := -G_{kj}^0 K_0{}^j{}_0$ and $\Phi_{hk} := G_{hj}^0 K_k{}^j{}_0 - G_{kj}^0 K_h{}^j{}_0$.

Proposition 2.4 *A horizontal potential Θ yields, in a covariant way, the* classical Lagrangian $\mathcal{L} : J_1 \boldsymbol{E} \to T^* \boldsymbol{T}$, *with coordinate expression*

$$\mathcal{L} = \left(\tfrac{1}{2} G_{ij}^0 x_0^i x_0^j + x_0^i A_i + A_0 \right) d^0 ,$$

where A_λ are the components of the potential A observed by the observer o associated with the chart. The Lagrangian is defined locally and up to a gauge, but does not depend on any observer. The Poincaré–Cartan form associated with the Lagrangian $\mathcal{L}$ turns out to be just Θ. The Euler–Lagrange equation associated with $\mathcal{L}$ turns out to coincide with the generalised Newton law.

Proposition 2.5 *A horizontal potential* Θ *and an observer o yield the* classical *Hamiltonian* $\mathcal{H} : J_1 E \to T^* T$ *and the* classical momentum $\mathcal{P} : J_1 E \to T^* E$, *defined as the negative of the o-horizontal component and the o-vertical components of* Θ, *respectively. Thus, we can write*

$$\Theta = -\mathcal{H} + \mathcal{P}.$$

In an adapted chart, we have the coordinate expressions

$$\mathcal{H} = \left(\tfrac{1}{2} G_{ij}^0 x_0^i x_0^j - A_0\right) d^0, \quad \mathcal{P} = \left(G_{ij}^0 x_0^j + A_i\right) d^i.$$

They are defined locally and up to a gauge, and depend on the choice of the observer.

The Newton law can be achieved also through $\mathcal{H}$ and $\mathcal{P}$, by means of a Hamiltonian formalism; but this procedure is non covariant, as it depends on the choice of an observer.

Classical Lie algebras.   Additionally, our structures yield further results on Lie algebras of functions and lifts of functions.

First of all, we obtain the *Poisson Lie bracket* $\{f, g\} := \Lambda^\sharp (df \wedge dg)$ for the functions of phase space.

A function f of phase space is *conserved* along the solutions of the Newton law if and only if $\gamma . f = 0$. We denote the space of conserved functions by $\mathrm{Con}(J_1 E, \mathbb{R})$. This space turns out to be a subalgebra of the Poisson algebra.

The time fibring and the spacelike metric yield, in a covariant way, a distinguished subset of the set of functions of phase space [3]. Namely, we define a *special quadratic function* to be a function of phase space, whose second fibre derivative (with respect to the affine fibres of phase space over spacetime) is proportional to the spacelike metric. In other words, in coordinates, the special quadratic functions are the functions of the type

$$f = \tfrac{1}{2} f^0 G_{ij}^0 x_0^i x_0^j + f^i G_{ij}^0 x_0^j + \overset{o}{f},$$

with f^0, f^i, $\overset{o}{f} \in \mathrm{Map}(E, \mathbb{R})$. The *time component* of a special quadratic function f as above is defined to be the (coordinate independent) map $f'' := f^0 u_0 : E \to \bar{\mathbb{T}}$.

Proposition 2.6 *The space of special quadratic functions* $\mathrm{Spec}(J_1 E, \mathbb{R})$ *turns out to be a Lie algebra through the* special *Lie bracket*

$$[\![f, g]\!] := \{f, g\} + \gamma(f'').g - \gamma(g'').f,$$

with coordinate expression

$$[\![f, g]\!]^0 = f^0 \partial_0 g^0 - g^0 \partial_0 f^0 - f^h \partial_h g^0 + g^h \partial_h f^0,$$

$$[\![f, g]\!]^i = f^0 \partial_0 g^i - g^0 \partial_0 f^i - f^h \partial_h g^i + g^h \partial_h f^i,$$

$$[\![f, g]\!] = f^0 \partial_0 \overset{o}{g} - g^0 \partial_0 \overset{o}{f} - f^h \partial_h \overset{o}{g} + g^h \partial_h \overset{o}{f}$$
$$- (f^0 g^k - g^0 f^k) \Phi_{0k} + f^h g^k \Phi_{hk}.$$

Corollary 2.1 *We have the following distinguished subalgebras of the special Lie algebra:*

- *The subalgebra* $\mathrm{Quan}(J_1 E, \mathbb{R}) \subset \mathrm{Spec}(J_1 E, \mathbb{R})$ *of* quantisable functions f, *whose time components* f'' *depend only on time;*
- *the subalgebra* $\mathrm{Time}(J_1 E, \mathbb{R}) \subset \mathrm{Quan}(J_1 E, \mathbb{R})$ *of time functions* f, *whose time components* f'' *are constant;*
- *the subalgebra* $\mathrm{Aff}(J_1 E, \mathbb{R}) \subset \mathrm{Time}(J_1 E, \mathbb{R})$ *of* affine functions f, *whose time components* f'' *vanish;*
- *the subalgebra* $\mathrm{Map}(E, \mathbb{R}) \subset \mathrm{Aff}(J_1 E, \mathbb{R})$ *of* spacetime functions.

Example 2.1 *We obtain*

$$\mathcal{L}_0, \mathcal{H}_0 \in \mathrm{Time}(J_1 E, \mathbb{R}), \quad \mathcal{P}_i \in \mathrm{Aff}(J_1 E, \mathbb{R}), \quad x^\lambda \in \mathrm{Map}(E, \mathbb{R}).$$

Clearly, the special bracket and the Poisson bracket coincide on $\mathrm{Aff}(J_1 E, \mathbb{R})$.

We have distinguished lifts of special quadratic functions to vector fields of spacetime and of phase space. Let us denote by $\mathrm{Pro}(E, TE) \subset \mathrm{Sec}(E, TE)$ the subalgebra of vector fields of E which are projectable on T.

Proposition 2.7 *The time fibring and the spacelike metric yield, in a covariant way, for each* $f \in \mathrm{Spec}(J_1 E, \mathbb{R})$, *the tangent lift* $X[f] : E \to TE$, *whose coordinate expression is*

$$X[f] = f^0 \partial_0 - f^i \partial_i.$$

The lift $\mathrm{Spec}(J_1 E, \mathbb{R}) \to \mathrm{Sec}(E, TE) : f \mapsto X[f]$ *turns out to be a Lie algebra morphism (with respect to the special bracket and the standard Lie bracket, respectively); its kernel is* $\mathrm{Map}(E, \mathbb{R})$.

Example 2.2 *We obtain*

$$X[\mathcal{L}_0] = \partial_0 - A_0^i \partial_i, \quad X[\mathcal{H}_0] = \partial_0, \quad X[\mathcal{P}_i] = -\partial_i, \quad X[x^\lambda] = 0,$$

where $A_0^i := G_0^{ij} A_j$. *We observe that* $X[\mathcal{L}] := u^0 \otimes X[\mathcal{L}_0]$ *turns out to be the unique observer for which the spacelike component of the observed potential A vanishes. Moreover,* $X[\mathcal{H}] := u^0 \otimes X[\mathcal{H}_0]$ *turns out to be just the observer by which we have defined the Hamiltonian.*

Proposition 2.8 *For each vector field X of E projectable on* T, *the spacetime fibring yields, in a covariant way* [39], *the* holonomic prolongation

$$X_{\mathrm{hol}}^\uparrow := X_{(1)} : J_1 E \to T J_1 E,$$

whose coordinate expression is

$$X_{\mathrm{hol}}^\uparrow = X^\lambda \partial_\lambda + (\partial_0 X^i + \partial_j X^i x_0^j - \partial_0 X^0 x_0^i) \partial_i^0.$$

This prolongation turns out to be an injective Lie algebra morphism.

Corollary 2.2 *For each $f \in \mathrm{Quan}(J_1 E, \mathbb{R})$, the time fibring yields, in a covariant way, the* holonomic lift

$$X_{\mathrm{hol}}^{\uparrow}[f] := \big(X[f]\big)_{(1)} : J_1 E \to T J_1 E,$$

whose coordinate expression is

$$X_{\mathrm{hol}}^{\uparrow}[f] = f^0 \partial_0 - f^i \partial_i - (\partial_0 f^i + \partial_j f^i x_0^j + \partial_0 f^0 x_0^i) \partial_i^0.$$

The lift $\mathrm{Quan}(J_1 E, \mathbb{R}) \to \mathrm{Sec}(J_1 E, T J_1 E) : f \mapsto X_{\mathrm{hol}}^{\uparrow}[f]$ *turns out to be a Lie algebra morphism (with respect to the special bracket and the standard Lie bracket, respectively); its kernel is* $\mathrm{Map}(E, \mathbb{R})$.

Example 2.3 *We obtain*

$$X_{\mathrm{hol}}^{\uparrow}[\mathcal{L}_0] = \partial_0 - A_0^i \partial_i - (\partial_0 A_0^i + \partial_j A_0^i x_0^j) \partial_i^0,$$
$$X_{\mathrm{hol}}^{\uparrow}[\mathcal{H}_0] = \partial_0, \quad X_{\mathrm{hol}}^{\uparrow}[\mathcal{P}_i] = -\partial_i, \quad X_{\mathrm{hol}}^{\uparrow}[x^\lambda] = 0.$$

For each function f of phase space, we obtain, in a covariant way, the dt-vertical *Hamiltonian lift* $\Lambda^\sharp(df) : J_1 E \to V J_1 E$. More generally, for each function f of phase space and for each *time scale* $\tau : J_1 E \to \bar{\mathbb{T}}$, we obtain the τ-*Hamiltonian lift* $\gamma(\tau) + \Lambda^\sharp(df) : J_1 E \to T J_1 E$. In particular, we obtain the following result.

Proposition 2.9 *For each $f \in \mathrm{Spec}(J_1 E, \mathbb{R})$, the cosymplectic structure yields, in a covariant way, the* Hamiltonian lift

$$X_{\mathrm{Ham}}^{\uparrow}[f] := \gamma(f'') + \Lambda^\sharp(df) : J_1 E \to T J_1 E,$$

whose coordinate expression is

$$X_{\mathrm{Ham}}^{\uparrow}[f] = f^0 \partial_0 - f^i \partial_i + X_0^i \partial_i^0,$$

where

$$X_0^i = G_0^{ij} \Big(\tfrac{1}{2} \partial_j f^0 G_{hk}^0 x_0^h x_0^k + (-f^0 \partial_0 G_{jh}^0 + f^k \partial_k G_{jh}^0 + G_{hk}^0 \partial_j f^k) x_0^h$$
$$+ \partial_j \overset{o}{f} + \Phi_{hj} f^h + f^0 \Phi_{j0} \Big).$$

The lift $\mathrm{Quan}(J_1 E, \mathbb{R}) \to \mathrm{Sec}(J_1 E, T J_1 E) : f \mapsto X_{\mathrm{Ham}}^{\uparrow}[f]$ *turns out to be a Lie algebra morphism (with respect to the special bracket and the standard Lie bracket, respectively); its kernel is* $\mathrm{Map}(T, \mathbb{R})$.

Example 2.4 *We obtain*

$$X_{\mathrm{Ham}}^{\uparrow}[\mathcal{H}_0] = \partial_0 - G_0^{ij} \partial_0 \mathcal{P}_j \partial_i^0, \quad X_{\mathrm{Ham}}^{\uparrow}[\mathcal{P}_i] = -\partial_i + G_0^{hj} \partial_i \mathcal{P}_h \partial_j^0,$$
$$X_{\mathrm{Ham}}^{\uparrow}[x^0] = 0, \qquad\qquad X_{\mathrm{Ham}}^{\uparrow}[x^i] = G_0^{ij} \partial_j^0.$$

The interest of the above Hamiltonian lift is due to the following result concerning the projectability, which will play an important role in quantum mechanics.

Proposition 2.10 ([3]) *The τ-Hamiltonian lift of a function f of $J_1 E$ is projectable on a vector field of spacetime if and only if $f \in \mathrm{Spec}(J_1 E, \mathbb{R})$ and $\tau = f''$. Moreover, if these conditions are fulfilled, then the τ-Hamiltonian lift projects on the tangent lift of f.*

2.2 Covariant quantum mechanics

We proceed by sketching our covariant model of quantum mechanics on a curved spacetime fibred over absolute time. We recall the basic elements of the model and present new results as well.

Quantum structure. According to [3,46], for quantum mechanics of a charged spinless particle in the above classical background (including the given gravitational and electromagnetic external fields), we postulate:

(Q.1) A *quantum bundle* $Q \to E$, which is a one dimensional complex vector bundle over spacetime;

(Q.2) a *Hermitian metric* $\mathfrak{h} : E \to \mathbb{C} \otimes (Q^* \otimes_E Q^*) \otimes_E \Lambda^3 V^* E$ of the quantum bundle, with values in the complexified space of spacelike volume forms of spacetime.

Locally, we shall refer to a scaled complex *quantum basis* $(\mathfrak{b})$ normalised by the condition $h(\mathfrak{b}, \mathfrak{b}) = \eta$. The associated scaled complex chart is denoted by (z). Then, we obtain the scaled real basis $(\mathfrak{b}_1, \mathfrak{b}_2) := (\mathfrak{b}, i\mathfrak{b})$ and the associated scaled real chart (w^1, w^2). If $\Psi \in \mathrm{Sec}(E, Q)$, then we write $\Psi = \Psi^1 \mathfrak{b}_1 + \Psi^2 \mathfrak{b}_2 = \psi \mathfrak{b}$, where Ψ^1, Ψ^2 and ψ are, respectively, the scaled real and complex components of Ψ. Moreover, we consider the *extended quantum bundle*, $Q^\uparrow \to J_1 E$, obtained by extending the base space of the quantum bundle to the classical phase space, which here plays the role of space of classical observers.

Each system of connections $\{\overset{o}{\mathsf{q}}\}$ of the quantum bundle parametrised by the classical observers induces, in a covariant way, a connection q of the extended quantum bundle, which is said to be *universal* [46,47]. The universal connections are characterised in coordinates by the condition $\mathsf{q}^0_i = 0$.

Then, we postulate:

(Q.3) a *quantum connection* q of the extended quantum bundle, which is Hermitian, universal and whose curvature is $R[\mathsf{q}] = i\Omega$.

We recall that Ω incorporates the mass m of the particle and the Planck constant $\hbar$.

190   J. Janyška, M. Modugno, D. Saller

Proposition 2.11 *The coordinate expression of the quantum connection, with respect to a quantum basis and a spacetime chart, turns out to be locally of the type*

$$\text{Ч}_0 = -i\mathcal{H}_0, \quad \text{Ч}_i = i\mathcal{P}_i, \quad \text{Ч}_i^0 = 0.$$

The above classical Hamiltonian $\mathcal{H}$ and momentum $\mathcal{P}$ are referred to the observer o associated with the spacetime chart (x^λ) and to a classical horizontal potential Θ of Ω, which is locally determined by the quantum connection Ч and the quantum basis $\mathfrak{b}$. Then, the gauge of the classical potential $A := o^*\Theta$ is determined by the quantum connection and the quantum basis. Moreover, we recall that A includes both the gravitational and the electromagnetic potential.

These minimal geometric objects Q.1,..., Q.3 constitute the only source, in a covariant way, of all further objects of quantum mechanics.

Actually, the quantum connection lives on the extended quantum bundle, whose base space is the phase space; on the other hand, the covariance of the theory requires that the significant physical objects be independent from observers. This fact suggests a method of projectability, in order to get rid of the observers encoded in the phase space. Actually, we have already used this method in the classical theory, just in view of these developments of quantum mechanics. Indeed, this method turns out to be fruitful.

Quantum dynamics. The *quantum dynamics* can be obtained in the following way: the method of projectability yields, in a covariant way, a distinguished quantum Lagrangian (hence, the generalised Schroedinger equation, the quantum momentum and the probability current) [3,46].

Even more, the covariance implies the essential uniqueness of the above Lagrangian and of the Schroedinger equation [6,14].

Proposition 2.12 *The coordinate expression of the quantum Lagrangian is*

$$\mathfrak{L}[\Psi] = \tfrac{1}{2}\big(i(\bar{\psi}\partial_0\psi - \psi\partial_0\bar{\psi}) + 2A_0\bar{\psi}\psi - G_0^{ij}(\partial_i\bar{\psi}\partial_j\psi + A_iA_j\bar{\psi}\psi)$$
$$- iG_0^{ij}A_j(\bar{\psi}\partial_i\psi - \psi\partial_i\bar{\psi}) + k\rho_0\bar{\psi}\psi\big)\sqrt{|g|}d^0 \wedge d^1 \wedge d^2 \wedge d^3,$$

where ρ is the scalar curvature of the fibres of spacetime determined by the spacelike metric and $k \in \mathbb{R}$ is a real constant (which is not determined by the covariance).

Corollary 2.3 *The coordinate expression of the generalised Schroedinger equation turns out to be*

$$\left(\partial_0 - iA_0 + \tfrac{1}{2}\frac{\partial_0\sqrt{|g|}}{\sqrt{|g|}} - \tfrac{1}{2}i(\overset{o}{\Delta}_0 + k\rho_0)\right)\psi = 0,$$

where

$$\overset{o}{\Delta}_0 := G_0^{hk}(\partial_h - iA_h)(\partial_k - iA_k) + \frac{\partial_h(G_0^{hk}\sqrt{|g|})}{\sqrt{|g|}}(\partial_k - iA_k)$$

is the spacelike quantum Laplacian.

Corollary 2.4 *We obtain the conserved* probability current *with coordinate expression*

$$\Psi^* j = (\bar\psi\psi)\upsilon_0^0 - G_0^{hk}\left(i\tfrac{1}{2}(\bar\psi\,\partial_h\psi - \psi\,\partial_h\bar\psi) + A_h\bar\psi\psi\right)\upsilon_k^0,$$

where $\upsilon_\lambda^0 := i(\partial_\lambda)\sqrt{|g|}\,d^0 \wedge d^1 \wedge d^2 \wedge d^3$.

Quantum operators. We obtain distinguished operators acting on the sections of the quantum bundle in the following covariant way.

First of all, we have a distinguished family of *second order pre-quantum operators*.

Proposition 2.13 *The Schroedinger operator yields, for each time scale* $\tau : E \to \bar{\mathbb{T}}$, *the second order linear operator* $\mathfrak{S}(\tau) : J_2 Q \to Q$, *which acts on the sections* Ψ *of the quantum bundle, according to the coordinate expression*

$$\mathfrak{S}(\tau)[\Psi] = i\tau^0 \left(\partial_0 - iA_0 + \tfrac{1}{2}\frac{\partial_0\sqrt{|g|}}{\sqrt{|g|}} - \tfrac{1}{2}i\left(\overset{o}{\Delta}_0 + k\rho_0\right)\right)\psi\,\flat.$$

In particular, each $f \in \mathrm{Spec}(J_1 E, \mathbb{R})$ *yields, in a covariant way, the second order pre-quantum operator* $\mathfrak{S}[f] := \mathfrak{S}(f'')$.

Then, we obtain a distinguished family of first order operators, by classifying the vector fields of the quantum bundle which preserve the Hermitian metric. A vector field Y of Q is said to be *Hermitian* if it is projectable on E and on T, is real linear over its projection on E and fulfills $L[Y]\mathfrak{h} = 0$.

We denote the space of Hermitian vector fields of Q by $\mathrm{Her}(Q, TQ)$.

Proposition 2.14 *A vector field* Y *of* Q *is Hermitian if and only if its coordinate expression is of the type*

$$Y \equiv Y[f] = f^0\partial_0 - f^i\partial_i + \left(i(\overset{o}{f} + A_0 f^0 - A_i f^i) - \tfrac{1}{2}\mathrm{div}_\eta X[f]\right)\mathbb{I},$$

where $f \in \mathrm{Quan}(J_1 E, \mathbb{R})$ *and where* $\mathbb{I} = (w^1\partial w_1 + w^2\partial w_2)$ *denotes the identity vertical vector field of the quantum bundle. The above expression of* $Y[f]$ *turns out to be independent of the choice of coordinates.*

The space of Hermitian vector fields $\mathrm{Her}(Q, TQ)$ *is closed with respect to the Lie bracket. Moreover, the map*

$$\mathrm{Quan}(J_1 E, \mathbb{R}) \to \mathrm{Her}(Q, TQ) : f \mapsto Y[f]$$

is an isomorphism of Lie algebras (with respect to the special bracket and the standard Lie bracket, respectively). Furthermore, the map $\mathrm{Her}(Q, TQ) \to \mathrm{Pro}(E, TE) :$ $Y[f] \mapsto X[f]$ *turns out to be a central extension of Lie algebras by* $\mathrm{Map}(E, i\mathbb{R})\otimes\mathbb{I}$.

For each $f \in \mathrm{Quan}(J_1 E, \mathbb{R})$, the vector field $Y[f] : Q \to TQ$ is said to be the *quantum lift* of f .

Example 2.5 *We obtain*

$$Y[\mathcal{L}_0] = \partial_0 - A_0^i \partial_i - \left(iA_i A_0^i + \tfrac{1}{2} \left(\frac{\partial_0 \sqrt{|g|}}{\sqrt{|g|}} - \frac{\partial_i (A_0^i \sqrt{|g|})}{\sqrt{|g|}} \right) \right) \mathcal{I},$$

$$Y[\mathcal{H}_0] = \partial_0 - \tfrac{1}{2} \frac{\partial_0 \sqrt{|g|}}{\sqrt{|g|}} \mathcal{I}, \quad Y[\mathcal{P}_i] = -\partial_i + \tfrac{1}{2} \frac{\partial_i \sqrt{|g|}}{\sqrt{|g|}} \mathcal{I},$$

$$Y[x^\lambda] = ix^\lambda \mathbb{I}.$$

Corollary 2.5 *Each quantisable function f yields, in a covariant way, the first order operator acting on the sections of the quantum bundle*

$$Z[f] := iL[Y[f]],$$

whose coordinate expression is, for each $\Psi \in \mathrm{Sec}(E, Q)$,

$$Z[f].\Psi = i\left(f^0 \partial_0 \psi - f^i \partial_i \psi - \left(i(\overset{o}{f} + A_0 f^0 - A_i f^i) \right.\right.$$
$$\left.\left. - \tfrac{1}{2} \mathrm{div}_\eta X[f] \right) \psi \right) \flat.$$

For each quantisable function f, we say $Z[f]$ to be the associated *first order pre-quantum operator*. We denote the space of the first order pre-quantum operators by $\mathrm{Oper}_1(Q)$.

Proposition 2.15 *The space $\mathrm{Oper}_1(Q)$ turns out to be a Lie algebra through the bracket*

$$\big[Z[f], Z[g] \big] := -i\big(Z[f] \circ Z[g] - Z[g] \circ Z[f] \big).$$

Moreover, the map $\mathrm{Quan}(J_1 E, \mathbb{R}) \to \mathrm{Oper}_1(Q) : f \mapsto Z[f]$ turns out to be an isomorphism of Lie algebras (with respect to the special bracket and the above Lie bracket, respectively).

Example 2.6 *We obtain*

$$Z[\mathcal{H}_0].\Psi = i\left(\partial_0 \psi + \tfrac{1}{2} \frac{\partial_0 \sqrt{|g|}}{\sqrt{|g|}} \psi \right) \flat,$$

$$Z[\mathcal{P}_i].\Psi = -i\left(\partial_i \psi + \tfrac{1}{2} \frac{\partial_i \sqrt{|g|}}{\sqrt{|g|}} \psi \right) \flat,$$

$$Z[x^\lambda].\Psi = x^\lambda \psi \flat.$$

The above results appear to be a covariant "correspondence principle" yielding *pre-quantum operators* associated with quantisable functions. However, we still need to introduce the Hilbert stuff carrying the standard probabilistic interpretation of quantum mechanics. It can be done in the following covariant way [3,46].

Let us restrict our postulate (C.3), by requiring that the fibring of spacetime over time makes spacetime a bundle. Thus, we postulate that the fibres of spacetime are each other isomorphic. Then, we consider the infinite dimensional *functional quantum bundle* $\boldsymbol{H}_c \to \boldsymbol{T}$, whose fibres are constituted by the compact support smooth sections, at fixed time, of the quantum bundle ("regular sections"). The Hermitian metric $\mathfrak{h}$ equips this bundle with a pre-Hilbert metric $\langle\,,\,\rangle$. Then, a true Hilbert bundle $\boldsymbol{H} \to \boldsymbol{T}$ can be obtained by a completion procedure. This bundle has no distinguished splittings into time and type Hilbert fibre; such a splitting can be obtained by choosing a classical observer.

Each regular section Ψ of the quantum bundle can be regarded as a section $\widehat{\Psi}$ of the functional quantum bundle. Accordingly, each "regular" operator $\mathfrak{O}$ acting on sections of the quantum bundle can be regarded as an operator $\widehat{\mathfrak{O}}$ acting on the sections of the functional quantum bundle.

Our previous results yield, for each quantisable function f, two distinguished operators acting on the sections of the functional quantum bundle, namely $\widehat{Z[f]}$ and $\widehat{\mathfrak{S}[f]}$. Actually, in general, both operators do not act on the fibres of the functional bundle (at fixed time), because they involve the partial derivative ∂_0. On the other hand, we have the following results [3,15,46].

Proposition 2.16 *Let* $f \in \mathrm{Quan}(J_1 E, \mathbb{R})$. *Then, the combination*

$$\widehat{f} := \widehat{Z[f]} - \widehat{\mathfrak{S}[f]} \tag{1}$$

acts on the fibres of the functional bundle. We have the following coordinate expression

$$\widehat{f}(\widehat{\Psi}) = \Big(-\tfrac{1}{2} f^0 (\overset{o}{\Delta}_0 + k\rho_0) - \mathrm{i} f^j (\partial_j - \mathrm{i} A_j) + \overset{o}{f}$$
$$- \mathrm{i}\tfrac{1}{2}\frac{\partial_j (f^j \sqrt{|g|})}{\sqrt{|g|}} \Big) \psi\widehat{\mathfrak{b}}.$$

Moreover, $\widehat{f}$ is symmetric with respect to the Hermitian metric $\langle\,,\,\rangle$.

For the self-adjointness of $\widehat{f}$ further global conditions on f are needed.

For each $f \in \mathrm{Quan}(J_1 E, \mathbb{R})$, we say $\widehat{f}$ to be the *quantum operator* associated with f.

Example 2.7 *We obtain the following distinguished quantum operators*

$$\widehat{\mathcal{H}_0}(\widehat{\Psi}) = -\Big(\tfrac{1}{2}\overset{o}{\Delta}_0 + \tfrac{1}{2} k\rho_0 - A_0 \Big) \psi\widehat{\mathfrak{b}},$$

$$\widehat{\mathcal{P}_j}(\widehat{\Psi}) = -\mathrm{i}\Big(\partial_j + \tfrac{1}{2}\frac{\partial_j \sqrt{|g|}}{\sqrt{|g|}} \Big) \psi\widehat{\mathfrak{b}},$$

$$\widehat{x^\lambda}(\widehat{\Psi}) = x^\lambda \psi\widehat{\mathfrak{b}}.$$

The space of the fibre preserving maps of the functional quantum bundle into itself becomes a Lie algebra through the bracket $[h, k] := -\mathrm{i}(h \circ k - k \circ h)$.

Proposition 2.17 *For each $f, g \in \mathrm{Quan}(J_1 E, \mathbb{R})$, we obtain*

$$[\hat{f}, \hat{g}] = \widehat{[\![f, g]\!]} - \mathrm{i}\widehat{[S[f], Z[g]]} + \mathrm{i}\widehat{[S[g], Z[f]]}. \tag{2}$$

In particular, for each $f, g \in \mathrm{Aff}(J_1 E, \mathbb{R})$, we obtain

$$[\hat{f}, \hat{g}] = \widehat{[\![f, g]\!]} = \widehat{\{f, g\}}. \tag{3}$$

Thus, the above results suggest our covariant "equivalence principle". The Feynmann path integral approach can be nicely formulated in our framework [3]. In fact, the quantum connection Ч yields, in a covariant way, a non linear connection of the extended quantum bundle over time; moreover, this connection allows us to interpret the Feynmann amplitudes through the parallel transport of this connection. However, unfortunately, our theory does not contribute so far to the hard problem of the measure arising in the Feynmann theory.

The case of a spin particle (generalised Pauli equation) can be approached in an analogous way, by considering a further quantum bundle of dimension two, with the only additional postulate of a suitable soldering form [48].

3   Symmetries

Next, we classify the infinitesimal symmetries of the classical and quantum structures. We show that these symmetries are controlled by the Lie algebra of quantisable functions and its distinguished subalgebras. Moreover, we discuss the strict relations between classical and quantum symmetries.

3.1   Classical symmetries

We start by discussing the main results concerning symmetries of the classical structure.

Subalgebras.   First we analyse further distinguished subalgebras of the algebra of quantisable functions.

Proposition 3.1 *We have the following distinguished subalgebras of the algebra of quantisable functions:*

- *the subalgebra $\mathrm{Hol}(J_1 E, \mathbb{R}) \subset \mathrm{Quan}(J_1 E, \mathbb{R})$, which is constituted by the functions f such that $X^{\uparrow}_{\mathrm{hol}}[f] = X^{\uparrow}_{\mathrm{Ham}}[f]$;*
- *the subalgebra $\mathrm{Unim}(J_1 E, \mathbb{R}) \subset \mathrm{Quan}(J_1 E, \mathbb{R})$, which is constituted by the functions f such that $\mathrm{div}_\eta X[f] = 0$;*
- *the subalgebra $\mathrm{Self}(J_1 E, \mathbb{R}) \subset \mathrm{Quan}(J_1 E, \mathbb{R})$, which is constituted by the functions f such that $i(X^{\uparrow}_{\mathrm{hol}}[f])\Omega = df$.*

If $f \in \mathrm{Hol}(J_1 E, \mathbb{R})$, then we set

$$X^{\uparrow}[f] := X^{\uparrow}_{\mathrm{Ham}}[f] = X^{\uparrow}_{\mathrm{hol}}[f].$$

Proposition 3.2 *We have*

$$\mathrm{Time}(J_1 E, \mathbb{R}) \cap \mathrm{Con}(J_1 E, \mathbb{R}) = \mathrm{Time}(J_1 E, \mathbb{R}) \cap \mathrm{Self}(J_1 E, \mathbb{R}).$$

Then, we set

$$\mathrm{Clas}(J_1 E, \mathbb{R}) := \mathrm{Time}(J_1 E, \mathbb{R}) \cap \mathrm{Con}(J_1 E, \mathbb{R})$$
$$= \mathrm{Time}(J_1 E, \mathbb{R}) \cap \mathrm{Self}(J_1 E, \mathbb{R})$$

and denote the space of the tangent lifts of elements of $\mathrm{Clas}(J_1 E, \mathbb{R})$ by

$$\mathrm{Clas}(E, TE) \subset \mathrm{Pro}(E, TE). \tag{4}$$

Proposition 3.3 *We have*

$$\mathrm{Time}(J_1 E, \mathbb{R}) \cap \mathrm{Con}(J_1 E, \mathbb{R}) \subset \mathrm{Hol}(J_1 E, \mathbb{R})$$
$$\mathrm{Time}(J_1 E, \mathbb{R}) \cap \mathrm{Con}(J_1 E, \mathbb{R}) \subset \mathrm{Unim}(J_1 E, \mathbb{R}).$$

Proposition 3.4 *The special and the Poisson brackets coincide in* $\mathrm{Clas}(J_1 E, \mathbb{R})$. *Hence, this space turns out to be a subalgebra of the Poisson and of the special algebras. Moreover,* $\mathrm{Clas}(E, TE)$ *turns out to be closed with respect to the standard Lie bracket.*

We call the elements of $\mathrm{Clas}(J_1 E, \mathbb{R})$ *classical generators*. This name will be justified by Proposition 3.5, Corollary 3.1 and Corollary 3.2.

Symmetries. A vector field $X^\uparrow \in \mathrm{Sec}(J_1 E, T J_1 E)$ is said to be a *symmetry of the classical structure* if it is projectable on E and T and fulfills

$$L[X^\uparrow]dt = 0, \quad L[X^\uparrow]\Omega = 0.$$

We denote the space of the symmetries of the classical structure by $\mathrm{Clas}(J_1 E, T J_1 E)$.

Proposition 3.5 ([16]) *A vector field* $X^\uparrow$ *of* $J_1 E$ *projectable on* E *fulfills* $L[X^\uparrow]dt = 0$ *and* $L[X^\uparrow]\Omega = 0$ *if and only if, locally,*

$$X^\uparrow = X^\uparrow[f], \quad with \quad f \in \mathrm{Clas}(J_1 E, \mathbb{R}),$$

where f *is defined up to a constant.*

Corollary 3.1 *If* $f \in \mathrm{Clas}(J_1 E, \mathbb{R})$, *then we obtain*

$$L[X[f]]G = 0, \quad L[X[f]]\eta = 0,$$
$$L[X^\uparrow[f]]\gamma = 0, \quad L[X^\uparrow[f]]K = 0.$$

Corollary 3.2 *If X is a vector field of* E *projectable on* T, *such that* $L[X^\uparrow_{\mathrm{hol}}]\mathcal{L} = 0$, *then we obtain locally*

$$X = X[f], \quad X^\uparrow_{\mathrm{hol}} = X^\uparrow[f], \quad with \quad f \in \mathrm{Clas}(J_1 E, \mathbb{R}),$$

where f *is defined up to a constant.*

3.2 Quantum symmetries

Eventually, we classify the vector fields of the extended quantum bundle which preserve the full quantum structure: all fibrings (on quantum bundle, on phase space, on spacetime, on time), the Hermitian metric, the quantum connection. Moreover, we compare the symmetries of the quantum structure with the symmetries of the quantum Lagrangian.

Symmetries of the quantum structure. A vector field $Y^\uparrow$ of $\boldsymbol{Q}^\uparrow$ is said to be a *symmetry of the quantum structure* if it is projectable on $\boldsymbol{Q}$, $J_1\boldsymbol{E}$, $\boldsymbol{E}$, $\boldsymbol{T}$, is real linear over $J_1\boldsymbol{E}$ and fulfills

$$L[Y^\uparrow]dt = 0, \quad L[Y^\uparrow]\hbar = 0, \quad L[Y^\uparrow]\text{ч} = 0.$$

We denote the space of symmetries of the quantum structure by $\mathrm{Quan}(\boldsymbol{Q}^\uparrow, T\boldsymbol{Q}^\uparrow)$.

For each $f \in \mathrm{Hol}(J_1\boldsymbol{E}, \mathbb{R})$, we define its *extended quantum lift* to be the vector field of the extended quantum bundle

$$Y^\uparrow[f] := \text{ч}\big(X^\uparrow[f]\big) + \big(\mathrm{i}f - \tfrac{1}{2}\,\mathrm{div}_\eta\, X[f]\big)\mathbb{I}.$$

Proposition 3.6 *A vector field $Y^\uparrow$ of $\boldsymbol{Q}^\uparrow$ is a symmetry of the quantum structure if and only if it is of the type*

$$Y^\uparrow = Y^\uparrow[f], \quad with \quad f \in \mathrm{Clas}(J_1\boldsymbol{E}, \mathbb{R}).$$

The space $\mathrm{Quan}(\boldsymbol{Q}^\uparrow, T\boldsymbol{Q}^\uparrow)$ is closed with respect to the Lie bracket. Moreover, the map $\mathrm{Clas}(J_1\boldsymbol{E}, \mathbb{R}) \to \mathrm{Quan}(\boldsymbol{Q}^\uparrow, T\boldsymbol{Q}^\uparrow) : f \mapsto Y^\uparrow[f]$ is an isomorphism of Lie algebras (with respect to the special bracket and the standard Lie bracket, respectively). Furthermore, the map $\mathrm{Quan}(\boldsymbol{Q}^\uparrow, T\boldsymbol{Q}^\uparrow) \to \mathrm{Clas}(\boldsymbol{E}, T\boldsymbol{E}) : Y^\uparrow[f] \mapsto X[f]$ turns out to be a central extension of Lie algebras by $\mathrm{i}\mathbb{R} \otimes \mathbb{I}$.

Symmetries of the quantum dynamics. Next, we compare the symmetries of the quantum connection and the symmetries of the quantum Lagrangian.

Proposition 3.7 *For each $f \in \mathrm{Quan}(J_1\boldsymbol{E}, \mathbb{R})$, we obtain, in a covariant way, the holonomic quantum lift of f, defined as the holonomic prolongation* [39]

$$Y_{\mathrm{hol}}[f] := \big(Y[f]\big)_{(1)} : J_1\boldsymbol{Q} \to TJ_1\boldsymbol{Q}$$

of the quantum lift $Y[f]$, whose coordinate expression is

$$\begin{aligned}
Y_{\mathrm{hol}}[f] = {}& f^0\partial_0 - f^i\partial_i \\
& - \tfrac{1}{2}\,\mathrm{div}_\eta\, X[f](w^1\partial_1 + w^2\partial_2 - w^1_\lambda\partial^\lambda_1 - w^2_\lambda\partial^\lambda_2) \\
& - \tfrac{1}{2}\partial_\lambda\,\mathrm{div}_\eta\, X[f](w^1\partial^\lambda_1 + w^2\partial^\lambda_2) \\
& + (f^0 A_0 - f^i A_i + \overset{o}{f})(w^1\partial_2 - w^2\partial_1 + w^1_\lambda\partial^\lambda_2 - w^2_\lambda\partial^\lambda_1) \\
& + \partial_\lambda(f^0 A_0 - f^i A_i + \overset{o}{f})(w^1\partial^\lambda_2 - w^2\partial^\lambda_1) \\
& - \partial_0 f^0(w^1_0\partial^0_1 + w^2_0\partial^0_2) + \partial_\lambda f^i(w^1_i\partial^\lambda_1 + w^2_i\partial^\lambda_2).
\end{aligned}$$

Proposition 3.8 *Let* $f \in \text{Time}(J_1 E, \mathbb{R})$. *Then, the following conditions are equivalent:*

1) $L[Y_{\text{hol}}^{\uparrow}[f]]\Psi = 0$,   2) $L[Y_{\text{hol}}[f]]\mathcal{L} = 0$,

3) $i(X_{\text{hol}}^{\uparrow}[f])\Omega = df$,   4) $\gamma.f = 0$,   5) $f \in \text{Clas}(J_1 E, \mathbb{R})$.

Eventually, we consider the conserved currents associated with symmetries of the quantum Lagrangian, according to the standard Noether theorem. Additionally, our results allow us to associate such currents with classical quantisable functions.

For each $f \in \text{Quan}(J_1 E, \mathbb{R})$, we define the associated *quantum current* to be the 3-form

$$\mathfrak{J}[f] := -i(Y[f])\Pi : J_1 Q \to \Lambda^3 T^* Q,$$

where Π is the Poincaré–Cartan form [7] associated with the quantum Lagrangian.

Corollary 3.3 *For each* $f \in \text{Clas}(J_1 E, \mathbb{R})$, *the current* $\mathfrak{J}[f]$ *is conserved along the solutions* $\Psi : E \to Q$ *of the Schroedinger equation.*

Example 3.1 *The current associated with the constant function* $1 \in \text{Clas}(J_1 E, \mathbb{R})$ *is just the conserved probability current.*

Moreover, for each affine function and quantum section, we obtain, in a covariant way, a spacelike 3-form (which can be integrated on the fibres of spacetime), according to the following result.

Proposition 3.9 *Let* $f \in \text{Aff}(J_1 E, \mathbb{R})$. *Then, for each* $\Psi \in \text{Sec}(E, Q)$ *we obtain*

$$\left(\Psi^*(\mathfrak{J}[f])\right)^{\vee} = \tfrac{1}{2}\left(\mathfrak{h}(Z[f].\Psi, \Psi) - \mathfrak{h}(\Psi, Z[f].\Psi)\right)$$

where $^{\vee}$ *denotes the vertical restriction. We have the coordinate expression*

$$\left(\Psi^*(\mathfrak{J}[f])\right)^{\vee} = \left(f^i(\Psi^1 \partial_i \Psi^2 - \Psi^2 \partial_i \Psi^1) + \overset{o}{f}(\Psi^1 \Psi^1 + \Psi^2 \Psi^2)\right) \\ \sqrt{|g|}\check{d}^1 \wedge \check{d}^2 \wedge \check{d}^3.$$

Acknowledgements. This paper has been partially supported by Ministry of Education of the Czech Republic under the Project MSM 143100009, Grant of the GA ČR No. 201/99/0296 (Czech Republic), Department of Applied Mathematics of University of Florence (Italy), Department of Mathematics of University of Mannheim (Germany), GNFM of INDAM (Italy), Project of cooperation N. 19/35 "Differential equation and differential geometry" between Czech Republic and Italy.

Marco Modugno would like to thank the organizers of the meeting for invitation and warm hospitality.

References

1. Jadczyk, A., Modugno, M. (1992): An outline of a new geometric approach to Galilei general relativistic quantum mechanics. in *Differential geometric methods in theoretical physics*, ed. by C. N. Yang, M. L. Ge and X. W. Zhou, World Scientific, Singapore, pp. 543–556
2. Jadczyk, A., Modugno, M. (1993): A scheme for Galilei general relativistic quantum mechanics, in *Proceedings of the* 10th *Italian Conference on general relativity and gravitational physics*, Bardonecchia, 1–5 September, ed. by M Cerdonio, R. D'Auria, M. Francariglia, G. Magnano, World Scientific, New York
3. Jadczyk, A., Modugno, M. (1994): Galilei general relativistic quantum mechanics. Report Dept. Appl. Math, Univ. of Florence, pp. 215
4. Abraham, R., Marsden, J.E. (1978): Foundations of Mechanics. 2nd ed., Benjamin-Cummings
5. Janyška, J. (1995): Remarks on symplectic and contact 2-forms in relativistic theories. Bollettino U.M.I. **7**, 9–B, 587–616
6. Janyška, J. (1995): Natural quantum Lagrangians in Galilei general relativistic quantum Lagrangians. Rendiconti di Matematica, S. VII, Vol. **15**, Roma, 457–468
7. Modugno, M., Vitolo, M. (1996): Quantum connection and Poincaré–Cartan form, in *Gravitation, electromagnetism and geometrical structures*, ed. by G. Ferrarese , Pitagora, Bologna, 237–279
8. Janyška, J., Modugno, M. (1996): Relations between linear connections on the tangent bundle and connections on the jet bundle of a fibred manifold. Arch. Math. (Brno), **32**, 281–288; http://www.emis.de/journals
9. Vitolo, R. (1996): Spherical symmetry in classical and quantum Galilei general relativity. Annales de l'Institut Henri Poincaré, **64**, (2), 177–203
10. Vitolo, R. (1996): Quantum structures in general relativistic theories. In *Proceedings of the XII Italian Conference on General Relativity and Gravitational Physics*, Roma, 1996; World Scientific, Singapore
11. Janyška, J., Modugno, M. (1999): On the graded Lie algebra of quantisable forms, in *Differential Geometry and Applications*, ed. by I. Kolář, O. Kowalski, D. Krupka, J. Slovak, Proceedings of the 7th International Conference, Brno, 10–14 August 1998, Masaryk University, 601–620
12. Vitolo, R. (1998): Quantising the rigid body. In: Proceedings of the VII Conference on Differential Geometry and Applications, Brno 1998, 653–664
13. Vitolo, R. (1999): Quantum structures in Galilei general relativity. Ann. Inst. 'H. Poinc. **70**, (3), 239–257
14. Janyška, J. (2001): A remark on natural quantum Lagrangians and natural generalized Schrödinger operators in Galilei quantum mechanics, in Proceedings of the 20th Winter *School of geometry and physics*, Srni, January 15–22, 2000, Supplemento ai rendiconti del Circolo Matematico di Palermo, Serie II, Numero **66**, pp. 117–128
15. Modugno, M., Tejero Prieto, C., Vitolo, R. (2000): Comparison between geometric quantisation and covariant quantum mechanic, in *Proceedings Lie Theory and Its Applications in Physics* – Lie III, 11–14 July 1999, Clausthal, Germany, ed. by H.-D. Doebner, V.K. Dobrev, J. Hilgert, World Scientific, Singapore, 155–175
16. Saller, D., Vitolo, R. (2000): Symmetries in covariant classical mechanics. J. Math. Phys. **41** (10), 6824–6842
17. Trautman, A. (1963): Sur la théorie Newtonienne de la gravitation. C. R. Acad. Sc. Paris t. **257**, 617–620

18. Trautman, A. (1966): Comparison of Newtonian and relativistic theories of space-time, in *Perspectives in geometry and relativity*, N. **42**, Indiana University press, 413–425
19. Dombrowski, H.D., Horneffer, K. (1964): Die Differentialgeometrie des Galileischen Relativitätsprinzips. Math. Z. **86**, 291–311
20. Duval, C. (1985): The Dirac & Levy-Leblond equations and geometric quantization, in *Diff. Geom. Meth. in Math. Phys.*, Proceedings of the 14th International Conference held in Salamanca, Spain, June 24–29, ed. by P.L. García, A. Pérez-Rendón, L.N.M. **1251**, Springer-Verlag, Berlin, pp. 205–221
21. Duval, C. (1993): On Galilean isometries. Clas. Quant. Grav. **10**, 2217–2221
22. Duval, C., Burdet, G., Künzle, H.P., Perrin, M. (1985): Bargmann structures and Newton–Cartan theory. Phys. Rev. D **31** (8), 1841–1853
23. Duval, C., Gibbons, G., Horvaty, P. (1991): Celestial mechanics, conformal structures, and gravitational waves. Phys. Rev. D **43** (12), 3907–3921
24. Duval, C., Künzle, H.P. (1984): Minimal gravitational coupling in the Newtonian theory and the covariant Schr'odinger equation. G.R.G. **16** (4), 333–347
25. Ehlers, J. (1989): The Newtonian limit of general relativity, in *Fisica matematica classica e relatività*, Rapporti e Compatiilitá, Elba 9-13 giugno 1989, pp. 95–106
26. Havas, P. (1964): Four-dimensional formulation of Newtonian mechanics and their relation to the special and general theory of relativity. Rev. Modern Phys. **36**, 938–965
27. Kuchař, K. (1980): Gravitation, geometry and nonrelativistic quantum theory. Phys. Rev. D **22** (6), 1285–1299
28. Künzle, H.P. (1972): Galilei and Lorentz structures on space-time: comparison of the corresponding geometry and physics. Ann. Inst. H. Poinc. **17** (4), 337–362
29. Künzle, H.P. (1974): Galilei and Lorentz invariance of classical particle interaction. Symposia Mathematica **14**, 53–84
30. Künzle, H.P. (1976): Covariant Newtonian limit of Lorentz space-times. G.R.G. **7** (5), 445–457
31. Künzle, H.P. (1984): General covariance and minimal gravitational coupling in Newtonian spacetime, in *Geometrodynamics Proceedings*, ed by A. Prastaro, Tecnoprint, Bologna, pp. 37–48
32. Künzle, H.P., Duval, C. (1984): Dirac field on Newtonian space-time. Ann. Inst. H. Poinc. **41** (4), 363–384
33. Le Bellac, M., Levy-Leblond, J.M. (1973): Galilean electromagnetism. Nuovo Cim. B **14** (2), 217–233
34. Levy-Leblond, J.M. (1971): Galilei group and Galilean invariance, in *Group theory and its applications*, ed. by E. M. Loebl, Vol. **2**, Academic, New York, pp. 221–299
35. Mangiarotti, L. (1979): Mechanics on a Galilean manifold. Riv. Mat. Univ. Parma **5** (4) , 1–14
36. Schmutzer, E., Plebanski, E. (1977): Quantum mechanics in non inertial frames of reference. Fortschritte der Physik **25**, 37–82
37. Tulczyjew W.M. (1981): Classical and quantum mechanics of particles in external gauge fields. Rend. Sem. Mat. Univ. Torino **39**, 111–124
38. Tulczyjew, W.M. (1985): An intrinsic formulation of nonrelativistic analytical mechanics and wawe mechanics. J. Geom. Phys. **2**, (3), 93–105
39. Kolář, I., Michor, P., Slovák, J. (1993): Natural operations in differential geometry. Springer-Verlag, Berlin
40. Libermann, P., Marle, Ch.M. (1987): Symplectic geometry and analytical mechanics. Reidel Publ., Dordrecht
41. Woodhouse N. (1992): Geometric quantization. Second Edn, Clarendon Press, Oxford

42. Janyška, J., Modugno, M. (1996): Classical particle phase space in general relativity, in Proc. Conf. Diff. Geom. Appl., Brno 28 August –1 September 1995, Masaryk University, 1996, ed. by J. anyška, I. Kolar, J. Slovak, pp. 573–602; http://www.emis.de/proceedings

43. Janyška, J. (1998): Natural Lagrangians for quantum structures over 4-dimensional spaces. Rendiconti di Matematica, S. VII, Vol. **18**, Roma, 623–648

44. Janyška, J., Modugno, M. (2000): Quantisable functions in general relativity, in *Opérateurs différentiels et Physique Mathématique*, ed. by J. Vaillant, J. Carvalho e Silva, Textos Mat. Ser. B, **24**, 161–181

45. Janyška, J., Modugno, M. (1997): On quantum vector fields in general relativistic quantum mechanics. General Mathematics **5**, Proceeding of the 3rd International Workshop on Differential Geometry and its Applications, Sibiu (Romania) 1997, 199–217

46. Jadczyk, A., Janyška, J., Modugno, M. (1998): Galilei general relativistic quantum mechanics revisited, in *Geometria, física-matemática e outros ensaios*, ed. by A.S. Alves, F.J. Craveiro de Carvalho, J.A. Pereira da Silva, Departamento de Matematica, Universidade de Coimbra, Coimbra, pp. 253–313

47. García, P.L. (1972): Connections and 1-jet fibre bundle. Rendic. Sem. Mat. Univ. Padova **47**, 227–242

48. Canarutto, D., Jadczyk, A., Modugno, M. (1995): Quantum mechanics of a spin particle in a curved spacetime with absolute time. Rep. on Math. Phys. **36**, 1, 95–140

The following additional references are useful for a comparison with the current literature:

Albert, C. (1989): Le théorème de reduction de Marsden–Weinstein en géométrie cosymplectique et de contact. J. Geom. Phys. **6** (4), 627–649

Balachandran, A.P., Gromm, H., Sorkin, R.D. (1987): Quantum symmetries from quantum phases. Fermions from Bosons, a Z_2 Anomaly and Galileian Invariance. Nucl. Phys. B **281**, 573–583

Cattaneo, V. (1970): Invariance Relativiste, Symetries Internes et Extensions d'Algébre de Lie. Thesis Université Catholique de Louvain

de Leon, M., Marrero, J.C., Padron, E. (1997): On the geometric quantization of Jacobi manifolds. J. Math. Phys. **38**, (12), 6185–6213

Fanchi, J.R. (1993): Review of invariant time formulations of relativistic quantum theories. Found. Phys. **23**, 487–548

Fanchi, J.R. (1994): Evaluating the validity of parametrized relativistic wave equations. Found. Phys. **24**, 543–562

Fernández, M., Ibañez, R., de Leon, M. (1996): Poisson cohomology and canonical cohomology of Poisson manifolds. Archivium Mathematicum (Brno) **32**, 29–56

Gotay, M.J. (1986): Constraints, reduction and quantization. J. Math. Phys. **27** (8), 2051–2066

Horwitz, L.P. (1992): On the definition and evolution of states in relativistic classical and quantum mechanics. Foun. Phys. **22**, 421–448

Horwitz, L.P., Rotbart, F.C. (1981): Non relativistic limit of relativistic quantum mechanics. Phys. Rev. D **24**, 2127–2131

Kyprianidis, A. (1987): Scalar time parametrization of relativistic quantum mechanics: The covariant Schr'odinger formalism. Phys. Rep **155**, 1–27

Marmo, G., Morandi, G., Simoni. A. (1988): Quasi-invariance and central extensions. Phys. Rev. D **37**, p. 2196–2206

Marsden, J.E., Ratiu, T. (1995): Introduction to mechanics and symmetry. Texts in Appl. Math. **17**, Springer, New York

Michel, L. (1965): Invariance in quantum mechanics and group extensions, in *Group Theoretical Concepts and Methods in Elememtary Particle Physics*, ed. by F. Gürsey, Gordon and Breach, New York

Peres, A. (1995): Relativistic quantum measurements, in *Fundamental problems of quantum theory*, Ann. N. Y. Acad. Sci., **755**

Piron, C., Reuse, F. (1979): On classical and quantum relativistic dynamics. Found. Phys. **9**, 865–882

Simms, D.J. (1968): Lie groups and quantum mechanics, in *Lect. Notes Math.*, Vol. **52**, Springer, Berlin Heidelberg New York

Simms, D.J., Woodhouse, N. (1977): Lectures on Geometric Quantization. Lect. Notes Phys. **53**, Springer, Berlin Heidelberg New York

Sniaticki, J. (1980): Geometric quantization and quantum mechanics. Springer–Verlag, New York

Tuynman, G.M., Wigerinck, W.A.J.J. (1987): Central extensions and physics. J. Geom. Phys. **4**,(3), 207–258

Anti-de Sitter Quantum Field Theory and the AdS-CFT Correspondence

U. Moschella

Abstract. We give a short account of a new approach to anti-de Sitter Quantum Field Theory that is based on the assumption of certain analyticity properties of the n-point correlation functions. We then discuss the application of this formalism to the construction of conformal field theories that are naturally obtained on the covering of the cone asymptotic to the AdS manifold, and that satisfy the axioms of Luscher an Mack.

1 Introduction

Anti-de Sitter (AdS) Quantum Field Theory (QFT) has recently come again to the general attention because of the by-now famous AdS/CFT (Conformal Field Theory) correspondence that conjectures a duality between type IIB superstring theory on $AdS_5 \times S_5$ and super Yang–Mills field theory on the flat four-dimensional boundary of AdS_5 (times the five sphere S_5) [1].

AdS QFT is rendered difficult by the existence of closed timelike curves and by the fact that the AdS manifold is not globally hyperbolic but has a boundary at spacelike infinity. While the first problem is easily solved by passing to the covering of the manifold, the second is typical either of the AdS manifold or of its covering and cannot be avoided. On the other hand, it is the existence of that boundary at spacelike infinity that allows the AdS/CFT correspondence to be formulated.

There two main approaches to AdS QFT: the first and older one is based on group-theoretical methods [2] following ideas that can be traced back to Dirac [3]. The main concern in the second and subsequent approach [4] to this subject has been to specify boundary conditions such that the difficulties arising by lack of global hyperbolicity could be circumvented. Both these approaches have naturally influenced very much the recent research on this subject. However, their applicability is more or less limited to free AdS QFT's.

We introduce a general framework for the study of AdS QFT which can be characterized in terms of locality, AdS covariance and a spectral condition formulated in terms of analyticity properties of the n-point functions; the latter can be interpreted as the energy spectrum condition and express more conveniently the boundary conditions one should impose to get a well-defined AdS QFT. A CFT on the boundary is naturally obtained by requiring a certain asymptotic behaviour for the n-point functions. The limiting procedure we will use directly constructs CFT's on the universal covering of the asymptotic cone of AdS space-time in the sense of Lüscher and Mack [5]; the conformal invariance of the corresponding Minkowskian (interacting) field theories on the boundary of AdS is thereby clearly exhibited without making use of any field equation, while field equations enter crucially in all other proposals to make effective the AdS/CFT conjecture [6–8]. For two-point functions, we are able to give a more complete treatment which is based on stronger analyticity properties [9] which are closely similar to that enjoyed by two-point functions in flat spacetime.

This paper is a short account of results that have been obtained in [9–11] together with M. Bertola, J. Bros, H. Epstein, V. Gorini and R. Schaeffer. I thank all of them for their collaboration and friendship.

2   Geometry of the AdS manifold

From a geometrical point of view, the AdS manifold is a simple example of curved space-time. Indeed, the AdS geometry is a solution of the cosmological Einstein's equations with the same degree of symmetry as the flat Minkowski solution. Apart from other physical reasons (i.e. its relations with supersymmetry and string theory, which recently culminated in the AdS/CFT correspondence), this is already a good explanation why there has been a lot of work on AdS QFT. Let us start by giving a short account of the AdS geometry. We consider the vector space $\mathbb{R}^{d+2}$ equipped with the following pseudo-scalar product:

$$X \cdot X' = X^0 X'^0 + X^{d+1} X'^{d+1} - X^1 X'^1 - \cdots - X^d X'^d. \tag{1}$$

The $(d+1)$-dimensional AdS universe can then be identified with the manifold $\mathrm{AdS}_{d+1} = \{X \in \mathbb{R}^{d+2}, \ X^2 = 1\}$, where $X^2 = X \cdot X$, endowed with the induced metric. The AdS relativity group is $G = SO_0(2, d)$. Two events X, X' of AdS_{d+1} are space-like separated if $(X - X')^2 < 0$, i.e. if $X \cdot X' > 1$. The complexification of AdS_{d+1}: $\mathrm{AdS}_{d+1}^{(c)} = \{Z = X + iY \in \mathbb{C}^{d+2}, \ Z^2 = 1\}$ will be also crucially used.

There are two parametrization of the AdS manifold which are relevant to us. The *"covering parametrization"* $X = X[r, \tau, e]$: it is obtained by intersecting AdS_{d+1} with the cylinders with equation $\{X^{0^2} + X^{d+1^2} = r^2 + 1\}$, and is given by

$$\begin{cases} X^0 = \sqrt{r^2 + 1} \sin \tau, \\ X^i = r \, e^i, & i = 1, \dots, d \\ X^{d+1} = \sqrt{r^2 + 1} \cos \tau, \end{cases} \tag{2}$$

with $e^2 \equiv e^{1^2} + \dots + e^{d^2} = 1$ and $r \geq 0$. For each fixed value of r, the corresponding "slice"

$$C_r = \mathrm{AdS}_{d+1} \cap \{X^{0^2} + X^{d+1^2} = r^2 + 1\}, \tag{3}$$

of AdS_{d+1} is a Lorentz manifold $\mathbb{S}_1 \times \mathbb{S}_{d-1}$. The complexified space $\mathrm{AdS}_{d+1}^{(c)}$ is obtained by giving arbitrary complex values to r, τ and to the coordinates $e = (e^i)$ on the unit $(d-1)$-sphere.

The parametrization (2) allows one to introduce relevant coverings of AdS_{d+1} and $\mathrm{AdS}_{d+1}^{(c)}$ by unfolding the 2π-periodic coordinate τ (resp. $\Re\tau$), interpreted as a time-parameter: these coverings are denoted respectively by $\widehat{\mathrm{AdS}}_{d+1}$ and $\widehat{\mathrm{AdS}}^{(c)}_{d+1}$. A privileged "fundamental sheet" is defined on these coverings by imposing the condition $-\pi < \Re\tau < \pi$. Similarly one introduces a covering $\hat{G}$ of the group G. By transitivity, AdS_{d+1} and $\widehat{\mathrm{AdS}}_{d+1}$ are respectively generated by the action of G and $\hat{G}$ on the base point $B = (0, \ldots, 0, 1)$. The notion of space-like separation in $\widehat{\mathrm{AdS}}_{d+1}$ can be specified as follows: let $X, X' \in \widehat{\mathrm{AdS}}_{d+1}$ and let g an element of $\hat{G}$ such that $X' = gB$; define $X_g = g^{-1}X$. X and X' are spacelike separated if X_g is in the fundamental sheet of $\widehat{\mathrm{AdS}}_{d+1}$ and $(X - X')^2 \equiv (g^{-1}X - g^{-1}X')^2 < 0$. This implies that $X_g = X_g[r, \tau, e]$ with $-\pi < \tau < \pi$ and $\sqrt{r^2 + 1}\cos\tau > 1$. It is also interesting to note that on each manifold C_r the condition of space-like separation between two points $X = X[r, \tau, e]$ and $X' = X'[r, \tau', e']$ reads (in view of (2):

$$(X - X')^2 = 2(r^2 + 1)(1 - \cos(\tau - \tau')) - r^2(e - e')^2 < 0, \tag{4}$$

and that the corresponding covering manifold $\hat{C}_r$ therefore admits a global causal ordering which is specified as follows:

$$(\tau, e) > (\tau', e') \quad \text{iff} \quad \tau - \tau' > 2\,\mathrm{Arcsin}\left(\frac{(e - e')^2}{4}\frac{r^2}{r^2 + 1}\right)^{\frac{1}{2}}. \tag{5}$$

The *"Poincaré parametrization"* $X = X(v, x)$: it only covers the part Π of the AdS manifold which belongs to the half-space $\{X^d + X^{d+1} > 0\}$ of the ambient space and is obtained by intersecting AdS_{d+1} with the hyperplanes $\{X^d + X^{d+1} = e^v\}$, each slice Π_v (or "horosphere") being an hyperbolic paraboloid:

$$\begin{cases} X^\mu = e^v x^\mu, & \mu = 0, 1, \ldots, d - 1 \\ X^d = \sinh v + \frac{1}{2}e^v x^2, \\ X^{d+1} = \cosh v - \frac{1}{2}e^v x^2, \end{cases} \tag{6}$$

In each slice Π_v, $x^0, \ldots, x^{d-1}$ can be seen as coordinates of an event of a d-dimensional Minkowski spacetime $\mathbb{M}^d$ with metric $ds_M^2 = dx^{0\,2} - dx^{1\,2} - \cdots - dx^{d-1\,2}$ and $x^2 = x^{0\,2} - x^{1\,2} - \cdots - x^{d-1\,2}$ (here and in the following where it appears, an index M stands for Minkowski). The scalar product (1) and the AdS metric can then be rewritten as follows:

$$X \cdot X' = \cosh(v - v') - \frac{1}{2}e^{v+v'}\left(x - x'\right)^2, \quad ds_{\mathrm{AdS}}^2 = e^{2v}ds_M^2 - dv^2. \tag{7}$$

Equation (7) implies that $(X(v, x) - X(v, x'))^2 = e^{2v}(x - x')^2$. This in turn implies that space-like separation in any slice Π_v can be understood equivalently in the Minkowskian sense of the slice itself or in the sense of the ambient AdS universe.

The *Euclidean submanifold* E_{d+1} of $\widehat{\mathrm{AdS}}^{(c)}_{d+1}$ is the set of all points $Z = X + iY$ in $\widehat{\mathrm{AdS}}^{(c)}_{d+1}$ such that $X = (0, X^1, \ldots, X^{d+1})$, $Y = (Y^0, 0, \ldots, 0)$ and $X^{d+1} > 0$. It is therefore represented by the upper sheet (characterized by the condition $X^{d+1} > 0$)

of the two-sheeted hyperboloid with equation $X^{d+1^2} - Y^{0^2} - X^{1^2} - \cdots - X^{d^2} = 1$. E_{d+1} is equally well represented in both parametrizations (2) and (6) as follows:

$$Z = Z[r, \tau = i\sigma, e]; \quad (r, \sigma, e) \in \mathbb{R} \times \mathbb{R} \times \mathbb{S}_{d-1}, \tag{8}$$

or

$$Z = Z(v, (iy^0, x^1, \ldots, x^{d-1})); \quad v \in \mathbb{R}, \ (y^0, x^1, \ldots, x^{d-1})) \in \mathbb{R}^d. \tag{9}$$

In view of (12), E_{d+1} is contained in the fundamental sheet of $\widehat{\mathrm{AdS}}_{d+1}^{(c)}$.

3 Quantum field theory

Let us consider now a free AdS QFT. Canonical quantization of fields on curved space-times goes as follows. Given a field equation, say the Klein–Gordon equation, one introduces the scalar product

$$(\varphi_1, \varphi_2) = -i \int_\Sigma \bar\varphi_1(x) \overleftrightarrow{\partial}_\mu \varphi_2(x) d\Sigma^\mu, \tag{10}$$

in the space of (classical) solutions of the equation, where Σ is a spacelike Cauchy hypersurface and $d\Sigma$ is the associated volume element, and looks for a complete set (in the sense of the given scalar product) of mode solutions $u_i(x)$. The field φ is then given by mode expansion

$$\varphi(x) = \sum_i [a_i u_i(x) + a_i^\dagger \bar u_i(x)], \tag{11}$$

and canonical quantization is achieved by assuming the commutation rules (CCR) $[a_i, a_j^\dagger] = \delta_{ij}$, $[a_i, a_j] = [a_i^\dagger, a_j^\dagger] = 0$ and by choosing the corresponding vacuum.

The ambiguity inherent the quantization of fields on a gravitational background appears here clearly: in fact the previous mode expansion is generally based on an arbitrary choice of local coordinates (which may or may not extend to the whole space). Moreover it is in general impossible to characterize the physically relevant vacuum states as the fundamental states for the energy in the usual sense and the analogue of a spectral condition is in general lacking. For the AdS case the situation is even worse: the lack of global hyperbolicity renders the procedure useless from the start, since there does not exist a global Cauchy surface and information can enter by spacelike infinity[1].

We will now describe an approach based on the properties of the analytic continuation of correlation functions that is closely similar to what one does in Minkowski [12] and de Sitter [13] QFT's and that gives solution to both problems.

---

[1] The brilliant idea in [4] was to use the conformal embedding of the AdS manifold in the Einstein Static Universe which is globally hyperbolic. Then it has been possible for certain values of the field's mass to pull back a well defined QFT on the ESU produce the boundary conditions that make the resulting QFT well defined. Unfortunately, the procedure is very special and tricky and can work at most for free field theories.

3.1 Wightman functions

We consider a general scalar QFT on $\widehat{\mathrm{AdS}}_{d+1}$. According to the general reconstruction procedure [12], a theory is completely determined by the set of all n-point vacuum expectation values of the field Φ, given as distributions on the corresponding product manifolds $(\widehat{\mathrm{AdS}}_{d+1})^n$: $\mathcal{W}_n(X_1, \ldots X_n) = \langle \Omega, \Phi(X_1) \ldots \Phi(X_n)\Omega \rangle$. These distributions are supposed to be tempered when represented in the variables of the covering parametrization $X_j = X_j[r_j, \tau_j, e_j]$ and to satisfy the following general requirements:

AdS invariance:

$$\mathcal{W}_n(gX_1, \ldots gX_n) = \mathcal{W}_n(X_1, \ldots X_n), \quad \text{for any } g \in \hat{G}. \tag{12}$$

Local commutativity:

$$\mathcal{W}_n(X_1, \ldots X_j, X_{j+1}, \ldots, X_n) = \mathcal{W}_n(X_1, \ldots X_{j+1}, X_j, \ldots, X_n), \tag{13}$$

for X_j, X_{j+1} space-like separated in the sense of the covering space $\widehat{\mathrm{AdS}}_{d+1}$, as defined above.

The usual *positive-definiteness* and *hermiticity* properties are formulated as in any spacetime [12]. Let us discuss now the physical content of the theory. The infinitesimal generator $J_{0,d+1}$ of the covering of the subgroup of the rotations in the plane $(0, d + 1)$ can be interpreted as the generator of time translations. Therefore it is natural to assume that it be represented by a self-adjoint operator whose spectrum is bounded from below. By using the standard Laplace transform argument [12] in the corresponding time-variables $\tau_1, \ldots, \tau_n$, one is led to formulate this spectral condition by the following analyticity property of the Wightman functions:

Spectral condition: each distribution $\mathcal{W}_n(X_1[r_1, \tau_1, e_1], \ldots, X_n[r_n, \tau_n, e_n])$ is the boundary value of a holomorphic function $W_n(Z_1, \ldots, Z_n)$ which is defined in a complex neighborhood of the set $\{Z = (Z_1, \ldots, Z_n); Z_j = X_j + iY_j \in \widehat{\mathrm{AdS}}_{d+1}^{(c)}; Z_j = Z_j[r_j, \tau_j, e_j]; \Im\tau_1 < \Im\tau_2 < \cdots < \Im\tau_n\}$.

As a by-product, the Schwinger function S_n, that is the restriction of each W_n to the Euclidean submanifold $\{(Z_1, \ldots, Z_n) \in (E_{d+1})^n; \sigma_1 < \sigma_2 < \cdots < \sigma_n\}$, is well-defined. For a more refined and satisfactory formulation of the spectral condition see [9].

4 AdS/CFT correspondence

4.1 Dimensional boundary condition. Fields restricted to $\hat{C}_r$

In order to obtain relevant QFT's on the boundary of the AdS spacetime we are led to assume the following power-decrease at infinity for the Wightman functions (expressed with the help of the coordinates (2)):

208   U. Moschella

Dimensional boundary condition at infinity: a scalar QFT on $\widehat{\text{AdS}}_{d+1}$ is said to be of asymptotic dimension Δ if the following limits exist in the sense of distributions:

$$\lim_{\min(r_1,\dots,r_n)\to+\infty} (r_1\cdots r_n)^{\Delta} \mathcal{W}_n(X_1[r_1,\tau_1,\underset{1}{\text{e}}],\dots,X_n[r_n,\tau_n,\underset{n}{\text{e}}])$$

$$= \mathcal{W}_n^{\infty}\left([\tau_1,\underset{1}{\text{e}}],\dots,[\tau_n,\text{e}_n]\right). \quad (14)$$

This condition calls for the following comment. Generally speaking distributions cannot be restricted to lower dimensional manifold. In our framework the spectral condition implies that the above condition is well-defined. In fact, for each fixed $r_1,\dots,r_n$ and $\text{e}_1,\dots,\text{e}_n$, the existence of an analytic continuation W_n of $\mathcal{W}_n$ in the variables $\tau_1,\dots,\tau_n$ of the covering parametrization (2) in the tube domain $T_n = \{(\tau_1,\dots,\tau_n);\ \Im\tau_1 < \Im\tau_2 < \cdots < \Im\tau_n\}$ implies that the boundary value of W_n on the reals from this tube is a distribution in the variables $\tau_1,\dots,\tau_n$ on each submanifold obtained by fixing all the parameters r_j and e_j and that it is even a regular function of all these parameters. The limit in Eq. (14) is therefore also defined as a distribution in the variables $\tau_1,\dots,\tau_n$ with C^{∞} dependence with respect to the variables e_j. It is therefore also meaningful to consider the restrictions of the distributions $\mathcal{W}_n$ to the submanifolds $\left(\hat{C}_r\right)^n$ of $\left(\widehat{\text{AdS}}_{d+1}\right)^n$ (i.e. to the case when all variables r_j are equal to r). One then notices that the positivity conditions satisfied by assumption by the distributions $\mathcal{W}_n$ on $\widehat{\text{AdS}}_{d+1}$ can be extended to test-functions of the variables τ_j and e_j localized in these submanifolds $r_1 = \cdots = r_n = r$. In view of the standard reconstruction procedure [12], this allows one to say that in each slice $\hat{C}_r$ the given field on $\widehat{\text{AdS}}_{d+1}$ yields by restriction a well-defined quantum field $\Phi_r(\tau,\text{e})$. This field is obviously invariant under the product of the translation group with time-parameter τ by the orthogonal group $SO(d)$ of space transformations acting on the sphere $\mathbb{S}_{d-1}$ of the variables e. Moreover, it follows from the local commutativity postulate together with Eqs. (4) and (5) that the field Φ_r also satisfies local commutativity in the sense of the spacetime manifold $\hat{C}_r$. Finally, in view of the spectral condition, the n-point functions of Φ_r are (for each r) boundary values of holomorphic functions of the complex variables $\tau_1,\dots,\tau_n$ in the tube T_n, which shows that these theories satisfy a spectral condition with respect to the generator of time-translations.

4.2 Correspondence with conformal field theories on $\widehat{C}_{2,d}$ à la Lüscher–Mack

In a basic work by Lüscher and Mack [5] the concept of global conformal invariance in Minkowskian QFT has been associated in a deep and fruitful way with the general framework of QFT on the covering of a quadratic cone with signature $(+,+,-,\dots,-)$ in one dimension more. Since such a cone is precisely the asymptotic cone of the AdS quadric, it seems quite appropriate to try to formulate our general AdS/CFT correspondence in a way which exhibits as clearly as possible the connection between the previous AdS QFT framework CFT on the cone. Let us introduce therefore the asymptotic cone $C_{2,d}$ (resp. $C_{2,d}^{(c)}$) of AdS_{d+1} (resp. $\text{AdS}_{d+1}^{(c)}$). To do this, we first notice that by adapting the covering parametrization (2) of $\widehat{\mathrm{AdS}}_{d+1}$ to the case of its asymptotic cone, $C_{2,d} = \{\eta = (\eta^0, \ldots, \eta^{(d+1)}); \; \eta^{0^2} - \eta^{1^2} - \cdots - \eta^{d^2} + \eta^{d+1^2} = 0\}$, one readily obtains the following parametrization (with the same notations as [5], but in dimension $d + 2$):

$$\begin{cases} \eta^0 = r \sin \tau, \\ \eta^i = r\, \mathrm{e}^i, & i = 1, \ldots, d, \\ \eta^{d+1} = r \cos \tau, \end{cases} \tag{15}$$

with $\mathrm{e}^{1^2} + \cdots + \mathrm{e}^{d^2} = 1$ and $r \geq 0$, or in brief: $\eta = \eta[r, \tau, \mathrm{e}]$.

The parametrization (15) allows one to introduce the coverings $\widehat{C}_{2,d}$ and $\widehat{C}^{(c)}_{2,d}$ of $C_{2,d}$ and $C^{(c)}_{2,d}$ by again unfolding the 2π-periodic coordinate τ (resp. $\Re\tau$). A privileged "fundamental sheet" is defined on these coverings by imposing the condition $-\pi < \tau < \pi$ (resp. $-\pi < \Re\tau < \pi$).

We also note that the standard condition of space-like separation on $C_{2,d}$ is similar to the condition chosen on the AdS spacetime, namely

$$(\eta - \eta')^2 = r^2 \left[4 \left(\sin \left(\frac{\tau - \tau'}{2} \right) \right)^2 - (\mathrm{e} - \mathrm{e}')^2 \right] < 0, \tag{16}$$

and yields the corresponding global causal ordering on $\widehat{C}_{2,d}$

$$(\tau, \mathrm{e}) > (\tau', \mathrm{e}) \quad \text{iff} \quad \tau - \tau' > 2\mathrm{Arcsin} \left(\frac{(\mathrm{e} - \mathrm{e}')^2}{4} \right)^{\frac{1}{2}}, \tag{17}$$

equivalently written e.g. in [5] as $\tau - \tau' > \mathrm{Arccos}(\mathrm{e} \cdot \mathrm{e}')$. Note that in the space of variables $(\tau, \tau', \mathrm{e}, \mathrm{e}')$, the region described by Eq. (17) is exactly the limit of the region given by Eq. (5) when r tends to infinity. By taking the intersection of $C_{2,d}$ with the family of hyperplanes with equation $\eta^d + \eta^{d+1} = \mathrm{e}^v$, one obtains the analogue of the horocyclic parametrization (6), namely:

$$\begin{cases} \eta^\mu = \mathrm{e}^v x^\mu, & \mu = 0, 1, \ldots, d - 1, \\ \eta^d = \frac{1}{2}\mathrm{e}^v (1 + x^2), \\ \eta^{d+1} = \frac{1}{2}\mathrm{e}^v (1 - x^2), \end{cases} \tag{18}$$

which implies the following identity:

$$(\eta - \eta')^2 = \mathrm{e}^{v+v'} (x - x')^2. \tag{19}$$

By taking Eqs. (15) into account, one then sees that these formulae correspond (in dimension d) to the embedding of Minkowski space into the covering of the cone $C_{2,d}$ (see [14] and references therein), namely one has:

$$x^0 = \frac{\sin \tau}{\cos \tau + \mathrm{e}^d}, \quad x^i = \frac{\mathrm{e}^i}{\cos \tau + \mathrm{e}^d}, \tag{20}$$

with $\cos \tau + \mathrm{e}^d > 0$, $-\pi < \tau < \pi$.

Let us now consider a general QFT on $\widehat{\mathrm{AdS}}_{d+1}$ whose Wightman functions $\mathcal{W}_n$ satisfy AdS invariance together with the other properties described in the previous section. In view of Eq. (14) we can associate with the latter the following set of n-point distributions $\widetilde{\mathcal{W}}_n(\eta_1, \ldots, \eta_n)$ on $\widehat{\mathcal{C}}_{2,d}$:

$$\widetilde{\mathcal{W}}_n(\eta_1, \ldots, \eta_n) = (r_1 \cdots r_n)^{-\Delta} \mathcal{W}_n^{\infty}([\tau_1, \mathrm{e}_1], \ldots, [\tau_n, \mathrm{e}_n]). \tag{21}$$

First of all the set of distributions $\widetilde{\mathcal{W}}_n$ satisfy the required positive-definiteness condition for defining a QFT on $\widehat{\mathcal{C}}_{2,d}$. This is because the distributions $\mathcal{W}_n^{\infty}$ appear as the limits of the n-point functions of the QFT's on the spacetimes $\hat{C}_r$ when r tends to infinity. The positivity conditions satisfied by the latter are then preserved in the limit. It follows from the reconstruction procedure [12] that the set of distributions $\widetilde{\mathcal{W}}_n$ define a quantum field $\tilde{O}(\eta)$ on $\widehat{\mathcal{C}}_{2,d}$. $\tilde{O}(\eta)$ enjoys the following properties:

Local commutativity: Since the region (17) is the limit of (5) for r tending to infinity, it follows from the dimensional boundary condition and from the local commutativity of all fields Φ_r in the corresponding spacetimes $\hat{C}_r$ that the field $\tilde{O}(\eta)$ satisfies local commutativity on $\widehat{\mathcal{C}}_{2,d}$.

Spectral condition: In view of Eq. (14) extended to the complex domain T_n in the variables τ, we see that the n-point distributions $\widetilde{\mathcal{W}}_n(\eta_1, \ldots, \eta_n)$ are boundary values of holomorphic functions in the same analyticity domains of $\left(\widehat{\mathcal{C}}_{2,d}^{(c)}\right)^n$ as those of the Lüscher–Mack field theories [5].

The important result on which our AdS/CFT correspondence is based is that $\hat{G}$-invariance (12) of the AdS n-point functions, together with the other assumption we made imply the conformal invariance of the field $\tilde{O}(\eta)$; more precisely, the Wightman functions $\widetilde{\mathcal{W}}_n$ of this field are invariant under the action on $\widehat{\mathcal{C}}_{2,d}$ of the group $\hat{G}$, now interpreted as in [5] as the "quantum mechanical conformal group", namely that one has:

$$\widetilde{\mathcal{W}}_n(g\eta_1, \ldots, g\eta_n) = \widetilde{\mathcal{W}}_n(\eta_1, \ldots, \eta_n), \tag{22}$$

for all g in $\hat{G}$. A part of this invariance is trivial in view of how the limiting procedure is constructed: it is the invariance under the rotations in the $(0, d+1)$-plane (i.e. the translations in the time variables τ) and the invariance under the spatial orthogonal group of the subspace of variables $(\eta^1, \ldots, \eta^d)$ (acting on the sphere $\mathbb{S}_{d-1}$). For the proof of the nontrivial part we refer the reader to [10]. We can then summarize the main result of this section by the following statement:

Equations (14) and (21) display a general AdS/CFT correspondence for QFT's:

$$\Phi(X) \to \tilde{O}(\eta), \tag{23}$$

between a scalar quantum field $\Phi(X)$ on the covering $\widehat{\mathrm{AdS}}_{d+1}$ of AdS_{d+1} that is local and AdS invariant, they satisfy the spectral condition and the dimensional boundary *condition and a conformally invariant local field $\tilde{\mathcal{O}}(\eta)$ on the covering $\widehat{\mathcal{C}}_{2,d}$ of the cone $\mathcal{C}_{2,d}$, enjoying the Lüscher–Mack spectral condition; the degree of homogeneity (dimension) Δ of $\tilde{\mathcal{O}}(\eta)$ is equal to the asymptotic dimension of the AdS field $\Phi(X)$.*

Of course, from this general point of view, the correspondence may a priori be many-to-one. Finally, according to the formalism described in [5,14], the correspondence (23) can be completed by saying that there exists a unique conformal (Minkowskian) local field $\mathcal{O}(x)$ of dimension Δ whose n-point functions $\mathcal{W}_n^M$ are expressed in terms of those of $\tilde{\mathcal{O}}(\eta)$ by the following formulae:

$$\mathcal{W}_n^M(x_1, \ldots, x_n) = e^{(v_1 + \cdots + v_n)\Delta}\, \widetilde{\mathcal{W}}_n(\eta_1, \ldots, \eta_n)$$
$$= \Pi_{1 \le j \le n}(\eta_j^d + \eta_j^{d+1})^{\Delta}\, \widetilde{\mathcal{W}}_n(\eta_1, \ldots, \eta_n). \tag{24}$$

In this equation the Minkowskian variables x_j are expressed in terms of the cone variables η_j by inverting (18):

$$x_j^{\mu} = \frac{\eta_j^{\mu}}{\eta_j^d + \eta_j^{d+1}}. \tag{25}$$

5   Two-point functions

For two-point functions we are able to give a complete characterization of the spectral condition. There are indeed two distinguished complex domains [9] of $\mathrm{AdS}_{d+1}^{(c)}$, invariant under real AdS transformations, which are of crucial importance for a full understanding of the structures associated with two-point functions. They are given by

$$T^+ = \{Z = X + iY \in \mathrm{AdS}_{d+1}^{(c)};\ Y^2 > 0,\ \epsilon(Z) = +1\},$$
$$T^- = \{Z = X + iY \in \mathrm{AdS}_{d+1}^{(c)};\ Y^2 > 0,\ \epsilon(Z) = -1\}, \tag{26}$$

where

$$\epsilon(Z) = \mathrm{sign}(Y^0 X^{d+1} - X^0 Y^{d+1}). \tag{27}$$

T^+ and T^- are the AdS version of the usual forward and backward tubes T_M^+ and T_M^- of complex Minkowski spacetime, obtained in correspondence with the energy-momentum spectrum condition [12]; let us recall their definition (in arbitrary space-time dimension):

$$T_M^+ = \{z = x + iy \in \mathrm{M}^{(c)};\ y^2 > 0,\ y^0 > 0\},$$
$$T_M^- = \{z = x + iy \in \mathrm{M}^{(c)};\ y^2 > 0,\ y^0 < 0\}. \tag{28}$$

To understand the relation with the previous spectral condition we remark that, in the same way as these Minkowskian tubes are generated by the action of real Lorentz transcformations on the "flat" (one complex time-variable) domains $\{z = x + iy;\ y = (y^0, \mathbf{0});\ y^0 > 0$ (resp. $y^0 < 0)\}$, the domains (26) of $\mathrm{AdS}^{(c)}_{d+1}$ are generated by the action of the group G on the flat domains obtained by letting τ vary in the half-planes $\Im\tau > 0$ or $\Im\tau < 0$ and keeping r and e real in the covering parametrization (2) of the AdS quadric. We denote by $\hat{T}^+$ and $\hat{T}^-$ the covering of T^+ and T^-. It follows that AdS invariance and the spectral condition together give rise to the following

Normal analyticity condition for two-point functions: The two-point function $W(X, X')$ is the boundary value of a function $W(Z, Z')$ which is holomorphic in the domain $\hat{T}^- \times \hat{T}^+$ of $\widehat{\mathrm{AdS}}^{(c)}_{d+1} \times \widehat{\mathrm{AdS}}^{(c)}_{d+1}$.

A further use of AdS invariance implies that $W(Z, Z')$ is actually a function $w(\zeta)$ of a single complex variable ζ; this variable ζ can be identified with $Z \cdot Z'$ when Z and Z' are both in the fundamental sheet of $\widehat{\mathrm{AdS}}^{(c)}_{d+1}$; AdS invariance and the normal analyticity condition together imply the following

Maximal analyticity property: $w(\zeta)$ is analytic in the covering $\widehat{\Theta}$ of the cut-plane $\Theta = \{\mathbb{C} \setminus [-1, 1]\}$.

For special theories which are periodic in the time coordinate τ, $w(\zeta)$ is in fact analytic in Θ itself. One can now introduce all the usual Green functions. The "permuted Wightman function" $\mathcal{W}(X', X) = \langle \Omega, \Phi(X')\Phi(X)\Omega \rangle$ is the boundary value of $W(Z, Z')$ from the domain $\{(Z, Z') : Z \in \hat{T}^+,\ Z' \in \hat{T}^-\}$. The commutator function is then $\mathcal{C}(X, X') = \mathcal{W}(X, X') - \mathcal{W}(X', X)$. The retarded propagator $\mathcal{R}(X, X')$ is introduced by splitting the support of the commutator $\mathcal{C}(X, X')$ as follows

$$\mathcal{R}(X, X') = i\theta(\tau - \tau')\mathcal{C}(X, X'). \tag{29}$$

The other Green functions are then defined in terms of $\mathcal{R}$ by the usual formulae: the advanced propagator is given by $\mathcal{A} = \mathcal{R} - i\mathcal{C}$ while the chronological propagator is given by $\mathcal{F} = -i\mathcal{A} + \mathcal{W}$.

Note finally that, as a function of the single variable $\zeta = X \cdot X'$, the jump $i\delta w(\zeta)$ of $iw(\zeta)$ across its cut $(-\infty, +1]$ coincides with the retarded propagator $R(X, X')$ (or the advanced one); in the periodic (i.e. "true AdS") case, the support of δw reduces to the compact interval $[-1, +1]$.

5.1 Klein–Gordon fields and the AdS/CFT correspondence

The Wightman functions of fields satisfying the Klein–Gordon equation AdS_{d+1}

$$\Box_{\mathrm{AdS}}\Phi + m^2\Phi = 0, \tag{30}$$

display the simplest example of the previous analytic structure:

$$W_\nu(Z, Z') = w_\nu(\zeta) = \frac{e^{-i\pi\frac{d-1}{2}}}{(2\pi)^{\frac{d+1}{2}}}(\zeta^2 - 1)^{-\frac{d-1}{4}}\,Q^{\frac{d-1}{2}}_{\nu-\frac{1}{2}}(\zeta). \tag{31}$$

Here Q is a second-kind Legendre's function [15]; the parameter ν is linked to the field's mass by the relation

$$\nu^2 = \frac{d^2}{4} + m^2, \tag{32}$$

and the normalization of W_ν is chosen by the local Hadamard behavior. Since $W_\nu(Z, Z')$ and $W_{-\nu}(Z, Z')$ are solutions of the same Klein–Gordon equation (and share the same analyticity properties), the question arises if these Wightman function both define acceptable QFT's on AdS_{d+1}. The answer [16] is that only theories with $\nu \geq -1$ are acceptable and there are therefore two regimes: for $\nu > 1$ there is only one field theory corresponding to a given mass while for $|\nu| < 1$ there are two theories. The case $\nu = 1$ is a limit case. Equation (31) shows clearly that the only difference between the theories parametrized by opposite values of ν is in their large distance behavior. More precisely, in view of Eq. (3.3.1.4) of [15], we can write:

$$w_{-\nu}(\zeta) = w_\nu(\zeta)$$
$$+ \frac{\sin \pi \nu}{(2\pi)^{\frac{d+1}{2}}} \Gamma\left(\frac{d}{2} - \nu\right) \Gamma\left(\frac{d}{2} + \nu\right) (\zeta^2 - 1)^{-\frac{d-1}{4}} P_{-\frac{1}{2}-\nu}^{-\frac{d-1}{2}}(\zeta). \tag{33}$$

Now we notice that in this relation (where all terms are solutions of the same Klein–Gordon equation) the last term is *regular on the cut* $\zeta \in [-1, 1]$. This entails that, in the two theories, the c-number commutator $[\Phi(X), \Phi(X')]$ takes the same value for all (time-like separated) vectors (X, X') such that $|X \cdot X'| < 1$. Therefore we can say that *the two theories represent the same algebra of local observables at short distances (with respect to the radius R)*. But since the last term in the latter relation grows the faster the larger is $|\nu|$ (see [15], Eqs. (3.9.2)), we see that the two theories drastically differ by their long range behaviors.

The existence of the two regimes above has given rise to two distinct treatments of the AdS/CFT correspondence in the two cases [17] and symmetry breaking had been advocated to explain the difference.

In the present context, by applying the correspondence as given in Eq. (23), the two regimes can be treated in one stroke. Indeed, Eq. (3.9.2.21) of [15] reports the following large ζ behavior of the Legendre's function Q (valid for any complex ν):

$$Q_{\nu-\frac{1}{2}}^{\frac{d-1}{2}}(\zeta) \simeq e^{i\pi \frac{d-1}{2}} 2^{-\nu-\frac{1}{2}} \frac{\Gamma\left(\nu + \frac{d}{2}\right)}{\Gamma(\nu + 1)} \pi^{\frac{1}{2}} \zeta^{-\frac{1}{2}-\nu}. \tag{34}$$

It follows that the two-point function (31) and thereby all the n-point functions of the corresponding Klein–Gordon field satisfy the dimensional boundary conditions at infinity with dimension $\Delta = \frac{d}{2} + \nu$. Indeed, let τ and τ' be complex and such that $\Im\tau < \Im\tau'$. It follows that

$$W_\nu^\infty([\tau, e], [\tau', e']) = \lim_{r, r' \to \infty} (rr')^{\frac{d}{2}+\nu} W_\nu(Z[\tau, r, e], Z'[\tau', r', e'])$$
$$= \frac{2^{-\nu-1}}{(2\pi)^{\frac{d}{2}}} \frac{\Gamma(\nu + \frac{d}{2})}{\Gamma(\nu + 1)} \frac{1}{[\cos(\tau - \tau') - e \cdot e']^{\frac{d}{2}+\nu}}. \tag{35}$$

(see also [18]). This equation expresses nothing more than the behavior of the previous Legendre's function at infinity. Not only all the ν's are treated this way in one stroke but, also, one can study the boundary limit for theories corresponding to $\nu < -1$, even if the corresponding QFT may have no direct physical interpretation.

The two-point function of the conformal field $\tilde{\mathcal{O}}(\eta)$ on the cone $\widehat{\mathcal{C}}_{2,d}$ corresponding to (35) is then constructed by following the prescription of Eq. (21), which yields

$$
\begin{aligned}
\widetilde{W}_\nu(\eta, \eta') &= (rr')^{-\frac{d}{2}-\nu} W_\nu^\infty([\tau, e], [\tau', \overset{\prime}{e}]) \\
&= \frac{1}{2\pi^{\frac{d}{2}}} \frac{\Gamma(\nu + \frac{d}{2})}{\Gamma(\nu + 1)} \frac{1}{[-(\eta - \eta')^2]^{\frac{d}{2}+\nu}}.
\end{aligned}
\tag{36}
$$

Correspondingly, we can deduce from (36) the expression of the two-point function of the associated Minkowskian field on $\mathbb{M}^d$, given by formula (24); by taking Eq. (19) into account, we obtain:

$$
\begin{aligned}
W_\nu^M(z, z') &= e^{(\nu+\nu')(\frac{d}{2}+\nu)} \widetilde{W}_\nu\left(\eta\left(v, z\right), \eta'\left(v', z'\right)\right) \\
&= \frac{1}{2\pi^{\frac{d}{2}}} \frac{\Gamma(\nu + \frac{d}{2})}{\Gamma(\nu + 1)} \frac{1}{[-(z - z')^2]^{\frac{d}{2}+\nu}}.
\end{aligned}
\tag{37}
$$

In the latter, the Poincaré coordinates z and z' must be taken with the usual $i\epsilon$-prescription ($\Im z^0 < \Im z'^0$), which can be checked to be implied by the spectral condition b) of Sect. 2 through the previous limiting procedure.

Let us now describe how the previous limiting procedure looks like in the Poincaré coordinates (6) (see also [11]). These coordinates offer the possibility of studying directly the boundary behavior of the AdS Wightman functions in a larger domain of the complex AdS spacetime. This fact is based on the following simple observation: consider the parametrization (6) for two points with complex parameters specified by

$$
\begin{aligned}
Z &= Z(v, z), &\quad v \in \mathbb{R}, \ z \in T_M^-, \\
Z' &= Z'(v', z'), &\quad v' \in \mathbb{R}, \ z' \in T_M^+.
\end{aligned}
$$

It is easy to check that this choice of parameters implies that $Z \in T^-$ and $Z' \in T^+$. It follows that, given an AdS invariant two-point function satisfying locality and the normal analyticity condition, the following restriction automatically generates a local and (Poincaré) covariant two-point function on the slice Π_v, which satisfies the spectral condition [12] (in short: the two-point function of a general Wightman QFT):

$$
W_{\{v\}}^M(z, z') = W\left(Z(v, z), Z'(v, z')\right).
\tag{38}
$$

On the basis of the dimensional boundary condition (14), and of the fact (obtained by comparing (2) and (6)) that $\frac{e^v}{r} = \sqrt{1 + \frac{1}{r^2}} \cos \tau + e^d$ tends to the finite limit

$\cos \tau + e^d$ when r tends to infinity, one sees that the following limit exists and that it yields (in view of (21) and (24)):

$$\lim_{v \to +\infty} e^{2v\Delta} W^M_{\{v\}}(z, z') = W^M(z, z'). \tag{39}$$

The limiting two-point function $W^M(z, z')$ then automatically exhibits locality, Poincaré invariance and the spectral condition. (The invariance under special conformal transformations and scaling property would necessitate a special check, but they result from the general statement of conformal invariance of the limiting field $\tilde{\mathcal{O}}(\eta)$ completed by the analysis of [5]).

When applied to the Wightman functions of Klein–Gordon fields (i.e. with $\Delta = \frac{d}{2} + v$), the above discussion of the limiting procedure gives immediately the result obtained in Eq. (37) but in a larger complex domain:

$$\lim_{v \to \infty} e^{2v\left(\frac{d}{2}+v\right)} W_v(Z(v, z), Z'(v, z')) = \frac{1}{2\pi^{\frac{d}{2}}} \frac{\Gamma(v + \frac{d}{2})}{\Gamma(v + 1)} \frac{1}{[-(z - z')^2]^{\frac{d}{2}+v}} \tag{40}$$

In a completely similar way one can compute the bulk-to-boundary correlation function by considering a two-slice restriction $W_v(Z(v, z), Z'(v', z'))$ of W_v. The bulk-to-boundary correlation function is obtained by sending $v' \to \infty$ while keeping v fixed, by the following limit:

$$\lim_{v' \to \infty} e^{v'\left(\frac{d}{2}+v\right)} W_v(Z(v, z), Z'(v', z'))$$

$$= \frac{1}{2\pi^{\frac{d}{2}}} \frac{\Gamma\left(v + \frac{d}{2}\right)}{\Gamma(v + 1)} \frac{1}{\left(e^{-v} - e^v(z - z')^2\right)^{\frac{d}{2}+v}}. \tag{41}$$

6 Conclusions

We have shown a new framework to deal with AdS QFT which is closely similar to the standard framework of Wightman's QFT. This framework can be used either to prove general theorems [9] or to discuss applications. We have presented here how one can deal with the AdS/CFT correspondence. We hope to study in future how this formalism can be applied to the study of perturbation theory. This can be interesting from the AdS/CFT point of view but also in itself, since AdS QFT is believed to provide a useful curved spacetime infrared regularization to flat Minkowski (gauge) QFT's [19].

References

1. Maldacena, J.J.: Adv. Theor. Math. Phys. **2**, 231; hep-th/9711200
2. Fronsdal, C. (1974): Phys. Rev. D **10**, 589
3. Dirac, P.A.M. (1935): Ann. Math. **36**, 657

4. Avis, S.J., Isham, C.J., Storey, D. (1978): Phys. Rev. D **18**, 3565

5. Luscher, M., Mack, G. (1975): Commun. Math. Phys. **41**, 203

6. Gubser, S.S., Klebanov, I.R., Polyakov, A.M. (1998): Phys. Lett. B **428**, 105; hep-th/9802109

7. Witten, E. (1998): Adv. Theor. Math. Phys. **2**, 253; hep-th/9802150

8. Aharony, O., Gubser, S.S., Maldacena, J., Ooguri, H., Oz, Y. (2000): Phys. Rept. **323**, 183

9. Bros J., Epstein, H., Moschella, U. (1998): In preparation

10. Bertola, M., Bros, J., Moschella, U. and Schaeffer, R. (2000): Nucl. Phys. B **587**, 619; hep-th/9908140

11. Bertola, M., Bros, J., Gorini, V., Moschella, U., Schaeffer, R. (2000): Nucl. Phys. B **581**, 575; hep-th/0003098

12. Streater, R.F., Wightman, A.S. (1964): PCT, spin and statistics, and all that. W.A. Benjamin

13. Bros, J., Epstein, H., Moschella, U. (1998): Commun. Math. Phys. **196**, 535; gr-qc/9801099

14. Mack, G., Todorov, I.T. (1973): Phys. Rev. D **8**, 1764

15. Bateman, H. (1954): Higher transcendental functions. McGraw–Hill

16. Breitenlohner, P., Freedman, D.Z. (1982): Ann. Phys. **144**, 249

17. Klebanov, I.R., Witten, E. (1999): Nucl. Phys. B **556**, 89

18. Giddings, S.B. (1999): Phys. Rev. D **61**, 106008

19. Callan, C.G., Wilczek, F. (1990): Nucl. Phys. B **340**, 366

Geodetic Contributions to Gravitational Experiments in Space

E.C. Pavlis

Abstract. Geodesy has been traditionally a science that facilitated the testing of some of the most important laws of physics and their corollaries. In the past, these experiments were natural consequence of problems that geodesy had to solve in order to progress and refine its methods. In recent years, with a lot of the "geodetic problems" under control, geodesists have taken a closer look at problems that we can address keeping in mind that our current primary goal is to facilitate interdisciplinary research on global change and related topics (Fig. 1). One of the areas that geodesy can contribute the most, is the precise determination of the terrestrial gravity field and its temporal variations. This provides the precise, stable and free-of-gravitational-noise environment where very delicate experiments in gravitational physics can be conducted.

In what follows we will discuss recent and future developments and compare them to variations inferred from geophysical processes. Using this knowledge, we can model a priori a large number of variations that could affect any experiment requiring

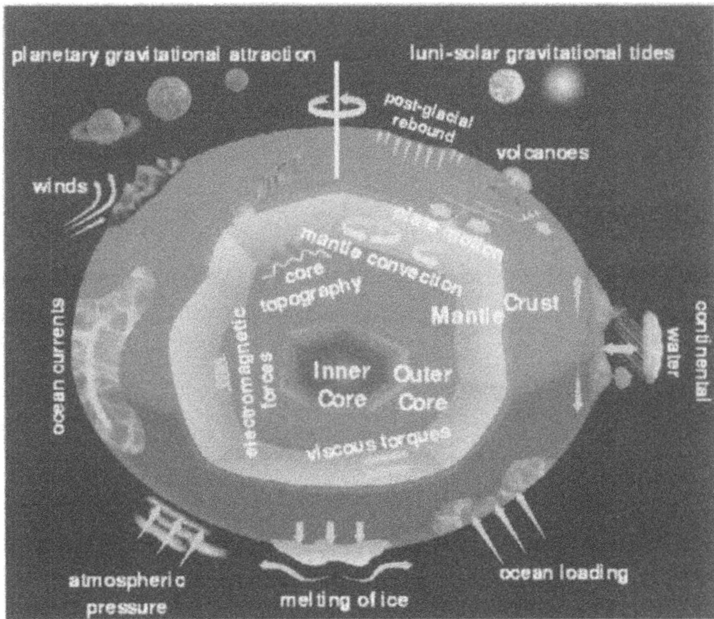

Fig. 1. Geophysical processes interactions sensed by near-Earth-orbiting spacecraft

ultra-precise measurement of extremely small gravity changes. Most gravitational experiments in space require isolation from such changes in the field, which can either mask the effects we are looking for, or even worse, disguise themselves as the effect itself! This contribution examines also the areas where current geodetic results and future research and proposed missions will make a profound contribution in testing physical laws in space.

Up to now, the attention of the community was centered on the static field, mainly because of the need to compute precise orbits for satellites carrying various instruments [1]. Since the emergence of climate change as one of the most important research topics at the international level [2], the temporal variations of the gravity field have become a prime topic and the focus of several space missions with clearly international nature. Some of these are already in orbit (*CHAMP*, launched July 15, 2000), others will be launched soon (*GRACE*, expected launch date February 2002) and some are still under development (*GOCE*, planned for 2005–2006). These missions however are only better (in terms of accuracy) and more complete (in terms of resolution) solutions of the problem. Long before they were even envisioned, geodesists had already identified on-going missions that are capable to address the temporal variations of at least the very long wavelength components of gravity.

The measuring technique for this, Satellite Laser Ranging (*SLR*), is one of the most widely used tracking techniques for precision orbit determination [3,4], and geophysical parameter estimation [5,6]. The technique has been around since the early 60s, and although the basic principle (Fig. 2) behind it has remained unchanged, the technique itself has evolved quite steadily in terms of precision and accuracy from the 1–2 meter level in the early days, to the few millimeters today, for a single range measurement [7].

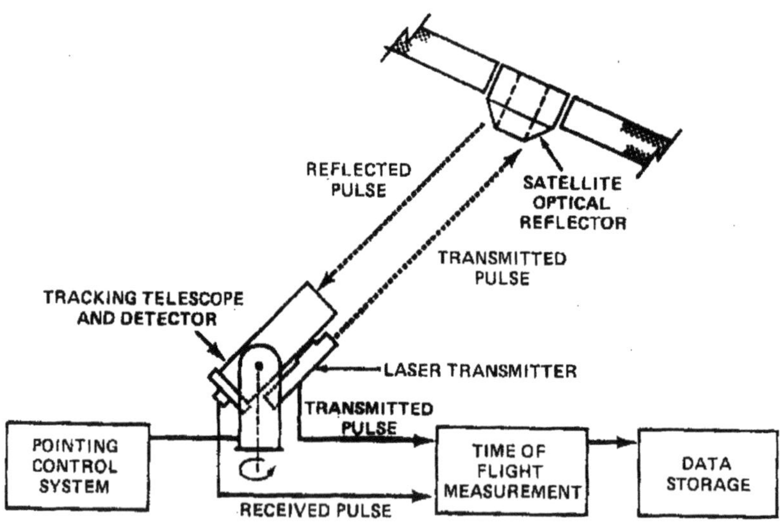

Fig. 2. Schematic explaining the principle of the satellite laser ranging technique

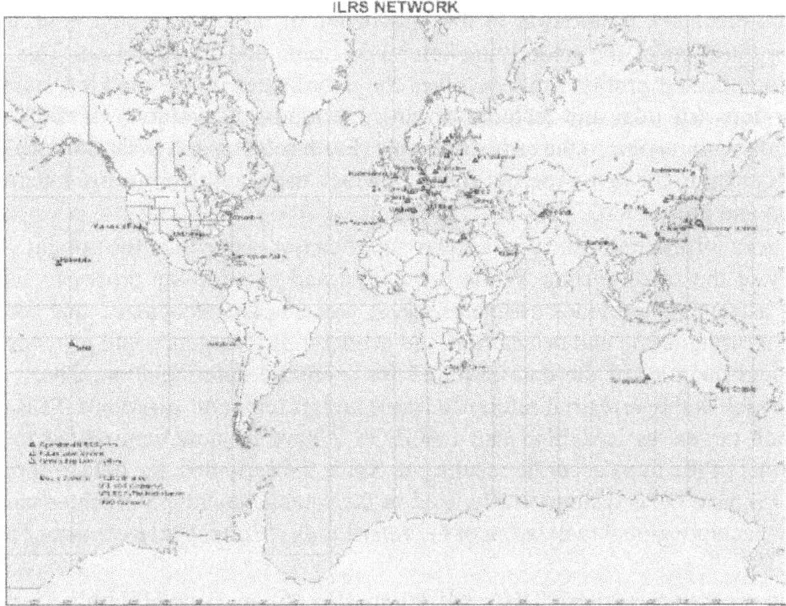

Fig. 3. The current international satellite laser ranging network

The network of tracking stations has also evolved over the decades with a similar pace and it numbers today about forty sites (Fig. 3). Some of these are mobile and can be easily re-deployed to achieve improved tracking configurations based on the community requirements. Over the years, the *client targets* have been primarily satellite missions that required very precise orbits, but not necessarily geodetic targets per se. To explain this better, one could make use of an analogy drawn from physics: if a *geodetic target* would be the equivalent of a particle in physics, then most of the SLR targets resemble school buses! There are exceptions of course, and these were developed specifically with the precise geodetic problems in mind, so they indeed approximate the concept of a *particle*. These are cannonball-shaped spacecraft, massive and dense, covered with high quality retro-reflectors, and placed in a variety of orbital altitudes and inclinations. The first such target was the French Starlette (launched in 1975), followed the year after by the US LAser GEOdynamics Satellite (LAGEOS), (Fig. 4).

These early geodetic targets validated the SLR technique and proved the original concept. Their example was followed by many more satellite missions with a need for precision orbits, irrespective of their primary goal. Over the years the global laser tracking network has serviced numerous missions, several tens of them simultaneously, and today there is a plethora of such targets being tracked for various applications (Fig. 5).

The majority of these targets provided useful data for the development of models of the terrestrial gravitational field and its tidal variations. However, to describe the orbits precisely, in addition to the knowledge of the gravitational field, there is also a question of the underlying reference frame and its evolution. This is a rather complicated problem that involves the coordinates of the tracking stations, their motions (of tidal and tectonic origin), the temporal evolution of the frame orientation with respect to the terrestrial crust (Earth rotation and polar motion), and its relationship to the quasi-inertial frame in which the satellite orbits are integrated (nutation and precession).

All these information must be known significantly better than the sought – for accuracy of the satellite rbits. While precession and nutation are processes which modern astrometry provides estimates for, it was quickly recognized that for the terrestrial origin, scale and orientation, the geodetic SLR targets could pro vide the quality and quantity of the data required for a precise determination. These data would establish the terrestrial reference frame and its temporal variations. SLR data from such targets as LAGEOS and LAGEOS 2 have by now contributed in this effort some of the most accurate results yet. Since the early 80s, the data of the first LAGEOS, have been systematically used in the establishment and maintenance of the most accurate global realization of the *International Terrestrial Reference Frame (ITRF)*, [8].

In the recent years, these TRF realizations (Fig. 6) are formed within a framework that accounts very precisely for the long wavelength temporal variations in the gravitational field. The reason behind this stems from the fact that the 0th, 1st, and 2nd degree harmonics of the spherical harmonic expansion that is traditionally used in modeling the field, are directly related to the scale, origin and orientation of the coordinate system underlying the TRF. This is in a way the geodetic linking of space, gravity and time.

Several SLR Analysis centers, including the Joint Center for Earth Systems Technology (*JCET*), are prime contributors to this effort. JCET submits an annual contribution of a combined analysis of the data set from both satellites for the period 1993-present [9]. A TRF realization comprises a set of precise tracking station posi-

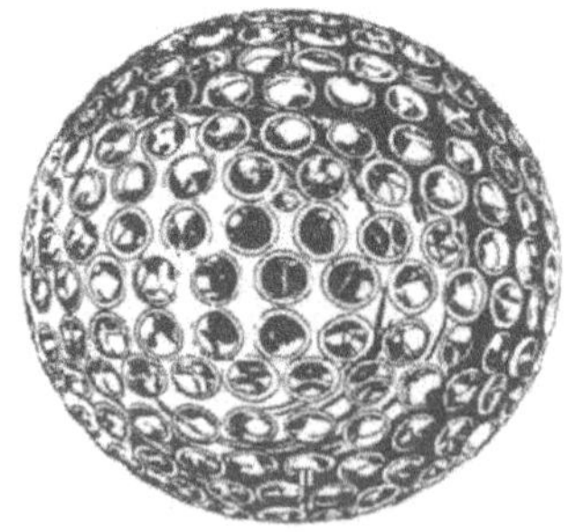

Fig. 4. The LAser GEOdynamics Satellite (LAGEOS)

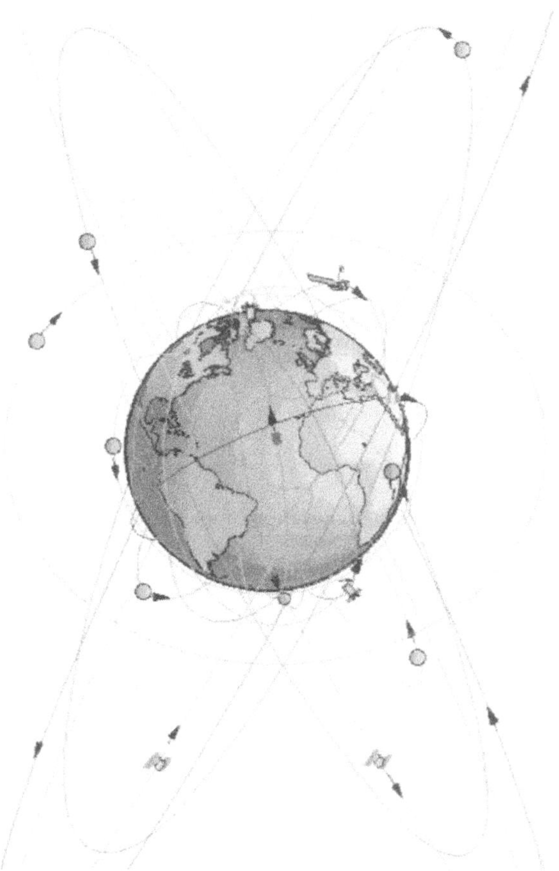

Fig. 5. The currently tracked missions by the international satellite laser ranging network

tions at some epoch and their precise linear velocities due to tectonic plate motion. The typical level of accuracy for SLR-derived TRF realizations is at the 2-5 millimeter level for the 3-D positions (Fig. 7), and 1-2 millimeter per year for the 3-D velocities (Fig. 8).

Along with this static TRF realization, a history of the temporal variations of its origin and its orientation are determined. The first is represented through a weekly time series of offsets of its *mean* origin with respect to the center of mass of the Earth system (geocenter), in the three cardinal directions (Fig. 9). The temporal variations in the orientation of the TRF with respect to the instantaneous rotational vector of Earth are described through a daily time series that define the offsets in polar motion (Fig. 10) and the excess in *length of day (LOD)* from the nominal value of 86400 s, (Fig. 11).

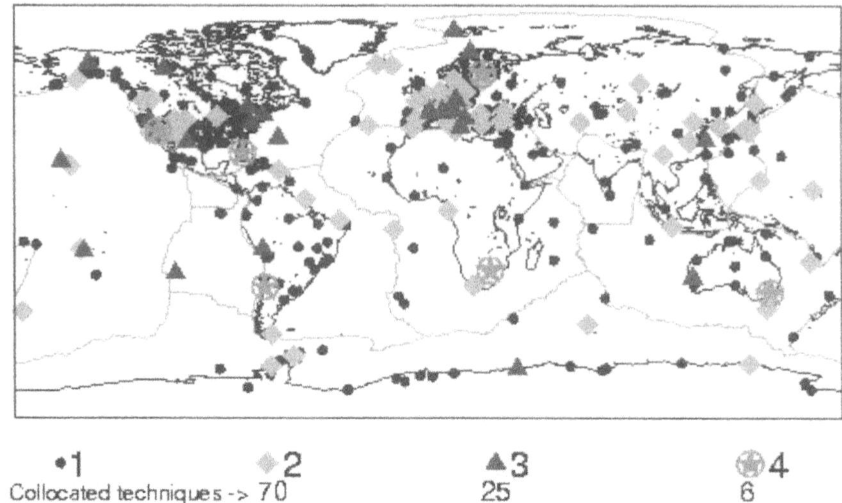

Fig. 6. The sites comprising the most recent realization of the ITRF: ITRF2000

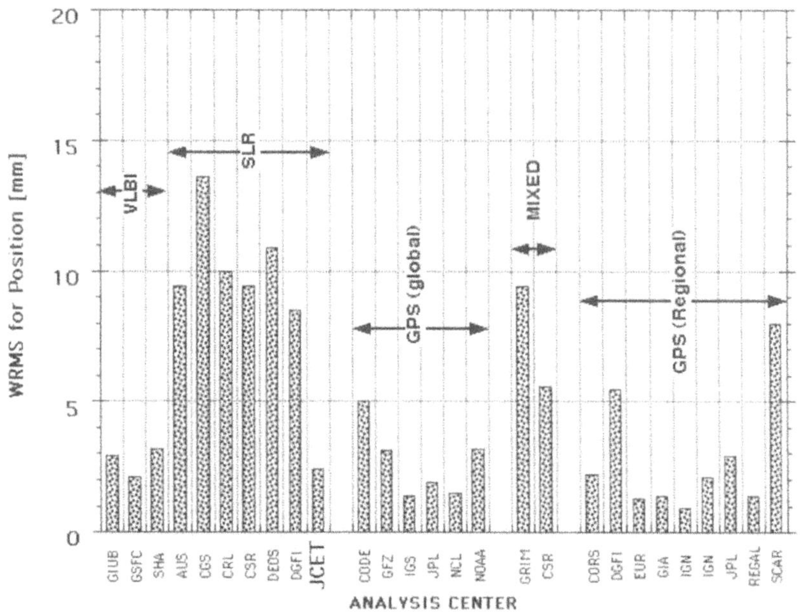

Fig. 7. Positional weighted root mean square (WRMS) error of the various contributions to ITRF2000

The results presented in Figs. 9 through 11 were obtained from the analysis of LA-GEOS and LAGEOS 2 SLR data, reduced at JCET [10,11], using NASA Goddard's GEODYN/SOLVE II software, [12].

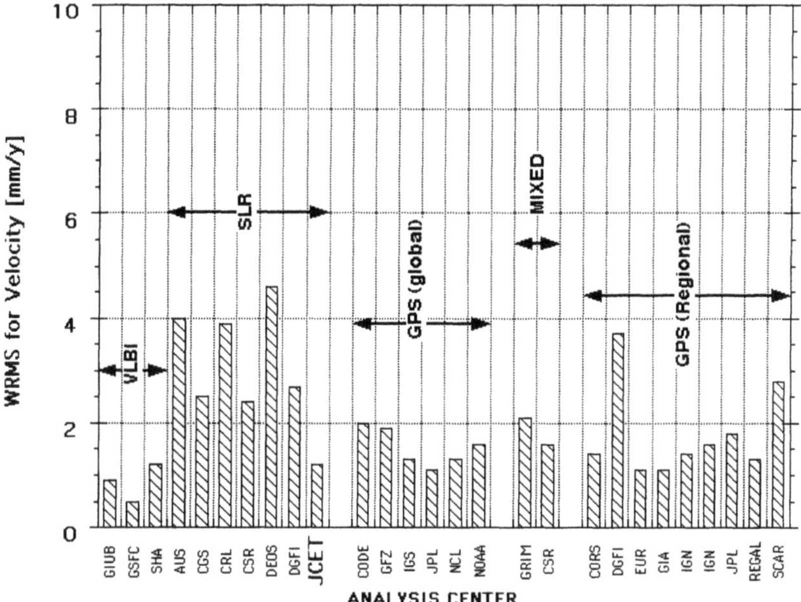

Fig. 8. Linear velocity weighted root mean square (WRMS) error of the various contributions to ITRF2000

The observed temporal variations in the gravity field are the manifestation of the redistribution of mass within the Earth system (solid Earth, oceans, atmosphere and cryosphere), as a result of the dynamics of the terrestrial environment driven by the incoming solar energy and the interactions between its various components. These variations cause concomitant changes in the Stokes' coefficients describing the terrestrial gravity field. Secular changes in J2 due to post-glacial relaxation have been observed since a many years. Similar changes in J3 have been attributed to changes in the ice sheets of Greenland and Antarctica. Seasonal changes in these coefficients have also been closely correlated with mass transfer in the atmosphere and oceans. The system is also affected severely by the hydrological cycle, which however is the most difficult to measure globally and accurately so far. This is reflected in the poorer correlation of the geodetic and geophysical signals in the z-direction, as illustrated in Fig. 9, the component for which the hydrological signal is the largest contribution in the geophysical series. The gist of all these observations is that geodetic estimation of the static and temporally varying gravitational field is an indispensable scientific tool reaching far beyond geodesy.

The initiation of studies for Earth Observing Systems in the mid-80s, and the launch of the first missions to contribute in the direction of observing *global change*, brought geodesy at a major cross-road. On the one hand, geodesy can contribute in the quest for such signals, and the possible contributions were enumerated earlier. On the other, the global synoptic fields that are now slowly becoming available from

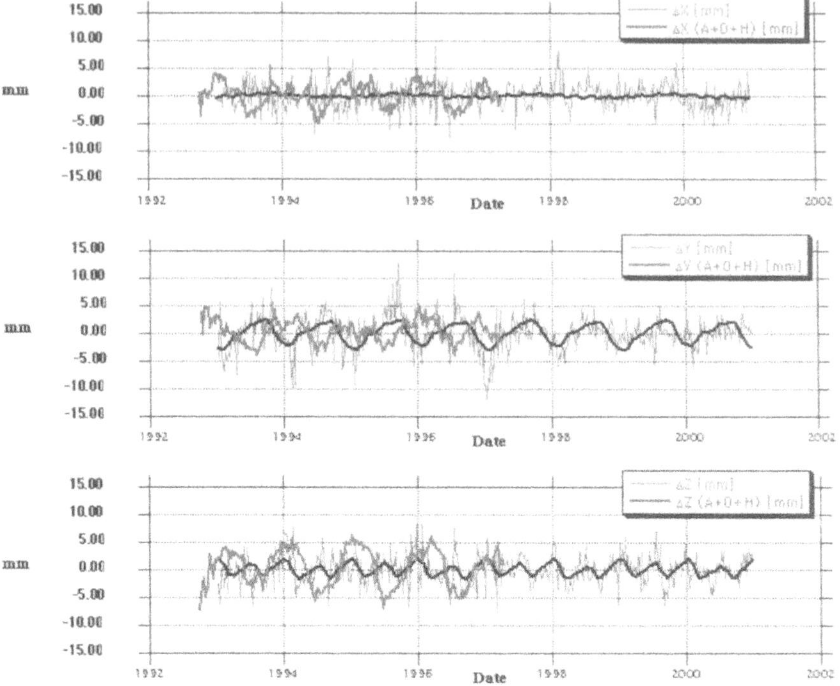

Fig. 9. The weekly motion of the geocenter from SLR compared to independently derived geophysical series (A+O+H)

these space missions (e.g. atmospheric and oceanic circulation, precipitation, snow and ice cover, water runoff, etc.), provide geodesists with unique opportunities for improved model development. By modeling a priori a large portion of the temporally varying gravitational signal on the basis of these observations, it should now be possible to derive new models of the static gravity field with unprecedented accuracy. That however presumed that geodesy had also an arsenal of similar tools to observe the gravitational field more precisely and with higher resolution. This was a geodesists' dream for over thirty years [13], and it is only now, that it is slowly being realized. Following a presentation of current geodetic results vis-à-vis gravitational experiments in relativistic physics, we will discuss the upcoming missions and their future role in the conduct of such experiments.

One of the predictions of General Relativity is the existence of a gravitomagnetic field near a massive, rotating body. In 1918, J. Lense and H. Thirring calculated the amount of *nodal dragging* that a moon of such a body would experience [14]. The dawn of space geodesy saw the first attempts of developing missions that would be able to measure this effect [15], henceforth to be called the *Lense–Thirring (L–T) effect*. The technology for most of these missions was very demanding and mostly yet to be developed, and the tracking capabilities were still quite poor. In this particular

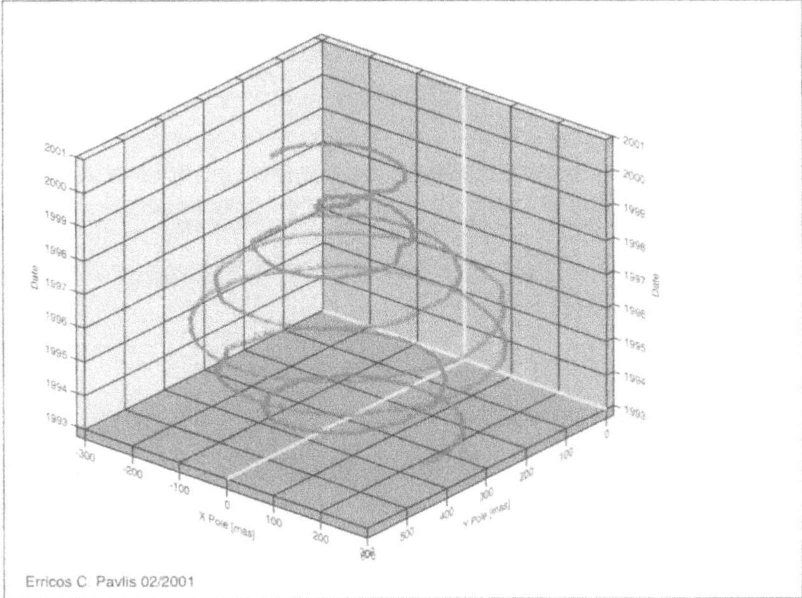

Fig. 10. The daily motion of the instantaneous terrestrial rotational axis with respect to the Conventional International Origin (CIO), determined from LAGEOS and LAGEOS 2 SLR data

case, the uncertainty of Newtonian gravitation effects would be neutralized by an ingenious mission design of two polar, drag-free, counter-orbiting spacecraft (Fig. 12). Eventually, the original initiative would lead to a dedicated mission, *Gravity Probe B (GP-B)*, at a very high cost and plagued by severe cost over-runs and launch delays. At present the launch date is set for 2002, but only a year ago it was spring of 2001, so the firmness of that projection therefore is rather doubtful.

On the other hand, the success of the LAGEOS mission in 1976 and the subsequent validation of SLR as a very precise tracking technique, opened new possibilities. By the late 80s SLR was mature enough in terms of both, hardware as well as an international tracking network, that the successful use of this technology was proposed and vigorously pursued by many groups from both communities [16]. Ciufolini modified slightly the original mission design, to take advantage of the already in space LAGEOS. The proposed counter-orbiting satellite, LAGEOS III (since there was already a LAGEOS II mission in advanced stages), would be placed in a similar to LAGEOS shape orbit but at the complementary inclination of 70°. The cost of an additional and identical LAGEOS spacecraft, the third to be build by then, would be comparatively insignificant. With the tracking from the existing SLR network being free and assured, the only cost to be of significance was that of the launch.

When this mission was not selected due to a change in the mission priorities of NASA, the efforts to use SLR were re-directed towards making the most out of the

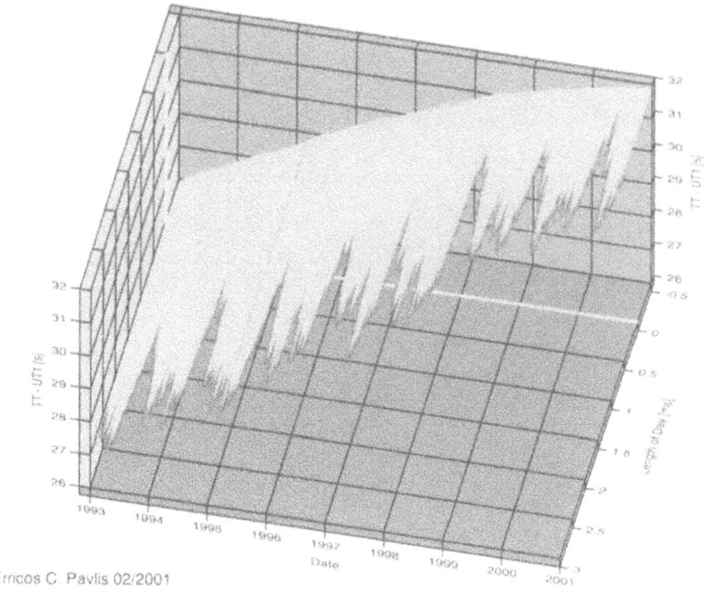

Fig. 11. The daily excess length of day (LOD) due to Earth's spin rate variations, determined from LAGEOS and LAGEOS 2 SLR data

existing LAGEOS and by now LAGEOS 2 (inclination $52°$) missions. The new attempt took advantage of the fact that in the case of LAGEOS 2, the orbit was significantly more elliptic than for LAGEOS, so one could use the L–T effect on the satellite's perigee in addition to the classical nodal effect (Fig. 13). Taking advantage of the latest improvements in modeling the static geopotential through the EGM96 model [1], the first results of this effort were reported in [17]. The reduction of four years of LAGEOS and LAGEOS 2 SLR data resulted in detecting and measuring L–T with a 20 % uncertainty and at a value only $+10\%$ off that predicted by GR (Fig. 14).

Although this first result is not nearly as accurate as GP-B promises (better than 1 %), the effort is continuing, extending the analysis to present. This would make use of eight years of data and it would more than halve the error estimate based on the error analysis presented in [18].

The success of this effort rekindled the idea of a third satellite with an orbital design optimized for the measurement of the L–T effect. In 1997 Ciufolini lead a large international team of scientists in a successful ASI proposal for a new dedicated laser mission, the *LAser RElativity Satellite* (*LARES*), Fig. 15. A Phase A study was conducted and a comprehensive report presented to ASI in late 1998. The proposed mission comprised a much smaller satellite than the LAGEOS spacecraft with fewer retro-reflectors and at a lower altitude. The inclination could be either polar or complementary to that of LAGEOS, depending on the launch options available. The mission was ranked second within the finally selected group of missions, out

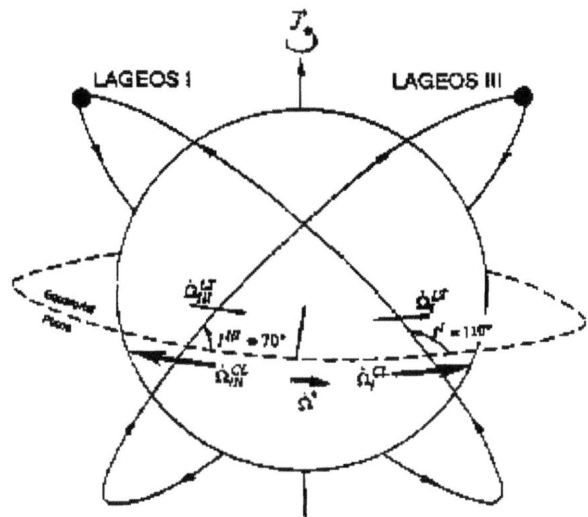

Fig. 12. The modified butterfly satellite mission configuration of Van Patten and Everitt [15] for the deployment of LAGEOS III as proposed by Ciufolini [16]

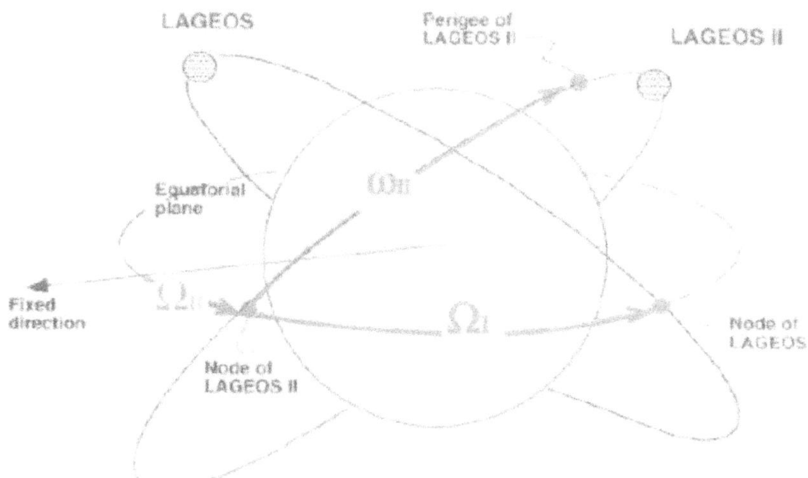

Fig. 13. The LAGEOS–LAGEOS 2 experiment to measure the Lense–Thirring effect with SLR

of some sixty original proposals. A year later, the same scenario was unsuccessfully proposed to NASA's *University Earth System Science (UnESS)* program. The Ciufolini group is currently seeking alternate means to launch this satellite in cooperation with various international groups.

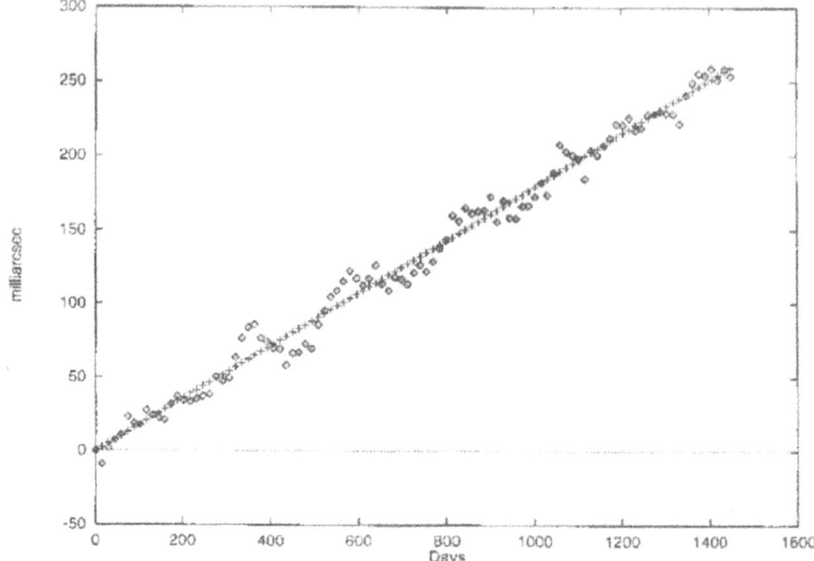

Fig. 14. Measurement of the L–T effect with SLR (o) vs. prediction (+), 1993–1996

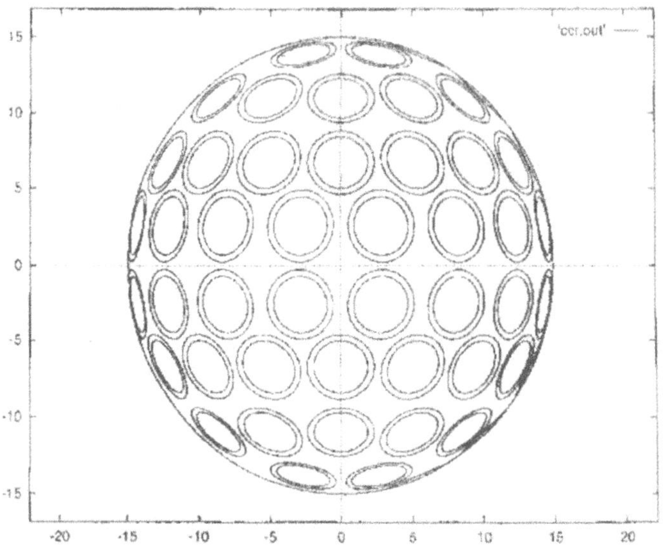

Fig. 15. The proposed design for the LARES satellite with 100 retro-reflectors

One reason behind the changes in the orbital configuration of the new LARES mission is the attempt to use the data from this experiment to test other relativistic theories, beyond GR. In particular, a large eccentricity in the orbit of LARES

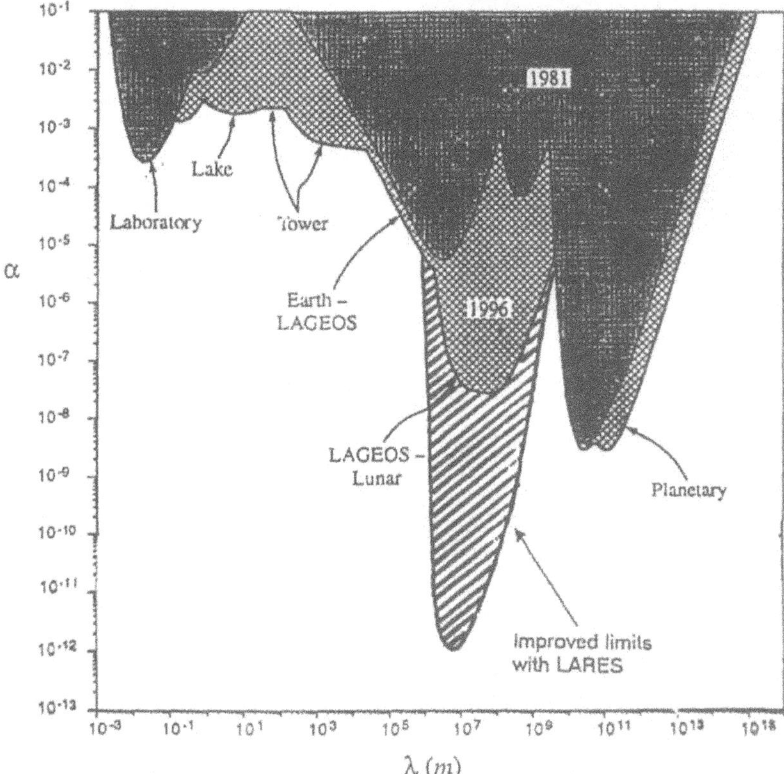

Fig. 16. Improved limits on Yukawa-type interactions at the 1–2 Earth radii range with LARES

would allow better observation of the evolution of its perigee. These observations are expected to place significant constraints on "new" Yukawa forces (Fig. 16) at the 1–2 Earth radii range. Furthermore, with improved gravitational modeling soon to become available from dedicated missions, it is now possible to envision the precise observation of along-track perturbations in the orbit of LARES with an annual period, leading to the detection of preferred frame effects [19].

Another reason behind the changes in the orbital configuration of the new mission LARES is the expected improvements in the static and temporally varying models of the geopotential from a number of gravity mapping missions that are nearing launch. We had alluded to these in the introductory sections: CHAMP, GRACE, and GOCE. The first mission, CHAMP, is already delivering data since last year. The aspect of this mission relevant to this area of research is its precise tracking from the *G*lobal *P*ositioning *S*ystem (*GPS*) constellation of satellites and the simultaneous measurement of non-gravitational forces from an on-board precise accelerometer. This is the first time that such an accelerometer is used for an extended mission.

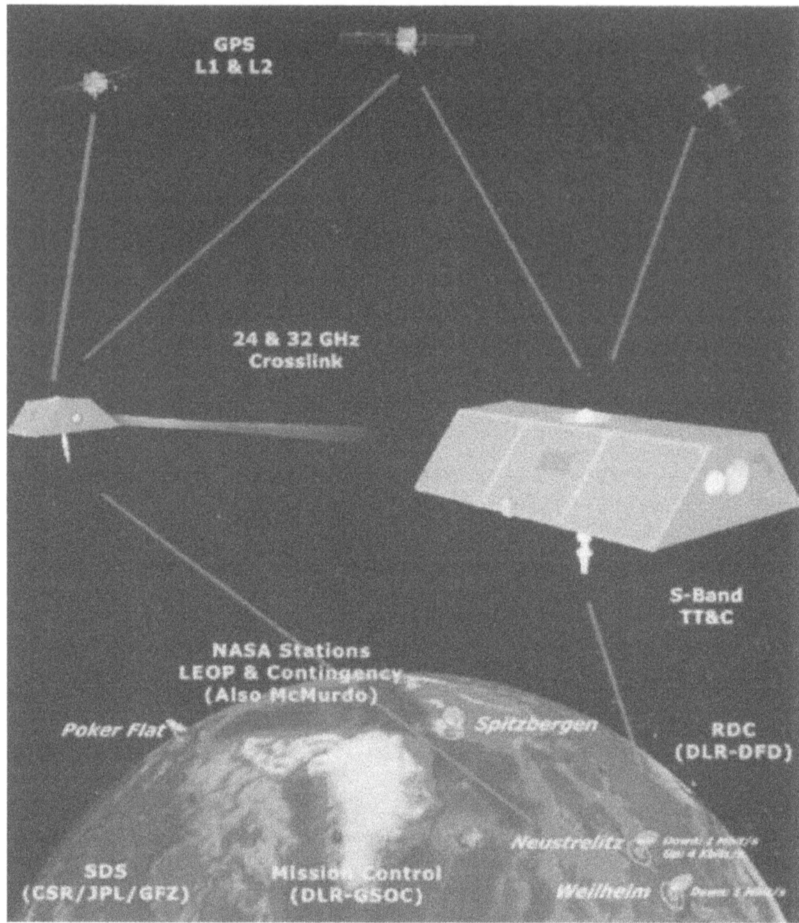

Fig. 17. A schematic of the GRACE mission concept

Due to the high altitude of the GPS spacecraft (12 hr orbits), and the low altitude of CHAMP ($\sim$ 450 km), the configuration is called "high-low" satellite-to-satellite tracking,
(H–L SST). The benefit of this configuration is that the entirety of the perturbations in the observed range-rate between the two spacecraft (s/c) can be now attributed to gravitation sensed by the low orbiter (non-gravitational forces accounted for using the observed accelerometry). The model enhancement expected from CHAMP is primarily the accuracy improvement of the very low degree terms (long wavelengths) up to about degree 40, to roughly two orders of magnitude better than what EGM96 provides today.

By early 2002, the second mission, GRACE, will be launched (Fig. 17). This is sort of a pair of "CHAMPs" flying in tandem (nominally $\sim$ 250 km apart), enhanced with higher quality accelerometers and a one-way radio frequency range-difference measuring system on both of them. This configuration of satellite-to-satellite tracking between two low orbiters is called "low–low" (L–L SST). The L–L configuration is characterized from its ability to resolve features proportional to the distance between the two s/c. GRACE therefore, since it also carries GPS receivers on both s/c, will contribute to the observation of the long, as well as the medium wavelengths (up to 120) of the gravitational spectrum. Moreover, the mission is polar ($i = 89.5\,^\circ$) and it will sample the field well enough to produce monthly snapshots. With a projected lifetime of five years, it is hoped that we will acquire enough such *monthly snapshots* to be able to determine the temporal variations in the long wavelengths down to at least seasonal frequencies. This achievement alone will be a unique contribution compared to today's status quo. Only the secular variations of the first few zonal terms up to degree 6 have been studied so far. Annual and semi-annual signals have been determined only for J2 and J3, and these are not as definitive results as we would like to have. The third mission, GOCE, is really the true "geopotential mapping" mission that geodesists always hoped for. Instead of relying on perturbations in some linear observable from which to infer the gravitational signal (as in H–L and L–L), GOCE will carry a gradiometer, a device that will measure directly the gravitational tensor components in space. Along with precise knowledge of the time and location of these measurements (GOCE will also be tracked with GPS), we can then construct a gravitational model solving a geodetic boundary problem by means of one of the standard geodetic techniques. Although this task is not as simple as it sounds, gradiometry is definitely the cleanest and most direct measurement of gravitation that we could possibly make in space. With a two six-month observing period scenario,

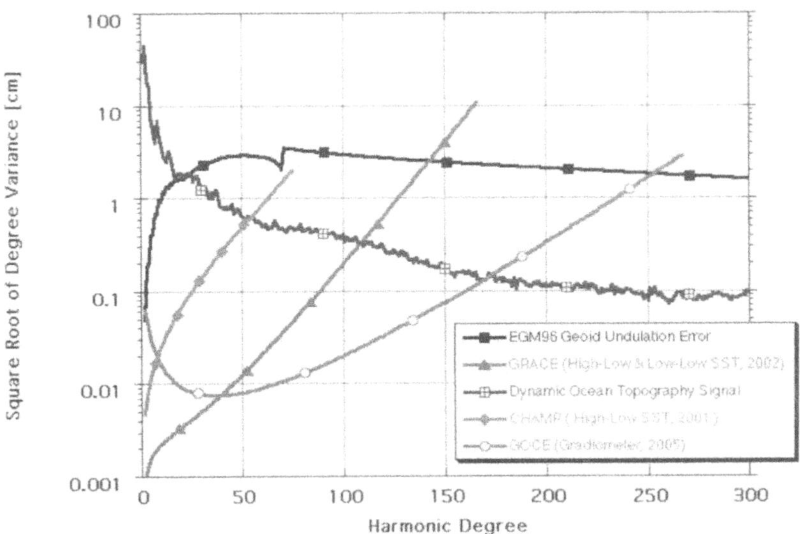

Fig. 18. A comparison of geophysical signals and gravitational error spectra from current and future geopotential mapping missions

GOCE promises to deliver a model that will extend our precise knowledge to much higher resolution, somewhere in the neighborhood of maximum degree 250 or even higher (Fig. 18). It is expected that by that time, the release of additional observations on land and over the oceans from national data bases and terrestrial campaigns (airborne and ship-borne), will allow the development of global models to maximum degree ~ 2000.

While these models will not have the same uniform quality for the very high degree portions (above ~ 250) on a global scale, their overall improvement will remove all of the ambiguities associated with the static long and medium wavelengths. Furthermore, all of the significant temporal signals which may alias themselves in gravitational experiments in fundamental physics, will also be precisely observed. It is not surprising then that the present decade has been called the International Decade for Geopotential Research. Once these data have been reduced to the model level, the ball will be in the observers' court. They will need to improve the precision and sensitivity of their measurements, to take advantage of these geopotential modeling improvements. And the struggle continues...

References

1. Lemoine, F.G., Kenyon, S.C., Factor, J. K., Trimmer, R. G., Pavlis, N.K., Chinn, D.S., Cox, C.M., Klosko, S.M., Luthcke, S.B., Torrence, M.H., Wang, Y.M., Williamson, R.G., Pavlis, E.C., Rapp, R.H., Olson, T.R. (1998): The Development of the Joint NASA GSFC and the National Imagery and Mapping Agency (NIMA) Geopotential Model EGM96. NASA/TP-1998-206861. Goddard Space Flight Center, Greenbelt, Maryland, July 1998
2. NRC (1997): Satellite Gravity and the Geosphere: Contributions to the Study of the Solid Earth and Its Fluid Envelope. Washington D.C., National Academy Press
3. Nerem, R.S. et al. (1993): Expected Orbit Determination Performance for the TOPEX/Poseidon Mission. IEEE Transactions on Geoscience and Remote Sensing, Vol. **31**, 2
4. Pavlis, E.C. (1995): Comparison of GPS s/c Orbits Determined from GPS and SLR Tracking Data. Adv. Space Res. **16**, 12, 55–58
5. Smith, D.E. et al. (1994): Contemporary global horizontal crustal motion. Geophys. J. Int. **119**, 511–520
6. Nerem, R.S. et al. (1994): Gravity Model Development for TOPEX/POSEIDON: Joint Gravity Models 1 and 2. J. of Geophys. Res. **99**, C12
7. Degnan, J.J. (1993): Millimeter Accuracy Satellite Laser Ranging: A Review. AGU Monograph Series, Contributions of Space Geodesy to Geodynamics: Technology. Geodynamics **25**, 133–162
8. Altamimi, Z. et al. (2001): The Terrestrial Reference Frame and the Dynamic Earth. Eos Trans. AGU **82** (25), 273, 278–279
9. Pavlis, E.C. (2000): JCET's Contribution to the IERS Terrestrial Reference Frame 2000. Eos Trans. AGU **81** (48), Fall Meet. Suppl., F312
10. Pavlis, E.C. (2001): Satellite laser ranging constraints on global mass transport in the Earth system. Geophysical Res. Abstracts (CD) **3**, EGS, Nice, France
11. Pavlis, E.C. (2001): Earth orientation from satellite laser ranging (SLR): quality, content and resolution. Geophysical Res. Abstracts (CD), **3**, EGS, Nice, France

12. Pavlis, D.E., et al. (1998): GEODYN systems description. Vol. **3**, Greenbelt, MD: NASA GSFC

13. MIT (1970): The Terrestrial Environment: Solid Earth and Ocean Physics. NASA Contractor Report CR-1579. Cambridge, MA: Massachusetts Inst. of Technology

14. Lense, J., Thirring, H. (1984): Translation of the German original (Phy. Z. **19**, 156, 1918), by B. Mashhoon, F.W. Hehl, D.S. Theiss. Gen. Relativ. Gravitation **16**, 711

15. Van Patten, R.A., Everitt, C.W.F. (1976): A possible experiment with two counter-orbiting drag-free satellites to obtain a new test of Einstein's General theory of Relativity and improved measurements in Geodesy. Celestial Mechanics **13**, 429–447

16. Ciufolini, I. (1989): A comprehensive introduction to the LAGEOS gravitomagnetic experiment: From the importance of the gravitomagnetic field in physics to preliminary error analysis and error budget. Int. J. of Mod. Physics A **4** (13), 3083–3145

17. Ciufolini, I., Pavlis, E.C., Chieppa, F., Fernandes-Vieira, E., Pérez-Mercader, J. (1998): Test of General Relativity and Measurement of the Lense-Thirring Effect with Two Earth Satellites. Science **279**, 2100–2103

18. Pavlis, E.C., Iorio, L. (2001): The impact of tidal errors on the determination of the Lense–Thirring effect from satellite laser ranging. Accepted in the International Journal of Modern Physics D

19. Nordtvedt, K. (1999): Gravitational preferred frames and Earth satellite orbits. Class. Quantum Grav. **16**, L19

The r-modes Oscillations and Instability:
Surprises from Magnetized Neutron Stars

L. Rezzolla

Abstract. The instability of r-mode oscillations in rapidly rotating neutron stars has been so far a source of surprises. The analyses carried to date have revealed the surprising existence of this instability and, perhaps even more surprisingly, have shown that such instability has a rather large parameter space in which it can survive against the damping produced by shear and bulk viscosity, as well as against the interaction with a solid star crust. The magnetohydrodynamic coupling of the modes with a stellar magnetic field, which is likely to be present, has recently been shown to provide another surprising aspect of the instability. We here review the relevance of a stellar magnetic field, its modifications under the action of the r-mode instability, and how the interaction between r-mode oscillations and a magnetic field might limit the onset and duration of the instability.

1 Introduction

Almost twenty years after their basic properties were first investigated in Newtonian rotating stars [1], r-mode oscillations in rotating relativistic stars have "surprised" the community interested in sources of gravitational waves when they were shown to be unstable to the emission of gravitational radiation. Perhaps more surprising is that for perfect fluid stellar models, the instability sets in with arbitrarily slow rotation rates [2,3]. This is a significant difference from previously investigated mode-instabilities, whose onset does depend on a specific (and usually high) rate of rotation. The instability of r-modes oscillations is a manifestation of the more generic Chandrasekhar-Friedman-Schutz instability, but within this more general context it offers another surprise through its growth time. In all of the previously investigated instabilities, in fact, gravitational radiation couples to the time-varying mass multipole moments. For r-mode oscillations, however, gravitational radiation couples primarily through time-varying mass-current multipole moments. The growth time found in this case for hot, unmagnetized, and newly born neutron stars is sufficiently short that the instability has a large parameter space (in temperature and rotation velocities) where it can dominate viscous damping. Since this initial surprises, the literature on this subject has been rapidly growing (see Kokkotas' contribution to this volume and [4,5] for recent reviews and updated lists of references).

The energy budget governing the evolution of the r-mode instability is traditionally assumed to be regulated by the emission of gravitational waves and viscosity, which act as sources and sinks of energy, respectively [6–8]. However, other sources and sinks should be included in this budget, such as the loss of rotational and mode energy to electromagnetic radiation [9] or to magnetic energy. A new surprise in the evolution of the r-mode instability has emerged with the existence of a coupling between the mass-currents produced by the oscillations and the magnetic field present in the neutron star [10] (expected to be in the range $10^{11} - 10^{13}$ G). What makes this surprise particularly relevant is twofold: firstly, because the generation of magnetic field is a generic feature of shearing flows perpendicular to magnetic field lines in a highly conducting plasma, such as hot neutron star matter. Secondly, because the generation of magnetic field represents an important channel into which the energy of the mode can be funneled. Indeed, this channeling can be so efficient that, as pointed out by Spruit [11], extremely intense magnetic fields could be created during the instability, which could then become buoyant-unstable and generate powerful flashes of γ-rays.

In the following we review our present understanding on the complex magneto-hydrodynamical (MHD) coupling between r-mode oscillations and a pre-existing magnetic field, paying special attention to the stages when the oscillations reach a nonlinear development. In doing so we will make use of the linear results which are presently available and exploit phenomenological considerations to extend these results to the nonlinear regime. Nonlinear physical phenomena analogous to the r-mode oscillations and occurring either in Earth's atmosphere or fluid-dynamics laboratories, will be our guides. Using them to deduce nonlinear effects, we will investigate the interaction of the r-mode oscillations with a magnetic field and the modifications introduced to the onset and evolution of the instability. In particular, we will show that the dynamics of the r-mode oscillations inevitably couples to any pre-existent magnetic field and results in a net amplification of the latter. This generation of magnetic field is accompanied by a conversion of the energy of the mode into magnetic energy and might therefore impede the onset or the growth of the r-modes instability. In particular, if the initial magnetic field is sufficiently strong, it will prevent gravitational radiation from exciting the r-mode instability. An initially weak magnetic field will be, however, amplified during the instability and could cause the instability to die out as the star spins down.

This review is organized as follows: In Sect. 2 we take a careful look at the equations of motions for fluid elements undergoing r-mode oscillations and discuss how the nonlinear behaviour can be deduced from the linear equations. In Sect. 3 we review the phenomenological model we have used for the background evolution of the instability, while in Sect. 4 we present an effective Lagrangian treatment of the MHD equations. Numerical results from our model are presented in Sect. 5 and they are then used in Sect. 6 to deduce their impact on the evolution of the instability. Finally, Sect. 7 contains our conclusions and considerations on the impact of magnetic fields on the gravitational wave emission from the r-mode instability.

2 Kinematic properties

A key role in the coupling between r-mode oscillations and the magnetic field is played by the kinematic properties of the modes. A careful look at the equations of motion will reveal nonlinear effects which manifest themselves mostly in secular velocity fields. Within this Section we will assume that the the nonlinear coupling among different modes is negligible[1] and that there is a total absence of magnetic field.

2.1 The Eulerian perturbation velocity field

Consider some representative fluid elements moving on an isobaric surface of a rotating star and experiencing r-mode oscillations. For a Newtonian inviscid star, these are solutions of the perturbed hydrodynamic equations having Eulerian velocity perturbations of "axial type" [6,3]. In an orthonormal basis such perturbations may be written as

$$\delta \mathbf{v}_1(r, \theta, \varphi, t) = \alpha \Omega R \left(\frac{r}{R}\right)^{\ell} \mathbf{Y}^B_{\ell m} e^{i\sigma t},\tag{1}$$

where R and Ω are the radius and angular velocity of the unperturbed star, α is a dimensionless coefficient that describes the amplitude of the perturbation, and σ is the frequency of the mode in the inertial frame. In equation (1), $\mathbf{Y}^B_{\ell m}$ is the magnetic-type vector spherical harmonic. Consider now a frame instantaneously corotating with the star. In this frame, the differential equations governing the motion of fiducial fluid elements of an $\ell = m$ mode[2] in the coordinate basis $(t,\ r,\ \theta,\ \varphi)$ are

$$\dot{\theta}(t,\theta,\varphi) = \alpha \Omega \left(\frac{r}{R}\right)^{\ell-1} c_{\ell} \sin\theta (\sin\theta)^{\ell-2} \cos\left[\ell\varphi + \left(\frac{2\Omega}{\ell+1}\right)t\right],\tag{2}$$

$$\dot{\varphi}(t,\theta,\varphi) = -\alpha \Omega \left(\frac{r}{R}\right)^{\ell-1} c_{\ell} \cos\theta (\sin\theta)^{\ell-2} \sin\left[\ell\varphi + \left(\frac{2\Omega}{\ell+1}\right)t\right],\tag{3}$$

where c_{ℓ} is a constant coefficient [10], and the dot refers to the total derivative with respect to the time coordinate (Note that $\dot{r} = 0$ at first order in Ω). The angular frequency σ can be related to its angular frequency ω in the corotating frame and to the stellar angular velocity; at the lowest order in Ω, this relation is $\sigma = \omega - \ell\Omega$.

Note that hereafter we will consider the r-mode instability retaining only the the lowest order term in Ω. Nevertheless, we will often consider neutron stars spinning very near the break-up limit. While this approximation might be a reasonable one, it hides the possibly significant modifications that could appear when the general relativistic rotational effects are fully taken into account.

2.2 Nonlinear motions from linear equations

The linear-order equations (2)–(3) do not generate a significant secular magnetic field since they lead to unitary strain tensors [cf. Eq. (11)]. However, Eqs. (2)–(3) can provide important information about the nonlinear motions of fluid elements and, in particular, about whether they lead to a secular drift velocity. When the nonlinear

---

[1] As shown in [12], this is a rather good approximation.
[2] Hereafter we will always refer to modes for which $m = \ell$.

expressions are not available, in fact, a rather standard technique [13,14] allows to calculate second-order quantities from linear results. This is an approximation but in some relevant examples, such as sound waves and shallow water waves, one finds that the lowest-order *nonlinear* corrections to the linear velocity field make no contribution to the estimated velocity field: i.e. the drift velocity is given exactly to $\mathcal{O}(\alpha^2)$ by the linear velocity field.

We have used of this technique and obtained analytical expressions for the values of the velocity perturbations at second-order in α. In particular, we have expanded the equations of motion in powers of α, averaged over a gyration, and retained only the lowest-order non-vanishing term. Interestingly, we find that a second-order drift velocity exists in the φ-direction and the total displacement in φ from the onset of the oscillation at t_0 up to time t is then found to be[3]

$$
\begin{aligned}
\Delta \tilde{x}^{\varphi}(r, t) &\equiv \int_{t_0}^{t} \delta v_1^{\varphi}(t')dt' \\
&= \frac{2}{\ell+1}\left(\frac{r}{R}\right)^{\ell-1} \kappa_{\ell}(\theta) \int_{t_0}^{t} \alpha^2(t')\Omega(t')dt' + \mathcal{O}(\alpha^3).
\end{aligned}
\tag{4}
$$

with $t \gg P$, with P being the oscillation period. The matching coefficient $\kappa_{\ell}(\theta)$ is introduced to relate the instantaneous and secular velocities and is dependent on the mode number ℓ and on the θ position on an isobaric surface of the star. For the $\ell = 2$, $\kappa_2(\theta) \equiv (1/2)^7(5!/\pi)(\sin^2\theta - 2\cos^2\theta)$ and hence the net displacement in the azimuthal direction after an oscillation is approximately $2\pi\alpha$ times the radius of gyration.

The azimuthal drift velocity of a given fluid element is readily calculated from (4). In an orthonormal basis and for the $\ell = 2$ mode, this is

$$
\mathbf{v}_{\mathrm{d}}(r, \theta, t) = \frac{2}{3}\kappa_2(\theta)\alpha^2(t)\Omega(t)R\left(\frac{r}{R}\right)^2 \mathbf{e}_{\hat{\varphi}},
\tag{5}
$$

where $\mathbf{e}_{\hat{\varphi}}$ is the unit vector in the φ direction. Note that the secular velocity field (5) is also responsible for a differential rotation in both the radial and polar directions.

It is important to emphasize that using Eqs. (2)–(3) to compute the displacement of an element of fluid expanding $\dot{\theta}$ and $\dot{\varphi}$ in powers of α is not equivalent to considering nonlinear effects in the fluid equations (see [15] for a detailed discussion of this). However, analogous fluid-dynamical processes whose nonlinear behaviour is known suggest the existence of a secular drift velocity of $\mathcal{O}(\alpha^2)$. Moreover, we expect the drift of fluid elements given by the velocity field $\mathbf{v}_{\mathrm{d}}$ to be qualitatively correct and perhaps exact to $\mathcal{O}(\alpha^2)$. After this prediction was first made [10], the existence of an $\mathcal{O}(\alpha^2)$ drift velocity and differential rotation has been verified both in analytical toy models [16] and, at least qualitatively, through nonlinear numerical simulations both in the general relativistic Cowling approximation [17] and in Newtonian simulations [18] implementing postNewtonian radiation-reaction forces [19].

In Fig. 1 we show numerical integrations of the equations of motion (2) and (3) on the northern hemisphere of the rotating star. In particular we show the projected

---

[3] At second-order in α there is no secular motion in the $\theta-$direction.

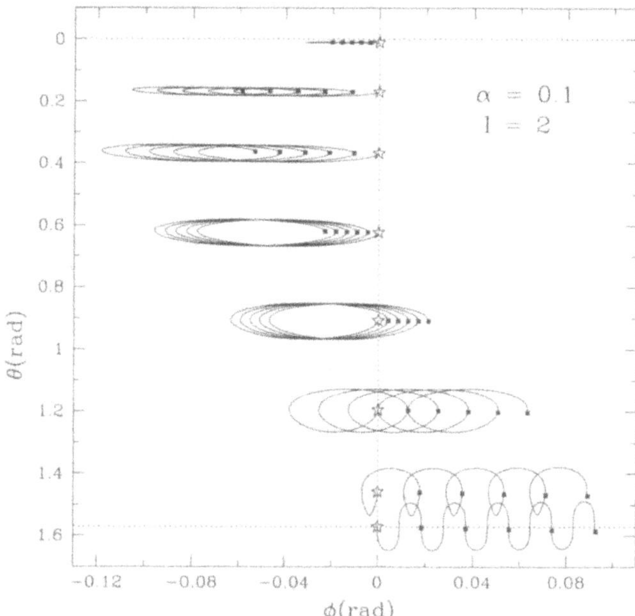

Fig. 1. Projected trajectories $\theta(t)\sin\theta(t)\cos\varphi(t)$, and $\varphi(t)\sin\theta(t)\cos\varphi(t)$ of seven fiducial fluid elements as seen in the corotating frame. The projected trajectories refer to five periods for an r-mode oscillation with $\ell = 2$, $\alpha = 0.1$, and $\Omega = \Omega_B$. All of the fluid elements are initially positioned on the $\varphi_0 = 0$ meridian and have different latitudes (indicated with stars), while the positions at the end of each oscillation are indicated with filled squares

(θ, φ) trajectories over five periods for fiducial fluid elements subject to an r-mode oscillation with $\ell = 2$ (the coordinates are those of a reference frame corotating with the star). All of the fluid elements are initially positioned on the $\varphi_0 = 0$ meridian, but at different latitudes and these positions are indicated with stars. The positions at the end of an oscillation are instead indicated with filled squares. In solving equations (2) and (3) we have assumed that α has the constant value 0.1 and that the star is rotating at the break-up limit $\Omega_B \equiv (2/3)\sqrt{\pi G \bar{\rho}}$, with $\bar{\rho} = 3M/(4\pi R^3)$.

The existence of a secular velocity field is of great importance for the generation of magnetic field and this will be discussed in more detail in Sect. 4. Before that, however, we need to define an evolutionary model for the instability and this will be briefly reviewed in the following section.

3   A phenomenological model for the evolution of the instability

In order to provide a direct comparison with results presented in the literature we have used the phenomenological model for the evolution of the instability presented by Owen et al. [7] in the case of a unmagnetized, fluid and rotating neutron star. In this model, the evolution of the r-mode instability is assumed to have three phases. The initial phase is the one during which any infinitesimal axial perturbation will grow exponentially over the timescale τ_{GW}, set by the gravitational radiation-reaction. For an $\ell = m = 2$ mode and a neutron star initially rotating at the "break-up" limit at a temperature $\sim 10^{10}$ K, this timescale has been estimated to be of the order of a few tens of seconds. The exponential growth is followed by an intermediate phase in which the mode's amplitude reaches its saturation value α_{sat}, and the star is progressively spun down because of angular momentum loss via gravitational waves. This phase has been estimated to be of the order of ~ 1 year if conventional cooling rates and viscosity estimates are used and if no crust develops on the neutron star[4]. The final phase of the evolution is the one in which the star's angular velocity is so small that viscous dissipative effects dominate the radiation-reaction forces and the r-modes start to be damped out.

The evolution equations for the mode amplitude α and for the star's angular velocity Ω can be derived after requiring that the loss rates of energy and angular momentum are the same as those recorded at infinity in the form of gravitational waves [7]. These can then be synthesized as

before saturation $\qquad\qquad\qquad$ after saturation, before decay

$$\frac{d\alpha}{dt} = -\frac{\alpha}{\tau_{GW}} - \frac{\alpha}{\tau_V}\left(\frac{1-\alpha^2 Q}{1+\alpha^2 Q}\right), \qquad \frac{d\alpha}{dt} = 0; \quad \alpha = \alpha_{sat}, \tag{6}$$

$$\frac{d\Omega}{dt} = -\frac{2\Omega}{\tau_V}\left(\frac{\alpha^2 Q}{1+\alpha^2 Q}\right), \qquad \frac{d\Omega}{dt} = \frac{2\Omega}{\tau_{GW}}\left(\frac{Q\,\alpha_{sat}^2}{1-Q\,\alpha_{sat}^2}\right), \tag{7}$$

where τ_V is the viscous timescales comprising both bulk and shear viscosity, and Q is a constant.

The solutions to Eqs. (6) and (7) are shown in Fig. 2 where we have plotted the time evolution of the mode amplitude (dotted line) and of the star's angular velocity (solid lines), normalized to the saturation value and the break-up limit, respectively (The timescales used refer to an $\ell = 2$ mode.). The main panel of Fig. 2 shows the initial phases of the amplitude growth and angular velocity decay and is, for this reason, shown on a linear temporal scale. Different solid curves refer to different values of the saturation amplitude. Note that the decrease in the star's angular velocity and that its value after one year depends sensitively on the value used for α_{sat} and becomes very small only for $\alpha_{sat} = 1$. This is apparent from the inset which shows the decay of the star's angular velocity on a logarithmic time scale.

---

[4] The presence of a crust and the generation of a viscous boundary layer at the crust-core boundary could modify sensibly the viscous damping timescale [20]. See [21] for a recent discussion of the many aspects of this process.

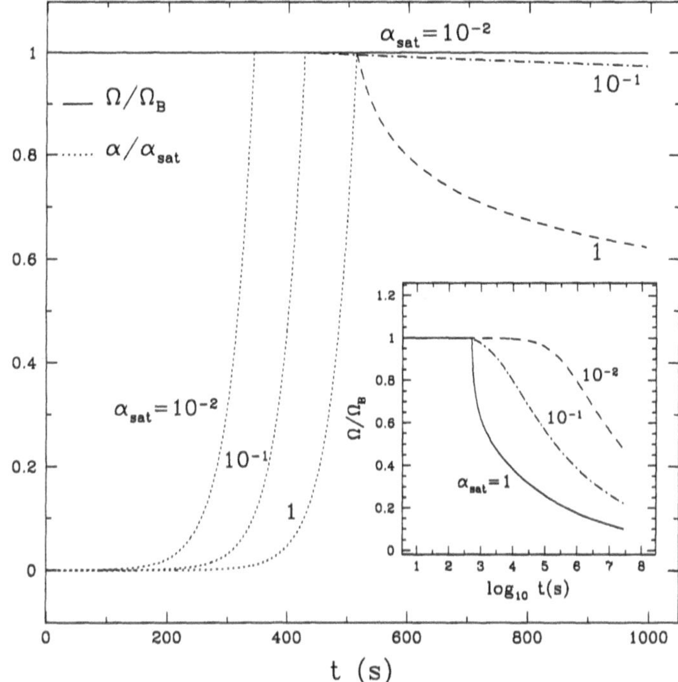

Fig. 2. Time evolution of the mode amplitude (dotted line) and of the star's angular velocity (solid lines) normalized to the saturation value and the break-up limit, respectively. Different curves refer to different values of the saturation amplitude α_{sat}. The small inset shows the decay of the star's angular velocity on a logarithmic time scale

4   Time evolution of the magnetic field: A Lagrangian approach

As mentioned in Sect. 2, secular velocity fields in a highly conducting plasma represent the necessary conditions for the generation of magnetic fields. To calculate the magnitude of these fields we have used the equations of ideal MHD (i.e. with infinite electrical conductivity) and expressed them in a Lagrangian formulation which is particularly suited to our problem. Because this is a rather unusual approach to the solution of the induction equations we briefly discuss it in what follows.

We start with combining the equation of mass conservation

$$\frac{d\rho}{dt} = -\rho \, (\nabla \cdot \mathbf{v}) \,, \tag{8}$$

where $d/dt \equiv (\partial/\partial t + \mathbf{v} \cdot \nabla)$, with the induction equation

$$\frac{\partial \mathbf{B}}{\partial t} = \nabla \times (\mathbf{v} \times \mathbf{B}) \,, \tag{9}$$

so as to obtain

$$\frac{D}{Dt}\left(\frac{\mathbf{B}}{\rho}\right) = \left(\frac{\mathbf{B}}{\rho}\cdot\nabla\right)\delta\mathbf{v}, \tag{10}$$

where we have decomposed the velocity $\mathbf{v}$ into $\mathbf{v} = \mathbf{v}_0 + \delta\mathbf{v}$, where $\mathbf{v}_0 \equiv \Omega\times\mathbf{r}$ is the uniform stellar rotation velocity and $\delta\mathbf{v}$ is the r-mode (axial) velocity perturbations[5]. Here, $D/Dt \equiv (\partial/\partial t + \mathbf{v}\cdot\nabla - \Omega\times)$ is the Lagrangian derivative for a fluid element moving at velocity $\mathbf{v}$ as seen in the corotating frame. Equation (10) can be integrated analytically to give [22,23]

$$\frac{B^j}{\rho}(\tilde{\mathbf{x}}, t) = \frac{B^k}{\rho}(\mathbf{x}, t_0)\frac{\partial\tilde{x}^j(t)}{\partial x^k(t_0)}. \tag{11}$$

Note that $\nabla\cdot\mathbf{v}_0 = 0$ at first order in the stellar angular velocity and $\nabla\cdot\delta\mathbf{v} = 0$ by definition of axial perturbations. To lowest order in Ω and α the flow is therefore incompressible and we can set $\rho(\tilde{\mathbf{x}}, t) = \rho(\mathbf{x}, t_0)$. The integral form (11) of the induction equation is particularly advantageous as it shows that the magnetic field at time t and position $\tilde{\mathbf{x}}$ can be computed from the magnetic field at time t_0 and position $\mathbf{x}$, using the tensorial coordinate strain $\partial\tilde{x}^j(t)/\partial x^k(t_0)$ that develops between t_0 and t. The advection of magnetic field lines is an obvious consequence of equation (11) and the problem of the magnetic field evolution is therefore transformed into the problem of determining the time evolution of the strain tensor $S_{jk}(t) \equiv \partial\tilde{x}^j(t)/\partial x^k$. While very compact and relatively simpler to solve numerically, equation (11) has the disadvantage of being sensitive to a correct evaluation of the strain tensor which might become inaccurate when the instability is fully developed. For this reason, and to verify the validity of the Lagrangian approach for very large saturation amplitudes, we have also implemented a more traditional Eulerian method, obtaining equivalent results (see [24]) for details).

5   Numerical results

We now present results from the numerical evolution of the magnetic field as obtained through the Lagrangian method of Sect. 4. The computations have been performed for a number of different values of the saturation amplitudes, mode numbers, and initial dipolar magnetic fields B_0.

The numerical strategy is based on the solution of the equations of motion (1) for a number of fiducial fluid elements suitably distributed on the surface of the star and calculated for several different mode numbers. The new positions after each oscillation are used to reconstruct the strain tensor $\partial\tilde{x}^j(t)/\partial x^k(t_0)$ from which the new magnetic field components are computed.

In Fig. 3 we show the time evolution of the the total secular magnetic field $\langle\Delta B\rangle$ scaled to the initial magnetic field and over a timescale of one year. The different

---

[5] In order to simplify our notation, we have dropped the index "1" for the linear velocity so that $\delta\mathbf{v} \equiv \delta\mathbf{v}_1$.

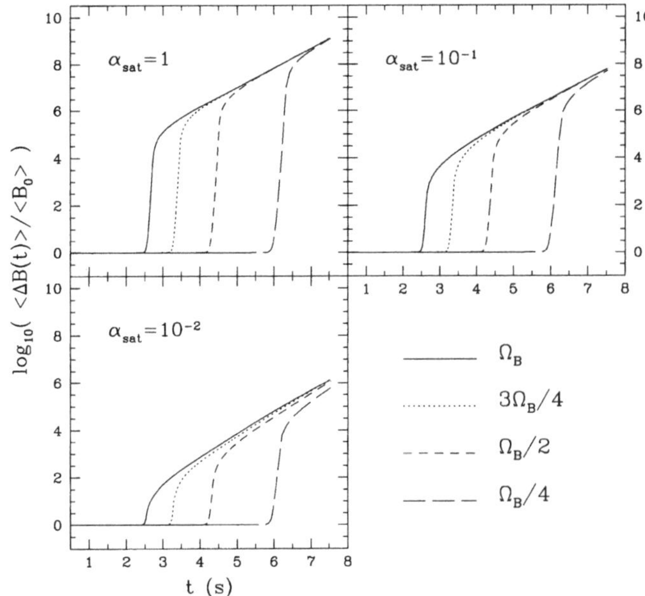

Fig. 3. Long-time evolution of the total secular magnetic field $\langle \Delta B \rangle$ scaled to the initial field for an $\ell = 2$ mode and different saturation amplitudes. The different lines refer to different values of the initial angular velocity

diagrams show how the evolution depends both on the final saturation amplitude and on the initial angular velocity of the star.

Most of the generation of the toroidal magnetic field takes place during the exponential growth of the mode and then settles onto a power law growth shortly thereafter. During this phase, any arbitrarily small toroidal magnetic field is amplified by the wrapping of the poloidal magnetic field produced by the (mostly) toroidal secular velocity field. The amplification is so large that it soon becomes comparable and larger than the seed poloidal field. Clearly, larger values of the saturation amplitude produce proportionally larger amplifications and the latter can be so dramatic in the case of $\alpha_{sat} = 1$, that the newly generated magnetic field at mode saturation has become four orders of magnitude larger than the pre-existing magnetic field. If not halted before, the toroidal magnetic field could grow to be nine orders of magnitude larger than the initial dipolar magnetic field over the timescale of one year. Note also that for stars rotating at slower angular velocities, the instability and the field growth set in at progressively later times. However, for any given saturation amplitude, the final magnetic field produced is rather insensitive of the rotation rate. Finally, it is worth mentioning that the poloidal components of the secular magnetic field do not show a significant growth and oscillate around their initial values.

6 Critical magnetic fields

We have seen how r-mode oscillations can produce secular drift velocity fields which, in a highly conducting conducting plasma such as neutron star matter, can generate intense toroidal magnetic field from pre-existing dipolar magnetic fields. It is now important to discuss the impact of these fields on the onset or growth of the r-mode oscillations. A fully self-consistent is beyond the scope of our approach. However, some important conclusions can still be drawn by considering the evolution of the instability and the generation of the magnetic fields as independent processes. In this case, the *nature* of the r-mode oscillations is unaffected by the generation of large magnetic fields and can, at any time, be described by the perturbative expressions (1). Under this hypothesis, it is then possible to calculate the strength of the magnetic field necessary to *prevent* the r-mode instability, or *suppress* the instability when this is free to develop.

6.1 Prevention of the instability

The shearing motions produced by r-mode oscillations will distort the magnetic field lines. For magnetic fields that are sufficiently intense, however, magnetic tension forces will tend to contrast these motions and distort the character of the oscillations. We will here refer to the *critical* magnetic field for the prevention of the instability it is convenient to consider the different energies in play since they can account for the global properties of the modes and of the magnetic field. Of course, during an oscillation the energy in the magnetic field will rise and fall. If there is not enough energy in the mode to supply the maximum magnetic energy increase required to complete an oscillation, a "full" r-mode oscillation will not occur (some other, smaller scale vibration might still occur). In other words, if the magnetic field exceeds a critical value, $B_{\mathrm{crit,P}}$, the magnetic stresses that build up during an oscillation will be so large that they will halt the fluid motion involved in the oscillation, i.e. the fluid momentum density will be brought to zero. The condition that defines $B_{\mathrm{crit,P}}$ is therefore $\delta E_{\mathrm{M}} = \widetilde{E}$, where $\widetilde{E}$ the energy in the mode as measured in the corotating frame of the equilibrium star

$$\widetilde{E} = \frac{1}{2}\frac{\alpha^2\Omega^2}{R^2}\int_0^R \rho\, r^6 dr \simeq 8.2 \times 10^{-3}\alpha^2 M\Omega^2 R^2, \tag{12}$$

and $\delta E_{\mathrm{M}} \equiv (1/8\pi)\int_{V_\infty} \delta B^2\, d^3\mathbf{x}$ is the change in magnetic energy density. Note that the fractional change in the mode energy produced by gravitational wave emission during a single oscillation period is $4\pi/\omega|\tau_{GW}|$. This is always $\ll 1$ and therefore can be neglected in this comparison. The expression for δE_{M} varies according to whether the neutron star core is made of a normal neutron fluid or it is superconducting, but hereafter we will refer to the first case only (see [10,24] for the superconducting case).

The induction equation can be integrated in time to estimate the variations in the magnetic field produced by the perturbation velocity field. Because of the periodic variation of the magnetic energy during an oscillation, the expression for the energy density should be averaged over half of an oscillation period, so that the magnetic energy produced is

$$\delta E_{\mathrm{M}} = \left(\frac{9\pi}{32}\right) \frac{1}{\Omega^2} \int_{V_*} \left[B_0 \frac{\delta v}{r}\right]^2 d^3\mathbf{x}, \tag{13}$$

In deriving (13) we have not included contributions to δE_{M} from changes in the magnetic field outside the star which are prevented by the high electric conductivity of the crust. Also, we have used the time average of the velocity perturbation as $\int_0^{P/2} \delta v(t')dt' \simeq (3\pi\Omega)/(2\delta v)$. Note that both energies (12) and (13) depend quadratically on the velocity perturbation δv, thus making the critical fields not sensitive to the amplitude development of the mode. The volume integral in (13) can be easily performed to give

$$\delta E_{\mathrm{M}} = \left[\frac{3\pi(1-p)}{32p}\right] \Lambda \alpha^2 B_S^2 R^3, \tag{14}$$

where $p \equiv R/R_{IN}$ and all of the angular dependence is contained in $\Lambda = \mathcal{O}(1)$. Using (14) with $p = 0.5$ we can now write the condition for mode prevention in terms of a critical field given by

$$B_{\mathrm{crit,P}} \simeq 2.5 \times 10^{16} \left(\frac{\Omega}{\Omega_B^*}\right) (M_{1.4})^{1/2} (R_{12.5})^{-1/2} \ \mathrm{G} \tag{15}$$

where $M_{1.4} \equiv M/1.4\,M_\odot$ and $R_{12.5} \equiv R/12.5\,\mathrm{km}$. The critical field (15) is larger than the one usually associated with newly-born neutron stars and it is therefore likely that for most of such stars the r-mode instability will be free to develop. Of course, an initially sub-critical magnetic field might well become critical as a result of the subsequent amplification; the determination of this new critical magnetic field is discussed in the following section.

6.2   Suppression of the instability

When the r-mode instability is not prevented by the initial magnetic field, it might be suppressed later on as a result of the newly produced magnetic field. In order to evaluate the *critical magnetic field for the suppression of the instability* we should ask whether the energy required to continue the oscillation in the next time interval Δt exceeds the energy that can be pumped into the mode by the emission of gravitational radiation. A critical point in the balance between the energy spent in producing magnetic field and the energy in the mode provided by the emission of gravitational waves is reached when the two rates are equal: i.e. $dE_{\mathrm{M}}/dt = (d\tilde{E}/dt)_{\mathrm{GW}}$, where

$$\left(\frac{d\tilde{E}}{dt}\right)_{\mathrm{GW}} = -\frac{32\pi G}{c^7} \left[\frac{4}{3(5)!!}\right]^2 \frac{\alpha^2 \Omega^4 \omega^4}{R^2} \left(\int_0^R \rho\, r^6 dr\right)^2, \tag{16}$$

is the rate of energy transferred to an $\ell = 2$ mode [7]. After the equality is reached, $dE_{\mathrm{M}}/dt > (d\widetilde{E}/dt)_{\mathrm{GW}}$. The reason for this is that the toroidal magnetic field, and hence dE_{M}/dt, will continue to grow whereas $(d\widetilde{E}/dt)_{\mathrm{GW}}$ remains the same or decreases significantly as gravitational radiation and viscous coupling to the star cause the star to spin down.

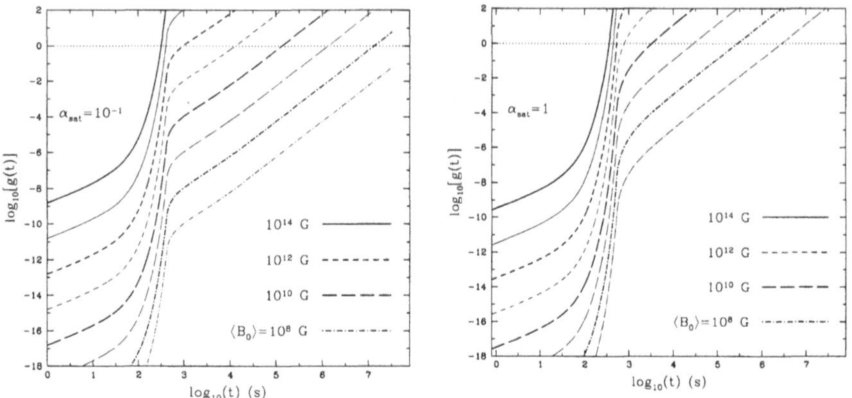

Fig. 4. Time evolution of the ratio $g(t)$. The calculation refers to our "standard" neutron star initially rotating at the break-up limit Ω_B^*. The **left** and **right** panel show the evolution for increasing values of the saturation amplitude and of the initial magnetic field $\langle B_0 \rangle$. Nearby **thick** and **thin** lines of the same type are used to compare the magnetic field evolution produced by the secular velocity (5) and a velocity field which is just a tenth of it

A measure of the relative importance of the two energy rates can be obtained through the time evolution of their ratio $g(t) \equiv (dE_{\mathrm{M}}/dt)(d\widetilde{E}/dt)_{\mathrm{GW}}^{-1}$, where the rate of production of magnetic energy takes the form

$$\frac{dE_{\mathrm{M}}(t)}{dt} = \left[\frac{(1-p)}{\pi p}\right] B_s^2 R^3 \Lambda' \alpha^2(t)\Omega(t) \int_0^t \alpha^2(t')\Omega(t')dt', \tag{17}$$

with again Λ' containing the angular dependence and being $\mathcal{O}(1)$.

Figure 4 shows the numerical computation of $g(t)$ for a "standard" neutron star initially rotating at the break-up limit Ω_B^* and four different values of the (volume averaged) initial magnetic field magnetic field $\langle B_0 \rangle$. The two panels refer $\alpha_{\mathrm{sat}} = 10^{-1}$ and 1, respectively. Note in each curve the phase of exponential growth of the magnetic field and the subsequent phase of power-law growth. Note also how different values of $\langle B_0 \rangle$ have the only effect of moving the curves in the vertical direction: the smaller the initial field, the longer it will take to reach a sufficiently strong magnetic field necessary for suppression. To account for the possible error in the estimate of the secular drift velocity and to show how the results are not very sensitive on this, we have shown with nearby thick and thin lines of the same type results with a drift velocity given by (5) and with a fictitious drift velocity which is just a tenth of the one given by (5). Finally, when read in terms of the times at which the suppression of the oscillations begins, Fig. 4 indicates that for an initial magnetic field $\langle B_0 \rangle = 10^{10} - 10^{14}$ G, the instability can be suppressed after a time between 4 days and 7 minutes for $\alpha_{\text{sat}} = 1$.

7   Conclusions

We have reviewed the onset and growth of the r-mode instability in rotating magnetized neutron stars. Because of the high conductivity of the hot neutron star matter and of the large scale mass currents produced by the instability, new surprises to the picture of the r-mode instability are brought in by the strong magnetic fields that are expected to accompany newly born neutron stars. We have here focussed our attention on the nature of the velocity field associated with the r-mode oscillations. Starting from linear expressions and working perturbatively on the mode's amplitude we have derived the expression for a toroidal secular velocity drift which is second-order in the mode's amplitude. This secular drift velocity field is often encountered in fluid-dynamical analogues of r-modes and the differential rotation it produces has been qualitatively confirmed both in simplified models [16] and in fully nonlinear simulations [17,18].

The coupling of the secular drift velocity with the pre-existing magnetic field is at the base of the generation of a toroidal magnetic field which can rapidly become many orders of magnitude larger than any seed magnetic field. The newly generated magnetic fields can influence the r-mode instability either preventing its onset (when sufficiently strong) or suppressing its saturated development (when they become so intense that the radiation-reaction cannot sustain their growth). We have also shown that it is likely that most newly born neutron stars will have an initial magnetic field sufficiently weak so as to trigger, at least for some time after their birth, the r-mode instability. We have in fact estimated that the critical initial magnetic field which would suppress even the first r-mode oscillation is $(B_0)_{\text{crit}} \sim 10^{16}$ G, a couple of orders of magnitude larger than the one usually associated with hot, young neutron stars. On the basis of our results, however, it also appears clear that the lifetime of such instability could effectively be very small. In particular, our calculations indicate that for an initial magnetic field $\langle B_0 \rangle = 10^{10} - 10^{14}$ G, the instability can be suppressed after a time between 4 days and 7 minutes for a saturation amplitude $\alpha_{\text{sat}} = 1$.

Acknowledgements. The work presented in this talk is in collaboration with F. Lamb, D. Marković and S. Shapiro. It is a pleasure to thank N. Andersson, K. Kokkotas, J. Miller, and N. Stergioulas for the numerous interesting discussions. Financial support for this research has been provided by the MURST and in part by the INFN.

References

1. Papaloizou, J., Pringle, J.E. (1978): MNRAS **182**, 423
2. Andersson, N. (1998): ApJ **502**, 708
3. Friedman, J.F., Morsink, S.M. (1998): ApJ **502**, 714
4. Andersson, N., Kokkotas, K.D. (2000): J. Mod. Phys. D **10**, 381
5. Friedman, J.L., Lockitch, K.H. (2001): Implications of the r-mode instability of rotating relativistic stars. In: Proceedings of the IX Marcel Grossman Meeting, Rome, July 2000, ed. by Gurzadyan V., Jantzen, R., Ruffini, R., World Scientific, in press
6. Lindblom, L., Owen, B.J., Morsink, S.M. (1998): Phys. Rev. Lett. **80**, 4843
7. Owen, B.J., Lindblom, L., Cutler, C., Schutz, B.F., Vecchio, A., Andersson, N. (1998): Phys. Rev. D **58**, 084020
8. Andersson, N., Kokkotas, K.D., Schutz, B.F. (1999): ApJ **510**, 846
9. Ho, W.C.G., Lai, D. (2000): ApJ, **543**, 386–394
10. Rezzolla, L., Lamb, F.K., Shapiro, S.L. (2000): ApJ **531**, L141
11. Spruit, H.C. (1999): Astron. and Astrophys. **341**, L1
12. Schenck, A.K., Arras, P., Flanagan, E.E., Teukolsky, S.A., Wasserman, I. (2000): astro-ph/000000
13. Landau, L.D., Lifshitz, E.M. (1987): Fluid Mechanics. Pergamon Press, Oxford, Great Britain. Sect. 65
14. Lighthill, J. (1980): Waves in Fluids. Cambridge University Press, Cambridge, Great Britain, p. 279
15. Rezzolla, L., Marković, D., Lamb, F.K., Shapiro, S.L. (2001): Phys. Rev. D, in press, Paper I
16. Levin, Y., Ushomirsky, G. (2000): MNRAS **322**, 515
17. Stergioulas, N., Font, J.A. (2001): Phys. Rev. Lett. **86**, 1148–1151
18. Lindblom, L., Tohline, J.E., Vallisneri, M. (2001): Phys Rev. Lett. **86**, 1152–1155
19. Rezzolla, L. (1999): ApJ **525**, 935
20. Bildsten, L., Ushomirsky, G. (2000): ApJ Lett. **529**, L33
21. Lindblom, L. (2001): Neutron star pulsations and instabilities, in Proceedings of the International Meeting "Gravitational Waves: A Challenge to Theoretical Astrophysics", Trieste June 2000, ed. by V. Ferrari, J.C. Miller, L. Rezzolla, ICTP Lecture Notes Series, Vol. **3**, p.257
22. Parker, E.N. (1979): Cosmical magnetic fields. Clarendon Press, Oxford, Great Britain, Sect. 4.3
23. Balbus, S.A., Hawley, J.F. (1998): Rev. Mod. Physics, **70**, 1
24. Rezzolla, L., Marković, D., Lamb, F.K., Shapiro, S.L. (2001): Phys. Rev. D, in press, Paper II

Strong Field Gravity and Quasi-Periodic Oscillations from Low-Mass X-ray Binaries

L. Stella, M. Vietri

Abstract. The relativistic precession model for quasi periodic oscillations, QPOs, in low mass X-ray binaries is reviewed. The behaviour of three simultaneous types of QPOs is well matched in terms of the fundamental frequencies for geodesic motion in the strong field of the accreting compact object for reasonable star masses and spin frequencies. The model works for neutron star as well as black hole candidate systems, as it ascribes the higher frequency kHz QPOs, the lower frequency kHz QPOs and the horizontal branch oscillations to the Keplerian, periastron precession and nodal precession frequencies of matter orbiting close to the inner edge of the accretion disk. The remarkable correlation between the centroid frequency of QPOs in both neutron star and black hole candidate low mass X-ray binaries is very well fit by the model. Some testable predictions are described. QPOs from low mass X-ray binaries might provide an unprecedented laboratory to test strong field general relativity.

1 Introduction

Old accreting neutron stars, NSs, in low mass X-ray binaries, LMXRBs, display a complex variety of quasi-periodic oscillations, QPOs, in their X-ray flux. These QPOs are revealed and studied through the broad peaks that they give rise to in the power spectra of the X-ray light curves. The *low frequency* QPOs ($\sim$ 1–100 Hz) that were discovered and studied from high luminosity "Z-sources" in the eighties are further classified into horizontal, normal and flaring branch oscillations (HBOs, NBOs and FBOs, respectively), depending on the simultaneous position occupied by a source in the X-ray colour-colour diagram (for a review see [1]). The kHz QPOs ($\sim$ 0.2 to $\sim$ 1.3 kHz) that were revealed and investigated with RXTE in a number of NS LMXRBs (see [2,3] and references therein) involve timescales similar to the dynamical timescales close to the NS. A common phenomenon is the presence of a pair of kHz QPOs (centroid frequencies of ν_1 and ν_2) which drift in frequency while mantaining their frequency difference $\Delta\nu \equiv \nu_2 - \nu_1 \approx$ 250–360 Hz roughly constant. Detailed studies showed that in four sources $\Delta\nu$ decreases significantly (by up to $\sim$ 100 Hz) as ν_2 increases; these are Sco X-1 [4], 4U1608-52 [5,6], 4U1735-44 [7] and 4U1728-34 [8]. Owing to poor statistics, a similar variation of $\Delta\nu$ in other sources would have remained undetected [9].

The kHz QPOs show remarkably similar properties across NS LMXRBs of the Z and Atoll groups, the luminosity of which differs by a factor of $\sim$ 10 on average. During type I bursts from six Atoll sources, a nearly coherent signal at a frequency of $\nu_{\text{burst}} \sim$ 290–580 Hz has also been detected (for a review see [10]). In a few cases ν_{burst} is consistent, to within the errors, with the frequency separation of the kHz QPO pair $\Delta\nu$ or twice its value $2\Delta\nu$. Yet in two sources (4U1636-53, [11], and 4U1728-34, [8]) ν_{burst} is significantly different from $\Delta\nu$ and its harmonics.

The presence of HBOs has been firmly established in both Atoll and Z-sources. Their frequency, ν_{HBO} (~ 15 to $\sim 60\,\text{Hz}$) shows a nearly quadratic dependence ($\sim \nu_2^2$) on the higher kHz QPO frequency. The frequency changes of the kHz QPOs and HBOs in individual sources are positively correlated with the instantaneous accretion rate.

QPO relative amplitudes are usually in the several percent (*rms*) range, although values as high as 10–15 % are not uncommon in the kHz QPOs of low luminosity Atoll sources. The energy spectrum of the QPOs must be somewhat harder than the source spectrum, as the relative QPO amplitude increases with photon energy. In most cases the coherence of the QPOs is limited to Q-values of ~ 1–10 (the quality factor Q is the ratio of the centroid frequency and width of the QPO power spectrum peak); on occasions Q of ~ 100–200 have been inferred for the kHz QPOs of a few Atoll sources.

A remarkable correlation between the centroid frequencies of QPOs (or peaked noise components) from LMXRBs was discovered ([12]). This correlation extends over nearly 3 decades in frequency and encompasses both NS and black hole candidate, BHC, systems.

The frequencies of these QPOs, despite their quasi-periodic nature, provide the most accurately measured observables of LMXRBs. A primary goal of any QPO model is therefore to explain the frequency range and dependence of the different QPO types of these sources. A variety of alternative QPO models has been proposed (see e.g. [13]). The basic features of the relativistic precession model, RPM, are reviewed here [14–16]. In the RPM the QPO signals arise from the fundamental frequencies

2   Periastron precession frequency and kHz QPOs

We consider here only infinitesimally eccentric and tilted orbits, under the assumption that the motion of matter in the innermost disk regions is dictated by the star's gravity alone. In the case of a circular geodesic in the equatorial plane ($\theta = \pi/2$) of a Kerr black hole of mass M and specific angular momentum a, the coordinate frequency measured by a static observer at infinity is

$$\nu_\varphi = \pm M^{1/2} r^{-3/2} [2\pi(1 \pm a M^{1/2} r^{-3/2})]^{-1}, \tag{1}$$

(we use units such that $G = c = 1$). The upper sign refers to prograde orbits. If we slightly perturb a circular orbit in the r and θ directions, the coordinate frequencies of the small amplitude oscillations within the plane (the epicyclic frequency ν_r) and in the perpendicular direction (the vertical frequency ν_θ) are given by (see [17] and references therein)

$$\nu_r^2 = \nu_\varphi^2 (1 - 6Mr^{-1} \pm 8a M^{1/2} r^{-3/2} - 3a^2 r^{-2}), \tag{2}$$

$$v_\theta^2 = v_\varphi^2(1 \mp 4aM^{1/2}r^{-3/2} + 3a^2r^{-2}). \tag{3}$$

In the Schwarzschild limit ($a = 0$) v_θ coincides with v_φ, such that the nodal precession frequency $v_{\mathrm{nod}} \equiv v_\varphi - v_\theta$ is identically zero. v_r, instead, is always lower than the other two frequencies, reaching a maximum for $r = 8M$ and going to zero at $r_{ms} = 6M$. This qualitative behaviour of v_r is preserved in the Kerr field ($a \neq 0$). Therefore the periastron precession frequency $v_{\mathrm{per}} \equiv v_\varphi - v_r$ is dominated by a "Schwarzschild" term over a wide range of parameters.

In the RPM the higher and lower frequency kHz QPOs are identified with $v_2 = v_\varphi$ and $v_1 = v_{\mathrm{per}}$, respectively. Therefore $\Delta v \equiv v_2 - v_1 = v_\varphi - (v_\varphi - v_r) = v_r$. For $a = 0$, Eqs. (1) and (2) give

$$v_r = v_\varphi(1 - 6M/r)^{1/2} = v_\varphi[1 - 6(2\pi v_\varphi M)^{2/3}]^{1/2}. \tag{4}$$

The curves in Fig. 1A show v_r vs. v_φ for $a = 0$ and selected values of M, the only free parameter in Eq. (4). The measured Δv vs. v_2 for 11 NS LMXRBs is also plotted. It is apparent that for NS masses of $\sim 2\ \mathrm{M}_\odot$, the simple model above is in qualitative agreement with the measured values, including the decrease of Δv for increasing v_2 seen in Sco X-1, 4U1608-52, 4U1735-44 and 4U1728-34. Remarkably, most points are close to the maximum of the epicyclic frequency.

The model above is only an approximation: first, the spacetime around a fast rotating NS is different from a Kerr spacetime (due to the star's oblateness induced by rotation); second, the effects of a finite (though small !) eccentricity are not taken into account.

Analytical formulae to partly correct for these effects were derived in [17]; Fig. 1B shows the fit to the observed Δv vs. v_2 relationship in Sco X-1 that was obtained through them. The orbital eccentricity, in particular, was varied in order to obtain different frequencies, while keeping the periastron distance $r_p = a(1 - e)$ fixed. The best model for a non-rotating NS is shown in Fig. 1B. The model reproduces fairly accurately the data with a minimum number of free parameters, the NS mass ($M \sim 1.9\ \mathrm{M}_\odot$) and periastron distance ($r_p \simeq 6.2\ M$). The latter value is close to the marginally stable orbit radius. When the NS spin is allowed a finite value (say $v_s \sim 300$–$600\ \mathrm{Hz}$), fits of very similar quality are obtained, the parameters of which differ only slightly from those given above. In essence, the effects induced by the NS rotation on v_r are small, though non-negligible.

The behaviour of the curves in Fig. 1 A,B, and therefore the ability of the model to match the observations, reflects the properties of the strong field Schwarzschild metric, since lower order expansions fail to reproduce the observed frequencies (see [17]).

Within the RPM, the maximum value of $v_r = \Delta v$ depends mainly on the mass of the compact object. The NS masses deduced from the simple modelling in Fig. 1 A,B are in the ~ 1.8–$2.0\ \mathrm{M}_\odot$ range, in agreement with the only relatively accurate mass measurement from optical spectro-photometry in any of these systems (Cyg X-2; $M = 1.78 \pm 0.23\ \mathrm{M}_\odot$, [18]). In general, within the RPM, Δv should not be obviously related to the NS spin frequency v_s. Therefore, it seems natural to identify v_s with v_{burst}, i.e. the relatively stable frequency seen during type I X-ray bursts (for a review

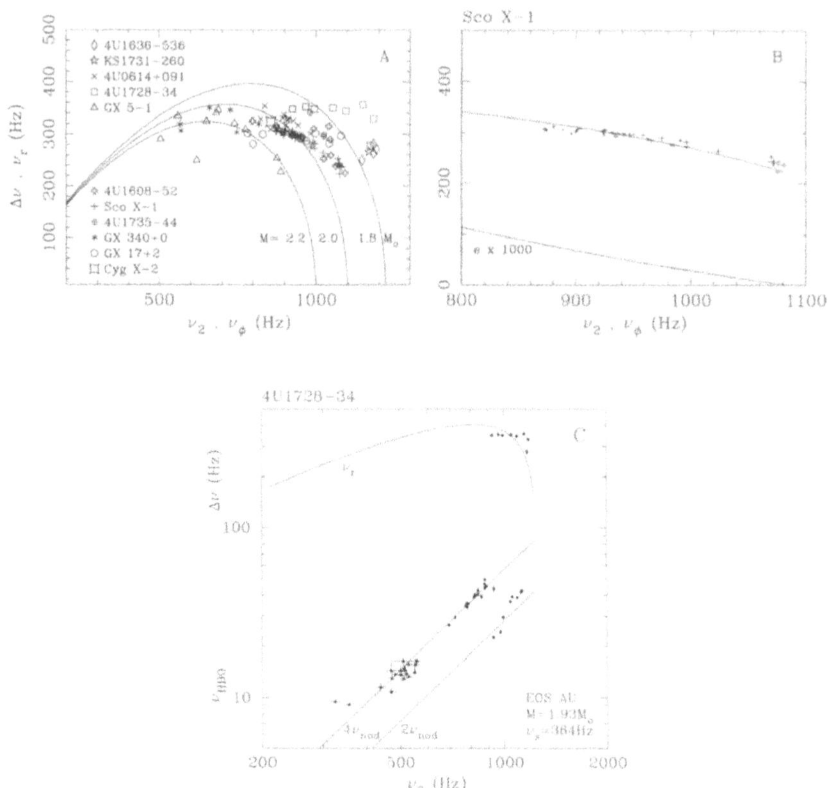

Fig. 1. (**A**) kHz QPO frequency difference $\Delta\nu$ versus higher QPO frequency ν_2 for 11 LMXRBs. Error bars are not plotted for the sake of clarity. The curves give the r- and φ-frequencies of matter in nearly circular orbit around a non-rotating neutron star, of mass 2.2, 2.0 and 1.8 $M_\odot$. (**B**) $\Delta\nu$ versus ν_2 in Sco X-1. The best fit model corresponds to the r- and φ-frequencies of matter orbiting a non-rotating 1.90 $M_\odot$ neutron star at a periastron distance of 6.25 M (17.5 km). The line marked with e gives the orbital eccentricity ($\times 1000$) (**C**) kHz QPO frequency difference $\Delta\nu$ and (double-branched) HBO frequency versus higher QPO frequency ν_2 in 4U1728-34 ([19,20,8]). The solid lines give the r-frequency and the 2nd and 4th harmonics of the nodal precession frequency ν_{nod} as a function of the φ-frequency for infinitesimally eccentric and tilted orbits in the spacetime of a 1.93 $M_\odot$ neutron star spinning at 364 Hz (EOS AU; [21])

see [10]; for a different point of view see [13]). The distribution of NS spins inferred in this way is fairly wide ($\sim$ 290–580 Hz) and compares well with that of millisecond radio pulsars, MSPs, in agreement with evolutionary scenarios in which LMXRBs are the progenitors of MSPs. The 401 Hz spin of SAX J1808.4-3658, the only bursting LMXRB displaying coherent pulsations in its persistent emission [22,23], is also in the range of spin frequencies deduced from ν_{burst}. None of the LMXRBs of the (high luminosity) Z-class has yet displayed burst oscillations; therefore their spin period is still to be measured. Cyg X-2 and GX 17+2, the only type I X-ray bursters in the group, might provide this information.

3 Nodal precession frequency and HBOs

In the RPM the HBO frequency is related to the nodal precession frequency, ν_{nod}, at the same radius at which the signals at ν_φ and ν_{per} are produced. From Eqs. (1) and (3) ν_{nod} can be written in the slow rotation limit ($a/M \ll 1$)

$$\nu_{\text{nod}} \simeq 4\pi a \nu_\varphi^2 \simeq 6.2 \times 10^{-5}(a/M)m\nu_\varphi^2 \, \text{Hz} \simeq 4.4 \times 10^{-8} \, I_{45}m^{-1}\nu_\varphi^2\nu_s \, \text{Hz},$$

$$(5)$$

where $M = m \, M_\odot$. This is the well known Lense–Thirring nodal precession formula. The latter equality refers to a rotating NS, where $aM = 2\pi\nu_s I$, with $I = 10^{45}I_{45}$ g cm^2 its moment of inertia.

If ν_φ and ν_s are measured, the only parameter in Eq. (5) that is not identified from observations is $I_{45}m^{-1}$; this can vary over a limited range, $0.5 < I_{45}m^{-1} < 2$, for virtually any mass and EOS (see the rotating NS models of refs. [24] and [25]). The stellar oblateness induced by the star's rotation gives rise to correction terms in the nodal precession frequency also (see [26] for a post-Newtonian formula). Their relative importance increases for high ν_s and ν_φ. Yet the Lense–Thirring term dominates over a wide range of parameters, such that a $\sim \nu_\varphi^2$ dependence is expected for ν_{nod}.

An approximately quadratic dependence of ν_{HBO} on the higher frequency kHz QPOs has been measured in a number of LMXRBs. This dependence was originally suggested on the basis of a few power spectra of the Atoll source 4U1728-34 [14]. Ref. [20] analysed a large set of power spectra from the same source and determined that the frequency ν_{low} of the ~ 10–50 Hz QPOs scales as $\nu_2^{2.11\pm0.06}$. (Note that the low frequency QPO vs. ν_2 relation of this source appears to be double-branched, with the centroid frequency shifting by a factor of ~ 2 across different observations; this suggests that on occasions the 2nd harmonic of ν_{HBO} is excited instead of the fundamental.) Ref. [15] first noticed that the HBO frequency of the Z-source GX 17+2 displays a nearly quadratic dependence on ν_2. [27] carried out a systematic study of Z-sources and determined that the HBO frequency is consistent with a ν_2^2 scaling (Cyg X-2 and Sco X-1 show evidence for a somewhat flatter dependence). In essence these results confirmed one of the basic features of the RPM, namely the nearly quadratic dependence of the nodal precession frequency on the φ-frequency.

If the NS spin frequency is measured, then for any value of ν_φ the model yields a predicted nodal precession frequency which is uncertain only by a factor of a few, mainly due to the allowed range of $I_{45}m^{-1}$ [14]. In the Atoll source 4U1728-34 burst oscillations and simultaneous kHz QPOs and HBOs were detected unambiguously [8,19,20]. Therefore its QPO frequencies can be used to test both the ν_{HBO} and $\Delta\nu$ versus ν_2 relationships predicted by the RPM, when the NS spin derived from burst oscillations is used ($\nu_{\text{burst}} \simeq 364$ Hz). In order to take fully into account of all the effects that contribute determining geodetic motion in the vicinity of the NS, we adopted a numerical approach and computed the spacetime metric of the star using Stergioulas' code, an equivalent of that of [25] (see [28]). From this, ν_r and ν_{nod} were derived as a function of ν_φ for geodesics with very small tilt angles and eccentricities [16,26].

Figure 1C shows the measured values of $\Delta\nu$ and ν_{HBO} versus ν_2 in 4U1728-34. Relatively high NS masses (see also Sect. 2) and stiff EOSs such as AU and UU [21] are required in this simple application of the RPM. The solid lines in Fig. 1C are for a 1.93 $M_\odot$ NS with EOS AU and $\nu_s = 364$ Hz. A good agreement is obtained if the HBO frequency, the lower of the two branches seen in 4U1728-34, is identified with the 2nd harmonics of ν_{nod} (i.e. $2\nu_{nod}$; see also [17,26]). Correspondingly the upper HBO branch is well fit by $4\nu_{nod}$. The geometry of the region producing the QPOs in the inner disk might be such that a stronger signal is produced at the even harmonics of the nodal precession frequency (e.g. [27]). The frequency range and trend of the epicyclic frequency ν_r in this model are also in reasonable agreement with the $\Delta\nu$ measurements; a more complex model is clearly required in order to fit these data more accurately.

In summary, the model presented here is capable of reproducing the salient features of both the $\Delta\nu$ versus ν_2 and ν_{HBO} versus ν_2 relationships of 4U1728-34, with just two free parameters (M and the EOS), the allowed range of which is very limited (moreover the EOS cannot even be varied continuously !). Concerning Z-sources and all other Atoll sources for which burst oscillations have not been detected yet, the NS spin can still be regarded as a free parameter. The application of the RPM is correlation (hereafter PBV correlation) involves both NS and BHC LMXRBs spanning different classes and a wide range of luminosities (see the points in Fig. 2). In kHz QPO NS systems, these components are the lower frequency kHz QPOs, ν_1, and the low frequency, HBO or HBO-like QPOs, ν_{HBO}. For BHC systems and lower luminosity NS LMXRBs the correlation involves either two QPOs, or a QPO and a peaked noise component. In all cases the frequency separation is about a decade and an approximate linear relationship ($\nu_{HBO} \sim \nu_1^{0.95}$) holds. The QPO frequencies from the peculiar NS system Cir X-1 varies over nearly a decade while closely following the PBV correlation and bridging its low and high frequency ends. Ref. [12] noted also that the ν_2 vs. ν_1 relations of different Atoll and Z-sources line-up with good accuracy.

The RPM matches precisely the PBV co. For all QPO sources, including BHCs, we use $\nu_{HBO} \simeq 2\nu_{nod}$ as in 4U1728-34 (see Sect.3). Figure 2A shows $2\nu_{nod}$ and ν_φ obtained from Eqs. (1)–(3) as a function of ν_{per} for corotating orbits and selected values of M and a/M. The high frequency end of each line is dictated by the orbital radius reaching the marginally stable orbit.

The separation of the lines in Fig. 2A testifies that while ν_{nod} depends weakly on the mass and more strongly on a/M, the opposite is true for ν_φ. By taking the weak field ($M/r \ll 1$) and slow rotation ($a/M \ll 1$) limit of Eqs. (1)–(3) the relevant first order dependence is made explicit,

$$\nu_\varphi \simeq (2\pi)^{-2/5} 3^{-3/5} M^{-2/5} \nu_{per}^{3/5} \simeq 33\, m^{-2/5} \nu_{per}^{3/5}\, \text{Hz}, \tag{6}$$

$$\nu_{nod} \simeq (2/3)^{6/5} \pi^{1/5} (a/M) M^{1/5} \nu_{per}^{6/5} \simeq 6.7 \times 10^{-2}\, (a/M) m^{1/5} \nu_{per}^{6/5}\, \text{Hz}. \tag{7}$$

For the case of rotating NSs we adopt the numerical approach outlined in Sect. 3. Results are shown in Fig. 2B for a NS mass of $1.95\,M_\odot$, EOS AU and $\nu_s = 300, 600, 900$ and $1200\,\mathrm{Hz}$ (corresponding to $a/M = 0.11, 0.22, 0.34$ and 0.47, respectively). Note that the approximate scalings in Eqs. (6) and (7) remain valid over a wide range of frequencies. Only for the largest values of ν_{per} and ν_s, ν_{nod} departs substantially from the $\sim \nu_{\mathrm{per}}^{6/5}$ dependence.

The measured QPO and peaked noise frequencies giving rise to the PBV correlation are also plotted in Fig. 2B. Higher kHz QPO frequencies from NS systems (ν_2) are included (for the sake of clarity NBOs and FBOs were excluded). The agreement over the range of frequencies spanned by each kHz QPO NS system should not be surprising: together with the accurate matching of the corresponding ν_2 vs. ν_1 relationship in Z-sources, this is indeed part of the evidence on which the RPM model was proposed. However the fact that the dependence of ν_{nod} on ν_{per} matches the observed $\nu_{\mathrm{HBO}} - \nu_1$ correlation to a good accuracy over ~ 3 decades in frequency (down to ν_1 of a few Hz), encompassing both NS and BHC systems, provides additional independent evidence in favor of the RPM. The observed variation of ν_{HBO} and ν_1 in individual sources (Cir X-1 is the most striking example, see Fig. 2B) further supports the scaling predicted by the RPM. The matching of the observed ν_2 vs. ν_1 relation in terms of ν_φ vs. ν_{per} is also quite accurate.

For EOS AU and $m = 1.95$, the ν_{HBO} vs. ν_1 values of most NS LMXRBs are best matched for ν_s in the ~ 600 to $900\,\mathrm{Hz}$ range. It is apparent from Fig. 2B that ν_s as low as $\sim 300\,\mathrm{Hz}$ are required for the Atoll sources with ν_{HBO} somewhat below the main PBV correlation: these are 4U1728-34 (see Sect. 3) and 4U1608-52 (from which burst oscillations have not been detected yet). The values above are close to the range of ν_s inferred from ν_{burst} in a number of other Atoll sources [3]. Z-type LMXRBs appear to require ν_s in the ~ 600 to $900\,\mathrm{Hz}$ range, a possibility that is still open since for none of these sources there exists yet a ν_s measurement. Note that the upper HBO branch of 4U1728-34 matches well the main PBV correlation. In the interpretation of Sect. 3 the lower branch corresponds to $2\nu_{\mathrm{nod}}$ and the upper branch to $4\nu_{\mathrm{nod}}$. One could further speculate that sources following the main PBV correlation, Z-sources in particular, are also in the upper HBO branch at $4\nu_{\mathrm{nod}}$; in this case their ν_s might be expected in the ~ 300–$400\,\mathrm{Hz}$ range.

For BHC LMXRBs the scatter around the PBV correlation implies values of a/M of ~ 0.1–0.3 (see Fig. 2A). The points from XTE J1550-564, while inconsistent with any single value of a/M, might lie along two distinct branches separated by a factor of ~ 2 in ν_{HBO}, similar to the case of 4U1728-34. In the RPM the high frequency BHC QPOs (e.g. the $\sim 300\,\mathrm{Hz}$ QPOs of GRO1655-40) are interpreted terms of $\nu_1 = \nu_{\mathrm{per}}$. This is at variance with the $\nu_1 = \nu_{\mathrm{nod}}$ interpretation of ref. [29], which requires high values of a/M (~ 0.95 in GRO1655-40), in contrast with BHC accretion-driven spin-up scenarios ([30]). From Fig. 2A it is apparent that the $\sim 300\,\mathrm{Hz}$ QPOs from GRO 1655-40 lie close to the high frequency end of the $a/M = 0.1$, $m = 7$ line. Since the mass of the BHC in GRO 1655-40 determined through optical observations is $\sim 7\,M_\odot$ ([31]), we conclude that, according to the RPM, $\nu_{\mathrm{per}} \simeq 300\,\mathrm{Hz}$ close to

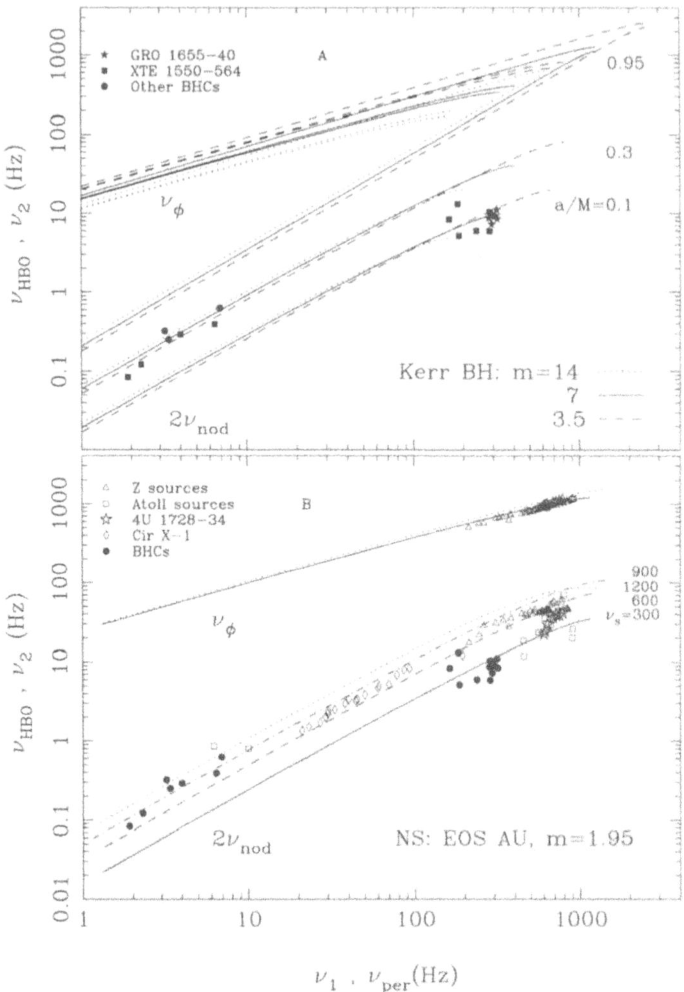

Fig. 2. Twice the nodal precession frequency, $2\nu_{nod} = \nu_{HBO}$, and φ-frequency, $\nu_\varphi = \nu_2$, vs. periastron precession frequency, $\nu_{per} = \nu_1$, for black hole candidates of various masses and angular momenta (panel A) and rotating neutron star models (EOS AU, $m = 1.95$) with selected spin frequencies (panel B). The measured QPO (or peaked noise) frequencies ν_1, ν_2 and ν_{HBO} giving rise to the PBV correlation are also shown in panel B for both BHC and NS LMXRBs and in panel A for BHC LMXRBs only; errors bars are not plotted (see [12] for a complete list of references). We included only those cases in which QPOs at ν_1 were unambiguously detected. NBO and FBO frequencies are not plotted

the marginally stable orbit, where by definition $\nu_\varphi = \nu_{per}$. This suggests that any additional QPO signal at $\nu_2 = \nu_\varphi$ would be very close to or even blended with the QPO peak at ν_1. The detection of two closeby or even partially overlapping QPO peaks close to $\sim 300\,\mathrm{Hz}$ in GRO 1655-40 would therefore provide further evidence in favor of the RPM interpretation.

4 Discussion

The RPM naturally explains the frequency range and dependence of the kHz QPOs and HBOs in NS LMXRBs, as well as the PBV correlation, which involves both NS and BHC systems. The model has a minimum number of free parameters. Testable predictions include:

(a) The frequency difference $\Delta\nu = \nu_r$ is expected to decrease also for low values of $\nu_2 = \nu_\varphi$ (see Eq. (4) and Fig. 1). Moreover $\Delta\nu$ should quickly decrease as ν_2 increases further and the orbital radius approaches the marginally stable orbit (see Fig. 1A).

(b) $\nu_2 = \nu_\varphi$ is expected to scale as $\nu_1^{3/5} = \nu_{\mathrm{per}}^{3/5}$. Extending the ν_2 vs. ν_1 correlation in NS systems toward lower frequencies and detecting the signal at ν_2 in BHC systems would provide important new tests.

In the RPM the QPO signals originate close to $r/M \sim 100\,m^{-2/5}\nu_{\mathrm{per}}^{-2/5}$ (see Eq. (6)). This radius, for individual sources, must decrease for increasing mass accretion rates (as $\dot{M}$ is positively correlated with e.g. ν_2). Inferred values range from close to the marginally stable orbit $(r/M \sim 6)$ to $r/M \sim 30$ over the frequency span covered by the PBV correlation (see Fig. 2). Many NS and BHC LMXRBs display X-ray spectra consisting of a soft component, usually interpreted in terms of emission from an optically thick accretion disk, and a harder, often power-law like component, likely due to a hot inner disk region. QPOs might originate at the transition radius between the optically thick disk and the hot region [16,32].

Since within the RPM the three QPO signals are produced by the same matter at the same radius, there should exists some coupling between the modulation at the three fundamental frequencies of motion. The characteristics and extent of such coupling will depend on the details of the mechanism giving rise to the modulation. Any coupling, however, will produce sidelobes around the main QPO peaks. Since, at first order, the amplitude of any sidelobe signal scales like the product of the amplitudes of the main signals, the QPO amplitudes being relatively small, the sidelobe amplitude should be smaller still. Ref. [33] recently discovered a low amplitude third kHz QPO peak tens of Hz longwards the lower frequency kHz QPO peak in three atoll sources. These authors report also that the frequency difference between the new peak and ν_1 is significantly higher than the frequency of the HBOs that are detected simultaneously in the same sources. Yet, it is premature to conclude that this third kHz QPO peak represents a different phenomenon than a sideband of the HBOs longwards of the lower frequency kHz QPOs. In fact the third kHz QPO peak was detected by averaging a large number of spectra with the using the so-called "shift and add" technique, where individual power spectra are added together after their frequency is shifted so as to keep the position of the peak at ν_1 fixed despite its frequency variations. Even if the systematic uncertainties of this technique were removed, it can be demonstrated that averaging the sidelobes over a range of ν_1 (and ν_{HBO}) may result in an average sidelobe peaking at a different frequency than the average HBO frequency. This is because of the amplitude variations that are known to accompany any QPO frequency variation. More extensive measurements with high signal to noise are required in order to settle this issue.

The success of the RPM is still largely based on frequency coincidence: a test-particle approach is adopted which is at best a very simple approximation. A complete and satisfactory theory cannot include blobs (they are sheared very quickly), and may contain a number of complex phenomena like magnetic coupling between the disk and the accretor, heating of the disk by the accretor's luminosity, radiation drag, and so on. Yet it seems reasonable that such theory is subject to the following constraints:

(1) the existence of clear correlations among all frequencies indicates that a coherent explanation for all of them is to be found simultaneously within a single framework;
(2) the extension of these correlations to black hole systems indicates that the phenomenon is either intrinsic to the accretion disk, or may be excited by the interaction of the disk with the accretor's (non-pulsing) luminosity;
(3) the considerable amplitude of the modulation (up to $\sim 15\,\%$) indicates that the phenomenon giving rise to QPOs does not originate in small regions of the accretion disk;
(4) the presence of strong damping due to viscosity clearly points to the existence of an external mechanism for excitation.

Theoretical efforts have so far concentrated on a specific subset of observations. For instance, ref. [34] has studied a new family of bending and precessing eigenmodes of a viscous accretion disk in its inner regions, reporting the existence of a new class of modes which are less promptly dissipated than expected, and which might explain the low–frequency QPOs. These modes are akin to test-particle motion (see also [35]), in that they are confined to very thin annuli surrounding the innermost disk radius, and thus they clearly fail to satisfy constraint no. (3). Furthermore, no plausible excitation mechanism is suggested. Constraint no. (3) plagues eigenmode solutions in general: the model by [36] where excitation of the Lense–Thirring peak is provided by magnetohydrodynamic coupling of the accretor's dipole field with the disk screening currents, suffers of this problem and, of course, of its inability to explain the phenomenology of black holes (constraint no. (2)).

A more promising approach by [37] leaves unaddressed the issue of excitation and considers just the response of a thin annular ring to a white-noise spectrum of small-amplitude perturbations. The authors find that the ring can respond only at selected frequencies close to the orbital frequencies for test-particle motion in the same gravitational potential, plus a red-noise spectrum similar to that observed in real sources. In this approach, it is difficult to understand how the excitation can be confined to a small annulus, in order not to spoil the small width of the kHz peaks thusly obtained. Constraint no. (3) also looms large.

Ideally, one would like to identify an eigenmode, or a coherent structure, covering a fairly wide range in radius (so as to circumvent constraint no. (3)), of hydrodynamical origin (to elude constraint no. (2)), giving rise simulataneously to all QPOs (see constraint no. (1)), and feeding on small-wavelength, small-amplitude perturbations which are always present in a turbulent hydrodynamical structure (see constraint no. (4)). We remark that vortices in accretion disks [38–40] have exactly these properties; we will explore this possibility in a future paper (Vietri & Stella, in preparation).

If confirmed, the RPM will provide an unprecedented opportunity to measure GR effects in the strong field regime, such as the periastron precession in the vicinity of the marginally stable orbit and the radial dependence of the Lense–Thirring nodal precession frequency. In principle, accurately measured kHz QPO and HBO frequencies would yield crucial information on the compact object such as its mass and angular momentum (e.g. by solving Eqs. (1)–(3) for m, a/M and r). Should suitable, additional observables be found, it might become possible to obtain a self-consistency check of the RPM, together with tests of GR in the strong field regime.

References

1. van der Klis, M. (1995): In: X-ray Binaries, Eds. W. H. G. Lewin, J. van Paradijs, E. P. J. van den Heuvel: Cambridge University Press, p. 252
2. van der Klis, M. (1998): In: Proc. NATO ASI, The many faces of neutron stars, Series C, Vol. **515**, p. 337
3. van der Klis, M. (2000): Ann. Rev. As. Ap. **38**, 717
4. van der Klis, M. (1997): ApJ **481**, L97
5. Mendez, M. (1998a): ApJ **494**, L65
6. Mendez, M. (1998b): ApJ **505**, L23
7. Ford, E.C. (1998): ApJ **508**, L155
8. Mendez, M., van der Klis, M. (1999): ApJ **517**, L51
9. Psaltis, D. (1998): ApJ **501**, L95
10. Strohmayer, T.E. (2000): Adv. Sp. Res. Proceedings of the 33rd COSPAR Scientific Assembly, in press; astro-ph/0012256
11. Mendez, M., van der Klis, M., Ford, E.C., Wijnands, R., van Paradijs, J. (1999): ApJ **511**, L49
12. Psaltis, D., Belloni, T., van der Klis, M. (1999): ApJ **520**, 262
13. Psaltis, D. (2000): Adv. Sp. Res. Proceedings of the 33rd COSPAR Scientific Assembly; astro-ph/0012251
14. Stella, L., Vietri, M. (1998a): ApJ **492**, L59
15. Stella, L., Vietri, M. (1998b): In: The active X-ray sky, Nucl. Phys. B [Proc. Suppl.] **69**, 135
16. Stella, L., Vietri, M., Morsink, S.M. (1999): ApJ **524**, L63
17. Stella, L., Vietri, M. (1999): Phys. Rev. Lett **82**, 17
18. Orosz, J.A., Kuulkers, E. (1999): MNRAS **305**, 132
19. Strohmayer, T.E. (1996): ApJ **469**, L9
20. Ford, E.C., van der Klis, M. (1998): ApJ **506**, L39, 1728–34
21. Wiringa, R.B., Fiks, V., Fabrocini, A. (1988): Phys. Rev. C **38**, 1010
22. Wijnands R.A.D., van der Klis M. (1998): Nature **394**, 344
23. Chakrabarty, D., Morgan, E.H. (1998): Nature **394**, 346
24. Friedman, J.L, Ipser, J.R., Parker, L. (1986): ApJ **304**, 115
25. Cook, G.B., Shapiro, S.L., Teukolsky, S.A. (1992): ApJ **398**, 203

26. Morsink, S. & Stella, L. (1999): ApJ **513**, 827
27. Psaltis, D. (1999): ApJ **520**, 763
28. Stergioulas, N., Friedman, J.L. (1995): ApJ **444**, 306
29. Cui, W., Zhang, S.N., Chen, W. (1998): ApJ **492**, L53
30. King, A.R., Kolb, U. (1999): MNRAS **305**, 654
31. Shahbaz, T. (1999): MNRAS **306**, 89
32. Di Matteo, T., Psaltis, D. (1999): ApJ **526**, L101
33. Jonker, P.G., Mendez, M., van der Klis, M. (2000): ApJ **540**, L29
34. Markovic, D., Lamb, F.K.: ApJ **507**, 316
35. Vietri, M., Stella, L. (1998): ApJ **503**, 350
36. Shirakawa, A., Lai, D. (2000): astro-ph 0012118
37. Psaltis, D., Norman, C. (2000): ApJ; astro-ph/0001391
38. Adams, F.C., Watkins (1995): ApJ **451**, 314
39. Godon, P., Livio, M. (1999): ApJ **523**, 350
40. Godon, P., Livio, M. (2000): ApJ **537**, 396

What Have We learned about Gamma Ray Bursts from Afterglows?

M. Vietri

Abstract. The discovery of GRBs' afterglows has allowed us to establish several facts: their distance and energy scales, the fact that they are due to explosions, that the explosions are relativistic, and that the afterglow emission mechanism is synchrotron radiation. On the other hand, recent data have shown that the fireball model is wrong when it comes to the emission mechanism of the true burst (which is unlikely to be synchrotron again) and that shocks are not external. Besides these relatively tame points, I will also discuss the less well established physics of the energy deposition mechanism, as well as the possible burst progenitors.

1 Introduction

Gamma ray bursts (GRBs) were discovered in 1969 [1] by American satellites of the *Vela* class aimed at verifying Russian compliance with the nuclear atmospheric test ban treaty. Though the discovery was made in 1969, the paper appeared only four years later because the authors had lingering doubts about the reality of the effects they had discovered. Since then, several thousands of bursts have been observed by a more than a dozen different satellites, but it is remarkable that the basic burst features outlined in the abstract of the 1969 paper (photons in the range 0.2–1.5 MeV, durations of 0.1–30 s, fluences in the range 10^{-5}–2×10^{-4} ergs cm^{-2}) have remained substantially unchanged.

Current evidence [2] has highlighted a wide (0.01–100 s) duration distribution, with hints of a bimodality which is claimed to correlate (at the 2.5σ level) with spectral properties. All bursts' spectra observed so far are strictly non-thermal, and there has never been any confirmation by BATSE of a supposed thermal component (nor of cyclotron lines or precursors, for this matter) claimed in previous reports. A remarkable feature reported by BATSE is the bewildering diversity of light curves, ranging from impulsive ones (a spike followed by a slower decay, nicknamed FREDs for Fast Rise-Exponential Decay), to smooth ones, to long ones with amazingly sharp fluctuations, including even some with a strongly periodic appearance (two such examples are the "hand" and the "comb", so nicknamed from the number of high-Q, regularly repeating sharp spikes).

The most exceptional result from BATSE, though, was the sky distribution of the bursts (Fig. 1). It was obvious from it that the bursts *had* to be extragalactic, as already discussed by theorists [3,4].

2000 BATSE Gamma–Ray Bursts

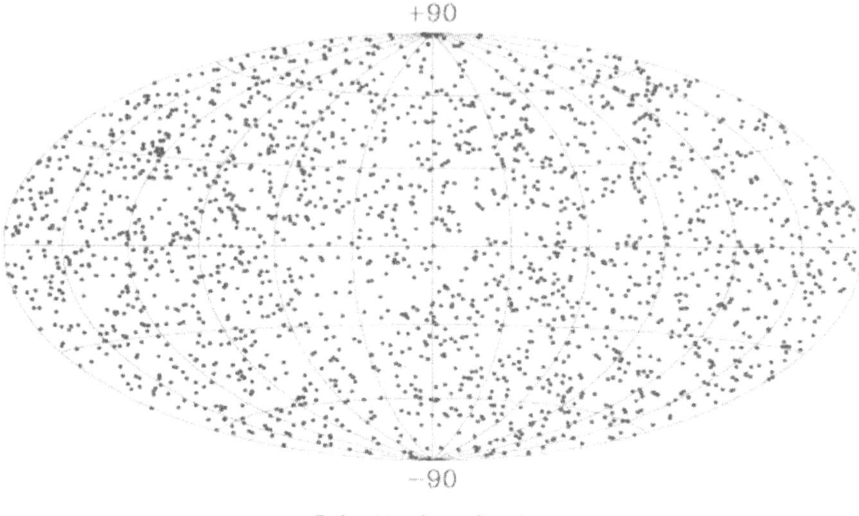

Fig. 1. Burst distribution on the plane of the sky

2 Afterglows

The next major step was triggered by BeppoSAX: in the summer of 1996, L. Piro and his coworkers located in archival data of the satellite the soft X-ray counterpart of a GRB (GRB 960720). They immediately conceived the idea of implementing a procedure to follow the next burst in real time, by re-orienting the whole satellite, after the initial detection by the Wide Field Cameras, so that the more sensitive Narrow Field Instruments could pinpoint the burst location to within 45 arcsecs, a feat never achieved in such short times, and by a single satellite. After an initial snafu (GRB 970111), the gigantic effort paid off with the discovery of the X-ray afterglow of GRB 970228 [5] immediately followed by the discovery of its fading optical counterpart [6], obtained through a search inside the WFC error box, in perfect agreement with theoretical predictions [7,8].

After the detection of the optical counterpart, the door was open to find the bursts' redshifts. Table 1 summarizes the status of our current knowledge: bursts' luminosities are for isotropic sources.

Two comments are in order. First, the bursts have *prima facie* a redshift distribution not unlike that of AGNs and of the Star Formation Rate (SFR). The initial hope that they might trace an even more distant and elusive Pop III, triggered by the fact that the second redshift detected was also the largest so far (GRB 971214, $z = 3.4$), has now vanished. Second, in order to place the energy release of GRB 990123 in context, one should notice that 4×10^{54} erg s is the energy obtained by converting the rest-mass of two solar masses, or, alternatively, the energy emitted by the whole

Table 1.

GRB	z	E_{iso}
970228	0.695	5×10^{51} erg
970508	0.835	2×10^{51} erg
971214	3.4	3×10^{53} erg
980703	0.93	3×10^{53} erg
990123	1.7	4×10^{54} erg
990510	1.6	2×10^{53} erg
990712	0.43	
991208	0.70	1.3×10^{53} erg
991216	≥ 1.02	

Universe out to $z \approx 1$ within the burst duration. So, a single (perhaps double) star outshines the whole Universe.

Besides the distance and energy scales, the major impact of the discovery of afterglows has been the establishment of some key features of the fireball model [9]:

1. bursts are due to explosions, as evidenced by their power-law behaviour;
2. the explosions are relativistic, as proved by the disappearence of radio flares;
3. the burst emission is due to synchrotron emission, as shown by the afterglow spectrum, and its optical polarization.

I will illustrate these points in the following, but, lest we become too proud, we should also remember that the fireball model has met some failures. The original version of the model [10] advocated the dissipation of the explosion energy at external shocks (i.e., those with the interstellar medium). Sari and Piran [11], following a point originally made by Ruderman [12] showed that these shocks smooth out millisecond timescale variability, which can only be maintained by the internal shocks proposed by [13]. Also, the fireball model originally ascribed even the emission from the burst proper (as opposed to the afterglow) to optically thin synchrotron processes; I will discuss in the section *Embarrassments* why this is exceedingly unlikely. Furthermore, even the last tenet of mid-90s common wisdom, i.e., that bursts are due to neutron binary mergers, does not look too promising at the moment (since some bursts seem to be located inside star forming regions, incompatible with the long spiral–in time), though of course it is by no means ruled out yet.

2.1 The fireball model

Here, one may assume that an unknown agent deposits 10^{51}–10^{54} erg s inside a small volume of linear dimension $\approx 10^6$–10^7 cm. The resulting typical energy density corresponds to a temperature of a few MeVs, so that electrons and positrons cannot be bound by any known gravitational field. In these conditions, optical depths for all known processes exceed 10^{10}. The fluid expands because of its purely thermal pressure, converting internal into bulk kinetic energy. Parametrizing the baryon component mass as $M_b \equiv E/\eta c^2$, it can be shown that, for $1 \leq \eta \leq 3 \times 10^5$ [14] the fluid achieves quickly (the fluid Lorenz factor increases as $\gamma \propto r$) a coasting Lorenz factor of $\gamma \approx \eta$.

The requisite asymptotic Lorenz factor is dictated by observations: photons up to $\epsilon_{\mathrm{ex}} \approx 18\,\mathrm{GeV}$ have been observed by EGRET from bursts [2]. For these photons to evade collisions with other photons, and thus electron/positron pair production, it is necessary that, in the reference frame in which a typical burst photon (with $\epsilon \approx 1\,\mathrm{MeV}$) and the exceptional photon are emitted, they appear as below pair production threshold: thus we must have $\epsilon' \epsilon'_{\mathrm{ex}} \leq 2m_e c^2$. Since $\epsilon' \approx \epsilon/\gamma$, and similarly for the other photons, we find [15]

$$\gamma \approx 300 \left(\frac{\epsilon}{1\,\mathrm{MeV}} \frac{\epsilon_{\mathrm{ex}}}{10\,\mathrm{GeV}} \right)^{1/2}. \tag{1}$$

From what we said above, we thus require a maximum baryon contamination, in an explosion of energy E, of $M_b < 10^{-6} M_\odot (E/10^{51}\,\mathrm{erg})(300/\eta)$.

The energy release is now assumed to be in the form of an inhomogeneous wind, with parts having a Lorenz factor larger than parts emitted previously. This leads to shell collisions (the internal shock model) at radii r_{sh} which allow a time-scale variablity $\delta t \approx r_{\mathrm{sh}}/2\gamma^2 c$; for $\delta t = 1\,\mathrm{ms}$, $r_{\mathrm{sh}} \approx 10^{13}\,\mathrm{cm}$, which fixes the internal shock radii. Particle acceleration at these internal shocks and ensuing non-thermal emission is thought to lead to the formation of the burst proper. At larger radii, a shock with the surrounding ISM forms, and shell deceleration begins at a radius $r_{\mathrm{ag}} = (3E/4\pi n m_p c^2 \gamma^2)^{1/3} \approx 10^{17}\,\mathrm{cm}$ for a $n = 1\,\mathrm{cm}^{-3}$ particle density typical of galactic disks. It is thought that the afterglow begins when the shell begins the slowdown, as this drives a marginally relativistic shock into the ejecta, thusly extracting a further fraction of their bulk kinetic energy.

2.2 Why explosions

The success of the fireball model lies in this, that it decouples the problem of the energy injection mechanism from the following evolution, which is, furthermore, an essentially hydrodynamical problem. It can be shown, in fact [16] that the evolution of the external shock is adiabatic, that the shock Lorenz factor decreases as $\gamma \propto r^{-3/2}$ because of the inertia of the swept–up matter, and thus r scales with observer's time as $t = r/\gamma^2 c \rightarrow \gamma \propto t^{-3/8}$ (for a radiative solution $\gamma \propto r^{-3/7}$, [17]). If afterglow emission is due to optically thin synchrotron in a magnetic field in near-equipartition with post-shock energy density, it can be shown that $B \propto \gamma$, that the typical synchrotron frequency at the spectral peak $\nu_m \propto \gamma B \gamma_e^2 \propto \gamma^4$ (where $\gamma_e \propto \gamma$ is the lowest post-shock electron Lorenz factor), and that $F(\nu_m) \propto t^{-3\beta/2}$, where β is the afterglow spectral slope. As it can be seen, these expectations are based exclusively upon the hydrodynamical evolution (and the synchrotron spectrum), and are thus reasonably robust.

We thus expect power–law time decays, a characteristic of strong explosions (see the Sedov analogy!), with time- and spectral-indices closely related. This is what is observed everywhere, from the X-ray through the optical to the radio, (see Piro and Fruchter, this volume), the few exceptions being discussed later on. In fact, the equality of the time-decay index of the X-ray and optical data in afterglows of individual sources has been taken as the key element to show that emission in the different bands is due to the same source. Time indices in the X-ray are in the range 0.7–2.2 (Frontera et al., in press).

2.3 Why synchrotron spectrum in the afterglow

After having established that bursts are due to explosions, we happily learn that afterglows emit through synchrotron processes. In Fig. 2 [18] we show the superposition of theoretical expectations for an optically thin synchrotron spectrum (including the cooling break at $\nu \approx 10^{14}$ Hz) with observations for GRB 970508. The remarkable agreement is even more exciting as we remark that observations are not truly simultaneous, but are scaled back to the same time by means of the theoretically expected laws for time-decay, thus simultaneoulsy testing the correctness of our hydro. Another piece of evidence comes from the discovery of polarization in the optical afterglow of GRB 990510 (Fig. 3), [19,20]. This polarization may appear small ($\approx 2\%$), but it is surely not due to Galactic effects: stars in the same field

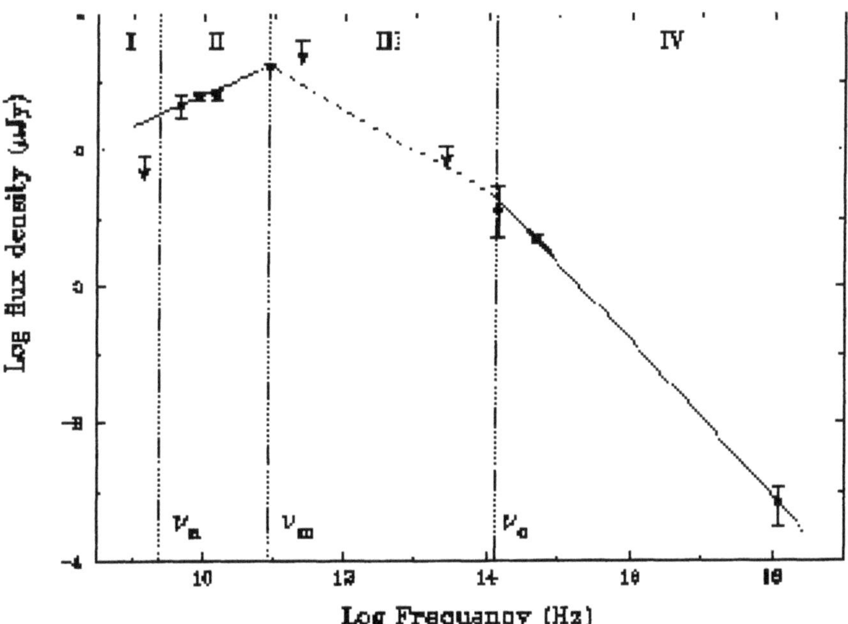

Fig. 2. Simultaneous spectrum of the afterglow of GRB 970508 in log flux density log frequency [Hz], from [18]

show a comparable degree of polarization, but along an axis different by about 50°. Also, polarization in the source galaxy is unlikely, because of a very stringent upper limit on the reddening due to this galaxy ([19]). The only remaining question mark is emission from an anisotropic source, but this would require a disk of 10^{18} cm to survive the intense γ ray (and X, and UV) flash: though not excluded, it does not look likely.

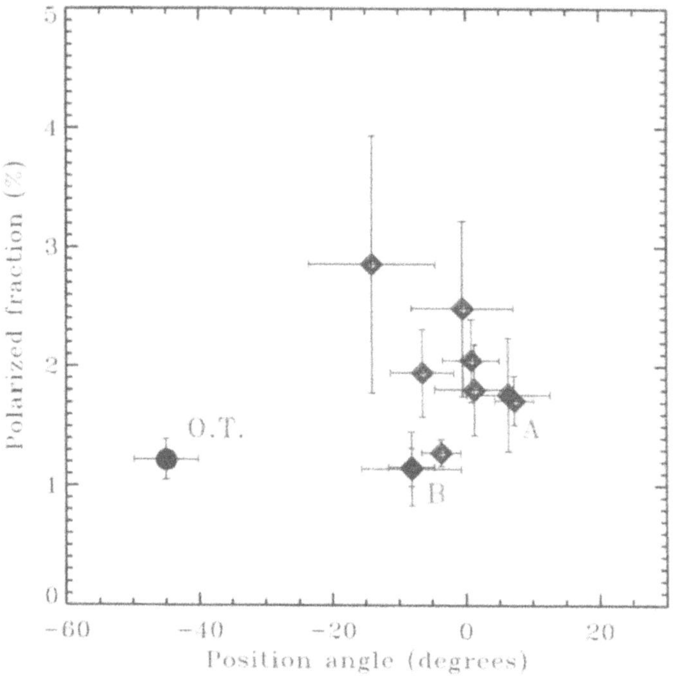

Fig. 3. Polarization amplitude and position angle for optical afterglow of GRB 990510, from [19]

2.4   Why relativistic expansion

Radio observations of the first burst observed so far (GRB 970508, [21]) showed puzzling fluctuations by about a factor of 2 in the flux, over a time-scale of days, disappearing after about 30 days from the burst (Fig. 4). This extreme, and unique behaviour, was explained in [22], where it was shown to be due to interference of rays travelling along different paths through the ISM, and randomly deflected by the spatially varying refractive index of the turbulent ISM. The wonderful upshot of this otherwise marginal phenomenon, is that these effects cease whenever the source

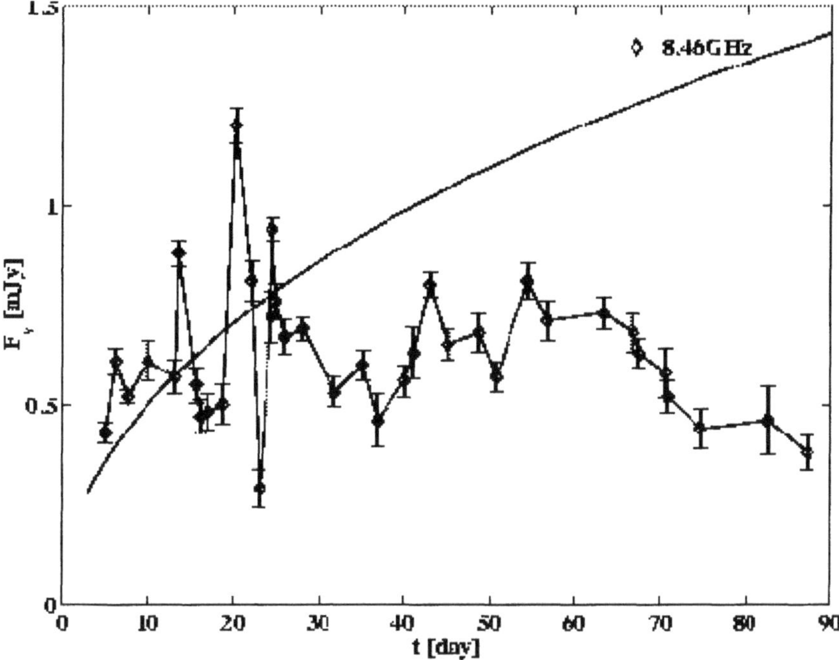

Fig. 4. VLA observatons at 8.46 GHz of the afterglow of GRB 970508, from [23]

expands beyond a radius

$$
R = 10^{17} \, \mathrm{cm} \frac{v_{10}^{6/5}}{d_{\mathrm{sc,kpc}} h_{75}} \left(\frac{SM}{10^{-2.5} m^{-20/3} \, \mathrm{kpc}} \right)^{-3/5} , \tag{2}
$$

where v_{10} is the radio observing frequency in units of 10^{10} Hz, $d_{\mathrm{sc,kpc}}$ is the distance of the ISM from us (assumed to be a uniform scattering screen), and SM is the Galactic scattering measure, scaled to a typical Galactic value.

The existence of interference effects is made more convincing by the amplitude of the average increase (a factor of 2, as observed), the correctness in the prediction of the time-interval between different peaks, and of the decorrelation bandwidth. Since flares disappear after about 30 days, it means that the average speed of the radio source is $R/30 \, \mathrm{days} = 3 \times 10^{10} \, \mathrm{cm \, s^{-1}}$. So we see directly that GRB 970508 expanded at an average speed of c over a whole month, giving us a direct observational proof that the source is highly relativistic. This proof is completely equivalent to superluminal motions in blazars, and is the strongest evidence in favor of the fireball model.

2.5   GRB 970508: our best case

The afterglow of GRB 970508 is our best case so far: it is in fact a burst for which not only do we know the redshift, but also a radio source that has been monitored for more than 400 days after the explosion [24]. Through these observations we can see the transition to a sub-relativistic regime at $t \approx 100\,$d, measure the total energetics of the following Sedov phase (unencumbered by relativistic effects!) $E_{\mathrm{New}} = 5 \times 10^{50}$ erg, determine two elusive parameters, $\epsilon_{\mathrm{eq}} = 0.5$ and $\epsilon_B = 0.5$ (the efficiencies with which energy is transfered to post-shock electrons by protons, and with which an equipartition field is built up), and the density of the surrounding medium $n \approx 0.4\,\mathrm{cm}^{-3}$. All of these values look reasonable (perhaps ϵ_{eq} and ϵ_B exceed our expectations by a factor of 10, a fact that could be remedied by introducing a slight density gradient which would keep the shock more efficient), so that our confidence in the external-shock-in-the-ISM model is boosted.

Another precious consequence of these late-time observations is that they yield information on beaming and energetics. In fact, GRB 970508 appeared to have a kinetic energy of $E_{\mathrm{rel}} = 5 \times 10^{51}$ erg when in the relativistic phase, a measurement which can be reconciled with E_{New} (remember that the expansion is adiabatic, so that we must have $E_{\mathrm{New}} = E_{\mathrm{rel}}$!) only if the unknown beaming angle, assumed $= 4\pi$ in deriving E_{rel}, is smaller than 4π by the factor $E_{\mathrm{New}}/E_{\mathrm{rel}}$; we thus have the only measurement of $\delta\Omega/4\pi = 0.1$, so far. This already rules out all classes of models requiring unplausibly large amounts of beaming, 10^{-8} or even beyond. Hopefully, more such measurements will come in the future, since this observationally heavy method is subject to many fewer uncertainties than the competing method of trying to locate breaks in the time-decay of afterglows. Also, the radiative efficiency of the burst can be estimated: correcting the observed burst energy release $E_{\mathrm{GRB}} = 2 \times 10^{51}$ erg for the same beaming factor, the radiative efficiency is $E_{\mathrm{GRB}}\delta\Omega/4\pi/(E_{\mathrm{New}} + E_{\mathrm{rel}}\delta\Omega/4\pi) = 0.3$, again a unique determination. Notice however that this figure is subject to a systematic uncertainty: we do not know whether the beaming fraction is the same for the burst proper and for the afterglow.

3   Embarrassments

Something is rotten in the fireball kingdom as well, namely, departures from pure power–law behaviours, and the spectra of the bursts proper.

3.1   Unpowerlawness

Departures from power–laws are expected when one considers the extremely idealized character of the solutions discussed so far: perfect spherical symmetry, uniform surrounding medium, smooth wind from the explosion, ϵ_{eq} and ϵ_B constant in space and time. The tricky point here is to disentangle these distinct factors. In GRB 970508 and GRB 970828 [25,26] a major departure was observed in the X-ray emission, within a couple of days from the burst; they constitute the single, largest violations observed so far, in terms of number of photons. It is remarkable that spectral variations were simultaneously observed, and that both bursts showed traces (at the 2.7σ significance level) of an iron emission line. The similarity of the bursts' behaviour argues in favor of the reality of these spectral features, which have been interpreted as thermal emission from a surrounding stellar–size leftover, pre-expelled by the burst's progenitor [27,28]. Clearly, these departures hold major pieces of information on the bursts' surroundings, and the nature of bursts' progenitors.

It has been argued [29] that, whenever the afterglow shell decelerates to below $\gamma \approx 1/\theta$, where θ is the beam semi-opening angle, emission should decrease because of the lack of emitting surface, compared to an isotropic source. But, in view of the existence of clear environmental effects (GRB 970508 and GRB 970828), it appears premature to put much stock in the interpretation of time-power-law breaks as due to beaming effects. And equally, it appears to this reviewer that the same comment applies to the interpretation of a resurgence of flux as due to the appearance of a SN remnant behind the shell. The major uncertainty here is the non-uniqueness of the interpretation: Waxman and Draine [30] have shown that effects due to dust can mimic the same phenomenon.

3.2 Bursts' spectra

A clear prediction of the emission of optically thin synchrotron is that the low-photon-energy spectra should scale like $dN_v/dv \propto v^\alpha$, with $\alpha = -3/2$, since the emission is in the fast cooling regime. Within thin synchrotron, there is no way to obtain $\alpha > -3/2$. This early-recognized requirement [31] is so inescapable that it has been dubbed the "line of death". Observations are notoriously discordant with this prediction. Preece et al. [32] have shown that, for more than 1000 bursts, α is distributed like a bell between -2 and 0, with mean $\bar{\alpha} \approx -1$. The tail of this distribution also contains a few tens of objects with $\alpha \approx +1$. An example of these can be found in [33] (GRB 970111), which is instructive since BeppoSAX has better coverage of the critical, low-photon-energy region. In particular, BATSE seems to loose sensitivity below $\approx 30\,\text{keV}$, but this is still not enough to explain away the discrepancy with the theory. Also, Preece et al. showed that the time-integrated spectral energy distribution has a peak at a photon energy $\epsilon_{\text{pk}} \approx 200\,\text{keV}$, and that ϵ_{pk} has a very small variance from burst to burst. Again, this does not seem dependent upon BATSE's lack of sensistivity above $700\,\text{keV}$, and again this has no explanation within the classic fireball model.

Many people agree that the neglect of Inverse Compton processes may be the root of the disagreement: the trick here is to devise a fireball model that smoothly incorporates them. One should remember that the details of the fireball evolution are *generic*, i.e., they do not depend upon any detailed property of the source, so that things like the radius at which the fireball becomes optically thin (to pairs or baryonic electrons), the radius at which acceleration ends, the equipartition magnetic field, and so on, are all reliably and inescapably fixed by the outflow's global properties. A step toward the solution has been made by [34] who remarked that at least some bursts have compactness parameters $l = 10(L/10^{53}\,\text{erg s}^{-1})(300/\gamma)^5 \gg 1$. Under these conditions, a pair plasma will form, nearly thermalized at $kT \approx m_e c^2$, and with Thomson optical depth $\tau_T \approx 10$. The modifications which this plasma will bring to the burst's spectrum are currently unknown, but it may be remarked that this configuration will be optically thick to both high-energy synchrotron photons due to non-thermal electrons accelerated at the internal shocks, and to low–energy cyclotron photons emitted by the thermal plasma, but it will be optically thin in the intermediate region reached by cyclotron photons upscattered via IC processes off non-thermal electrons. A model along this line (i.e., upscattering of cyclotron photons by highly relativistic electrons) is in preparation, but it remains to be seen whether it (like any other model, of course) can simultaneously explain the spectral shape and the narrow range of the spectral distribution peak energy ϵ_{pk}.

4   On the central engine

As remarked several times already, the fireball evolution is independent of the source nature. The only exisiting constraint is the maximum amount of baryon contamination, which is

$$M_b = \frac{E}{\eta c^2} = 10^{-6} M_\odot \frac{E}{10^{51}\,\mathrm{erg}} \frac{300}{\gamma}. \tag{3}$$

This is a remarkably small value: since the inferred luminosities exceed the Eddington luminosity by 13 orders of magnitude, they clearly have all it takes to disrupt a whole star, no matter how compact. Yet, the energy deposition must somehow occur outside the main mass, lest the explosion be slowed down to less relativistic, or even possibly Newtonian speeds. In order to satisfy this constraint, it has emerged that the most favorable configuration has a stellar-mass black hole ($M_{\mathrm{BH}} \approx 3$–$10 M_\odot$) surrounded by a thick torus of matter ($M_t \approx 0.01.-1 M_\odot$, with $\rho \approx 10^{10}$ g cm^{-3}). The presence of a black hole is *not* required by observations in any way: models involving neutron stars are still admissible, the advantage of having a black hole being only the deeper potential well: you may get more energy out per unit accreted mass. The configuration thusly envisaged has a cone surrounding the symmetry axis devoid of baryons, since all models leading to this configuration have large amounts of specific angular momentum, and thus baryons close to the rotation axis either are not there, or have accreted onto the black hole due to their lack of centrifugal support.

4.1   Energy release mechanism

There are two major mechanisms for energy release discussed in the literature, the first one to be proposed [35] being the reaction $\nu + \bar{\nu} \rightarrow e^- + e^+$. Neutrinos have non-negligible mean free paths in the tori envisaged here, so that this annihilation reaction will take place not inside the tori themselves, where they are preferentially generated because densities are highest, but in a larger volume surrounding the source. This is both a blessing and a disgrace: by occupying a larger volume, the probability that every neutrino finds its antiparticle to annihilate decreases, but then the energy is released in baryon–cleaner environments. The problem, though complex, is eminently suitable for numerical simulations, showing [36], and references therein) that about 10^{50} erg s can be released this way, above the poles of a black hole where less than $10^{-5} M_\odot$ are found.

Highly energetic bursts cannot be reproduced by this mechanism, due to its low efficiency: the second mechanism proposed involves the conversion of Poynting flux into a magnetized wind. The basic physical mechanisms are well-known [37] since they have been studied in the context of pulsar emission: electrons are accelerated by a motional electric field $E = v \wedge B/c$ due to the rotation of a sufficiently strong magnetic dipole, attached either to a black hole, or to the torus. Photons are then produced by synchrotron or curvature radiation, and photon/photon collisions produce pairs, to close the circle and allow looping. In order to carry away 10^{51} erg s^{-1}, a magnetic field of $\approx 10^{15}$ G is required. This is not excessive, since it is about three orders of magnitude below equipartition with torus matter, and because such fields already exist in nature, see SGR 1806-20 and SGR 1900+140: the key point is to understand whether some kind of dynamo effect can lead to these high values within the short allotted time.

Depending upon whether the open magnetic field lines extending to infinity are connected to the black hole or to the torus, the source of the energy of the outflow will be the rotational energy of the black hole (the so-called Blandford–Znajek effect) or of the torus. The first case is traditionally discussed in the context of AGNs [38], but it is harshly disputed whether the energy outflow may be actually dominated by the black hole rather than by the disk [39,40]. On the other hand, the torus looks ideal as the source of a dynamo: its large shear rate, the presence of the Balbus–Hawley instability to convert poloidal into toroidal flux, and the possible presence of the anti-floating mechanism inhibiting ballooning of the magnetic field [41], all seem to favor the existence of a fast dynamo. It should also be remarked that the configuration of the magnetic field in this problem is known: in fact, the configuration discussed in [42] for black holes, only uses the assumptions of steady-state and axial symmetry, and is thus immediately extended to magnetic fields anchored to the torus. What is really required here is a first order study, of the sort published by [43] on angular momentum removal from young, pre-main-sequence stars via magnetic stresses, and on the associated $\alpha - \omega$ dynamo. Until such studies are made, it will be premature to claim that neutrino annihilations are responsible for the powering of GRBs.

4.2 Progenitors

There is no lack of proposed progenitors, but I will discuss only binary neutron mergers [44], collapsars [45,46] and SupraNovae [47,48].

Clearly, NS/NS mergers is the best model on paper: it involves objects which have been detected already, orbital decay induced by gravitational wave emission is shown by observations to work as per the theory, and numerical simulations by Janka's group show that a neutrino-powered outflow in baryon-poor matter can be initiated. The major theoretical uncertainties here concern bursts' durations and energetics: all numerical models produce short bursts (≈ 0.1 s) with modest energetics, $E < 10^{51}$ erg. This is a direct consequence of the mechanism for powering the burst: large, super-Eddington luminosities are carried away by neutrinos, leading to a large mass influx, but only a small fraction, 1–3%, can be harnessed for the production of the burst. Furthermore, we cannot invoke large beaming factors in this case: the outflow is only marginally collimated, in agreement with expectations that an accretion disk with inner and outer radii $R_{out}/R_{in} \approx$ a few (for the case at hand) can only produce a beam semi-opening angle of R_{in}/R_{out}. So, perhaps, this model may account for the short bursts, but it should be remembered that nothing of what was discussed above pertains to this subclass: BeppoSAX (and thus all BeppoSAX-triggered observations) can only detect long bursts.

On the other hand, future space missions, whether or not able to locate short bursts, can provide a decisive test of this model, provided they can follow with sufficient sensitivity a given burst for several hours. This model, in fact, is the only one proposed so far according to which some explosions should take place outside galaxies: according to [49], about 50% of all bursts will be located more than 8 kpc from a galaxy, and 15% in the IGM. This characteristic is testable without recourse to optical observations. In fact, the afterglow begins with a delay (as seen by an outside observer) of $t_d = (r_{ag} - r_{sh})/\gamma^2 c \approx r_{ag}/\gamma^2 c$, which varies greatly depending upon the environment in which the burst takes place:

$$
t_d = \begin{cases} 15\,\text{s} & \text{ISM, n} = 1\,\text{cm}^{-3} \\ 5\,\text{min} & \text{galactic halo, n} = 10^{-4}\,\text{cm}^{-3} \\ 4\,\text{h} & \text{IGM, n} = 10^{-8}\,\text{cm}^{-3}. \end{cases} \tag{4}
$$

Between the burst proper and the beginning of the power-law-like afterglow, thus a silence of recognizable duration is expected.

Collapsars are currently in great vogue as a possible source of GRBs: the large amount of energy available as the core of a supermassive star collapses directly to a black hole is in fact very attractive, even though (again!) the limited efficiency of the reaction $v + \bar{v} \rightarrow e^- + e^+$ makes most of this energy unavailable. Here too there is some evidence that these objects must exist [46] and numerical simulations again showing energy preferentially deposited along the hole rotation axis are also available [50]. Here however, what is truly puzzling is how the outflow can pierce the star's outer layers without loading itself with baryons: we should remember that at most $10^{-6} M_\odot$ can be added to 10^{51} erg: more baryons imply a proportionately slower outflow. The argument is that the dynamical timescale of the outer layers of a massive stars is of order of a few hours, so that, even if the core collapses and pressure support is removed, nothing will happen during the energy release phase: the outflow must pierce its way through. Two processes seem especially dangerous: Rayleigh–Taylor instability of the fluid heated-up by neutrino annihilations as it is weighed upon by the colder, denser outer layers, and Kelvin–Helmholtz instability after the hot fluid has pierced the outer layers and is passing through the hole. It is well-known that the non-linear development of these instabilities leads to mass entrainment, and that the time-scale for the development of these instabilities is very fast. Furthermore, the baryon-free outflow may be "poisoned" by baryons to a deadly extent, even if numerical simulations, with their finite resolution, were to detect nothing of the kind.

The third class of models, SupraNovae, concerns supramassive neutron stars which are stabilized against self-gravity by fast rotation, to such an extent that they cannot be spun down to $\omega = 0$ because they implode to a black hole. As the star's residual magnetic dipole sheds angular momentum, this is exactly the fate to be expected for the whole star, except for a small equatorial belt, whose later accretion will power the burst. It is easy to show that this implosion must take place in a very baryon-clean environment. The major uncertainties here concern the channels of formation and the existence of this equatorial belt. Two channels of formation have been proposed: direct collapse to a supramassive configuration [47] and slow mass accretion in a low-mass X-ray binary [48]. Both are possible, though none yet is supported by observations. The existence of the left-over belt has recently been questioned by [51], who however simulated the collapse of neutron stars with intermediate equations of state, which are entirely (or nearly exactly so) contained inside the marginally stable orbit even before collapse: clearly, these must be swallowed whole by the resulting black hole. Soft equations of state are free of this objection, and are thus much more likely to leave behind an equatorial belt. The soft EoSs are especially favored since the neutron stars must survive the r-mode instability, and thus soft EoSs [52] would be in any case required. So one might say that the existence of these stars hinges on onc uncertainty only, the EoS of nuclear matter. Besides the baryon-clean environments, SupraNovae have another advantage over rival models: only the lowest density regions would be left behind, precisely those with the smallest neutrino losses. The powering of the burst can thus occur through accretion caused by removal of angular momentum by magnetic stresses, without the parallel, unproductive, neutrino generation.

5 Conclusions

It is difficult to end on an upbeat note: we cannot expect in the near future a rate of progress similar to the one we witnessed in the past three years. In particular, it may be expected that the next flurry of excitement will come with the beginning of the SWIFT mission, which promises to collect relevant data (redshifts, galaxy types, location within or without galaxies, absorption or emission features in the optical and in the X-ray) for a few hundred bursts. This data will nail the major characteristics of the environment (at large) in which bursts take place, and we may be able to rule out a few models. On the other hand, the energy release process, shrouded as it is in optical depths $> 10^{10}$, will remain mysterious, our only hope in this direction being gravitational waves.

Judging by the analogy with radio pulsars, this will correspond to the flattening of the learning curve. Aside from this, we may hope to locate the equivalent of the binary radio pulsar, but, differently from Jo Taylor, we have to be awfully quick in grabbing it.

References

1. Klebesadel, R.W., Strong, I.B., Olson, R.A. (1973): Ap. J. L. **182**, L85
2. Fishman, G.J., Meegan, C.A. (1995): ARAA **33**, 415
3. Usov, V.V., Chibisov, G.V.: Soviet Astr. **19**, 115

4. Paczyński, B. (1986): Ap. J. L. **308**, L43
5. Costa, E., Frontera, F., Heise, J., Feroci, M., Zand, J.I., Fiore, F., Cinti, M.N., Dal Fiume, D., Nicastro, L., Orlandini, M., Palazzi, E., Rapisarda, M., Zavattini, G., Jager, R., Parmer, A., Owens, A., Molendi, S., Cosumano, G., Maccarone, M. C., Giarrusso, S., Coletta, A., Antonelli, L.A., Giommi, P., Muller, J.H., Piro, L., Buttler, R.C. (1997): Nature **387**, 783
6. van Paradijs, J., (1997): Nature **386**, 686
7. Vietri, M. (1997a): Ap. J. L. **478**, L9
8. Mészáros, P., Rees, M.J. (1997): Ap. J. **476**, 232
9. Rees, M.J., Mèszàros, P. (1992): MNRAS **258**, P41
10. Mészáros, P., Rees, M.J. (1993): Ap. J. **405**, 278
11. Sari, R., Piran, T. (1997): Ap. J. **485**, 270
12. Ruderman, R. (1975): Ann. NY. Acad. Sci. **262**, 164
13. Paczyński, B., Xu, G. (1994): Ap. J. **427**, 708
14. Mészáros, P., Laguna, P., Rees, M.J. (1993): Ap. J. **415**, 181
15. Baring, M. (1993): Ap. J. **418**, 391
16. Waxman, E. (1997): Ap. J. L. **489**, L33
17. Vietri, M. (1997b): Ap. J. L. **488**, L105
18. Galama, T., et al. (1998): Ap. J. L. **500**, L97
19. Covino, S., Lazzati, D., Ghisellini, G., Saracco, P., Campana, S., Chincarine, G., Di Serego, S., Cimatti, A., Vanzi, L., Pasquini, L., Haardt, F., Israel, G.L., Stella, L., Vietri, M. (1999): Astr. Ap. **348**, L1
20. Wijers, R.A.M.J., Vreeswijk, P.M., Galama, T.J., Rol, E., van Paradijs, J., Konvelioton, C., Giblin, T., Masetti, N., Palazzi, E., Pion, E., Frontera, F., Nicastro, L., Falomo, R., Soffitta, P., Piro, L. (1999): Ap. J. L. **523**, L33
21. Frail, D., et al. (1997) Nature **389**, 261
22. Goodman, J.J. (1997): New Astr. **2**, 449
23. Waxman, E., Frail, D., Kulkarni, D. (1998): Ap. J. **497**, 288
24. Frail, D., Waxman, E., Kulkarni, S. (2000): Ap. J.bf 537, 191; astro-ph 9910319
25. Piro, L., costa, E., Feroci, M., Frontera, F., Amati, L., Dal Fiume, D., Antonelli, L.A., Heise, J. (1999): Ap. J. L. **514**, L73–L77
26. Yoshida, A., Namiki, M., Otani, C., Kawai, N., Murakami, T., Ueda, Y., Shibata, R., Uno, S. (1999): Astr. Ap. S. **138**, 433–434
27. Lazzati, D., Campana, S., Ghisellini, G. (1999): MNRAS **304**, L31
28. Vietri, M., Perola, G.C., Piro, L., Stella, L. (1999): MNRAS **308**, P29
29. Rhoads, J. (1997): Ap. J. L. **487**, L1
30. Waxman, E., Draine, B.T. (2000): Ap. J. **537**, 796
31. Katz, J. (1994): Ap. J. **422**, 248
32. Preece, R.D., Briggs, M.S., Mallozzi, R.S., Pendleton, G.N., Paciesas, W.S., Band, D.L. (2000): Ap. J. S. **126** (1), 19–36
33. Frontera, F., et al. (1999): Ap. J. S., in press, astro-ph 9911228
34. Ghisellini, G., Celotti, A. (1998): Ap. J. L. **511**, L93
35. Berezinsky, V.S., Prilutskii, O.F. (1986): Astr. Ap. **175**, 309
36. Janka, T., Eberl, T., Ruffert, M., Fryer, C.L. (1999): Ap. J. L. **527**, L39
37. Usov, V.V. (1992): Nature **357**, 472
38. Rees, M.J., Begelman, M.C., Blandford, R.D., Phinney E.S. (1984): Nature **295**, 17
39. Ghosh, P., Abramowicz, M.A. (1997): MNRAS **292**, 887
40. Livio, M., Ogilvie, G.I., Pringle, J.E. (1999): Ap. J. **512**, 100
41. Kluzniak, W., Ruderman, M. (1998): Ap. J. L. **505**, L113

42. Thorne, K.S., Price, R.H., MacDonald, D.A. (1986): Black holes: The membrane paradigm. Yale University Press, New Haven
43. Tout, C.A., Pringle, J.E. (1992): MNRAS **256**, 269
44. Narayan, R., Paczynski, B., Piran, T. (1992): Ap. J. L. **395**, L83
45. Woosley, S. (1993): Ap. J. **405**, 273
46. Paczyǹski, B. (1998): Ap. J. L. **494**, L45
47. Vietri, M., Stella, L. (1998): Ap. J. L. **507**, L45
48. Vietri, M., Stella, L. (1999): Ap. J. L. **527**, L43
49. Bloom, J.S., Sigurdsson, S., Pols, O.R. (1999): MNRAS **305**, 763
50. Mc Fayden, A., Woosley, S.E. (1999): Ap. J. **524**, 262
51. Shibata, M., Baumgarte, T.W., Shapiro, S.L. (2000): Phys. Rev. D **6104** (4), 4012
52. Weber, F. (1999): In Pulsars as astrophysical laboratories for nuclear and particle physics, Institute of Physics Bristol, UK

Gravitational Waves
and the Death-Spiral of Compact Binaries

C. M. Will

Abstract. The completion of a network of advanced laser-interferometric gravitational-wave observatories will make possible the study of the inspiral and coalescence of binary systems of compact objects (neutron stars and black holes), using gravitational radiation. To extract useful information from the waves, theoretical general relativistic gravitational waveform templates of extremely high accuracy will be needed for filtering the data, probably as accurate as $O[(v/c)^6]$ beyond the predictions of the quadrupole formula. We review theoretical methods for calculating accurate waveforms, and focus on one known as Direct Integration of the Relaxed Einstein Equations (DIRE). The new method is free of divergences or undefined integrals, correctly predicts all gravitational wave "tail" effects caused by backscatter of the outgoing radiation off the background curved spacetime, and yields radiation that propagates asymptotically along true null cones of the curved spacetime. The method also yields equations of motion through $O[(v/c)^4]$, radiation-reaction terms at $O[(v/c)^5]$ and $O[(v/c)^7]$, and gravitational waveforms and energy flux through $O[(v/c)^4]$, in agreement with other approaches. We report on progress in evaluating the $O[(v/c)^6]$ contributions.

1 Introduction

Some time in this decade, a new window for astronomy and relativistic gravity may be realized, with the completion and operation of kilometer-scale, laser interferometric gravitational-wave observatories in the U.S. (LIGO project), Europe (VIRGO and GEO-600 projects) and Japan (TAMA-300 project). Gravitational-wave searches at these observatories are scheduled to commence around 2002. The LIGO broad-band antennae will have the capability of detecting and measuring the gravitational waveforms from astronomical sources in a frequency band between about 10 Hz (the seismic noise cutoff) and 500 Hz (the photon counting noise cutoff), with a maximum sensitivity to strain at around 100 Hz of $\Delta l/l \sim 10^{-22}$ (rms). The most promising source for detection and study of the gravitational-wave signal is the "inspiralling compact binary" – a binary system of neutron stars or black holes (or one of each) in the final minutes of a death spiral leading to a violent merger. Such is the fate, for example of the Hulse–Taylor binary pulsar PSR 1913+16 in about 240 million years. Given the expected sensitivity of the "advanced LIGO" (around 2007), which could see such sources out to hundreds of megaparsecs, it has been estimated that from 3 to 100 annual inspiral events could be detectable. Other sources, such as supernova core collapse events, instabilities in rapidly rotating nascent neutron stars, signals from non-axisymmetric pulsars, and a stochastic background of waves, may be detectable (for reviews, see Ref. [1] and other articles in this volume).

The analysis of gravitational-wave data from such inspiral sources will involve some form of matched filtering of the noisy detector output against an ensemble of theoretical "template" waveforms which depend on the intrinsic parameters of the inspiralling binary, such as the component masses, spins, and so on, and on its inspiral evolution. How accurate must a template be in order to "match" the waveform from a given source (where by a match we mean maximizing the cross-correlation or the signal-to-noise ratio)? In the total accumulated phase of the wave detected in the sensitive bandwidth, the template must match the signal to a fraction of a cycle. For two inspiralling neutron stars, around 16,000 cycles should be detected; this implies a phasing accuracy of 10^{-5} or better. Since $v/c \sim 1/10$ during the late inspiral, this means that correction terms in the phasing at the level of $(v/c)^5$ or higher are needed. More formal analyses confirm this intuition [2].

Because it is a slow-motion system ($v/c \sim 10^{-3}$), the binary pulsar is sensitive only to the lowest-order effects of gravitational radiation as predicted by the quadrupole formula. Nevertheless, the first correction terms of order v/c and $(v/c)^2$ to the quadrupole formula, were calculated as early as 1976 [3]. These are now conventionally called "post-Newtonian" (PN) corrections, with each power of v/c corresponding to half a post-Newtonian order (0.5 PN), in analogy with post-Newtonian corrections to the Newtonian equations of motion[1]. In 1976, the post-Newtonian corrections to the quadrupole formula were of purely academic, rather than observational interest.

But for laser-interferometric observations of gravitational waves, the bottom line is that, in order to measure the astrophysical parameters of the source and to test the properties of the gravitational waves, it is necessary to derive the gravitational waveform and the resulting radiation back-reaction on the orbit phasing at least to 2PN, or second post-Newtonian order, $O[(v/c)^4]$, beyond the quadrupole approximation, and probably to 3PN order.

2   Post-Newtonian generation of gravitational waves

The motion of isolated binary systems and the generation of gravitational radiation are long-standing problems that date back to the first years following the publication of GR, when Einstein calculated the gravitational radiation emitted by a laboratory-scale object using the linearized version of GR. Shortly after the discovery of the binary pulsar PSR 1913+16 in 1974, questions were raised about the foundations of the "quadrupole formula" for gravitational radiation damping (and in some quarters, even about its quantitative validity). These questions were answered in part by theoretical work designed to shore up the foundations of the quadrupole approximation [4], and in part (perhaps mostly) by the agreement between the predictions of the quadrupole formula and the *observed* rate of damping of the pulsar's orbit.

---

[1] This convention holds sway, despite the fact that pure Newtonian gravity predicts no gravitational radiation. It is often the source of confusion, since the energy carried by the lowest-order "Newtonian" quadrupole radiation manifests itself in a $(\text{post})^{5/2}$-Newtonian, or $O[(v/c)^5]$ correction in the equation of motion.

The challenge of providing accurate templates for LIGO-VIRGO data analysis has led to major efforts to calculate gravitational waves to high PN order. Three approaches have been developed.

The approach of Blanchet, Damour and Iyer is based on a mixed post-Newtonian and "post-Minkowskian" framework for solving Einstein's equations approximately, developed in a series of papers by Damour and colleagues [5]. The idea is to solve the vacuum Einstein equations in the exterior of the material sources extending out to the radiation zone in an expansion (post-Minkowskian) in "nonlinearity" (effectively an expansion in powers of Newton's constant G), and to express the asymptotic solutions in terms of a set of formal, time-dependent, symmetric and trace-free (STF) multipole moments [6]. Then, in a near zone within one characteristic wavelength of the radiation, the equations including the material source are solved in a slow-motion approximation (expansion in powers of $1/c$) that yields a set of STF source multipole moments expressed as integrals over the "effective" source, including both matter and gravitational field contributions. The solutions involving the two sets of moments are then matched in an intermediate zone, resulting in a connection between the formal radiative moments and the source moments. The matching also provides a natural way, using analytic continuation, to regularize integrals involving the non-compact contributions of gravitational stress-energy, that might otherwise be divergent.

An approach called DIRE is based on a framework developed by Epstein and Wagoner (EW) [7], and extended by Will, Wiseman and Pati. We shall describe DIRE briefly below.

A third approach, valid only in the limit in which one mass is much smaller than the other, is that of black-hole perturbation theory. This method provides numerical results that are exact in v/c, as well as analytical results expressed as series in powers of v/c, both for non-rotating and for rotating black holes. For non-rotating holes, the analytical expansions have been carried to 5.5 PN order (for a review see [8]). In all cases of suitable overlap, the results of all three methods agree precisely.

3   Direct integration of the relaxed Einstein equations (DIRE)

Like the post-Minkowskian approach, DIRE involves rewriting the Einstein equations in their "relaxed" form, namely as an inhomogeneous, flat-spacetime wave equation for a field $h^{\alpha\beta}$, whose formal solution can be written

$$h^{\alpha\beta}(t, \mathbf{x}) = 4 \int_C \frac{\tau^{\alpha\beta}(t - |\mathbf{x} - \mathbf{x}'|, \mathbf{x}')}{|\mathbf{x} - \mathbf{x}'|} d^3x', \tag{1}$$

where the source $\tau^{\alpha\beta}$ consists of both the material stress-energy, and a "gravitational stress-energy" made up of all the terms non-linear in $h^{\alpha\beta}$, and the integration is over the past flat-spacetime null cone C of the field point $(t, \mathbf{x})$. The wave equation is accompanied by a harmonic or deDonder gauge condition $h^{\alpha\beta}{}_{,\beta} = 0$, which serves to specify a coordinate system, and also imposes equations of motion on the sources. Unlike the BDI approach, a *single* formal solution is written down, valid everywhere in spacetime. This formal solution, is then iterated in a slow-motion ($v/c < 1$), weak-field ($||h^{\alpha\beta}|| < 1$) approximation, that is very similar to the corresponding procedure in electromagnetism. However, because the integrand of this retarded integral is not compact by virtue of the non-linear field contributions, the original EW formalism quickly runs up against integrals that are not well defined, or worse, are divergent. Although at the lowest quadrupole and first few PN orders, various arguments can be given to justify sweeping such problems under the rug [3], they are not very rigorous, and provide no guarantee that the divergences do not become insurmountable at higher orders. As a consequence, despite efforts to cure the problem, the EW formalism fell into some disfavor as a route to higher orders, although an extension to 1.5 PN order was accomplished [9].

The resolution of this problem involves taking literally the statement that the solution is a *retarded* integral, i.e. an integral over the *entire* past null cone C of the field point [10]. To be sure, that part of the integral that extends over the intersection $\mathcal{N}$ between the past null cone and the material source and the near zone is still approximated as usual by a slow-motion expansion involving spatial integrals over a constant-time hypersurface $\mathcal{M}$ of moments of the source, including the non-compact gravitational contributions, just as in the post-Minkowskian framework. But instead of cavalierly extending the spatial integrals to infinity as was implicit in the original EW framework, and risking undefined or divergent integrals, we terminate the integrals at the boundary of the near zone, chosen to be at a radius $\mathcal{R}$ given roughly by one wavelength of the gravitational radiation. For the integral over the rest of the past null cone $C - \mathcal{N}$ exterior to the near zone ("radiation zone"), we neither make a slow-motion expansion nor continue to integrate over a spatial hypersurface, instead we use a coordinate transformation in the integral from the spatial coordinates d^3x' to quasi-null coordinates du', $d\theta'$, $d\varphi'$, where

$$t - u' = r' + |\mathbf{x} - \mathbf{x}'|, \tag{2}$$

to convert the integral into a convenient, easy-to-calculate form, that is manifestly convergent, subject only to reasonable assumptions about the past behavior of the source:

$$h_{C-\mathcal{N}}^{\alpha\beta}(t, x) = 4 \int_{-\infty}^{u} du' \int \frac{\tau^{\alpha\beta}(u' + r', \mathbf{x}')}{t - u' - \mathbf{n}' \cdot \mathbf{x}} [r'(u', \Omega')]^2 d^2\Omega'. \tag{3}$$

This transformation was suggested by earlier work on a non-linear gravitational-wave phenomenon called the Christodoulou memory [11]. Not only are all integrations now explicitly finite and convergent, one can show that all contributions from the finite, near-zone spatial integrals that depend upon $\mathcal{R}$ are actually *cancelled* by corresponding terms from the radiation-zone integrals, valid for both positive and negative powers of $\mathcal{R}$ and for terms logarithmic in $\mathcal{R}$ [12]. Thus the procedure, as expected, has no dependence on the artificially chosen boundary radius $\mathcal{R}$ of the near-zone. In addition, the method can be carried to higher orders in a straightforward manner. The result is a manifestly finite, well-defined procedure for calculating gravitational radiation to high orders.

Thus, for field points in the far zone, the integral over the near zone takes the standard form of a multipole expansion,

$$h_{\mathcal{N}}^{\alpha\beta}(t,\mathbf{x}) = 4 \sum_{m=0}^{\infty} \frac{(-1)^m}{m!} \frac{\partial^m}{\partial x^{k_1} \dots \partial x^{k_m}} \left(\frac{1}{r} M^{\alpha\beta k_1 \dots k_m}(t-r) \right), \qquad (4)$$

$$M^{\alpha\beta k_1 \dots k_m}(t-r) \equiv \int_{\mathcal{M}} \tau^{\alpha\beta}(t-r,x')x'^{k_1} \dots x'^{k_m} d^3 x', \qquad (5)$$

where integrals are over the *finite* hypersurface $\mathcal{M}$. For field points in the near zone, the integral over the near zone can be expanded in the form of a sequence of "Poisson"-like potentials and superpotentials (and "superduper"-potentials) evaluated at a fixed time t:

$$h_{\mathcal{N}}^{\alpha\beta}(t,\mathbf{x}) = 4 \sum_{m=0}^{\infty} \frac{(-1)^m}{m!} \frac{\partial^m}{\partial t^m} \int_{\mathcal{M}} \tau^{\alpha\beta}(t,x')|\mathbf{x}-\mathbf{x}'|^{m-1} d^3 x'. \qquad (6)$$

In each case, the integrals must be combined with the corresponding integral over the rest of the past light cone, Eq. (3). Because of the aforementioned general proof, it is not necessary to keep any terms in these integrals that depend explicitly on the radius $\mathcal{R}$; this simplifies calculations considerably.

4   Equations of motion to 3.5PN order

We assume that the orbiting bodies are sufficiently small compared to their separation that tidal effects, or effects due to their finite size, can be ignored. For inspiralling compact binaries, this is believed to be a good approximation until the final few orbits. This amounts to replacing a perfect fluid stress-energy tensor,

$$T^{\mu\nu} = (\rho + p)u^\mu u^\nu + p g^{\mu\nu}, \qquad (7)$$

with that of "point-masses":

$$T^{\mu\nu} = \sum_A m_A \delta^3(\mathbf{x} - \mathbf{x}_A) u^\mu u^\nu / u^0 \sqrt{-g}, \qquad (8)$$

where ρ, p and u^μ are the density, pressure and four-velocity of the fluid, respectively. However, because of gravitational non-linearities, such a stress-energy tensor will lead to infinities at the location of each body, hence one must find a way to regularize in order to isolate the physically relevant terms. Blanchet et al. use a regularization procedure based on the Hadamard "partie fini". Our approach, which is less formal, though probably equivalent, is to isolate those terms in any integral of fields over a body that neither vanish nor blow up as D, the size of the body, shrinks to zero. Terms that vanish as D^N represent tidal and spin effects (and their relativistic generalizations), which we are ignoring. Terms that diverge as D^{-N} are "self-energy" terms; we assume that these can be uniformly absorbed into renormalized masses for the bodies. It is important to stress that this is an assumption, whose validity has been checked in general only to 1PN order (no Nordtvedt effect in GR) and under restricted circumstances to 2PN order. The result is a well-defined procedure for keeping "finite, point-mass" terms.

With these assumptions, the equations of motion for each body take the form of a geodesic equation,

$$\rho^* \frac{d^2 x^j}{dt^2} + \rho^* \Gamma^j_{\mu\nu} v^\mu v^\nu - \rho^* \Gamma^0_{\mu\nu} v^j v^\mu v^\nu = 0, \tag{9}$$

where $\rho^* = \rho u^0 \sqrt{-g} \to \sum_A m_A \delta^3(\mathbf{x} - \mathbf{x}_A)$, and $v^\mu = dx^\mu/dt$.

To obtain equations of motion valid through 3.5PN order, it is necessary to iterate the relaxed Einstein equation four times. Evaluating the resulting Poisson-like potentials for two fluid balls, integrating the equation of motion (9) over one of the bodies, and keeping only terms that are finite as the bodies shrink in size, one obtains equations of motion of the schematic form

$$a_1^i = -\frac{m_2}{r^2}[n^i + O(\epsilon) + O(\epsilon^2) + O(\epsilon^{5/2}) + O(\epsilon^3) + O(\epsilon^{7/2}) + \dots], \tag{10}$$

$$a_2^i = \frac{m_1}{r^2}[n^i + (1 \rightleftharpoons 2)], \tag{11}$$

where r is the distance between the bodies and $n^i \equiv (x_1^i - x_2^i)/r$. The expansion parameter ϵ is related to the orbital variables by $\epsilon \sim m/r \sim v^2$, v is the relative velocity, and $m = m_1 + m_2$ is the total mass ($G = c = 1$).

We have evaluated all contributions to $h^{\mu\nu}$ formally through 3.5PN order in terms of Poisson-like potentials [12], and have calculated them explicitly for two compact bodies through 2.5PN order and at 3.5PN order. At 2PN order, we obtain equations of motion in complete agreement with those of Damour and Deruelle (Eqs. (154)–(160) of Ref. [13]) and Blanchet et al. (Eq. (8.4) of Ref. [14]). Contributions at 3PN order using the post-Minkowski approach have been reported [15,16], while evaluation of these contributions in the DIRE approach is in progress.

The contributions at 2.5PN and 3.5PN order represent gravitional-radiation reaction and its post-Newtonian corrections. Iyer and Will [17] have shown that, assuming energy and angular momentum balance, the relative two-body equations of motion at 2.5PN order can be written in the form

$$\mathbf{a} = -\frac{8}{5}\eta(m/r^2)(m/r)\left[-(A_{2.5} + A_{3.5})\dot{r}\mathbf{n} + (B_{2.5} + B_{3.5})\mathbf{v}\right], \tag{12}$$

where $\eta = m_1 m_2/m^2$, $\dot{r} = dr/dt$, and

$$A_{2.5} = (3 + 3\beta)v^2 + (\frac{23}{3} + 2\alpha - 3\beta)\frac{m}{r} - 5\beta\dot{r}^2, \tag{13}$$

$$B_{2.5} = (2 + \alpha)v^2 + (2 - \alpha)\frac{m}{r} - 3(1 + \alpha)\dot{r}^2, \tag{14}$$

where α and β are arbitrary, and reflect the effects of coordinate freedom on the equations of motion. The values $\alpha = 4$, $\beta = 5$ correspond to the so-called "Burke–Thorne" gauge, in which the radiation reaction is expressed solely as a quasi-Newtonian potential $\Phi_{RR} = -\frac{1}{5}M_{ij}^{(5)}(t)x^i x^j$, where $M_{ij}(t)$ is the traceless moment of inertia tensor of the system and the superscript (5) denotes five time derivatives. This also corresponds to the gauge used by Blanchet [18]. Our 2.5PN equations of motion yield the values $\alpha = -1$, $\beta = 0$, which corresponds to the gauge used by Damour and Deruelle (Eq. (161) of Ref. [13]).

At 3.5PN order, the expressions for $A_{3.5}$ and $B_{3.5}$ are

$$A_{3.5} = a_1 v^4 + a_2 v^2 m/r + a_3 v^2 \dot{r}^2 + a_4 \dot{r}^2 m/r + a_5 \dot{r}^4 + a_6 (m/r)^2, \tag{15}$$

$$B_{3.5} = b_1 v^4 + b_2 v^2 m/r + b_3 v^2 \dot{r}^2 + b_4 \dot{r}^2 m/r + b_5 \dot{r}^4 + b_6 (m/r)^2. \tag{16}$$

Energy and angular momentum balance yield values for the 12 coefficients modulo 6 arbitrary gauge parameters. Our 3.5PN equations of motion yield the values

$$a_1 = -\frac{183}{28} - \frac{15}{2}\eta, \qquad b_1 = -\frac{313}{28} - \frac{3}{2}\eta, \tag{17}$$

$$a_2 = -\frac{173}{14} - \frac{186}{6}\eta, \qquad b_2 = \frac{205}{42} + \frac{37}{2}\eta, \tag{18}$$

$$a_3 = \frac{285}{4} + \frac{15}{2}\eta, \qquad b_3 = \frac{339}{4} + \frac{3}{2}\eta, \tag{19}$$

$$a_4 = -\frac{147}{4} - 47\eta, \qquad b_4 = -\frac{205}{12} - \frac{106}{3}\eta, \tag{20}$$

$$a_5 = -70, \qquad b_5 = -75, \tag{21}$$

$$a_6 = -\frac{989}{14} - 23\eta, \qquad b_6 = -\frac{1325}{42} - 13\eta. \tag{22}$$

By comparing these values with the Iyer and Will expressions (Eqs. (2.18) of Ref. [17]), we obtain values for the 6 arbitrary gauge parameters. The fact that 12 constraints yield a consistent solution for the 6 parameters is a useful check of the method and algebra. The result yields radiation reaction equations to 3.5PN order in the generalization of Damour-Deruelle gauge. Blanchet's multipole expressions [18] for radiation reaction at 3.5PN order yield a different set of coefficients, which correspond to the generalization of the Burke–Thorne gauge.

5 Gravitational radiation waveform and energy flux

By iterating the relaxed Einstein equations for field points in the far zone, and substituting the two-body equations of motion to the appropriate order, we obtain an explicit formula for the transverse-traceless (TT) part of the radiation-zone field, denoted h^{ij}, which is the waveform to be detected in laser interferometric systems. In terms of an expansion beyond the quadrupole formula, it has the schematic form,

$$h^{ij} = \frac{2\mu}{R}\left\{ \tilde{Q}^{ij}[1 + O(\epsilon^{1/2}) + O(\epsilon) + O(\epsilon^{3/2}) + O(\epsilon^2)\dots] \right\}_{TT}, \tag{23}$$

where μ is the reduced mass, R is the distance to the source, and $\tilde{Q}^{ij}$ represents two time derivatives of the mass quadrupole moment tensor (the series actually contains multipole orders beyond quadrupole). The 0.5PN and 1PN terms were derived by Wagoner and Will [3], the 1.5PN terms by Wiseman [9]. The contribution of gravitational-wave "tails", caused by backscatter of the outgoing radiation off the background spacetime curvature, at $O(\epsilon^{3/2})$, were derived and studied by several authors. The 2PN terms including 2PN tail contributions were derived by two independent groups and are in complete agreement (see Refs. [10] and [19] for details and references to earlier work). The 2.5PN terms and various specific higher-order terms, such as "tails-of-tails" have been derived by Blanchet and collaborators, while the formidable job of evaluating the 3PN terms is still in progress.

To illustrate some of the results of the DIRE method, we note that, in calculating the gravitational waveform through 2PN order, we explicitly retained all terms that depend on positive powers of the radius $\mathcal{R}$ of the near zone [10]. The multipole expansion of the integral over the near zone yielded

$$
h^{ij}_{\mathcal{N}}(t, \mathbf{x}) = \mathcal{R}- \text{independent terms} + 1/\mathcal{R} \text{ terms}
$$
$$
- \frac{1912}{315} \frac{m}{R} \,^{(4)}Q^{ij}(u)\mathcal{R}, \tag{24}
$$

where $^{(4)}Q^{ij}(u)$ represents four time derivative of the quadrupole moment of the source. The $\mathcal{R}$-dependent term is of 2PN order; this explains why sweeping "divergent" terms under the rug at 1PN order was successful, if not fully justified. The integral over the rest of the null cone yields

$$
\begin{aligned}
h^{ij}_{\mathcal{C}-\mathcal{N}}(t, \mathbf{x}) =\ & \frac{4m}{R} \int_0^\infty ds \,^{(4)}Q^{ij}(u - s) \left[\ln\left(\frac{s}{2R + s} \right) + \frac{11}{12} \right] \\
& + \frac{4m}{3R} \hat{N}^k \int_0^\infty ds \,^{(5)}Q^{ijk}(u - s) \left[\ln\left(\frac{s}{2R + s} \right) + \frac{97}{60} \right] \\
& - \frac{16m}{3R} \epsilon^{(i|ka} \hat{N}^k \int_0^\infty ds \,^{(4)}J^{a|j)}(u - s) \left[\ln\left(\frac{s}{2R + s} \right) + \frac{7}{6} \right] \\
& + \frac{1912}{315} \frac{m}{R} \,^{(4)}Q^{ij}(u)\mathcal{R}.
\end{aligned} \tag{25}
$$

Notice that the $\mathcal{R}$-dependent term from the outer integral exactly cancels that from the near-zone integral. The other terms, involving integrals over the past history of the source are the "tails", and are in complete agreement with tail terms derived by other methods. In addition by suitably combining the tail terms with the leading order quadrupole (Q^{ij}), octopole (Q^{ijk}), and current quadrupole (J^{aj}) terms in the waveform, one can show that part of the effect of these terms, say for a circular orbit of frequency ω, is to convert the phase of the wave from $\psi = \omega(t - R)$ to $\psi = \omega(t - R - 2m \ln R - \text{const})$. This demonstrates that, despite the use of a flat-spacetime wave equation to embody Einstein's equations, the propagation of the radiation follows the true, Schwarzschild-like null cones far from the source.

There are also contributions to the waveform due to intrinsic spin of the bodies, which, for compact bodies, occur at $O(\epsilon^{3/2})$ (spin-orbit) and $O(\epsilon^2)$ (spin-spin); these have been calculated elsewhere [20].

Given the gravitational waveform, one can compute the rate at which energy is carried off by the radiation (schematically $\int \dot{h}h\,d\Omega$, the gravitational analog of the Poynting flux). For the special case of non-spinning bodies moving on quasi-circular orbits (*i.e.* circular apart from a slow inspiral), the energy flux through 2PN order has the form

$$\frac{dE}{dt} = \frac{32}{5}\eta^2\left(\frac{m}{r}\right)^5\left[1 - \frac{m}{r}\left(\frac{2927}{336} + \frac{5}{4}\eta\right) + 4\pi\left(\frac{m}{r}\right)^{3/2} + \left(\frac{m}{r}\right)^2\left(\frac{293383}{9072} + \frac{380}{9}\eta\right)\right],$$

(26)

where $\eta = m_1 m_2/m^2$. Assuming that energy radiated to infinity is balanced by an equal loss of orbital energy, we can translate this into a formula for the evolution of the orbital frequency f_b, and thence the gravitational wave frequency $f = 2f_b$. Through 2PN order, it has the form

$$\dot{f} = \frac{96\pi}{5}f^2(\pi\mathcal{M}f)^{5/3}\left[1 - \left(\frac{743}{336} + \frac{11}{4}\eta\right)(\pi mf)^{2/3} + 4\pi(\pi mf) + \left(\frac{34103}{18144} + \frac{13661}{2016}\eta + \frac{59}{18}\eta^2\right)(\pi mf)^{4/3} + O[(\pi mf)^{5/3}]\right],$$

(27)

where $\mathcal{M} = \eta^{3/5}m$ is the "chirp" mass. In Eqs. (26) and (27), the first term is the quadrupole contribution, the second term is the 1PN contribution, the third term, with the coefficient 4π, is the "tail" contribution, and the fourth term is the 2PN contribution first reported jointly by Blanchet et al. [21].

Similar expressions can be derived for the loss of angular momentum and linear momentum. These losses react back on the orbit to circularize it and cause it to inspiral. Radiation of linear momentum can also cause "radiation recoil", a phenomenon which is probably unobservable. The result is that the orbital phase (and consequently the gravitational-wave phase) evolves non-linearly with time. It is the sensitivity of the broad-band LIGO and VIRGO-type detectors to phase that makes the higher-order contributions to dE/dt so observationally relevant. A ready-to-use set of formulae for the 2PN gravitational waveform template, including the non-linear evolution of the gravitational-wave frequency (not including spin effects) have been published [22] and incorporated into the Gravitational Radiation Analysis and Simulation Package (GRASP), a software toolkit used in LIGO.

6   Gravitational-wave tests of general relativity

In addition to opening a new astronomical window, the detailed observation of gravitational waves by such observatories may provide the means to test general relativistic predictions for the polarization and speed of the waves, and for gravitational radiation damping. This subject has been reviewed elsewhere [23,24].

Acknowledgements. This work was supported in part by the National Science Foundation, Grant Number PHY 96-00049. This paper is based in part on a paper given at the 9th Yukawa International Seminar in July 1999 [25].

References

1. Thorne, K.S. 1995): Gravitational Waves, in *Proceedings of the Snowmass 95 Summer Study on Particle and Nuclear Astrophysics and Cosmology*, ed. by E. W. Kolb, R. Peccei, World Scientific, Singapore pp. 398–425; gr-qc/9506086
2. Finn, L.S., Chernoff, D.F. (1993): Phys. Rev. D **47**, 2198; gr-qc/9301003;
   Cutler, C., Flanagan, É.E. (1994): Phys. Rev. D **49**, 2658; gr-qc/9402014;
   Poisson, E. and Will, C.M. (1995): Phys. Rev. D**52**, 848; gr-qc/9502040;
   Damour, T., Iyer, B.R., Sathyaprakash, B.S. (1998): Phys. Rev. D**57**, 885;
   Poisson, E.: In: Proceedings of the Second Gravitational-Wave Data Analysis Workshop, in press
3. Wagoner, R.V., Will, C.M. (1976): Astrophys. J. **210**, 764
4. Walker, M., Will, C.M. (1980): Phys. Rev. Lett. **45**, 1741;
   Anderson, J.L. (1980): Phys. Rev. Lett. **45**, 1745;
   Damour, T. (1983): Phys. Rev. Lett. **51**, 1019;
   D. Christodoulou, B. G. Schmidt, (1979): Commun. Math. Phys. **68**, 275;
   Isaacson, R.A., Welling, J.S., Winicour, J. (1984): Phys. Rev. Lett. **53**, 1870
5. Blanchet, L., Damour, T. (1986): Phil. Trans. R. Soc. London A **320**, 379;
   Blanchet, L., Damour, T. (1988): Phys. Rev. D **37**, 1410;
   Blanchet, L., Damour, T. (1989): Ann. Inst. H. Poincaré (Phys. Theorique) **50**, 377;
   Damour, T., Iyer, B.R. (1991): Ann. Inst. H. Poincaré (Phys. Theorique) **54**, 115;
   Blanchet, L. (1995): Phys. Rev. D **51**, 2559; gr-qc/9501030;
   Blanchet, L., Damour, T. (1992): Phys. Rev. D **46**, 4304
6. Thorne, K.S. (1980): Rev. Mod. Phys. **52**, 299
7. Epstein, R., Wagoner, R.V. (1975): Astrophys. J. **197**, 717 (EW)
8. Mino, Y., Sasaki, M., Shibata, M., Tagoshi, H. and Tanaka, T. (1997): Prog. Theor. Phys. [Suppl.] **128**, 1; gr-qc/9712057
9. Wiseman, A.G. (1992): Phys. Rev. D **46**, 1517
10. Will, C.M., Wiseman, A.G. (1996): Phys. Rev. D **54**, 4813; gr-qc/9608012
11. Wiseman, A.G., Will, C.M. (1991): Phys. Rev. D**44**, R2945
12. Pati, M.E., Will, C.M. (2001): Phys. Rev. D **62**, 124015; gr-qc/0007087
13. Damour, T. (1987): The problem of motion in Newtonian and Einsteinian gravity, in *300 Years of Gravitation*, ed. by S.W. Hawking, W. Israel, Cambridge University Press, Cambridge, pp. 128–198
14. Blanchet, L., Faye, G., Ponsot, B. (1998): Phys. Rev. D **58**, 124002; gr-qc/9804079)
15. Jaranowski, P., Schäfer, G. (1998): Phys. Rev. D **57**, 7274; gr-qc/9712075;
    Jaranowski, P., Schäfer, G. (1999): Phys. Rev. D **60**, 124003; gr-qc/9906092
16. Blanchet, L., Faye, G. (2000): Phys. Lett. **A271**, 58;
    Blanchet, L., Faye, G. (2001): Phys. Rev. **D63**, 062005
17. Iyer, B., Will, C.M. (1995): Phys. Rev. D **52**, 6882

18. Blanchet, L. (1993): Phys. Rev. D**47**, 4392
19. Blanchet, L., Damour, T., Iyer, B.R. (1995): Phys. Rev. D **51**, 5360; gr-qc/9501029
20. Kidder, L.E., Will, C.M., Wiseman, A.G. (1993): Phys. Rev. D **47**, R4183; gr-qc/9211025; Kidder, L.E. (1995): Phys. Rev. D **52**, 821; gr-qc/9506022
21. Blanchet, L., Damour, T., Iyer, B.R., Will, C.M., Wiseman, A.G. : Phys. Rev. Lett. **74**, 3515; gr-qc/9501027
22. Blanchet, L., Iyer, B.R., Will, C.M., Wiseman, A.G. (1996): Class. Quantum Grav. **13**, 575; gr-qc/9602024
23. Will, C.M. (1999): Physics Today **52**, 38 (October)
24. Will, C.M. (2001): Living Reviews in Relativity. **4**, 4; gr-qc/0103036
25. Will, C.M. (1999): Prog. Theor. Phys. [Suppl.] **136**, 158

Gravity in Anti-de-Sitter Space and Quantum Field Theory

A. Zaffaroni

Abstract. We discuss some aspects of gravitational theories in Anti-de-Sitter spacetime, from the AdS/CFT correspondence to the phenomenology of extra-dimensions and brane-worlds.

1 Introduction

Classical gravitational theories with an Anti-de-Sitter (AdS) vacuum have been recently studied in relation to Quantum Field Theory (QFT) and phenomenology. Such studies originated from the discovery of D-branes in string theory. D-branes are solitonic objects, topological defects in spacetime, whose worldvolume supports non-abelian gauge theory degrees of freedom. Their discovery stimulated the investigation of gauge theory properties using string theory. It also reinforced the idea that we can live on branes fluctuating in spacetimes of dimension greater than four.

The unconventional behavior of gravity in AdS spacetime is at the basis of many related results. Some of the results on AdS are stringy–inspired. However, the stringy description is useful when it is weakly coupled, i.e. when it reduces to a classical action for gravity, or supergravity. In such cases, we can use a classical theory to describe strong coupled phases of gauge theories or to model suggestive scenarios for phenomenology. For the purpose of our talk, we do not need string theory. General Relativity is sufficient. We will include at most a cosmological constant or extra scalar fields. One noticeably circumstance is that, for the description of physical systems in dimension d, we will need AdS spacetimes of dimension $d + 1$.

Section 2 discusses some aspects of the string duality known as AdS/CFT, which is briefly reviewed. It will clarify the behavior of gravity in an AdS vacuum. Contrary to naive expectations, it can be shown that all of the degrees of freedom of a gravitational theory in AdS can be re-expressed in terms of the degrees of freedom (without gravity!) of a CFT with one dimension less. Section 3 describes the Randall–Sundrum model and related brane–world scenarios. These models contain examples of non-compact *compactifications*, where gravity is localized on topological defects. In these models a slice of AdS is used. A deep connection to AdS/CFT still exists and it will be discussed.

In our brief journey through different models, we will focus in particular on few selected topics. We will discuss the AdS/CFT interpretation of Renormalization Group flows in QFT and the holographic interpretation of the compact Randall–Sundrum model. We will discuss the role of a mode, the radion, which is important both for conceptual and phenomenological reasons.

2 The AdS/CFT correspondence: a concise review

The relation between gravity in AdS and QFT is a recent result of string theory. There is by now evidence that all string backgrounds with AdS_{d+1} space-time provide string duals for d-dimensional conformal field theories (CFT) [1]. This is the explicit realization of an old idea, emerged in the seventies after the work of G. t'Hooft. Studying the properties of gauge theories with a large number of color N, it was realized that the quantum theory can be conveniently described in terms of stringy variables. For confining theories, the strings just represent the non-abelian flux tubes connecting quarks. For theories that do not confine the interpretation of the strings is less obvious, but a stringy description is still suggested by the large N expansion. What kind of string theory do we need? It was suspected for a while that unconventional string theories were needed. Since conventional string theory is a possible consistent formulation of quantum gravity, it seemed not plausible to use it for describing a d-dimensional QFT that does not include gravity. Recent results of string theory however indicate that conventional closed string theories with an AdS_{d+1} vacuum may serve the purpose. The reason for this lies in the existence of string vacua (AdS_{d+1} backgrounds for example) where gravity is an illusion: the same physics can be described by a set of gauge theory degrees of freedom. We can understand the fictitious gravity as an artifact of holography. AdS_{d+1} has a conformal boundary at infinity which is isomorphic to d-dimensional Minkowski spacetime. All the physics in AdS_{d+1} is uniquely determined by boundary data. Remarkably, a gravity theory in AdS has an equivalent description as a gauge theory on the boundary.

2.1 Conformal theories

The correspondence between AdS gravity and QFT works at the best in the case of gauge theories that are conformal. This is known as AdS/CFT correspondence. Gauge theories living on the worldvolume of type II string or M-theory d-dimensional branes are dual to string or M-theory backgrounds $AdS_{d+1} \times X$ [1], where X is a suitably chosen compact manifold which will not play any role in our discussion. Here, we will only consider the case $d = 4$. String theory has two expansion parameters, the string coupling g_s determining the loop expansion and the string tension $1/\alpha'$ determining the higher derivative expansion in the effective action for gravity and stringy states. These two parameters correspond to the two gauge theory parameters: N, the number of colors, and g_{YM}, the coupling constant. We will focus on the weak coupling limit of string theory, described by a classical effective Lagrangian, which is dual to a strongly coupled gauge theory at large N.

The basic relation between a five dimensional relativistic theory (including gravity) with AdS_5 vacuum and a four dimensional CFT is given by the symmetry group of the two theories. The isometry group $O(4, 2)$ of AdS_5 is also the group of conformal transformations in four dimensions. The content of the AdS/CFT correspondence consists in a one-to-one map between observables and a prescription for computing correlation functions for CFT composite operators using the AdS_5 theory.

The source for a CFT operators O, defined via the coupling

$$S_{\text{CFT}} + \int dx^4 O(x)\omega(x), \tag{1}$$

is identified with the boundary value of an AdS field ω. This identification makes sense since the AdS$_5$ boundary is isomorphic to Minkowski spacetime. The metric in AdS is naturally associated to the CFT stress-energy tensor via the coupling $\int dx^4 \sqrt{-g}\, g_{\mu\nu} T^{\mu\nu}$. More generally, the symmetries (and currents) of field theories can be read from those of the gravity theory according to the rule: a gauge symmetry in AdS corresponds to a global symmetry in field theory. The identification of the other observables depends on the details of the background. The rules of AdS/CFT state that the CFT partition function in presence of the source $\omega(x)$ is given by the classical 5d action $S[\hat{\omega}]$ for a solution of the 5d equations of motion. We choose the solution $\hat{\omega}$ that converges to ω on the AdS boundary and behaves well at the AdS horizon. We have the identification [1]

$$\langle e^{i \int dx^4 \omega(x) O(x)} \rangle_{\text{CFT}} = e^{iS[\hat{\omega}]} . \tag{2}$$

As an example, consider a free field ω of mass m in AdS. Consider coordinates where AdS looks like (we use signature $(-, +, +, +, +)$)

$$ds^2 = dy^2 + e^{2\varphi(y)} dx^\mu dx_\mu, \quad \mu = 0, 1, 2, 3, \tag{3}$$

with $\varphi = -y/R$. $r = -\infty$ is the boundary and $r = \infty$ the horizon. The appropriate boundary condition for the field ω, following from the asymptotic solution of the equations of motion, is $\hat{\omega}(x, y) \rightarrow e^{(4-\Delta)r} \omega(x)$, where $m^2 = \Delta(\Delta - 4)$. Δ is identified with the conformal dimension of the dual operator O. From (2) we can extract the two point functions $< O(x)O(y) > \sim |x - y|^{-2\Delta}$ [1], as expected in a conformal theory.

2.2 Non conformal theories

We also expect that general backgrounds (3), with an arbitrary warp factor φ, but still preserving 4d Poincaré invariance, are dual to non-conformal theories. In particular, backgrounds that are asymptotically AdS have a natural interpretation as deformations of CFT's. This framework is extremely useful for studying the gravitational description of QFT Renormalization Group (RG) flows and, ultimately, the non-perturbative dynamics of non-conformal gauge theories.

Consider a certain four-dimensional CFT. As discussed in the previous Section, it exists a dual 5d Lagrangian. In general, for many purposes, this effective 5d Lagrangian can be taken as a general Lagrangian for scalars coupled to gravity

$$L = \sqrt{-g}\left[-\frac{R}{4} + \frac{1}{2} g^{IJ} \partial_I \lambda_a \partial_J \lambda_b G^{ab} + V(\lambda) \right]. \tag{4}$$

The scalar fields correspond to the CFT operators. We only retain in the Lagrangian the modes we are interested in. The form of the potential depends on the particular case we are considering.

For every critical point of the potential V at constant $\hat{\lambda}_a$, we have an AdS$_5$ solution with radius determined by the value of the potential $V(\hat{\lambda}_a)$. This corresponds to a CFT, as discussed in the previous Section. But we can also consider more general solutions. The equations of motion for the scalars and the metric read

$$\ddot{\lambda}_a + 4\dot{\varphi}\dot{\lambda} = \frac{\partial V}{\partial \lambda_a}, \quad 6(\dot{\varphi})^2 = \sum_a (\dot{\lambda}_a)^2 - 2V. \tag{5}$$

Solutions with non zero scalar fields, which are asymptotically AdS, describe RG flows for a four dimensional CFT [2,3]. Such flows may describe deformations of CFT or choice of different vacua of the same theory [4]. Equations (5) represent the gravity description of the QFT RG flow, y representing the energy scale along the flow.

Many solutions corresponding to non-conformal gauge theories have been studied [5–8]. Moreover, many general QFT theorems have been proved in the gravity context [2,3,9]. We will quote only a particular example, the c theorem [2,3]. In a gauge field theory in 4d, the trace anomaly is given by

$$T^\mu_\mu = \frac{\tilde{\beta}}{2g^2} F^2_{\mu\nu} + \frac{c}{16\pi^2} W^2_{\mu\nu\rho\sigma} - \frac{a}{16\pi^2} \tilde{R}^2_{\mu\nu\rho\sigma}. \tag{6}$$

Here $W_{\mu\nu\rho\sigma}$ and $R_{\mu\nu\rho\sigma}$ are the Weyl and curvature tensors for an external metric $g_{\mu\nu}$ that couples to the energy-momentum tensor $T_{\mu\nu}$. In two dimensions, it is known that the c function always decreases along a RG flow. In four dimensions, the situation is much less clear and there is no general QFT theorem for c. We now will see that the dual gravitational picture helps in proving a c theorem in four dimensions.

The external anomaly coefficients a and c have a straightforward interpretation in the dual gravity theory. c is always equal to a and it is associated with the cosmological constant Λ_{CFT} at the critical points [10]. More interestingly, one can prove that for the class of field theories that have a gravity dual, a c theorem exists. Indeed we can exhibit a c function that is monotonically decreasing along the flow [2,3]. The c function $c(y) \sim (T_{yy})^{-3/2}$ is constructed with the y component of the stress-energy tensor

$$T_{yy} = 6(\dot{\varphi})^2 = \sum_a (\dot{\lambda}_a)^2 - 2V. \tag{7}$$

At the critical points, where $\dot{\lambda}_a = 0$,

$$c(y) = c_{\text{UV,IR}} \sim (-V)^{-3/2}_{\text{UV,IR}} \sim \Lambda^{-3/2}_{\text{UV,IR}}, \tag{8}$$

and using the equations of motion one can easily check that $c(y)$ is monotonic [2,3].

Let us discuss the meaning of the monotonicity of c. It is determined by the convexity of the function φ, which follows from (5)

$$4\ddot{\varphi} = -6 \sum_a (\dot{\lambda}_a)^2. \tag{9}$$

In a well defined 5d theory, the kinetic term for the scalars is positive definite and φ is convex. We see that the monotonicity of c is a consequence of positivity of energy. The c theorem can be indeed related to the weak energy dominance condition [3].

Notice that the concept of the c function is not uniquely defined. Central charges are unambiguously defined only at the conformal fixed points. Outside conformality, there is more than one interpolating and decreasing function $c(y)$. In QFT, for example, $c(y)$ depends on the renormalization scheme. Even in the AdS context, there is a second natural way of defining a c function: $c(y)$ appears in the two-point function of the stress-energy tensor. We may call it the *canonical c* function and it is not necessarily equal to the above defined *holographic c* function [11]. The *canonical c* function can be computed using the rules of AdS/CFT [11]. This is generally a non-trivial computation and, at the best of our knowledge, there is no proof (using gravity) that the *canonical c* function decreases. However the decreasing can be explicitly checked in the few analytically solvable examples of flows [11]. We know at least three (or four) different definitions of c functions using gravity [2,3,9,11,12].

We should notice that classical gravity in AdS only describes a very restricted class of gauge theories or, at least, of their phases. First of all these theories are at strong coupling. Moreover, they always have $a = c$, which is in general not the case in field theory. Finally, being dual to pure gravity, all operators with spin greater than two should decouple. In this regime, theories are simplified and gravity defines a *holographic* scheme, where general QFT results, like the c theorem [2,3] and the Callan–Symanzick equation [9], can be efficiently proved. For studying the most general gauge theory in the most general phase the full string theory is required.

3 Brane worlds

The relation between AdS and phenomenology emerged in an apparently different context. The idea that the existence of extra spacetime dimensions could explain four-dimensional physics goes back to the twenties and the work of Kaluza and Klein. In this scenario, conventional four-dimensional Einstein gravity is recovered only at large distances. At distances of the order of the radius of the internal dimensions, physics becomes effectively five dimensional. This is not in contradiction with experiments if the threshold is smaller than current results on gravitational forces, which only tested Einstein gravity above the millimeter. We can formulate three basic possible scenarios.

- We live localized in a four-dimensional plane in a $4 + n$ dimensional spacetime with n extra compact dimensions.
- The n extra dimensions are not compact but 4d Einstein gravity is still recovered at large distances.

- The extra dimensions are not compact and physics is effectively five-dimensional both at large and short distances, but 4d Einstein gravity is recovered at intermediate distances.

In all these scenarios, we are supposed to live in the dimensions remaining after compactification or on four-dimensional branes embedded in the larger spacetime. The first scenario is well known and it is certainly conceptually consistent. It is the standard scenario in the KK (Kaluza–Klein) literature as well as in string theory, where the critical ten dimensional vacuum is compactified to four dimensions. It was recently pointed out that it can also shed light on the hierarchy problem [13,14]. The second scenario is less conservative, but it was recently pointed out that it is consistent [15]. Consider for simplicity the case of one extra dimension. The generic five dimensional spacetime of interest has the form (3). Start with a five dimensional gravity theory which includes at low energies the standard Einstein term $2M_{(5)}^3 \int \sqrt{-g} R_{(5)}$, where $2M_{(5)}^3 = 1/(16\pi G_{(5)})$, and dimensionally reduce to four dimensions using

$$ds^2 = dy^2 + e^{2\varphi(y)} g_{\mu\nu} dx^\mu dx_\nu, \quad \mu, \nu = 0, \ldots, 3. \tag{10}$$

We obtain

$$\left(2M_{(5)} \int dy\, e^{2\varphi(y)} \right) \int d^4 \sqrt{-g} R_{(4)}. \tag{11}$$

The term in round brackets in (2) determines the effective four dimensional Newton constant

$$\frac{1}{16\pi G_{(4)}} = 2M_{(5)} \int dy\, e^{2\varphi(y)}. \tag{12}$$

A 4d observer would say to live in four dimensions provided that the expression in (12) is finite, that is the warp factor $\exp(2\varphi(y))$ is integrable. This is easily obtained, in the Randall-Sundrum (RS) model [15] for example, with $\varphi(y) = -2k|y|$, which is the patching of two truncated AdS$_5$ spaces. This result could be spoiled by the tower of KK modes, which, with a non-compact dimension, form a continuum. Contrary to naive expectations, the KK modes only give negligible corrections to Einstein theory. The third scenario is much less conservative and its consistency is still controversial.

 We will now discuss a series of models that exemplify the various mentioned scenarios.

3.1 The set-up

The metric (3) is the most general 5d metric with four-dimensional Poincaré invariance. We now assume that it is a solution of a five-dimensional gravity theory with three-brane sources,

$$\int dx^5 \sqrt{-g} (2M^3 R - \Lambda_a) - \sum_i \delta(y - y_i) \tau_i \int dx^4 \sqrt{-\hat{g}}, \tag{13}$$

where $\hat{g}$ is the induced metric on each brane. The 5d cosmological constant Λ_a may vary in the different domains of spacetime, labelled by a, delimited by three-branes. We shall consider the case $\Lambda_a \leq 0$. The derivative of the warp factor jumps at the three-brane positions by an amount related to the tensions of the branes $\Delta\varphi'(y_c) = -\tau_i/(12M^3)$. Tensions and cosmological constants in the bulk must satisfy these jump conditions. As a consequence there is a fine-tuning in order to get flat four-dimensional space.

The non-compact RS model [15] has $\varphi(y) = -k|y|$. The bulk metric is AdS_5 with cosmological constant $\Lambda = -24M^3k^2$ and there is a brane of positive tension $\tau_1 = 24M^3k$ at the origin. An orbifold Z_2 symmetry $y \to -y$ is also imposed. The compact RS model [15] is obtained by compactifying the y direction on a circle and introducing a second brane at the other orbifold point $y = r$ [15]. The tension of the second brane is negative $\tau_2 = -\tau_1$. Since the negative tension brane is sitting at an orbifold fixed point, the negative energy mode associated with its fluctuations in the transverse direction is projected out.

The compact RS model belong to the first class of scenarios mentioned at the beginning of Sect. 3. It is a compactification with a non trivial warp factor. The four dimensional effective theory contains a massless graviton and a massless scalar field r, named radion, whose vacuum expectation value determines the distance between the branes. For our purpose, it will be convenient to put the two branes at $y = r_0$ and $y = r_1$. An orbifold projection will be always understood, even if, strictly speaking, it is not necessary. The four dimensional Planck scale is given by

$$M_P^2 = \frac{M^3}{k}\left(e^{-2kr_0} - e^{-2kr_1}\right). \tag{14}$$

3.2   The non-compact RS model

From (14) we see that, in the limit $r_1 \to \infty$, which defines the non-compact RS model, the four dimensional Planck scale stays finite. A four dimensional graviton has been localized on the brane and its effective theory at large distances is just Einstein gravity with Plack mass $M_P^2 = M^3\,e^{-2kr_0}/k$. We could worry about the effect of the KK modes, which form a continuum due to the non compactness of the fifth direction. They certainly correct the effective theory, but it can be shown that they give small $1/r^4$ corrections to the Newton law [15]. These corrections are negligible at very large distances.

The model has an intriguing relation with the AdS/CFT correspondence [16]. The truncation of the AdS boundary corresponds, on the CFT side, to the coupling to gravity. The RS brane acts as an UV cut-off. We then obtain a new type of duality: the RS model can be equivalently described by a strongly coupled CFT coupled to 4d gravity. It can be shown that the effect of integrating out the CFT matter fields gives $1/r^4$ corrections to the Newton law identical to the effects of the KK modes in the RS model [16].

Should we consider the non-compact RS model as a 4d theory in disguise? This depends on taste. Certainly all of the effects of the KK modes can be taken into account by replacing them with a CFT. However, this CFT is strongly coupled. The classical gravitational description in 5d is useful for many purposes.

3.3 The compact RS model

We now discuss the holographic interpretation of the compact RS model [17,18]. As discussed in Sect. (2.2), we interpret the coordinate y as an energy scale. The region between the two branes represents the energy regime where the 4d theory is well approximated by a CFT. The brane at r_0 (UV or Planck brane) represents the UV cut-off. The brane at r_1 (IR brane) abruptly ends AdS space. This represents a breakdown of conformal invariance in the IR. We claim that this is a spontaneous breaking and that the compact RS model is dual to a 4d field theory with non-linearly realized conformal symmetry. The massless radion field is an exact modulus of the model, so it is naturally interpreted as the Goldstone boson of broken dilatation invariance [17,18]. We can consider the compact RS model as an idealized description of a CFT (coupled to gravity) along an exactly flat direction, parameterized by the radion. For example, Coulomb or Higgs branches typically exist in gauge theories with AdS duals, which are obtained from D-branes.

In order to simplify things we can decouple 4d gravity by sending the UV brane all the way to the AdS boundary $r_0 \to -\infty$. Using the variable $\mu = k\exp(-ky)$, which is suitable for our purposes, the IR brane is at the position labelled by μ_1. In AdS/CFT, scale invariance is identified with the isometry $x \to \lambda x, \mu \to \mu/\lambda$. The conformal symmetry is non-linearly realized in the non compact RS model, since the position of the IR brane is changed $\mu_1 \to \mu_1/\lambda$. The physics is however unchanged, so that μ_1 parameterizes a manifold of equivalent vacua. The associated Goldstone boson is the radion μ. We can strengthen the interpretation by considering the effective Lagrangian for radion and 4d gravity calculated in [19]

$$L = \sqrt{-g}\,\frac{M^3}{k^3}\left\{2(\mu_0^2 - \mu^2)R(g) - 12(\partial\mu)^2\right\}. \tag{15}$$

When $\mu_0 \to \infty$, 4d gravity decouples and g is just a background probing our CFT. The μ dependent terms are Weyl invariant, as expected. With the rules of AdS/CFT one can also check that the the dilatation current two-point function has a pole, as dictated by Goldstone's theorem [18]. Moreover, the residue at the pole computed using AdS/CFT coincides with that extracted from (15) [18].

Corresponding to a radion value $\langle\mu\rangle = \mu_1$ there should be some CFT operators O_i of dimensions d_i getting VEVs. Since the local geometry between the two branes is *exactly* AdS, we expect that the operator that spontaneously breaks conformal invariance has formally infinite dimension [18].

A massless or light scalar field has disastrous phenomenological consequences. In the RS literature, the radion is stabilized by the Goldberger–Wise mechanism [20]. A nearly massless scalar field is introduced in the bulk, with boundary conditions on the UV and IR branes that make it run. This generates a potential for the radion field. In the holographic description, as discussed in Sect. (2.2), a non trivial profile for bulk scalar fields corresponds to a deformation of the original CFT, which indeed would generically lift the flat direction [18,17]. To give phenomenologically acceptable results, the deforming operator should have dimension close to 4.

3.4 The GRS model

We saw consistent examples of the first two scenarios mentioned at the beginning of Sect. 3. The third scenario is much less conservative and more exotic. We can now provide an example of this scenario, the Gregory–Rubakov–Sibiryakov (GRS) model [21]. Consider a modification of the compact RS model where at the right hand side of the IR brane we take flat five dimensional space. The warp factor for the GRS model is

$$
\varphi(y) = \begin{cases} -ky & 0 \le y \le r, \\ -kr & y \ge r. \end{cases} \tag{16}
$$

The cosmological constant equals respectively $\Lambda_1 = -24Mk^2$ and $\Lambda_2 = 0$ to the left and to the right of the second brane. The matching conditions at the branes fix $\tau_1 = 24M^3k$, $\tau_2 = -24M^3k$. Notice that the IR brane has negative tension.

The model has many attractive features. The theory is effectively five-dimensional both at small and at large scales while it localizes gravity at intermediate scales. This can be understood by considering r as an IR regulator. For $r \to \infty$ we should recover the non-compact RS model and a localized 4d graviton. For finite r, the 4d graviton still exists as a metastable state [21–23]. If the life-time of the graviton is greater than the age of the universe, we could reasonably live in such scenario. At very large distances, which we could never probe, the physics is that of flat five dimensional spacetime. However, the presence of a freely fluctuating brane with negative tension violates positivity of the energy and makes the model unstable. The kinetic term of the 4d radion field, determining the position of the IR brane, can be explicitly computed [24] and it is negative. As noticed in [25], there is no modification of the model by including bulk fields which cures the problem. The requirement of flat space at infinity indeed requires a warp factor with $\ddot{\varphi} \ge 0$, which is in contradiction with positivity of the energy, as discussed in Sect. (2.2).

It is intriguing that *at the classical level* the GRS model exactly reproduces Einstein gravity [21,24]. With a massless scalar field we would rather expect a Brans-Dicke theory. To correctly reproduce the Einstein theory, the graviton propagator must have the form

$$
\langle h_{\mu\nu}h_{\rho\sigma}\rangle \sim \left(\frac{\eta_{\mu\rho}\eta_{\nu\sigma} + \eta_{\nu\rho}\eta_{\mu\sigma}}{2} - \frac{\eta_{\mu\nu}\eta_{\rho\sigma}}{2} \right) \frac{1}{q^2} = \left(\frac{P_2}{2} - \frac{P_0}{2} \right) \frac{1}{q^2}, \tag{17}
$$

where we have neglected terms involving q_μ in the tensor structure. However, the graviton bound state, being made of massive KK graviton, has a propagator

$$
\left(\frac{P_2}{2} - \frac{P_0}{3} \right) \frac{1}{q^2}, \tag{18}
$$

which has a crucial $1/3$ factor. The radion field contributes a P_0/q^2 term to the propagator, with the right coefficient for reproducing Einstein gravity as in (17) [24]. Since, due to five dimensional coordinate invariance, brane matter always couples to both the graviton bound state and the radion, Einstein gravity is exactly reproduced at the classical level. Notice that the fact that the radion has negative kinetic term is crucial for this agreement [23,24]. Therefore, the very same ingredient that allows to reproduce Einstein gravity, makes the model unstable.

It is not known if there exists consistent modifications of the GRS model which realize the third of the scenarios mentioned at the beginning of Sect. 3.

Acknowledgements. I would like to thank all the collaborators with whom some of the results reported here were obtained. A particular thank to D. Anselmi, A. Kehagias, P. Fré, S. Ferrara, L. Girardello, M. Petrini, M. Porrati and R. Rattazzi for a continuous collaboration on the AdS/CFT and the RS scenario.

References

1. Maldacena: J. (1998): ATMP **2**, 231; hep-th/9711200;
   Gubser, S.S., Klebanov, I.R., Polyakov, A.M. (1998): Phys. Lett. B **428**, 105; hep-th/9802109;
   Witten, E. (1998): Adv. Theor. Math. Phys. **2**, 253; hep-th/9802150
2. Girardello, L., Petrini, M., Porrati, M., Zaffaroni A. (1998): JHEP **12**, 022; hep-th/9810126;
   (1999): JHEP **05**, 026; hep-th/9903026;
   (2000): Nucl. Phys. B **569**, 451; hep-th/9909047
3. Freedman, D.Z., Gubser, S.S., Pilch, K., Warner, N.P.: hep-th/9904017;
   (2000): JHEP **0007**, 038; hep-th/9906194
4. Balasubramanian, V., Kraus, P., Lawrence, A., Trivedi, S. (1999): Phys. Rev. D **59**, 104021; hep-th/9808017;
   Klebanov, I.R., Witten, E. (1999): Nucl. Phys. B **556**, 89; hep-th/9905104
5. Warner, N.P. (2000): Class. Quant. Grav. **17**, 1287; hep-th/9911240
   Petrini, M., Zaffaroni, A.: hep-th/0002172 (in press)
6. Pilch, K., Warner, N.P. (2001): Nucl. Phys. B **594**, 209, hep-th/0004063; hep-th/0006066 (in press)
7. Polchinski, J., Strassler, M.J.: hep-th/0003136
8. Klebanov, I.R., Strassler, M.J. (2000): JHEP **0008**, 052; hep-th/0007191
9. Porrati, M., Starinets, A. (1999): Phys. Lett B **454**, 77; hep-th/9903085;
   Balasubramanian, V., Kraus, P. (1999): Phys. Rev. Lett **83**, 3605; hep-th/9903190;
   de Boer, J., Verlinde, E., Verlinde, H. (2000): JHEP **0008**, 003, hep-th/9912012
10. Henningson, M., Skenderis, K. (1998): JHEP **07**, 023; hep-th/9806087;
    Gubser, S.S. 1999: Phys. Rev. D **59**, 025006; hep-th/9807164
11. Anselmi, D., Girardello, L., Porrati, M., Zaffaroni, A. (2000): Phys. Lett. B **481** 346, hep-th/0002066
12. Alvarez, E., Gomez, C. (1998): Nucl. Phys. B **541**, 441; hep-th/9807226
13. Arkani-Hamed, N., Dimopoulos, S., Dvali, G. (1998): Phys. Lett. B **429**, 263; hep-ph/9803315;

(1999): Phys. Rev. D **59**, 086004; hep-ph/9807344;
Antoniadis, I., Arkani-Hamed, N., Dimopoulos, S., Dvali, G. (1998): Phys. Lett. B **436**, 257; hep-ph/9804398

14. Randall, L., Sundrum, R. (1999): Phys. Rev. Lett. **83**, 3370; hep-ph/9905221
15. Randall, L., Sundrum, R. (1999): Phys. Rev. Lett. **83**, 4690; hep-th/9906064
16. Witten: Talk at ITP conference *New Dimensions in Field Theory and String Theory*, Santa Barbara, November 1999; http://www.itp.ucsb.edu/online/susy_c99/discussion;
    Gubser, S.S.: hep-th/9912001
17. Arkani-Hamed, N., Porrati, M., Randall, L. (2001): JHEP **0108**, 017, hep-th/0012148
18. R. Rattazzi, A. Zaffaroni (2001): JHEP **0104**, 021; hep-th/0012248
19. Csaki, C., Graesser, M., Randall, L., Terning, J. (2000): Phys. Rev. **D62**h, 045015; hep-ph/9911406;
    Goldberger, W.D., Wise, M.B. (2000): Phys. Lett. B **475**, 275; hep-ph/9911457
20. Goldberger, W.D., Wise, M.B. (1999): Phys. Rev. Lett. **83**, 4922; hep-ph/9907447
21. Gregory, R., Rubakov, V.A., Sibiryakov, S.M. (2000): Phys. Rev. Lett. **84**, 5928; hep-th/0002072;
    (2000): Phys. Lett. B **489**, 203; hep-th/0003045
22. Csaki, C., Erlich, J., Hollowood, T.J. (2000): Phys. Rev. Lett. **84**, 5932; hep-th/0002161;
    (2000): Phys. Lett. B **481** 107; hep-th/0003020;
    Csaki, C., Erlich, J., Hollowood, T.J., Terning, J. (2001): Phys. Rev. **D63**, 065019; hep-th/0003076
23. Dvali, G., Gabadadze, G., Porrati, M. (2000): Phys. Lett. B **484**, 112; hep-th/0002190;
    (2000): Phys. Lett. B **484**, 129; hep-th/0003054
24. Rattazzi, R., Zaffaroni, A. (2000): JHEP **0007**, 056, hep-th/0004028
25. Witten, E. (2001): 4th UCLA International Symposium on sources and detection of dark matter and dark energy in the univere, Marina del Rey, February 2000. Ed. by Cline, D.B., Springer, Berlin; hep-ph/0002297

Black Hole Formation in Supernovae:
Prospects of Unveiling Fallback Emission

L. Zampieri

Abstract. We review the formation mechanism of neutron stars and black holes in core-collapse supernovae and the conditions for establishing a copious fallback of stellar material onto the newly born compact remnant. Accretion from the base of the envelope is capable both to turn the neutron star into a black hole and to give rise to detectable emission of radiation. The potential of unveiling the emission from fallback onto a black hole and its observational implications are discussed in detail.

1 Introduction

Neutron Stars (NSs) and solar-mass Black Holes (BHs) are the evolutionary end-points of massive stars. Trying to identify these exotic objects has been one of the most fascinating challenges for Astronomers over the last 30 years. Not only they are a prediction of Einstein's theory of General Relativity, but also they provide unique "laboratories" to test strong gravity effects and our theories of nuclear interactions under extreme conditions. Because of the absence of direct emission of radiation, gathering observational evidence for the existence of astrophysical BHs is difficult. Observational searches for BHs are mainly aimed to find evidence of their gravitational effects on the surrounding environment. In Galactic X-ray binary systems the gravitational pull of the remnant can strip material away from the companion star and accrete it via the formation of a disk, producing the observed X-ray emission. Measurements of the mass function for a number of systems end up with a lower limit of $3M_\odot$ for the mass of the compact object implying that it must be a BH. On the other hand, in nuclei of several active and regular galaxies high precision Hubble Space Telescope measurements of the motion of stars and gas show that a large concentration of mass (10^6–$10^8 M_\odot$) must reside in a very small central region ($\approx 0.1\,\mathrm{pc}$), thus providing evidence for the presence of supermassive BHs.

Although there is not unanimous agreement on the mechanism leading to the formation of supermassive BHs, the theoretical scenario for the formation of NSs and solar-mass BHs, as those inferred to be present in Galactic X-ray binary systems, appears to be quite well established and envisages the collapse of the core during the late evolutionary stages of massive ($> 8M_\odot$) progenitor stars. While clear evidence for the formation of NSs in supernova explosions was found long ago through the discovery of radio pulsars in supernova remnants, similar evidence for the presence of solar-mass BHs in the site of their formation is still lacking. Recent observations of an overabundance of Nitrogen and Oxygen in the atmosphere of the companion star in the BH binary system GRO J1655-40 support indirectly the BH-forming supernova scenario [1], but a convincing direct proof of BHs in otherwise successful supernovae has not been gathered as yet. An entirely new prospect for identifying BHs in supernovae may come from the potential of detecting the characteristic emission from ongoing accretion fueled by the stellar envelope after the explosion. In the following, we review the results of recent studies aimed to uncover the emission from fallback onto a BH formed in the aftermath of the explosion and clarify the analogies and differences with respect to the fallback luminosity produced by a NS.

2   Core-collapse supernovae and formation of BHs

2.1   NS or BH formation ?

Detailed radiation hydrodynamic simulations of supernova explosions show that, after core-collapse, a hot protoneutron star forms, cooling down primarily by neutrino emission. These elusive particles have a very small cross section and can easily escape in a few seconds without interacting significantly with the surrounding stellar material (as confirmed by the detection of a burst of neutrinos at the time of the explosion of SN 1987A, the brightest supernova since that observed in 1604 by Kepler). While only a tiny fraction ($\sim 1\%$) of the energy carried away by neutrinos ($\sim 10^{53}$ erg) is effectively absorbed by the stellar material, it may nevertheless be sufficient to unbind all the envelope and cause the explosion of the whole star [2]. The exact evolution depends on the rate of early infall of stellar material on the collapsed core and on the binding energy of the envelope (that depend in turn on the progenitor mass M_*).

- *Stars with $8M_\odot < M_* < 19M_\odot$* – Despite difficulties in understanding the requirements of a successful explosion (see, e.g., [2]), it appears that neutrino re-heating is capable of restoring the pressure support behind the prompt shock (generated by the sudden halt of the collapse when nuclear matter density is reached), reviving it. When the shock breaks out, it originates the spectacular optical display of the supernova. The energy deposited by neutrinos is larger than the binding energy of the envelope, that is then expelled. The collapsed core has a typical mass $\sim 1.4M_\odot$ and gives birth to a NS [3]. If, during core-collapse, the remnant acquires a large magnetic field and becomes rapidly rotating, it may be revealed relatively soon after the explosion (10–100 years) through the energy output from its radio emission.
- *Stars with $19M_\odot < M_* < 25M_\odot$* – The early evolution is similar to that of less massive stars. After neutrino re-heating and shock revival, the star explodes producing a successful supernova with the formation of a NS. However, the total binding energy of the stellar envelope tends to increase rapidly with progenitor mass. It reaches a maximum ($\sim 10^{51}$ erg) for $M_* \sim 25M_\odot$, becoming comparable to the total explosion energy of the supernova [4]. Therefore, after the passage of the shock (and possibly the reverse shock formed at the H–He interface) a variable amount of matter ($0.1 - 0.3M_\odot$) may remain gravitationally bound to the collapsed remnant and *fall back* onto it [3]. Typically, these stars have core

masses around 1.6–1.8 $M_\odot$ and their fate depends critically on the equation of state at nuclear matter densities that fixes the maximum stable mass M_{crit} for NSs. For soft equations of state (as the one proposed by [5] as a consequence of K^- condensation), M_{crit} may be as low as 1.5 $M_\odot$ (gravitational mass) and therefore early collapse of the hot protoneutron star to a BH might be expected even before that significant fallback establishes [6], as early suggested by [7]. More recent models of nuclear forces and accurate Monte Carlo modeling of nucleon interactions indicate that $M_{crit} = 1.8$–2.2 $M_\odot$ [8]. The fallback of material from the envelope may increase their mass up to 1.7–2.1 $M_\odot$ (baryonic mass) and then turn the newly formed NS into a BH [9], as already emphasized by [10,11].

- *Stars with $25M_\odot < M_* < 40M_\odot$* – The lack of internal pressure support after core collapse causes an early accretion of gas from the base of the envelope before that neutrinos can deposit their energy and revive the shock. This effect becomes progressively more important with increasing progenitor mass. A lot of energy must be spent in order for the delayed shock to overcome the ram pressure of the infalling stellar material and to reverse the motion of the envelope producing a successful supernova explosion [4]. This has the effect to decrease the explosion energy, increase the amount of mass that remains gravitationally bound to the central remnant after the passage of the shock, and give rise to significant fallback at later times. Typically, the collapsed core has a mass of $\sim 2M_\odot$ and more than $1M_\odot$ of envelope material may fallback onto it on a dynamical timescale (1–10 hours). The mass of the central remnant becomes larger than $3M_\odot$, sufficient to turn it into a (*fallback*) BH [10]. Most of the heavy elements synthetized during the explosion are advected into the BH and the expanding supernova remnant is depleted of heavy elements.
- *Stars with $M_* > 40M_\odot$* – The amount of mass accreted from the base of the envelope during the phase in which the shock stalls is so large to induce the direct collapse of the hot protoneutron star into a BH before that neutrino re-heating becomes effective. No explosion takes place and the whole star collapses and forms a BH of few tens of $M_\odot$, event known as *failed supernova* [12].

2.2 Shock propagation and fallback

The propagation of a supernova shock is rather complex and determines how the explosion energy is distributed in the envelope of the progenitor star. This has important consequences for the onset of significant fallback. The shock may increase or decrease its velocity depending on the progenitor density profile. Because of hydrodynamic interactions, the post-shock material decelerates in regions where the shock slows down (ρr^3 increases with r [13]). This effect is particularly noticeable when the shock reaches the H–He interface. The deceleration is so strong that a reverse shock may form, causing a further slow-down of the post-shock material. However, despite following the details of the evolution of the envelope during the explosion is rather complex, the pushing and shoving of (post-shock) material to gain elbow room away from the energy release that originated the explosion results approximately in a characteristic homologous velocity profile ($V \propto r$) [14]. Gas at the outer edge gets shoved the most and, hence, has the highest velocity, while matter near the center is pushed and pulled backwards and forwards and, therefore, has a smaller velocity. It is precisely this low-velocity, inner part of the expanding envelope (inside the helium layer) that may eventually remain bound and fall back onto the collapsed remnant. Assuming that, after the passage of the shock (and, possibly, the reverse shock) the motion is ballistic and spherically symmetric, an initially expanding element of fluid with positive binding energy E_b will fall back on a central point mass M in a characteristic time $t_* \sim (\pi/\sqrt{2})GME_b^{-3/2}$. Gas with smaller and smaller binding energy will turn around and fall back after a progressively larger time, fueling on ongoing accretion of stellar material onto the central object.

The accretion of material from the base of the envelope and the early fallback fueled by gravitationally bound gas have a fundamental role in determining the mass and therefore the nature of the collapsed remnant formed in the aftermath of the explosion. This has implications for the number and the distribution of NS e BH remnants in our Galaxy [4,6]. Here we will not elaborate further on this issue, but will instead turn to discuss in more detail the character and the properties of the accretion flow induced by fallback.

3   Radial fallback at hypercritical rates

Although angular momentum may eventually lead to the formation of an accretion disk, detailed calculations of fallback have been carried out only for radial flows. The analysis of fallback in spherical symmetry has allowed us to reach a basic understanding of several physical effects and of the hierarchy of timescales that characterize this phenomenon [15–17]. Investigations of hypercritical accretion disks performed up to now either do not reach the same level of consistency [18] or are not directly relevant for the regimes of interest here [19]. We will briefly discuss fallback disks later.

As mentioned above, the amount of material that falls back becomes significant only for stars with $M_* > 19M_\odot$. The typical peak accretion rate during early fallback amounts to $\approx 0.5M_\odot/20\text{hrs} \sim 10M_\odot \text{ yr}^{-1}$. Therefore, accretion is highly hypercritical and photons are completely advected with the flow. The radius below which photons are advected inward faster than they can diffuse outwards is called trapping radius and is given by [20,21])

$$r_{\text{tr}} = \kappa \dot{M}/4\pi c, \tag{1}$$

where $\dot{M}$ is the accretion rate and κ is the gas opacity. Because initially r_{tr} is very large, radiation produced in the vicinity of the compact object cannot escape and the properties of the early accretion induced by fallback do not depend significantly on the nature of the central compact object (NS or BH). Furthermore, we expect that the evolution be quasi-stationary, because all the typical timescales increase with radius and then the inner regions of the flow respond quickly to the secular changes happening in the outer zones. During the early stages, both shock and radioactive heating increase the gas temperature to $\approx 10^7$ K, so that the internal (gas+radiation) pressure is dynamically important.

Consequently, the early fallback can be well approximated through a sequence of stationary solutions of spherically accreting, polytropic, Bondi-like flows [22]. This is shown clearly by detailed numerical radiation-hydrodynamic simulations in spherical symmetry [16,17] (see below). The characteristic time for the Bondi hydrodynamic accretion is the accretion timescale $t_a = r_a/c_s$, where $r_a = GM/c_s^2$ is the accretion radius (M is the mass of the central remnant) and c_s the gas sound velocity. The peak accretion rate is reached at the initial accretion timescale $t_{a,0} \approx 20$ hrs (see Fig. 1). Because of expansion, the density in the outer envelope decreases secularly with time. As density decreases, the effects of radiation diffusion at the accretion radius r_a can no longer be neglected. Accretion is not adiabatic and the Bondi formula does not strictly apply. The accretion rate can be approximated by the expression $\dot{M} = Q\dot{M}_B$, where $\dot{M}_B$ is the Bondi accretion rate and $Q = t_a/t_{\mathrm{diff}}$ is the ratio of the accretion timescale to the diffusion timescale $t_{\mathrm{diff}} = \tau r/c$ (τ is the optical thickness at radius r) [11,20]. The time at which $t_a \sim t_{\mathrm{diff}}$ is typically of ~ 10 days. During this phase $\dot{M}$ increases with respect to $\dot{M}_B$ (Fig. 1) owing to a slight flattening of the pressure gradient at the accretion radius caused by radiation diffusion. This continues until eventually the accretion rate becomes limited by the supernova expansion, when t_a becomes larger than the expansion timescale $t_0 = r/V$ (V is the velocity at radius r) at r_a. This happens also after a typical time of the order of 10 days, so that the modified Bondi accretion stage is very short. When $t_a > t_0$ pressure effects become dynamically negligible and the motion of the gas can be well approximated by the radial ballistic motion of test particles. The accretion rate shows a characteristic power-law decay with time [15] (see Fig. 1)

$$\dot{M}_{\mathrm{dust}} = \frac{16\pi^{5/3}}{9} GM\rho_0 t_0 \left(\frac{t}{t_0}\right)^{-5/3} . \tag{2}$$

ρ_0 and t_0 denote the gas density and expansion time at the onset of the ballistic phase. While for simplicity we have assumed that the envelope is uniform, in reality the

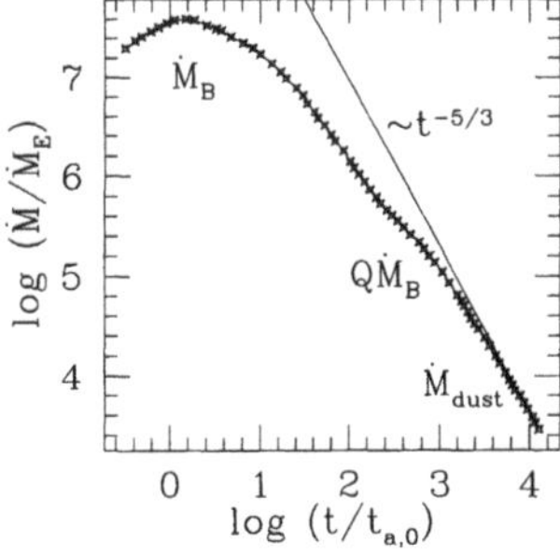

Fig. 1. Accretion rate (in units of the Eddington rate $\dot{M}_E$ defined below) vs time (in units of the initial accr. timescale $t_{a,0}$) for a model computed by Zampieri et al. [16]

situation is more complex and both ρ_0 and t_0 vary from the inner helium layer to the outer hydrogen envelope [17].

In the following we will refer to the ongoing accretion of bound envelope material following the transition to ballistic motion as to the *late-time or secondary fallback*, as opposed to the preceding *early or primary fallback* stage. Clearly, much of the mass is being accreted during primary fallback, in a time comparable to the initial accretion timescale (see again Fig. 1), and may be sufficient to turn a newly formed NS into a BH (as discussed in the previous section). A rough estimate of the amount of mass accreted in this phase is given by $M_{\rm acc} \approx \dot{M}_B(t_{a,0})\, t_{a,0} \approx 8(GM)^3\rho/c_s^6$ (ρ and c_s are the initial post-shock density and sound velocity of the envelope). This quantity is strongly dependent on the initial temperature. The total mass accreted during secondary fallback is comparatively much smaller. However, this late time accretion phase is overwhelmingly important for the possibility of giving rise to detectable emission of radiation.

3.1 Radiation-hydrodynamic simulations of fallback onto a BH

Self-consistent fully relativistic computations of the radiation-hydrodynamic evolution of spherically symmetric supernova fallback in presence of a BH is a respectable task, involving the simultaneous solution of the equations of relativistic radiation hydrodynamics for a self-gravitating matter fluid which is interacting with radiation coupled to the moments of the relativistic transfer equations [16,23]. Tackling the simulation of realistic envelopes is even more difficult because one must deal with the effects of radioactive decay and the opacities of heavy elements. The equations can be cast in the form ($G = c = 1$) [16,17]

$$\epsilon_{,t} + a\kappa_{\rm P}(B - w_0) + p\left(\frac{1}{\rho}\right)_{,t} = Q_\gamma, \tag{3}$$

$$u_{,t} + a\left\{\frac{\Gamma}{b}\left(\frac{p_\mu - b\kappa_{\rm R}\rho w_1}{\rho h}\right) + 4\pi r\left[p + \left(\frac{1}{3} + f\right)w_0\right] + \frac{M_e}{r^2}\right\} = 0, \tag{4}$$

$$\frac{(\rho r^2)_{,t}}{\rho r^2} + a\left(\frac{u_{,\mu} - 4\pi b r w_1}{r_{,\mu}}\right) = 0, \tag{5}$$

$$(M_e)_{,\mu} = 4\pi r^2 r_{,\mu}\left(e + w_0 + \frac{u}{\Gamma}w_1\right), \tag{6}$$

$$(w_0)_{,t} - a\kappa_{\rm P}\rho(B - w_0) + \left[\frac{4}{3}\left(\frac{b_{,t}}{b} + 2\frac{r_{,t}}{r}\right) + \left(\frac{b_{,t}}{b} - \frac{r_{,t}}{r}\right)f\right]w_0$$
$$+ \frac{1}{abr^2}(w_1 a^2 r^2)_{,\mu} = 0, \tag{7}$$

$$\frac{(w_1)_{,t}}{w_1} = -a\kappa_R\rho - 2\left(\frac{b_{,t}}{b} + \frac{r_{,t}}{r}\right) - \frac{a}{w_1}\left[\frac{1}{3a^4b}(w_0 a^4)_{,\mu} \right.$$
$$\left. + \frac{1}{abr^3}(f w_0 a r^3)_{,\mu}\right], \quad (8)$$

where t and μ are the Lagrangian time and the comoving radial coordinate, r is the Schwarz-schild radial coordinate, a and b are the comoving-frame metric coefficients, u is the radial component of the fluid 4-velocity measured in the Eulerian frame, $\rho, e, p, \epsilon = (e - \rho)/\rho$ and $h = (e + p)/\rho$ are the rest-mass density, total mass-energy density, pressure, internal energy per unit mass and enthalpy of the gas flow (respectively) as measured in the comoving frame, M_e represents the effective gravitational mass-energy (for black hole + gas + radiation) contained within radius r, $\Gamma = (1 + u^2 - 2M_e/r)^{1/2}$, w_0 and w_1 are the radiation energy density and radiative flux (in units of $\mathrm{erg\,cm^{-3}\,s^{-1}}$), κ_P and κ_R are the Planck and Rosseland mean opacities, $B = a_R T^4$ is the Planck function, f is the closure Eddington factor for the moment equations, Q_γ is the energy input per unit mass and time from radioactive decay.

From the numerical point of view the greatest challenge in solving these equations is represented by the large dynamical range of characteristic times, going from the microsecond for the dynamical timescale in the accretion flow near the BH to the $\sim$ 100 days for the decay time of long-lived radioactive isotopes in the outer envelope. To make the numerical computation feasible, several different numerical procedures and acceleration measures have been employed [16,17]. Numerical solutions (subject to boundary conditions that simulate the presence of a BH) of the full relativistic radiation-hydrodynamic equations (3)–(8) for the expanding supernova envelope of a 26 $M_\odot$ progenitor and a 35 $M_\odot$ progenitor show a new and important result (see Fig. 2). After traversing all the phases that characterize a typical Type II supernova light curve (diffusion plateau and recombination peak) and reaching the stage at which heating from radioactive material becomes the dominant energy source, the luminosity deviates sharply from the characteristic exponential tail of heavy isotopes and relaxes on a distinguishable power-law decay: this marks the *emergence* of the luminosity from accretion onto the central BH. As shown in Fig. 2, more massive progenitors show an earlier emergence ($\sim$ 50 days) with respect to less massive ones ($\sim$ 1000 days) [17].

This important result is a consequence of the hydrodynamic evolution of the accretion flow at late times and the emission properties of BH accretion. By the time that accretion becomes ballistic, the dynamical effects of pressure gradients and of radioactive heating are negligible. The inner accreting region responds very quickly to the secular decrease in density in the outer expanding envelope, causing $\dot{M}$ to decrease. Although accretion is still occurring at hypercritical rates, the luminosity is well sub-Eddington because optically thick, spherical accretion onto BHs is very inefficient [24–27]. Therefore, radiation forces cannot affect significantly the inflow and the hydrodynamical expansion drives the evolution that preserves its ballistic and quasi-stationary character. As shown in Fig. 2 (panel c), the evolutionary track of the innermost accreting region of our computed models on the L–$\dot{M}$ (luminosity–

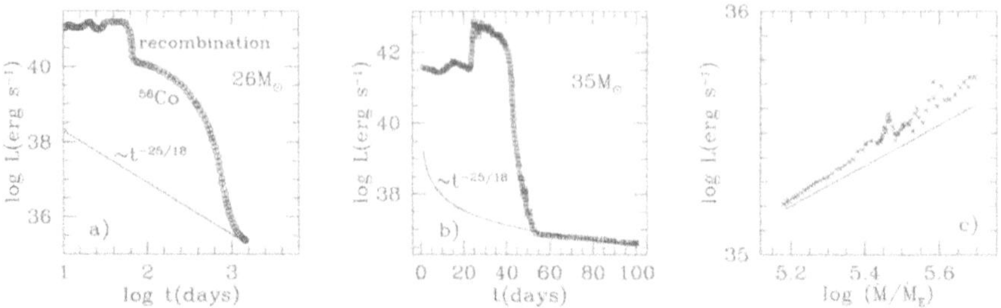

Fig. 2. Bolometric light curves for **(a)** a 26 $M_\odot$ progenitor and **(b)** a 35 $M_\odot$ progenitor (computed by Balberg et al. [17]). **(c)** L vs $\dot{M}$ (in units of $\dot{M}_E$) during secondary fallback (the solid line is the analytic approximation from Eq. [9])

accretion rate) plane follows closely the curve of the stationary solutions of spherical hypercritical accretion onto BHs, computed numerically by Nobili et al. [27]. For these solutions, Blondin [25] derived an approximate analytic expression by integrating the rate of compressional work done on the gas by gravitational forces from the trapping radius r_{tr} outwards. This expression relates the emerging luminosity to $\dot{M}$

$$\frac{L}{L_E} \simeq 4 \times 10^{-7} \left(\frac{\mu}{0.5}\right)^{-4/3} \left(\frac{k_{\mathrm{es}}}{0.4\,\mathrm{cm}^2\,\mathrm{g}^{-1}}\right)^{-1/3} \left(\frac{M}{M_\odot}\right)^{-1/3} \left(\frac{\dot{M}}{\dot{M}_E}\right)^{5/6} , \quad (9)$$

where μ and k_{es} are the mean molecular weight and electron scattering opacity of the bound gas, $L_E = 4\pi GMc/k_{\mathrm{es}}$ e $\dot{M}_E = L_E/c^2$ are the Eddington luminosity and accretion rate. We note that Eq. (9) was first used by [28] and [29] to estimate the luminosity emitted by accretion onto a putative BH in SN 1987A. Then, the decay with time of the accretion luminosity in the ballistic phase can be estimated inserting Eq. (2) in the expression for the luminosity (9). We obtain [16,30]

$$L = 2.1 \times 10^{36} \left(\frac{\mu}{0.5}\right)^{-4/3} \left(\frac{k_{\mathrm{es}}}{0.4\,\mathrm{cm}^2\,\mathrm{g}^{-1}}\right)^{-1/2} \left(\frac{M}{M_\odot}\right)^{2/3}$$
$$\left(\frac{\rho_0}{10^{-5}\,\mathrm{g\,cm}^{-3}}\right)^{5/6} \left(\frac{t_0}{20\,\mathrm{hr}}\right)^{20/9} \left(\frac{t}{1\,\mathrm{yr}}\right)^{-25/18} \mathrm{erg\,s}^{-1}. \quad (10)$$

Equation (10) is in good agreement with numerical results [16]. This luminosity is emitted in the IR-optical bands because the accreting envelope is largely opaque and the photosphere is located where the gas recombines at a temperature of ≈ 6000 K.

3.2   Radiation-hydrodynamic evolution of fallback onto NSs

At variance with accretion onto a BH, the structure of the hypercritical accretion flow associated with fallback onto a NS is strongly affected by the solid surface of the star and the effects of radiation pressure. Early hydrodynamic calculations of supercritical accretion onto NSs illustrated the important dynamical effects of radiation when the accretion energy is carried away by photons [31–33]. Subsequent calculations of hypercritical accretion flows showed the existence of steady state solutions in which, in the initial evolutionary stages, gravitational energy generated by subsonic infall through an extended envelope bounded by an accretion shock is effectively released by emission of neutrinos [11,28]. In these flows the optical depth is so large that photons cannot escape and radiation pressure dominates. The simplest approach to solve the envelope structure involves an integration of the steady state hydrodynamic equations from the shock radius $r_{\rm sh}$ (where strong shock conditions are applied) to the NS surface with a neutrino and pair emissivity term and a polytropic equation of state (with $\gamma = 4/3$) [11]. The accreting gas at large radius is basically unaffected by pressure gradients and is then in free-fall. Because the strong dependence of the cooling term on temperature causes neutrino emission to occur only very close to the NS, the position of the shock radius can be approximately derived by equating the rate of energy losses to the rate of gravitational potential energy release within a scale height $\sim r_{\rm ns}/4$ ($r_{\rm ns}$ is the NS radius) from the NS surface. This gives [11]

$$r_{\rm sh} = \left(\frac{7\pi\sqrt{2}}{3}\right)^{2/3} \frac{r_{\rm ns}^{8/5}\dot{M}^{-2/5}}{[(\pi A)^4 GM]^{1/15}}, \tag{11}$$

where $A = 3.1 \times 10^{-40}$ in cgs units. The effects of a central NS are hidden and the behavior of the accretion flow outside $r_{\rm sh}$ resembles closely that of a BH. Only photons above the trapping radius can emerge and, similarly to the BH case, the luminosity can be approximately calculated integrating the rate of compressional work from the trapping radius $r_{\rm tr}$ outwards [28]. Therefore, *during this phase, also the NS fallback luminosity is approximately given by Eq. (10).*

As the accretion rate diminishes, $r_{\rm sh}$ increases and eventually reaches the trapping radius $r_{\rm tr}$ (Eq. [1]) at the critical value

$$\dot{M}_c = 1.7 \times 10^{-4} (\kappa_{\rm es}/0.4\,{\rm cm}^2\,{\rm g}^{-1})^{-5/7}\,M_\odot\,{\rm yr}^{-1}.$$

From Eq. (2) this occurs at an age of about 6 months–1 year. When this happens, radiation from the shocked envelope can start to diffuse outwards. Because accretion is still proceeding at hypercritical rates, the luminosity overshoots the Eddington limit and the escape of radiation marks the emergence of the NS and has a dramatic effect on the subsequent hydrodynamic evolution of the accretion flow.

No completely consistent calculation of fallback onto a NS has been performed in this regime. The evolution depends strongly on the physical conditions of the flow and the outcome of the radiation-hydrodynamic interaction between the accreting gas and the intense radiation flux. Radiation diffusion may cut off the accretion flow at large radius leaving an extended, radiation-supported envelope bound to the NS [28]. This envelope is likely to be dynamically unstable but the timescale of this instability is not known. If this timescale is long, the luminosity may remain close to the Eddington limit for some time (few years) [11,28,34]. If free-bound transitions of heavy elements dominate the opacity, the resultant luminosity is limited to the local Eddington limit (computed with the local value of the opacity), that may be 3–4 orders of magnitude lower than the electron scattering Eddington luminosity. In SN 1987A, this process leads to the removal of the entire envelope in less than a year [35].

Irrespectively of the detailed radiation-hydrodynamic interaction between the accreting gas and the radiation field, it is conceivable that the evolution of the accretion flow during late-time fallback onto a NS proceeds in two stages. For $\dot{M}$ larger than $\sim 10^{-4} M_\odot \, \mathrm{yr}^{-1}$, the accretion efficiency is very low and equation (10) can be approximately applied to compute the fallback luminosity. After that $\dot{M}$ drops below $\sim 10^{-4} M_\odot \, \mathrm{yr}^{-1}$, the evolution becomes *qualitatively* different from that expected for secondary fallback onto a BH. At an age ≈ 1 year, the luminosity rapidly reaches the Eddington limit for scattering (line) opacity and remains at that level for ≈ 1 year, dropping abruptly when the radiation-supported envelope becomes dynamically unstable and accretion terminates.

4   Fallback via an hypercritical accretion disk

Very little is known about the structure and emission properties of hypercritically accreting fallback disks. In the typical physical conditions of the accreting gas during late-time fallback, the efficiency of the mechanism of angular momentum transport is not known. The centrifugal support may keep the gas from accreting, at least temporarily. If angular momentum transport occurs and a disk forms, its evolution may be self-similar with total angular momentum within the disk constant and mass flow rate decaying as a power law with time [18,36,37]. Physical conditions within the disk are very similar to those in the radial flows previously considered. The optical depth is so high that the vertical diffusion timescale is much larger than the "viscous" timescale and photon trapping dominates. If the central object is a NS it is likely that the innermost regions are somehow cooled down by neutrino emission [38]. On the other hand, in a hypercritical accretion disk onto a BH, neutrino cooling during late-time fallback is not important [19].

Because of photon trapping, it appears that, during the early evolutionary stages of late time fallback, most of the gravitational energy is either lost to neutrinos or advected into the BH. Thus, for both NSs and BHs the total bolometric luminosity will be low. As the accretion rate decreases, radiation diffusion effects become important. In the case of a NS, transition to an Eddington limited accretion stage might happen on a timescale shorter than that for the spherical case [38]. In the case of a BH, the evolution may continue along a sequence of steady states and the late-time accretion luminosity may still have a characteristic power law decay with time, determined by the secular decay of the disk accretion rate with t and the (at present unknown) functional relation between L and $\dot{M}$ for hypercritical accretion disks.

5   Observational prospects: can we unveil fallback emission from a BH?

As mentioned in Sect. 3.1, from an age of months the light curve of a typical Type II supernova is powered by the reprocessing of gamma ray photons emitted from the decay of radioactive isotopes. Because the decay time of long-lived isotopes is larger than the diffusion and expansion timescales, at this stage the behavior of the light curve follows the exponential decay of heavy elements. This behavior is qualitatively different and intrinsically steeper than the secular power law decay of the accretion luminosity generated by secondary fallback onto a BH [equation (10)]. Then the natural question is: Is it possible to unveil the BH fallback luminosity above the emission of radioactive isotopes? If observed, *the characteristic power law decay of the accretion luminosity could be used to reveal the presence of the BH and provide the first direct evidence of its formation in the aftermath of a supernova.* The detailed relativistic radiation hydrodynamic simulations reviewed in Sect. 3.1 have been undertaken to answer this fundamental question. At late times the accretion luminosity is always expected to dominate, but in a typical Type II supernova BH emergence may not be practically observable (see, e.g., the case of SN 1987A, in which the accretion tail produced by a putative BH would be unobservable for about 1000 years [16]). On the other hand, as shown in Fig. 2, under certain favorable conditions the fallback luminosity may emergence at an age ranging from about 50 days to 3 years when it is still at a detectable level. The favorable conditions are a large accretion rate induced by fallback and a low abundance of heavy elements. Neglecting the thermal and dynamical effects of radioactive heating (not important during secondary fallback), an estimate of the time of BH emergence t_e can be obtained simply by setting the power output from radioactive decay ($L = L_\gamma M_\gamma e^{-t/t_\gamma}$, where t_γ and M_γ are the decay time and the mass of an isotope and L_γ its characteristic luminosity) equal to the BH accretion luminosity (Eq. (10)). This gives [30]

$$\frac{t_e}{t_\gamma} - \frac{25}{18} \ln\left(\frac{t_e}{t_\gamma}\right) = 3.5 - \ln\left[\left(\frac{M}{M_\odot}\right)^{2/3}\left(\frac{M_\gamma}{10^{-3}M_\odot}\right)^{-1}\right.$$
$$\left.\left(\frac{\rho_0}{10^{-5}\,\mathrm{gcm}^{-3}}\right)^{5/6} \times \left(\frac{t_0}{20\,\mathrm{hr}}\right)^{20/9}\right], \quad (12)$$

where the relevant isotope to consider (at an age from months to few years) is ^{56}Co ($t_\gamma = 109$ days). For the values of the parameters in Eq. (12), the estimated time of emergence is $t_e \approx 3.5\,t_\gamma = 1$ yr, in fair agreement with numerical calculations. Equation (12) shows that the favorable conditions for BH emergence are a low abundance of ^{56}Co ($M_\gamma \sim 100$ times lower than typical) and a low expansion velocity ($t_0 \sim 5$ times larger than typical). Both conditions appear to be met in the supernova explosion of massive progenitors (in the range $25M_\odot < M_* < 40M_\odot$). The low abundance of heavy elements is due to the fact that the innermost part of the envelope, comprised of the heaviest elements, has been accreted by the hole, while the low expansion velocity is a consequence of a low energy explosion (see Sect. 2.2).

BH emergence might have been detectable in SN 1997D (as predicted by [30]). This supernova was discovered in January 1997 in a serendipitous observation of the parent galaxy NGC 1536 at a distance of 14 Mpc [39]. Although there is a different interpretation on the nature of the progenitor of this supernova [40], the observed light curve and nebular spectra can be explained as the result of an exceptionally low energy explosion of a 26 $M_\odot$ progenitor in which the collapse of the core and subsequent primary fallback may have originated a BH [41,42]. One of the light curves shown in Fig. 2, computed for the best fit 26 $M_\odot$ post-shock envelope structure of the SN 1997D progenitor, was aimed to model BH emergence for this supernova. The total luminosity at emergence was estimated to be marginally detectable with the Hubble Space Telescope STIS camera ($L(t_e) \approx 10^{36}$ erg s^{-1} corresponding to a visual magnitude $V \simeq 28$) with a 23000 s exposure (at a signal to noise ratio $S/N \simeq$ 10). However, no observation could be performed because the expected bolometric luminosity at emergence was grazing the detectability limit and the possibility of a successful detection was critically dependent on 100% uncertainties in the prediction of the model. In fact, although the specific time dependence expected for the persistent accretion luminosity ($\propto t^{-25/18}$) represents a distinct signature of BH emergence, gathering incontrovertible observational evidence for the presence of an accreting BH is made difficult by several effects, such as the exact value of the accretion luminosity (dependent on the mass of the central object and the physical properties of the initial bound material within the inner He mantle [17]), the precise contribution from radioactive luminosity (related to the actual late time opacity to gamma rays and to the unknown abundance of ^{57}Co and ^{44}Ti [43]), and the possible contamination from emission generated by circumstellar interaction.

Although SN 1997D has been the first BH-forming supernova candidate, it may not have offered the best chances for detectability of the accretion tail. Progenitors more massive than SN 1997D may provide earlier emergence and at a higher luminosity, as shown by the light curve of a 35 $M_\odot$ progenitor (Fig. 2). Emergence of the BH in these progenitors may be expected to happen soon after the clearing of the envelope by recombination at a luminosity $L(t_e) \approx 10^{37}$ erg s^{-1} [17], as suggested by [9]. At this luminosity, BH emergence would be detectable up to 20–25 Mpc (estimate not critically dependent on model uncertainties), unless cirmustellar emission and absorption become important (see, e.g., SN 1994W [44]). The rate of favorable BH-forming supernova events can be roughly estimated as [17]

$$R(d \le 20\,\text{Mpc}) = R_\text{SN} \times f_{\text{BH} \to \text{SN}} \times f_\text{CSM} \times N_\text{galaxies}(d \le 20\,\text{Mpc}), \qquad (13)$$

where R_SN is the rate of supernova explosion per galaxy, $f_{\text{BH} \to \text{SN}}$ the fraction of BH-forming events, f_CSM the fraction of events not contaminated by circumstellar interaction and $N_\text{galaxies}(d \le 20\,\text{Mpc})$ the number of galaxies within 20 Mpc. Assuming $R_\text{SN} \simeq 0.01$, $f_{\text{BH} \to \text{SN}} \simeq 0.1$ [45], $f_\text{CSM} \simeq 0.5$, $N_\text{CSM}(d \le 20\,\text{Mpc}) \simeq 1000$, we obtain $R \simeq 1$ every 2 years. This is only an upper limit because $f_{\text{BH} \to \text{SN}}$ includes also Type Ib/Ic events that are not good candidates (owing to a larger initial post-shock temperature and thus lower accretion rate [17]). A more conservative estimate, taking into account observational biases, may be 1 per several years [46].

As mentioned above, the fallback luminosity produced by a NS at an age of ≈ 1 year is *qualitatively* different with respect to that emitted by a BH. In a NS, when the shock radius overtakes the trapping radius (on a timescale of ≈ 1 year), the luminosity reaches the (local) Eddington limit and remains at that value for ≈ 1 year before dropping abruptly. In a BH, on the other hand, the late-time fallback luminosity continues to decay secularly as a power law with time [equation (10)]. Therefore the detection of a persistent power law tail in the late time (> 1 yr) light curve could not be mistaken with fallback emission from a NS. *The accretion tail must persist for more than ≈ 1 year in order to rule out the possibility that it is produced by fallback onto a NS during the advection dominated accretion stage.* The slope of the light curve can be measured by performing two or more observations, sufficiently spaced in time. If an hypercritical accretion disk forms, the qualitative discussion in the previous section suggest that the luminosity produced by a BH might be decaying as a power-law with time also in this case.

In this paper, we focused on the effects of late time fallback occurring at hypercritical rates in an attempt to unveil fallback emission early after the explosion. Although expansion leads to a steady decrease in $\dot{M}$, at an age of hundreds to thousands years fallback may still give rise to detectable emission of radiation. If a "standard" accretion disk (similar to those believed to be present in galactic X-ray binary systems) forms, its X-ray emission might be observable with present satellites [47]. Recently, it has been suggested that this emission may be relevant for point-like X-ray sources in supernova remnants (see, e.g., [37,48]).

6   Conclusions

We have reviewed the scenario for the formation of BHs in otherwise successful supernovae and the prospect of unveiling the emission from fallback of stellar material. If fallback was a spherically symmetric character, the early evolution is not dependent on the compact object formed during the explosion. *The presence of a BH may be inferred by the emergence of its accretion luminosity over emission from radioactive isotopes, with a characteristic and persistent power law decay with time* $(t^{-25/18})$, potentially detectable up to $\sim 20\,\mathrm{Mpc}$. Although NS fallback has not been investigated at the same level of consistency, the evolution of the NS accretion luminosity is qualitatively clear and, after ≈ 1 year, it is markedly different from that expected for a BH, making practically impossible to mistaken the presence of a NS for that of a BH.

Very little is known about the structure and emission properties of hypercritical fallback disks around both BHs and NSs. In light of its importance, this aspect deserves certainly further investigation in the future. It will be important to understand if an hypercritical disk around a BH may produce a persistent power-law decay of the accretion luminosity with time, providing a distinguishable signature similar to that found for radial fallback. Another aspect that will be very interesting to investigate is the character of fallback at an age of hundreds to thousands years. This will allow us to reach a better understanding of the possible connection between fallback and

X-ray emission from point-like X-ray sources in supernova remnants. Finally, it will be of the uttermost importance to undertake a systematic monitoring of supernovae that show favorable properties for unveiling BHs, as low ^{56}Co abundance and low expansion velocity. This would greatly improve the chances of detecting the first direct signature of BHs in supernovae.

Acknowledgement. I would like to thank Shmulik Balberg, Monica Colpi and Stu Shapiro for reading the manuscript.

References

1. Israelian, G. , Rebole, R., Basri, G., Casares, J., Martin, E.L. (1999): Nature **401**, 142
2. Janka, H.-Th. (2001): A & A **368**, 527
3. Woosley, S.E., Weaver, T.A. (1995): ApJS **101**, 181
4. Fryer, C.L. (1999): ApJ **522**, 413
5. Thorsson, V., Prakash, M., Lattimer, J.M. (1994): Nucl. Phys. A **572**, 693
6. Brown, G.E., Bethe, H.A. (1994): ApJ **423**, 659
7. Woosley, S.E., Weaver, T.A. (1986): ARA & A, **24**, 205
8. Akmal, A., Pandharipande, V.R., Ravenhall, D.G. (1998): Phys. Rev. D **58**, 1804
9. Woosley, S.E., Timmes, F.X. (1996): Nucl. Phys. A **606**, 137
10. Colgate, S.A. (1971): ApJ **163**, 221
11. Chevalier, R.A. (1989): ApJ **346**, 847
12. MacFadyen, A.I., Woosley, S.E. (1999): ApJ **524**, 262
13. Bethe, H.A. (1990): Rev. Mod. Phys. **62**, 801
14. Arnett, D. (1996): Supernovae and Nucleosynthesis. Princeton University Press, Princeton
15. Colpi, M., Shapiro, S.L., Wasserman, I. (1996): ApJ **470**, 1075
16. Zampieri, L., Colpi, M., Shapiro, S.L., Wasserman, I. (1998): ApJ **505**, 876
17. Balberg, S., Zampieri, L., Shapiro, S.L. (2000): ApJ **541**, 860
18. Mineshige, S., Nomura, H., Hirose, M., Nomoto, K., Suzuki, T. (1997): ApJ **489**, 227
19. Popham, R., Woosley, S.E., Fryer, C.L. (1999): ApJ **518**, 356
20. Begelman, M.C. (1978): A & A **70**, 583
21. Rees, M.J. (1978): Phys. Scr. **17**, 193
22. Bondi, H. (1952): MNRAS **112**, 195
23. Zampieri, L., Miller, J.C., Turolla, R. (1996): MNRAS **281**, 1183
24. Vitello, P.A.J. (1978): ApJ **225**, 694
25. Blondin, J.M. (1986): ApJ **308**, 755
26. Park, M.-G. (1990): ApJ **354**, 64
27. Nobili, L., Turolla, R., Zampieri, L. (1991): ApJ **383**, 250
28. Houck, J.C., Chevalier, R.A. (1991): ApJ **376**, 234
29. Brown, G.E., Weingartner, J.C. (1994): ApJ **436**, 843
30. Zampieri, L., Shapiro, S.L., Colpi, M. (1998): ApJ **502**, L149
31. Burger, H.L., Katz, J.I. (1980): ApJ **236**, 921
32. Burger, H.L., Katz, J.I. (1983): ApJ **265**, 393
33. Klein, R.I., Stockman, H.S., Chevalier, R.A. (1980): ApJ **237**, 912
34. Wu, H., Lin, X.B., Xu, H.G., You, J.H. (1998): A & A, **334**, 146
35. Fryer, C.L., Colgate, S.A., Pinto, P.A. (1999): ApJ **511**, 885
36. Mineshige, S., Nomoto, K., Shigeyama, T. (1993): A & A **267**, 95

37. Umeda, H., Nomoto, K., Tsuruta, S., Mineshige, S. (2000): ApJ **534**, L193
38. Chevalier, R.A. (1996): ApJ **459**, 322
39. De Mello, D., Benetti, S. (1997): IAU Circ. 6537
40. Chugai, N.N., Utrobin, V.P. (2000): A&A **354**, 557
41. Turatto, M., Mazzali, P.A., Young, T.R., Nomoto, K., Iwamoto, K., Benetti, S., Cappellaro, E., Danzinger, I.J., de Mello, D.F., Phillips, M.M., Suntzeff, N.B., Clocchiatti, A., Piemonte, A., Leibundgut, B., Cevarrubias, R., Meza, J., Sollerman, J. (1998): ApJ **498**, L129
42. Benetti S., Turatte, M., Balberg, S., Zampieri, L., Shapiro, S.L., Cappellaro, E., Nomoto, K., Nakamura, T., Mazzali, P.A., Patat, F. (2001): MNRAS **322**, 361
43. Timmes, F.X., Woosley, S.E., Hartmann, D.H., Hoffman, R.D. (1996): ApJ **464**, 332
44. Sollerman, J., Cumming, R.J., Lundqvist, P. (1998): ApJ **493**, 933
45. Fryer, C.L., Kalogera, V. (2001): ApJ **554**, 548
46. Balberg, S., Shapiro, S.L. (2001): ApJ **556**, 944
47. Perna, R., Hernquist, L., Narayan, R. (2000): ApJ **541**, 344
48. Chakrabarty, D., Pivovaroff, M.J., Hernquist, L.E., Heyl, J.S., Narayan, R. (2001): ApJ **548**, 800

Advanced Readout Configurations
for the Gravitational Wave Detector AURIGA

J.-P. Zendri, M. Bignotto, M. Bonaldi, M. Cerdonio, L. Conti, V. Crivelli Visconti,
M. De Rosa, P. Falferi, A. Marin, F. Marin, R. Mezzena, G.A. Prodi, M. Salviato,
G. Soranzo, L. Taffarello, A. Vinante, S. Vitale

Abstract. We report the status of the experimental effort devoted at improving the sensitivity
and widening the band of the gravitational wave detector AURIGA. The focus is on an optimized setup of the capacitive resonant transducer, read by an improved dc-SQUID amplifier
and on the implementation of an opto-mechanical resonant transducer. Both techniques, which
are complementary, should lead to an improvement of the detector performances of at least
two orders of magnitude in both energy sensitivity and bandwidth.

1 Introduction

The present global network of five resonant gravitational wave detectors [1–5] forms
an observatory for the gravitational astronomical events in our Galaxy [6]. Since the
rate of observable events in the Milky Way is unacceptably low (few per century), it
is required an improvement of the detector performances to extend the observed region to at least the local group of galaxies. Presently the sensitivity limit is set by the
transduction and amplification stage that typically operate a factor $10^4 \div 10^5$ above
the fundamental limit imposed by quantum mechanics (Standard Quantum Limit,
SQL). A substantial sensitivity improvement, which does not require any major detector reconfiguration, could thus be achieved by improving the readouts. According
to this point of view the AURIGA collaboration started a Research and Development program on two different transduction technologies: the well known capacitive
transducer and the innovative optical transducer. We report here the present state of
this program. In the first section we outline in a general way the parameters needed
to be maximized for a sensitive transducer chain. In the second section we describe
the facility we developed for testing transducers at ultracryogenic temperatures and
finally the last two sections are devoted to the description of the state of the two
transducer chains.

2 Transducer optimization

2.1 Monomode antenna

An acoustic gravitational wave detector consists of a massive body which reso-
nantly absorbs energy from an incoming gravitational wave. It can be schema-
tized as a spring-mass system with an equivalent mass M_{eq}, a resonant frequency

$\nu_R = \omega_R/(2\pi)$ and a quality factor Q_R. An incoming gravitational wave drives the oscillator displacement by the quantity $x(t)$ that depends on the wave amplitude, polarization and direction. For a resonator with a cylindrical shape (bar detector) we have:

$$M_{eq} = \frac{M_{bar}}{2}, \qquad F_{Grav} = \frac{M_{bar} L_{bar} \ddot{h}}{\pi^2}, \tag{1}$$

where L_{bar}, M_{bar} are respectively the length and the mass of the bar and $h(t)$ is the amplitude of an optimally oriented wave.

The resonator is also driven by some unavoidable noise forces as the thermal force F_{Ther} which, according to the fluctuation dissipation theorem, has the power spectrum:

$$S_{F_{Ther} F_{Ther}} = \frac{2k_B T M_{eq} \omega_R}{Q_R}, \tag{2}$$

where k_B is the Boltzman constant and T the resonator thermodynamic temperature. A second disturbance force, the so called "back-action force" F_{BA}, comes from the readout electronics which first transforms the oscillator displacement signal into an electrical signal and subsequently provides its amplification. Finally, the amplifier that here we assume for simplicity to be a voltage amplifier, introduces a broadband voltage noise described by the power spectrum $S_{V_n V_n}$. We define the transducer efficiency α as $V_{out} \equiv \alpha \cdot x$, where V_{out} is the output signal of the amplifier. The voltage noise power spectrum can thus be regarded as an equivalent displacement noise power spectrum $S_{x_n x_n} = S_{V_n V_n}/\alpha^2$. Quantum mechanics imposes for the transduction-amplification noise sources the lower limit

$$S_{x_n x_n} S_{F_{BA} F_{BA}} \geq \frac{\hbar^2}{4}, \tag{3}$$

where $\hbar$ is the Plank constant; the equality represents the amplifier SQL. All other noise sources can be, at least in principle, reduced to a negligible level. The total displacement noise power spectrum $S_{x_{n-tot} x_{n-tot}}$ is thus

$$S_{x_{n-tot} x_{n-tot}} = \frac{S_{F_{Ther} F_{Ther}} + S_{F_{BA} F_{BA}}}{M_{eq}^2 [(\omega_R^2 - \omega^2)^2 + \omega_R^2 \omega^2 / Q_R^2]} + S_{x_n x_n}. \tag{4}$$

It is useful here to introduce the total noise force F_{Tot} as the driving force which would produce at the detector output the measured noise displacement power spectrum $S_{x_{n-tot} x_{n-tot}}$. The expression for $S_{F_{Tot} F_{Tot}}$ is thus obtained multiplying eq.(4) by the square of the modulus of the harmonic oscillator force transfer function:

$$S_{F_{Tot} F_{Tot}} = S_{F_{Ther} F_{Ther}} + S_{F_{BA} F_{BA}} + S_{x_n x_n} M_{eq}^2 [(\omega_R^2 - \omega^2)^2 + \omega_R^2 \omega^2 / Q_R^2]. \tag{5}$$

A plot of the total noise power spectrum is presented in Fig. 1.

Let us suppose to have a signal force acting on the detector. In order to maximize the signal to noise ratio a filter has to be applied to the data. We expect that the

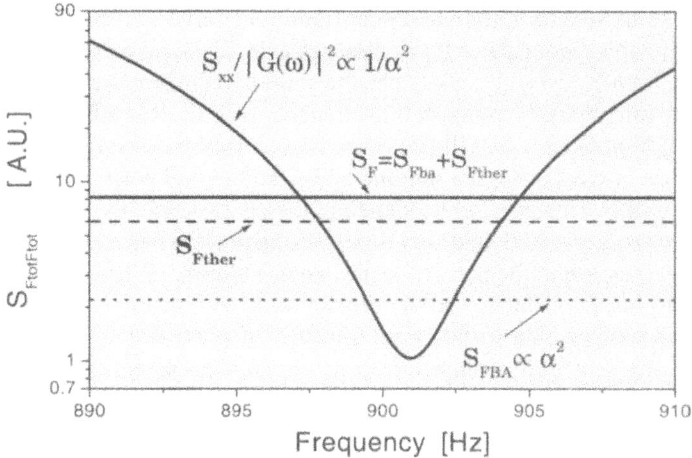

Fig. 1. Total noise force power spectrum as a function of the frequency. The broad band contribution $S_{F_{BA}F_{BA}} + S_{F_{Ther}F_{Ther}}$ and narrow band contribution $S_{x_n x_n}/|G(\omega)|^2$ are plotted separately. There $G(\omega) = [M_{eq}(\omega_R^2 - \omega^2 + i\omega\omega_R/Q_R)]^{-1}$ is the harmonic oscillator transfer function and α the transduction efficiency. The thermal noise (dashed line) and the back-action noise (dotted line) contribution to the broad band noise are also plotted

different frequency signal components need to be weighted with the noise level at the same frequency. Thus the weighting function should be proportional to the inverse of the the total noise. Indeed the Wiener filter theory predicts for the optimal filter $H(\omega)$:

$$H(\omega) = \frac{F[\omega]^* / S_{F_{Tot}F_{Tot}}}{1/2\pi \int_{-\infty}^{+\infty} d\omega |F[\omega]|^2 / S_{F_{Tot}F_{Tot}}}, \tag{6}$$

where $F[\omega]^*$ is the complex conjugate of the Fourier transform of the exciting signal force $F_{Grav}(t)$. The filter $H(\omega)$ removes the frequency intervals where the noise dominates over the signal and has to be multiplied by $F[\omega]$ to give the signal actually detected in the frequency domain. In particular, for a broad band signal like a Dirac delta function in time ($F[\omega] \approx$ const.) the filter reduces to a resonance centered at ω_R, with full width half maximum $\Delta\omega$:

$$\Delta\omega = \frac{1}{M_{eq}\omega_R} \sqrt{\frac{S_{F_{Ther}F_{Ther}} + S_{F_{BA}F_{BA}}}{S_{x_n x_n}}}. \tag{7}$$

The frequency components of the signal outside this region are in fact overwhelmed by the noise and thus $\Delta\omega$ represents the effective detector bandwidth. It is important to note that the detector bandwidth depends on its noise properties and not only on the harmonic oscillator dynamical properties. In particular, for a detector dominated by the thermal noise the bandwidth scales as $Q_R^{-1/2}$ while if the back-action dominates $\Delta\omega$ is independent of Q_R.

For a detector to be sensitive, the bandwidth should be large and the minima of Eq. (5) as low as possible (for instance for a broad band signal the expected sensitivity is approximately $\sqrt{\min\{S_{F_{\mathrm{Tot}}F_{\mathrm{Tot}}}\}/\Delta\omega}$). Given the amplifier, the best way to increase the bandwidth is to maximize the transduction efficiency α. Indeed as long as α is not too high ($\alpha \lesssim 10^9$ V/m using a SQUID amplifier) $S_{x_n x_n}$ decreases as α^{-2} while the minima of the noise curve Eq. (5) are dominated by the thermal noise contribution that is independent of α. However, as explained in the next section, the back-action contribution increases as α^2 and thus at very high transduction efficiencies an increase of α determines an increase of the power spectrum noise minima. When this regime, called the Giffard limit [7], is achieved the sensitivity is completely dominated by the amplifier noise sources. Above this level a further increase of the parameter α does not produce any sensitivity enhancement for broad band signals. However as the bandwidth still increases as α^2 an increment on α should help to discriminate the frequency details of the signal.

The first requirement for an optimal transduction chain is the use of linear amplifiers that operate close to the SQL. Moreover in order to full exploit the amplifier noise properties a very high transduction efficiency is needed. Unfortunately, all of the transducers presently integrated in resonant gravitational detectors have such physical limitations to the transduction efficiency that resonant transducer are required.

2.2 Resonant transducer

Resonant transducers were first proposed for inductive readouts [8] and subsequently used for all other techniques [9,10]. The basic idea is to connect to the main resonator a second lighter one and to monitor its motion. If both the oscillators have the same resonant frequency ω_R the maximal displacement of the lighter mass m_t is amplified by a factor $\mu^{-1/2} \equiv \sqrt{M_{\mathrm{eq}}/m_t}$. The transduction efficiency α is thus incremented of the same factor and becomes frequency-dependent, adding some complexity to the analyses of the sensitivity. The coupled resonances split the two normal modes by $\omega_R \cdot \sqrt{\mu}$ and their splitting increases with the transducer mass as $\sqrt{m_t}$. As suggested by Fig. 2, the overall system bandwidth cannot be much greater than this splitting. However, this gives only an upper bound on the detector bandwidth, since the displacement noise contribution, which scales as μ, can reduce the sensitivity between the two modes frequencies (see Fig. 2). The force noise contributions referred at the detector input acquire a frequency structure as well due to the resonant transducer. The amplitude of the thermal component is approximately independent of μ because the mass dependence of the transducer Nyquist force generator is canceled out by the transfer function. The back-action noise generator is instead mass independent and, when referred at the detector input, its contribution scales as μ^{-1}.

It is evident that there is an optimal choice for the transducer mass. Let us suppose to decrease m_t starting from a too high value, a situation similar to that described in Fig. 2. This operation would increase the sensitivity since it would widen the overall

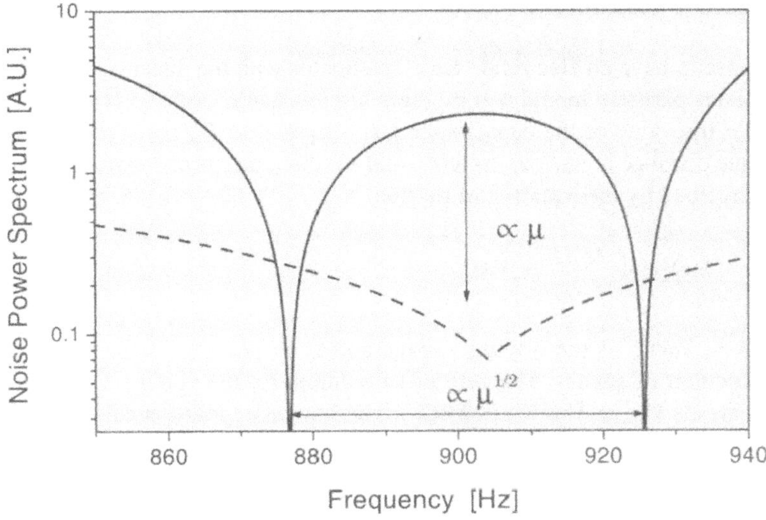

Fig. 2. Input noise force power spectrum for an acoustic detector with a resonant transducer. The thermal and back-action noise forces are represented by the dashed line while the amplifier broad band contribution is the solid line. The dependence on the mass ratio $\mu = m_t / M_{\mathrm{eq}}$ of the noise sources and the frequency split is underlined. The total noise source is the sum of these two contributions

bandwidth in spite of the decrease of the splitting of the modes. In fact, the corresponding decrease of the displacement noise would provide for useful additional bandwidth around and between the modes. At some mass value the displacement noise contribution within the modes would become comparable to the force noise (thermal plus back-action). Decreasing further m_t would reduce the sensitivity because the bandwidth is reduced proportionally to $\mu^{1/2}$ while the minima of the noise spectrum would be either stable or increasing, depending on what force noise would dominate, thermal or back-action. A lower bound on the optimal transducer mass is given by assuming that the force thermal noise dominates:

$$m_{t-\mathrm{min}} \approx \frac{1}{\omega_R} \sqrt{\frac{2 k_B T M_{\mathrm{eq}}}{\omega_R Q_R S_{x_n x_n}}}. \tag{8}$$

The optimal mass is higher if the back-action contribution dominates over the thermal noise, as is required for an optimized transducer readout at the Giffard limit. If this condition is not met by a simple resonant transducer, it should be convenient to use a transducer with two modes (i.e. with an intermediate mass resonator and a lighter mass resonator) and to monitor the displacement of the last lighter mass [11]. In this way, the modes splitting would still be proportional to μ (μ is now the mass ratio between two contiguous resonators) while the transduction efficiency would increase (proportional to μ^2) and correspondingly the back-action would show a stronger dependence (scales as μ^{-2}).

2.3 Parametric transducers

Let us suppose to have an electrical "RLC" resonator with the distance between the capacitor plates partially modulated by the main resonator motion $x(t) = x_0 e^{i\omega_R t}$. At first order the value of the capacitance $C(t)$ is equal to $C_0[1 - x(t)/d_0]$ where d_0 is the plate distance at rest ($d_0 \gg x(t)$) and C_0 the unperturbed capacitance. The circuit is described by the equation of motion:

$$L\ddot{Q}(t) + R\dot{Q}(t) + \frac{Q(t)}{C(t)} = V(t) \tag{9}$$

where $V(t) = V_0 e^{i\Omega t}$ is the driving voltage and $\Omega = (LC_0)^{-1/2}$ the unperturbed electrical resonant frequency. The zero order solution for the voltage V_C across the capacitor plates is $V_{C0} = V_0 e^{i\Omega t}/(i\Omega R C_0)$. The first order solution can be found by writing the charge as $Q(t) = V_{C0}C_0(1 + \epsilon(t))$ with $\epsilon(t) \ll 1$. The capacitor voltage thus becomes:

$$V_C = V_{C0}e^{i\Omega t}\left\{1 + \frac{x_0}{d_0}\frac{e^{i\omega_R t}}{\frac{i}{Q_{\mathrm{el}}} - \frac{2\omega_R}{\Omega}}\right\} \tag{10}$$

where $Q_{\mathrm{el}} = \Omega L/R$ is the electrical quality factor of the circuit. Comparing the above formula with the transduction efficiency α of a passive capacitive transducer ($\alpha = V_{C0}/d_0$), it becomes clear that the transduction efficiency is enhanced by a factor $\Omega/2\omega_R$ and therefore a parametric transducers usually operate at very high electrical frequency Ω. However when the frequency becomes too high, the efficiency is dominated by the term Q_{el} that usually ranges between 10^4 and 10^7. This last case is the one of the optical transducer in which the electrical resonator is replaced by an optical resonator (Fabry–Perot cavity) operating at very high frequency (about 10^{14}). The efficiency is thus dominated by the cavity Finesse that is the optical analogous of the electrical quality factor. Although parametric transducers produce a large transduction efficiency they have the disadvantage of a more complicate readout, the appearance of other noise sources related to the drive generator and in some case the need to reduce the dominant carrier signal (Eq. (10)) to a level compatible to the very sensitive signal amplifiers [12].

3 Transducer test facility

While the detector is taking data, the R&D effort is devoted to the realization of new transducers and amplifiers that would allow the required sensitivity improvement. It is obvious that the detector itself cannot be used as a bench test for these R&D devices. For this purpose we developed at the INFN *National Laboratories of Legnaro* a facility where transducers and amplifiers can be tested and set up in a configuration enabling a fast integration on the detector. The facility fulfills the following requirements:

- Working temperature below 100 mK. This requirement comes from the need to test all the equipment on the same physical environment as they would operate when employed in an ultracryogenic detector as AURIGA. The facility is thus equipped (see Fig. 3) with a commercial dilution refrigerator[1]. After installation, the dilution refrigerator with no thermal loads reached a temperature of approximately 15 mK with a measured cooling power of about 1 mW at 100 mK, well above our estimated thermal leaks at the same temperature.

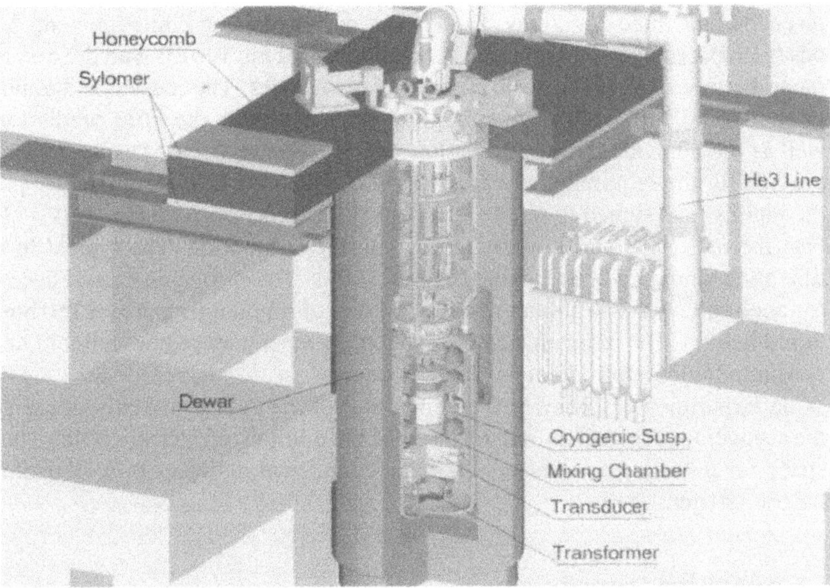

Fig. 3. Cross section of the Transducer Test Facility: The facility is hung to a honeycomb plate through 3 steel "legs" that provides an attenuation of 65 dB. The honeycomb plane is supported by dissipative sylomer® springs. The IVC houses the cryogenic suspensions which embrace the cold insert of the dilution refrigerator. The upper suspension spring is made of stainless steel to reduce the thermal leaks toward the ^{4}He bath. The material of the other springs is a high strength Al alloy; the last stage is a copper spring. Thermal link between transducer and the cold part of the DR is given by this last spring stage which is softly connected to the mixing chamber through annealed OHFC strips

- Seismic noise attenuation better than 140 dB at 900 Hz. In order to measure the transducer thermal noise at ultracryogenic temperatures the effect of seismic noise has to be made negligible. This goal is achieved using mechanical suspensions which hold the transducer providing an attenuation $\mathcal{A}$ of at least:

$$\mathcal{A} \leq \sqrt{\frac{2k_B T}{m_t Q_t \omega_R^3} \frac{1}{S_{\text{Seis}}}}, \tag{11}$$

---

[1] Leiden Cryogenics B.V., Tomatenstraat 4, 2321 HG Leiden (NL)

where ω_t, m_t and Q_t are respectively the resonant frequency, mass and quality factor of the transducer and T the thermodynamic temperature. For the typical transducer parameters $\omega_t = 2\pi \cdot 900\,\text{Hz}$, $m_t = 1\,\text{kg}$, $Q_t = 10^7$ and at a $T = 0.1\,\text{K}$, a mechanical attenuation of 140 dB is required, given the measured seismic noise power spectrum at 900 Hz of $S_{\text{Seis}}^{1/2} \approx 10^{-14}\,\text{m}/\sqrt{\text{Hz}}$.

In the facility this attenuation level is achieved using cryogenic suspensions installed inside the Internal Vacuum Chamber (IVC). They are a cascade of 4 annular masses connected by springs having a C-shape: this allows for a high attenuation on each of the three directions. The system, designed using Finite Elements Methods (FEM), provides a total vertical attenuation of about 190 dB with the frequency range between 160 Hz to 1900 Hz free of resonances. The measured attenuation of about 50 dB for each stage is in good agreement with the FEM predictions as well as the 30 mode frequencies below 160 Hz. The cryogenic suspensions alone provide for the required attenuation. However also the room temperature part of the facility is designed to reduce mechanical disturbances (see Fig. 3).

- Fast thermal cycle and low cryogenic liquid consumption. The typical thermal time for a whole cryogenic cycle is about 3 days. The cryogenic liquids necessary for cooling down to 4.2 K are about 200 liters of liquid nitrogen and 150 liters of liquid helium. The maintenance of cryogenic temperature require a liquid helium consumption of approximately 1 liter/hour.
- Wide Experimental Chamber. The inner part of the IVC is partially occupied by the cryogenic suspensions and the insert of the dilution refrigerator (DR). The free space for installing transducers is thus a cylinder of diameter $\varphi = 450\,\text{mm}$ and height 350 mm.

4 Capacitive transducer

The resonant capacitive transducer is a capacitor biased by a constant electric field E_0: one plate is connected to the bar end-face and the other is a light resonator (resonant transducer) attached to the main oscillator. Any relative displacement ΔL of the plates produces a voltage across the capacitor equal to $V_{\text{out}} = E_0 \Delta L$; the voltage signal is transformed into a magnetic flux signal and then read by a dc-SQUID. This is made by connecting the SQUID input coil to the transducer output after inserting between them an impedance matching transformer [13]. As the transduction efficiency E_0 is limited by the vacuum break-down field, approximately 10^7 V/m, the transducer needs to be resonant. Its optimal mass is determined using a numerical simulation that includes the system equations of motion and all the noise sources. The soundness of the result has been recently improved by using for the SQUID noise power spectrum the experimentally measured value [14] of the high sensitivity SQUID we employ. The computed optimal transducer mass of about 3 kg is almost ten times higher than the one used for the first AURIGA run in spite of the used SQUID amplifiers being very similar. This apparent contradiction comes from the fact that in the new version the electrical resonator formed by the transducer capacitance and the transformer primary coil is frequency matched to the mechanical oscillators. The transduction efficiency and thus the optimal mass (see Eq. (8)) is increased by inserting a resonant line between the transducer and the SQUID. Moreover an increase of the back action force $E_0 \cdot Q_n$ is expected as the noise charge Q_n due to the amplifier is amplified from the electrical resonance just at the mechanical resonance. Thus it is not surprising that the new expected AURIGA sensitivity curve shows a larger bandwidth (thanks to the higher transducer mass) but higher minima (due to higher back-action) than in the previous run (see Fig. 4). We expect as a result an improvement a factor 10 in the sensitivity for a gravitational wave burst.

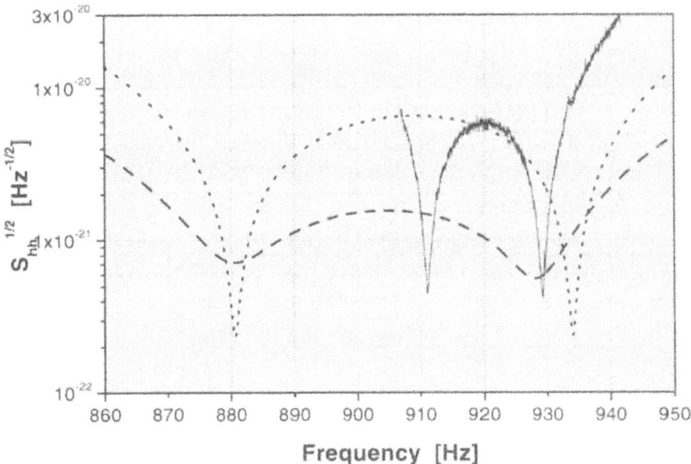

Fig. 4. Comparison between the equivalent strain noise power spectrum predicted for the next AURIGA run and the best (solid line) shown by AURIGA. The curves are obtained by transforming the input force power spectrum into a strain power spectrum through equation (1). All the used parameters have been measured on bench tests except for the bar resonant frequency which is the same of the previous AURIGA run. The dashed line correspond to a high transducer field $E_0 = 8 \cdot 10^6$ V/m. The dotted line is expected for low transducer field $E_0 = 2 \cdot 10^6$ V/m.

4.1 Experimental results

In order to reduce the electromagnetic interferences the new capacitive transducer is designed with the matching transformer located strictly close to the resonant plate. They are connected through 3 soft springs which allow the conservation of the high transducer mechanical quality factor in spite of the mechanical dissipations in the transformer box. The transducer resonant mass is designed using FEM in order to avoid spurious resonances and to distribute as much as possible the internal stresses. Its resonant plate shape is a variation of the classical "mushroom" with a diameter $\varphi = 250$ mm and weight 3.4 kg. The measured low-temperature mechanical quality factor is about $1.5 \cdot 10^6$ and does not depend on the transformer box being mounted or not. The matching transformer is located inside a copper box designed with FEM and lead coated. The box contains a superconducting primary coil of $L_p = 7.89$ H, two superconducting secondary coils, one strongly coupled $k = 0.86$ (where k is transformer geometrical coupling coefficient) and $L_s = 3.1\,\mu$H and one weakly coupled $k < 0.1$ to the primary, and finally the SQUID chip. The transformer box is vacuum sealed and can be filled with ^{3}He gas used for the low temperature thermalization. As the matching of the electrical circuit resonance to the mechanical one requires high electrical quality factor, the transformer is realized using low electrical loss materials as superconductors for the wires and the case walls and PTFE for the holders. During a preliminary test we observed that the measured value of Q_e is strongly affected by the dissipations coming from the transducer charge lines and it ranges between $2 \cdot 10^5$ and $7 \cdot 10^5$. The best result is obtained using a low temperature switch which disconnects the room temperature part of the charge line from the one at low temperature. All the calibration runs and the first noise measurements were done using the weakly coupled secondary coil because it doesn't affect the measurement of the quality factors and does not introduce any measurable back-action effect. We then found the configuration which allows the measurement, in the range 1.7 K ~ 4.2 K, of the thermal noise of the transducer and of the electrical resonator kept at about 80 Hz of distance. Up to now we have not explored lower temperatures as we have not yet connected the thermal links between the dilution refrigerator and the transducer.

The measured noise power spectrum of the commercial dc-SQUID[2] used for the AURIGA capacitive transducer exhibits a strong temperature independent component (at least half of the power spectral value at 4.2 K) [14]. As a consequence, a high sensitivity increase is not expected by cooling the SQUID chip at ultracryogenic temperature. Our strategy to obtain a dc SQUID operating with an energy sensitivity lower than $100\,\hbar$ involves the modification of a commercial dc SQUID system manufactured by Quantum Design[2]. We have replaced the conventional readout scheme (dc SQUID sensor – cold transformer – room temperature preamplifier) with a two stage configuration which employs a first dc-SQUID as a sensor and a second dc SQUID as preamlifier. In this way, the noise contribution of the room temperature electronics is negligible and the overall sensitivity, limited by the thermal noise of the first SQUID, can be further improved if the sensor is cooled to a temperature lower than 4.2 K.

Preliminary measurements have been carried out on the two stage SQUID system in the temperature range $(1.5 \div 4.2)$ K. The sensor SQUID was operated with open input coil; the flux noise Φ_n was sampled and it has been calculated the coupled energy sensitivity ϵ_n inferred as $\epsilon_n = L_{in}\Phi_n^2/2M^2$, where $L_{in} = 1.63\,\mu$H is the input coil inductance, and $M = 10.7$ nH is the mutual inductance between the input

---

[2] Quantum Design, 11578 Sorrento Valley Road, Suite 30, San Diego, CA 92121

coil and the SQUID loop. At 4.2 K the flux noise in the white noise frequency region was $1.16 \ \mu\Phi_0/\sqrt{Hz}$ and this corresponds to a coupled energy sensitivity of 388 $\hbar$. The coupled energy sensitivity is linear in temperature, in good agreement with the theory, down to 1.5 K, with a slope of 91 ± 1 $\hbar$/K. A non-thermal contribution to the noise energy was found to be 14 ± 2 $\hbar$. The energy sensitivity improvement obtained at 4.2 K with the two stage SQUID system is approximately a factor of 10, if we compare our result with that obtainable with the conventional readout electronics. Further experiments are in progress in order to measure the noise of the system in a dilution refrigerator and to couple the double SQUID to the capacitive transducer. The expected energy sensitivity at 100 mK should be better than 40 $\hbar$ which corresponds to a strain sensitivity of about 10^{-22}Hz$^{-1/2}$ and an overall bandwidth of 50 Hz.

5 Optical transducer

The fundamental idea for signal optical transduction [15] is to have a resonant optical cavity of length L formed by a mirror attached to a bar end face and a second mirror attached to a resonator having the same frequency as the bar (i.e. to a resonant transducer). Then a relative motion ΔL of the two mirrors, due to a bar vibration induced eventually by a g.w., is converted into a change $\Delta \nu$ of the optical resonant frequency ν according to:

$$\frac{\Delta L}{L} = \frac{\Delta \nu}{\nu}. \tag{12}$$

The idea expressed in Eq. (12) has been developed in the scheme shown in Fig. 5: this represents the working principle of the optical readout under development in AURIGA [2]. The optical resonant cavity cited above is here named *transducer cavity*. The naive approach of sending a light beam to this cavity and looking at its output as function of light frequency cannot work, essentially due to light frequency noise. In fact what one really needs are two cavities, one for stabilizing the light frequency and one for the signal: for that reason along with the transducer cavity there is also the *sensing cavity* or *reference cavity*. The basic components of the scheme are therefore a laser source and these two cavities.

The beam produced by the laser source is split in two: one is sent to the transducer cavity inside the cryostat that houses the bar, by means of optical fiber. The fraction of beam that is reflected is conveyed into another fiber cable and is then detected by a photodiode outside the cryostat: the current thus developed is used in order to obtain an error signal which is proportional to the frequency difference between laser and cavity, according to the FM sidebands technique [16,17]. For this purpose the light beam must be phase modulated: phase modulation is accomplished before the beam produced by the laser source is split. The error signal obtained is used by a servo system to frequency-lock the laser source to the transducer cavity: thus the light frequency carries information on bar motion.

In order to extract such information, the second beam is diverted into the sensing cavity, which acts as frequency reference: in other words, such cavity must be as

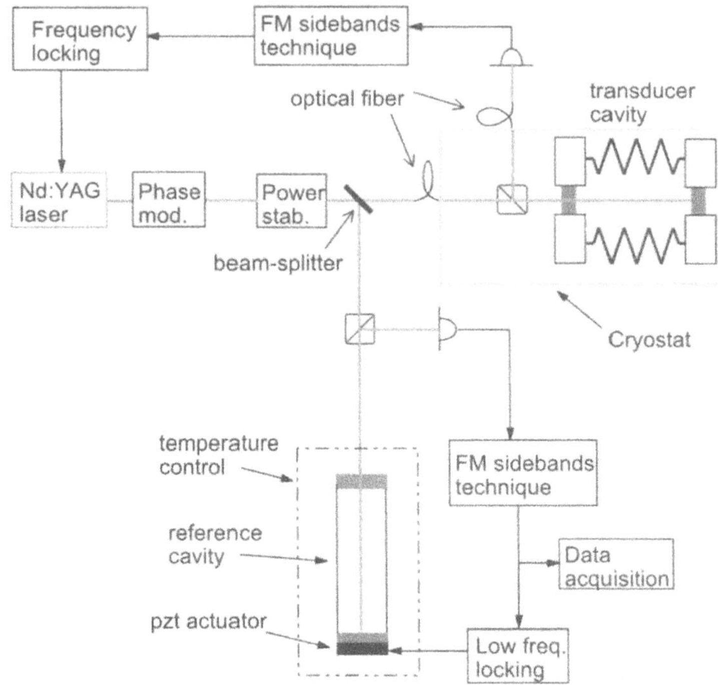

Fig. 5. Scheme of the optomechanical readout for resonant bar g.w. detectors

stable as possible (at least around detection frequency, i.e. around 1 kHz) and not to be affected by bar and transducer resonances. Again, the fraction of beam that is reflected is detected by a photodiode and its current used to obtain a signal which is proportional to the frequency difference between the sensing cavity (acting as length reference) and the laser frequency, i.e. the transducer cavity. Thus this error signal is that to be acquired as it depends only on bar and transducer motion. Anyway one must guarantee that the sensing cavity is in resonance condition with the light: therefore one should be able to correct its length at low frequency (i.e. in a frequency range lower than the kHz range, e.g. up to some tens of Hz) so to follow any slow drift of transducer cavity (due for instance to temperature variations).

The present configuration employs a room temperature sensing cavity, which can be temperature tuned to resonance and whose length can be controlled at very low frequency by temperature variation and at higher frequency by means of piezoelectric actuators. A detailed description of the sensing cavity has been given in ref. [18]. The error signal to be used for such controls is the same that, at higher frequencies, is acquired for g.w. detection.

An important point that should be underlined is that laser power noise acts as a *back-action noise,* in the sense that it determines fluctuations in the radiation pressure on transducer cavity mirrors, thus generating a noise force that back-acts on the detector. For this reason it is important that the laser power noise is reduced before entering the cryostat; in order to cancel out any phase fluctuations introduced eventually by the noise reduction, this is done before the laser beam is split.

With the parameters described in Table 5 one expects to reach a spectral strain noise at the level of $10^{-22}/\sqrt{Hz}$ within a bandwidth of 40 Hz [18].

Table 1. Parameters used for calculating the sensitivity of a bar detector equipped with the optical readout

bar temperature	0.1 K	transducer effective mass	6.5 kg
bar res. freq.	920 Hz	Q_{bar}	$5 \cdot 10^6$
transd. res. freq.	920 Hz	Q_{transd}	$5 \cdot 10^6$
trasd.-cav. length	0.01 m	sens.-cav. length	0.2 m
laser wavelength	1064 nm	mirror losses	1 ppm
sens.-cav. input mirror T	150 ppm	sens.-cav. output mirror T	2 ppm
transd.-cav. input mirror T	15 ppm	transd.-cav. output mirror T	2 ppm
photodiodes quantum efficiency	0.85	laser phase mod. amplitude	0.96 rad
transd.-cav. input power	2 mW	sens.-cav. input power	5 mW

5.1 Experimental results

The light source employed is a 100 mW diode-pumped Nd:YAG (wavelength $\lambda = 1\mu m$). The phase modulation is achieved by means of an electro-optic modulator resonating at 13.3 MHz. At such frequency a detected laser power of 10 mW shows a noise power density 1 dB above the shot-noise level. The phase modulation introduces a non-null residual amplitude modulation (RAM) which appears as a dc-offset of the photodetector signal demodulated at 13.3 MHz; the fluctuations of the RAM do not introduce any excess noise for frequencies above about 800 Hz. After propagation through a 12 m long single-mode polarization-maintaining optical fiber the the results are worse and a 30% decrease of the sensitivity is expected due to this effect.

The intensity noise reduction has been implemented and results are well within the requirement of achieving an almost shot-noise limit [19]. The gain of the feedback loop is maximized for frequencies around 1 kHz as this is the frequency band in which the detector is expected to be sensitive; at 1 kHz the residual noise is 5 dB above the quantum level for a 30 mW laser beam. The intensity noise reduction is also checked to be effective after the laser beams propagates along the single-mode polarization-maintaining optical fiber: the residual excess noise introduced by the waveguide is 2 dB for a 2-mW-laser beam [19] as that expected to be sent to the transducer cavity housed in the AURIGA cryostat.

We have built a feedback loop that frequency locks the laser source to the sensing cavity having as input signal the cavity error signal. The loop has two outputs: one is fed to the thermal control of the laser crystal, and is effective up to about 1 Hz, while the other is fed to a piezoelectric actuator mounted on the top of the laser crystal, and is effective between 1 Hz and few tens of kHz. Again, the loop gain is maximized for frequencies around 1 kHz. An *inloop* measurement shows that the frequency fluctuations of the laser beam with respect to the cavity it is locked to are below the shot-noise level in the frequency range between 250 Hz and 1000 Hz.

The true determination of the cavities noise can obviously be achieved only by comparison with a better reference: being this not available (we would otherwise have employed this better reference) we have constructed two identical sensing cavities (which we name *cav1* and *cav2*) and we thus measure the fluctuations of one with respect to the other, which are twice the fluctuations of a single one. At 1 kHz the total frequency noise is found to be at the level of $4 \cdot 10^{-3}$ Hz/$\sqrt{\text{Hz}}$, about 12 dB above the shot-noise level. The main culprit seems to be the piezoelectric actuators mounted on the sensing cavity, as above mentioned: work is presently devoted at the reduction of such noise and at the achievement of the shot-noise limit.

At the moment experimental effort is mainly devoted at mounting the whole readout to a room temperature bar identical to the one employed in AURIGA and thus demonstrating the effectiveness of the optical transducer.

References

1. Mauceli, E., Geng, Z.K., Hamilton, W.O., Johnson, W.W., Merkowitz, S., Morse, A., Price, B., Solomonson, N. (1996): Phys. Rev. D **54**, 1264
2. Prodi, G.A., Conti, L., Mezzena, R., Vitale, S., Taffarello, L., Zendri, J.P., Baggio, L., Cerdonio, M., Colombo, A., Visconti, V.C., Macchietto, R., Falferi, P., Bonaldi, M., Ortolan, A., Vedovato, G., Cavallini, E., Fortini, P. (1998): Initial Operation of the Gravitational Wave Detector Auriga, in *2nd E. Amaldi Conference on Gravitational Waves, CERN, CH, 1–4 July, 1997*, ed. by E. Coccia, G. Veneziano, G. Pizzella, World Scientific, pp. 148–158
3. Astone, P., Bassan, M., Bonifazzi, P., Carelli, P., Castellano, M.G., Cavallari, G., Coccia, E., Cosmelli, C., Fafone, V., Frasca, S., Majorana, E., Modena, I., Pallottino, G.V., Pizzella, G., Rapagnani, P., Ricci, F., Visco, M. (1993): Phys. Rev. D **47**, 362
4. Astone, P., Bassan, M., Bonifazzi, P., Carelli, P., Coccia, E., Cosmelli, C., Fafone, V., Frasca, S., Marini, A., Mazzitelli, G., Minenkov, Y., Modena, I., Modestino, G., Moleti, A., Pallottino, G.V., Papa, M.A., Pizzella, G., Rapagnani, P., Ricci, F., Ronga, F., Terenzi, R., Visco, M., Votano, L. (1997): Astroparticle Phys. **7**, 231
5. Blair, D.G., Ivanov, E.N., Tobar, M.E., Turner, P.J., van Kann, F., Heng, I.S. (1995): Phys. Rev. Lett. **74**, 1908
6. Allen, Z.A., Astone, P., Baggio, L., Busby, D., Bassan, M., Blair, D.G., Bonaldi, M., Bonifazzi, P., Carelli, P., Cerdonio, M., Coccia, E., Conti, L., Cosmelli, C., Visconti, V.C., D'Antonio, S., Fafone, V., Falferi, P., Fortini, P., Frasca, S., Hamilton, W.O., Heng, I.S., Ivanov, E.N., Johnson, W.W., Kingham, M., Locke, C.R., Marini, A., Martinucci, V., Mauceli, E., McHugh, M.P., Mezzena, R., Minenkov, Y., Modena, I., Modestino, G., Moleti, A., Ortolan, A., Pallottino, G.V., Pizzella, G., Prodi, G.A., Rocco, E., Ronga, F., Salemi, F., Santostase, G., Taffarello, L., Terenzi, R., Tobar, M.E., Vedovato, G., Vinante, A., Visco, M., Viale, S., Votano, L., Zendri, J.P. (2000): Phys. Rev Lett. **85**, 5046

7. Giffard, R.P. (1976): Phys. Rev. D **14**, 2478

8. Paik, H.J. (1976): J. Appl. Phys. **47**, 1168

9. Rapagnani, P. (1982): Il Nuovo Cimento C **5**, 385

10. Veitch, P.J., Blair, D.G., Linthorne, N.P., Mann, L.D., Ramm, D.K. (1987): Rev. Sci. Inst. **58**, 1910

11. Richard, J.P. (1984): Phys. Rev. Lett. **52**, 165

12. Ivanov, E.N., Turner, P.J., Blair, D.G. (1993): Rev. Sci. Instr. **64**, 3191

13. Carelli, P., Castellano, M.G., Cosmelli, C., Foglietti, V., Modena, I. (1985): Phys. Rev. A **32**, 3258

14. Falferi, P., Bonaldi, M., Cerdonio, M., Muck, M., Vinante, A., Mezzena, R., Prodi, G.A., Vitale, S. (2001): J. of Low Temp. **123**, 275

15. Conti, L., Cerdonio, M., Taffarello, L., Zendri, J.P., Ortolan, A., Rizzo, C., Ruoso, G., Prodi, G.A., Vitale, S., Cantatore, G., Zavattini, E. (1946): Rev. Sci. Instr. **69**, 554

16. Pound, R.V. (1946): Rev. Sci. Instr. **17**, 460

17. Drever, R.W.P., Hall, J.L., Kowalsky, F.V., Hough, J., Ford, G.M., Munley, A.J., Ward, H. (1983): Appl. Phys. B **31**, 97

18. Conti, L., Marin, F., De Rosa, M., Prodi, G.A., Taffarello, L., Zendri, J.P., Cerdonio, M., Vitale, S. (2000): An optical transduction chain for the AURIGA detector, in *Gravitational waves*, Proceedings of the Third E. Amaldi Conference, CalTech – California, 1999, ed. by S. Meshkov, AIP Conf. Proc., New York, p. 261

19. Conti, L., De Rosa, M., Marin, F. (2000): Appl. Opt. **39**, 5732–5738

Classical Relativity

Connections in Distributional Bundles and Field Theories

D. Canarutto

Abstract. Several standard notions of the differential geometry, such a tangent and jet paces, Frölicher–Nijenhuis bracket, connections and curvature, can be generalized to the case of bundles whose fibres are distribution spaces. Moreover a classical connection on a finite-dimensional bundle yields a connection on an associated distributional bundle. These idea can hed new light on field theories on a curved background.

1 Generalized half-densities

By a "classical manifold" we mean a smooth Hausdorff paracompact manifold of finite dimension. If Y is an oriented real classical manifold of dimension n, we set $\mathbb{V}Y := (\wedge^n TY)^+$ so that $\mathbb{V}^{-1/2}Y := (\mathbb{V}^{1/2})^* Y \to Y$ is the half-vector bundle of the so-called *half-densities*. Let $\mathcal{Y}_\circ$ be the vector space of all global smooth sections $Y \to \mathbb{C} \otimes \mathbb{V}^{-1/2} Y$ which have compact support. A topology on $\mathcal{Y}_\circ$ can be introduced by the standard procedure [1]; its topological dual $\mathcal{Y} := \mathcal{Y}_\circ'$ is called the space of *generalized complex half-densities*. Any sufficiently regular half-density θ can be seen as an element of $\mathcal{Y}$ by the rule

$$\langle \theta, u \rangle := \int_Y \theta(y) \otimes u(y).$$

We have $\mathcal{Y}_\circ \subset \mathcal{H} \subset \mathcal{Y}$, where $\mathcal{H}$ denotes the Hilbert space of $\mathcal{L}^2$ complex half-densities. This can be extended to half-densities valued in a vector bundle.

If Z is another oriented classical manifold, any orientation-preserving diffeomorphism $\varphi : Y \to Z$ determines an isomorphism $\varphi_* : \mathcal{Y}_\circ \to \mathcal{Z}_\circ$ of the corresponding half-density spaces, extended to generalized half-densities as

$$\langle \varphi_* \lambda, \varphi_* u \rangle = \langle \lambda, u \rangle, \quad \forall \lambda \in \mathcal{Y}, \ u \in \mathcal{Y}_\circ.$$

It is easy to see that $\varphi_* : \mathcal{Y} \to \mathcal{Z}$ is a continuous linear isomorphism, and that the correspondence $\varphi \mapsto \varphi_*$ behaves naturally with regard to compositions. In particular, if $\check{Y} \subset Y$ is an open subset, a local chart $y : \check{Y} \to \mathbb{R}^n$ yields a chart $y_* : \check{\mathcal{Y}} \to \mathcal{R}^n$, where $\mathcal{R}^n$ is the space of generalized complex half-densities on $\mathbb{R}^n$. For $\lambda \in \mathcal{Y}$ we set $\lambda_y := y_* \lambda \in \mathcal{R}^n$, which is analogous to the set of components of a vector in a finite dimensional space.

If $v : Y \to TY$ is a smooth vector field, then a Lie derivative is naturally defined by $\langle v.\lambda, u \rangle = -\langle \lambda, v.u \rangle$, and the map $\lambda \mapsto v.\lambda$ turns out to be a continuous linear operator in $\mathcal{Y}$. We have the coordinate expression

$$(v.\lambda)_y = v^i \partial_i \lambda_y + \tfrac{1}{2} (\partial_i v^i) \lambda_y.$$

2 Tangent prolongations

The notion of F-smoothness provides a general approach to calculus in infinite dimensional spaces [2,3]. It consists of the assignment of a set of curves and a set of functions, determining each other via the smoothness condition imposed on all compositions. For classical manifolds, and for manifolds modelled on Banach spaces, one recovers the usual smooth structure. Moreover there are important situations in which a simplified version of Frölicher's approach is sufficient to develop several basic ideas of differential geometry: one has only to consider a suitable reduced set of F-smooth curves, fulfilling a few consistency axioms. In this way, notions such as tangent spaces, jet spaces, connections and curvature, have been introduced and studied for a large class of functional bundles [4–6]. Everything can be formulated in terms of smooth classical maps, without getting involved in the topologies of functional spaces and other intricated questions (such as the characterizations of the whole sets of F-smooth curves and functions).

Here we adopt a similar but somewhat dual view: the space of "test" maps generates the F-smooth structure on its topological dual. Let $\mathbb{I} \subset \mathbb{R}$ be an open interval; a curve $\alpha : \mathbb{I} \to \mathcal{Y}$ will be called F-smooth if the map

$$\langle \alpha, u \rangle : \mathbb{I} \to \mathbb{C} : t \mapsto \langle \alpha(t), u \rangle,$$

is smooth for all $u \in \mathcal{Y}_\circ$. Two such curves α and β are said to coincide up to first order at, say, $t = 0$, if the maps $t \mapsto \langle \alpha(t), u \rangle$ and $t \mapsto \langle \beta(t), u \rangle$ coincide up to first order at $t = 0$, for all $u \in \mathcal{Y}_\circ$. This yields a natural definition of tangent bundle $\mathrm{T}\mathcal{Y}$, which turns out to be isomorphic to $\mathcal{Y} \times \mathcal{Y}$. The tangent vector to α is denoted by $\partial \alpha \equiv \frac{\partial \alpha}{\partial t} : \mathbb{I} \to \mathrm{T}\mathcal{Y}$.

Let M be a classical manifold. A map $\varphi : M \to \mathcal{Y}$ is called F-smooth if $\varphi \circ c : \mathbb{I} \to \mathcal{Y}$ is F-smooth for every smooth curve $c : \mathbb{I} \to M$ (equivalently, the maps $\varphi_u : M \to \mathbb{C} : x \mapsto \langle \varphi(x), u \rangle$ are smooth $\forall u \in \mathcal{Y}_\circ$). Then there is a unique map

$$\mathrm{T}\varphi \cong (\varphi \circ \pi_M , \mathcal{D}\varphi) : \mathrm{T}M \to \mathcal{Y} \times \mathcal{Y} = \mathrm{T}\mathcal{Y},$$

such that $\mathrm{T}\varphi \circ \partial c = \partial(\varphi \circ c)$ for every local smooth curve $c : \mathbb{R} \to M$. If $x = (x^a) : X \to \mathbb{R}^m$ is a local coordinate chart, $X \subset M$, and $x_a : \mathbb{R} \to M$ any coordinate curve, we define the partial derivative $\partial_a \varphi$ at the point $x_a(0)$ to be $\partial(\varphi \circ x_a)(0)$; we obtain a map $\partial_a \varphi : X \to \mathrm{T}\mathcal{Y}$. Locally we have

$$\mathrm{T}\varphi = \partial_a \varphi \, \mathcal{D}x^a, \quad \text{i.e.} \quad \langle \mathcal{D}\varphi, u \rangle = \langle \partial_a \varphi, u \rangle \, \mathcal{D}x^a, \quad u \in \mathcal{Y}_\circ.$$

Let N be a classical manifold; a map $f : \mathcal{Y} \to N$ is called F-smooth if $f \circ \alpha : \mathbb{I} \to \mathbb{R}$ is smooth for all F-smooth curves $\alpha : \mathbb{I} \to \mathcal{Y}$ (we remark that the F-smoothness of f does not allow any statement about its continuity with respect to the distribution space topology). Then there is a unique map $\mathrm{T}f : \mathrm{T}\mathcal{Y} \to \mathrm{T}N$, such that for any F-smooth curve α valued in $\mathcal{Y}$ one has

$$\mathrm{T}f(\partial \alpha(t)) = \mathrm{T}(f \circ \alpha)(t).$$

Finally, let $\Phi : \mathcal{Y} \to \mathcal{Z}$ be an F-smooth map. Then there is a unique map

$$\mathrm{T}\Phi \equiv (\Phi, \mathrm{D}\Phi) : \mathrm{T}\mathcal{Y} \to \mathrm{T}\mathcal{Z},$$

such that $\mathrm{T}\Phi(\partial\alpha(t_0)) = \partial(\Phi \circ \alpha)(t_0)$ for any F-smooth curve α valued in $\mathcal{Y}$. In particular, we obtain $\mathrm{T}\varphi_* = \varphi_* \times \varphi_*$.

The *tangent prolongations* $\mathrm{T}\varphi, \mathrm{T}f$ and $\mathrm{T}\Phi$ are F-smooth linear morphisms over φ, f and Φ, respectively. It is not difficult to see that all tangent prolongations behave naturally in terms of any compositions and cartesian products. Everything can be written in terms of chart expressions.

3   F-smooth bundles

By $p : E \to M$ we shall denote a smooth classical bundle ($\dim M = m$, $\dim E = m+n$) whose fibres are smoothly oriented manifolds. This means that $\wedge^n \mathrm{V}E \to E$ is a trivializable bundle with smoothly oriented fibres, so that we also have the smooth 2-fibred bundle $\mathbb{V}E \to E \to M$, where $(\mathbb{V}E)_x := \mathbb{V}(\mathrm{T}E_x)$, $x \in M$.

For each $x \in M$ we consider the space $\mathcal{E}_x$ of all complex generalized half densities on E_x. We introduce the fibred set

$$\wp : \mathcal{E} := \bigsqcup_{x \in M} \mathcal{E}_x \to M.$$

Let $X \subset M$ be an open submanifold. A local bundle trivialization $(x, y) : E_X \to X \times Y$ yields the local bundle trivialization

$$(x, y_*) : \mathcal{E}_X \to X \times \mathcal{Y} : \lambda_x \mapsto \big(x, (y_x)_*\lambda_x\big),$$

where $\lambda_x \in \mathcal{E}_x$, and y_x is the restriction of y to E_x. It is easy to see that a bundle atlas of E yields an F-smooth bundle atlas of $\mathcal{E}$.

Clearly, $\mathcal{E}$ turns out to be an F-smooth space in a natural way: a curve $\alpha : \mathbb{I} \to \mathcal{E}$ is defined to be F-smooth if $(x, y_*) \circ \alpha$ is such for any bundle trivialization (x, y); in other terms, $x \circ \alpha$ is a classical smooth curve and $y_* \circ \alpha$ is an F-smooth curve (in general, the F-smoothness of a map can be expressed via its trivialized expression). The definition of the tangent space $\mathrm{T}\mathcal{E}$ naturally follows. This is a fibred set over $\mathcal{E}$; a local bundle trivialization (x, y) on E yields the local bundle trivialization

$$\mathrm{T}(x, y_*) : \mathrm{T}\mathcal{E} \to \mathrm{T}X \times \mathrm{T}\mathcal{Y}.$$

A smooth atlas of E is seen to yield an F-smooth atlas of $\mathrm{T}\mathcal{E}$, so that $\mathrm{T}\mathcal{E} \to \mathcal{E}$, the tangent bundle of $\mathcal{E}$, is an F-smooth vector bundle. The definitions of the *vertical bundle* $\mathrm{V}\mathcal{E}$ and of the *first jet bundle* $\mathrm{J}\mathcal{E} \subset \mathrm{T}^*M \otimes_M \mathrm{T}\mathcal{E}$ are now analogous to those given in the finite-dimensional case. Replacing the base map x by a coordinate chart $x = (x^a)$ we have the fibred charts

$$(x^a, y_*) : \mathcal{E} \to \mathbb{R}^m \times \mathcal{Y},$$

$$(x^a, y_*, \dot{x}^a, \dot{y}_*) := \mathrm{T}(x, y_*) : \mathrm{T}\mathcal{E} \to \mathbb{R}^m \times \mathcal{Y} \times \mathbb{R}^m \times \mathcal{Y},$$

$$(x^a, y_*, y_{*a}) := \mathrm{J}(x, y_*) : \mathrm{J}\mathcal{E} \to \mathbb{R}^m \times \mathcal{Y} \times (\mathbb{R}^m \otimes \mathcal{Y}).$$

Tangent and jet prolongations of F-smooth maps $N \to \mathcal{E}$ and $\mathcal{E} \to N$, where N is a classical manifold, and of F-smooth maps between distributional bundles, can be defined in terms of the prolongations of their local expressions; all turn out to be F-smooth linear morphisms. In particular, let $\sigma : M \to \mathcal{E}$ be an F-smooth section and $\mathrm{j}\sigma : M \to \mathrm{J}\mathcal{E}$ its first jet prolongation. Setting $\sigma_y := y_* \circ \sigma : M \to \mathcal{Y}$, we obtain the local expression $y_{*a} \circ \mathrm{j}\sigma = \partial_a \sigma_y$. One can write local expressions for all kinds of prolongations of maps and fibred morphisms. With some caveats, they turn out to be similar to those of the classical finite-dimensional case. The same is true for transition maps between different charts.

4   F-smooth connections and field theories

A *connection* on the distributional bundle $\mathcal{E}$ is defined to be an F-smooth section $\Gamma : \mathcal{E} \to \mathrm{J}\mathcal{E}$. In the domain of a fibred chart (x^a, y) we have the local expression

$$\Gamma_{ay} := y_{*a} \circ \Gamma : \mathcal{E} \to \mathcal{Y}.$$

We shall only consider *linear* connections, that is connections Γ which are linear morphisms over M. Then $\Gamma_{ay} = \Gamma_{ayy} \circ y_*$ where $\Gamma_{ayy} : X \times \mathcal{Y} \to \mathcal{Y}$. The existence of global connections follows from standard arguments using the paracompactness of M.

As in the finite-dimensional case, the assignment of a connection is equivalent to that of other structures. First, Γ can be viewed as a linear map $\mathcal{E} \times_M \mathrm{T}M \to \mathrm{T}\mathcal{E}$; the image $\mathrm{H}_\Gamma \mathcal{E} := \Gamma(\mathcal{E} \times_M \mathrm{T}M)$ is a vector subbundle of $\mathrm{T}\mathcal{E} \to \mathcal{E}$, with m-dimensional fibres; if $v : M \to \mathrm{T}M$ is a smooth vector field, then $\Gamma_v : \mathcal{E} \to \mathrm{T}\mathcal{E}$ is an F-smooth vector field, called its *horizontal lift*. We also have the complementary map $\Omega := \mathbf{1} - \Gamma : \mathrm{T}\mathcal{E} \to \mathrm{V}\mathcal{E}$, which determines the decomposition $\mathrm{T}\mathcal{E} = \mathrm{H}_\Gamma \mathcal{E} \oplus_\mathcal{E} \mathrm{V}\mathcal{E}$.

Let $\sigma : M \to \mathcal{E}$ be an F-smooth section. The *covariant derivative* of σ is defined to be the linear morphsim over M

$$\nabla\sigma := \mathrm{pr}_2 \circ \Omega \circ \mathrm{T}\sigma : \mathrm{T}M \to \mathcal{E}.$$

If $v : M \to \mathrm{T}M$ is a vector field, then we also write $\nabla_v\sigma := \nabla\sigma \circ v$. The local expression is

$$(\nabla\sigma)_y := y_* \circ \nabla\sigma = \dot{x}^a \partial_a \sigma_y - \Gamma_{ay} \circ \sigma.$$

It can be seen that a classical smooth connection $\gamma : E \to \mathrm{J}E$ determines a connection Γ on $\mathcal{E} \to M$; if $y = (y^i)$ is a coordinate chart, we obtain the local expression

$$\Gamma_{ayy} = -\gamma_a^i \partial_i - \tfrac{1}{2}(\partial_i \gamma_a^i)\mathbf{1}y.$$

Hence, if $\sigma : M \to \mathcal{E}$ is an F-smooth section, we have

$$(\nabla\sigma)_y = \partial_a\sigma_y + \gamma_a^i\partial_i\sigma_y + \tfrac{1}{2}(\partial_i\gamma_a^i)\sigma_y.$$

Note that Γ is always linear, even if γ is not, and has a simple geometric interpretation. Suppose that, for sufficiently close $t, t_0 \in \mathbb{I}$, γ yields a diffeomorphism $\varphi_t : E_{c(t_0)} \to E_{c(t)}$ via parallel transport along a smooth curve $c : \mathbb{I} \to M$ (this is certainly the case if E is a vector bundle and γ is linear). Then, for each $\lambda \in \mathcal{E}_{c(t_0)}$ the smooth curve $\mathbb{I} \to \mathcal{E} : t \mapsto (\varphi_t)_*\lambda$ is exactly the *horizontal lift* of c (in $\mathcal{E}$) through λ. In general, the map φ_t is not defined on the whole fibre $\mathcal{E}_{c(t_0)}$ (even for t arbitrarily close to t_0), but the above interpretation applies, for example, whenever λ has compact support.

As further developments [7], it is possible to introduce the Lie bracket of fields $T\mathcal{E} \to TT\mathcal{E}$, the Frölicher-Nijenhuis bracket of tangent-valued forms (see also [5]) and the curvature of a connection. In particular, if Γ is the connection determined by the classical connection γ, then the chart expression of its curvature turns out to be

$$R_{ab\,yy} = -\rho_{ab}{}^i\partial_i - \tfrac{1}{2}(\partial_i\rho_{ab}{}^i)\mathbf{1}y,$$

where ρ denotes the classical curvature of γ.

5 Example: the Dirac equation

Connections in infinite-dimensional bundles of smooth maps have been used by Jadczyk and Modugno [4] in their Galileian "covariant quantization" setting (see also [8]). There one assumes a time fibration of spacetime, which yields a "Hilbert bundle" over time; time development of a particle's quantum history can be seen as parallel transport relatively to an F-smooth connection.

A similar approach (extended to distributions) for the Dirac equation on a curved Einstein spacetime M, provides the following example of a distributional connection which does *not* arise from a classical connection.

Fix a time fibration $t : M \to \mathbb{T}$, where $\mathbb{T}$ is a classical manifold diffeomorphic to $\mathbb{R}$, and write $\mathbb{V} := (\wedge^3 VM)^+ \subset \wedge^3 TM$ (inclusion is via the spacetime metric). For any $\tau \in \mathbb{T}$, $\mathbb{V}_\tau^*$ is the bundle of volume forms on the 3-manifold $M_\tau := t^{-1}(\tau)$. The metric determines a distinguished section $M \to \mathbb{V}$, so that, given a 4-spinor structure, a spinor field can be described as a section

$$\psi : M \to \mathbb{V}^{-1/2} \otimes W,$$

where W is the 4-spinor bundle.

Now the Dirac equation for ψ can be written in the form

$$\partial_0\psi^\alpha - \mathfrak{D}^\alpha(\psi) = 0,$$

with

$$\mathfrak{D}^\alpha(\psi) := \Xi_0{}^\alpha{}_\beta\,\psi^\beta + \tfrac{1}{2}\Gamma_0{}^h{}_h\,\psi^\alpha - (\gamma^0\gamma^j)^\alpha{}_\beta\,\partial_j\psi^\beta$$
$$+ (\gamma^0\gamma^j)^\alpha{}_\beta\,(\Xi_j{}^\beta{}_\rho + \tfrac{1}{2}\Gamma_j{}^h{}_h\,\delta^\beta{}_\rho)\psi^\rho - \mathrm{i}\,m\,\gamma^{0\alpha}{}_\beta\,\psi^\beta.$$

Here, γ denotes the Dirac map, Γ denotes the spacetime connection, and Ξ denotes the associated spinorial connection. Clearly, this can be extended to generalized sections, so that we obtain a connection $\mathfrak{D}$ on the bundle $\mathcal{W} \to \mathbb{T}$ of W-valued generalized densities over the classical bundle $M \to \mathbb{T}$. Generalized spinor fields $\psi : \mathbb{T} \to \mathcal{W}$ obeying the Dirac equation are then exactly the horizontal ones.

Present research regards the distributional bundles naturally arising in the 2-spinor formulation of Maxwell-Dirac field theory [9,10], and the applicability of these ideas to the study of quantum fields in curved background.

References

1. Schwartz, L. (1966): Théorie des distributions. Hermann, Paris
2. Frölicher, A., Kriegl, A. (1988): Linear spaces and differentiation theory. John Wiley & sons, Chichester
3. Kriegl, A., Michor, P. (1997): The convenient setting of global analysis. American Mathematical Society, Providence
4. Jadczyk, A., Janyska, J., Modugno, M. (1998): Galilei general relativistic quantum mechanics revisited, in *Geometria, Física-Matemática e Outros Ensaios, Homenagem a Antònio Ribeiro Gomes*, ed. by A.S. Alves, F.J. Craveiro de Carvalho, J.A. Pereira da Silva, pp. 253–313
5. Modugno, M., Kolář, I. (1998): Ann. Pol. Math. **LXVIII**, 2
6. Cabras, A., Kolář, A. (1995): Czech. Math. J. **45**, 120
7. Canarutto, D.(2000): Archivum Mathematicum Brno **36**, 2
8. Canarutto, D., Jadczyk, A., Modugno, M. (1995): Rep. Math. Phys. **36**
9. Canarutto, D. (1998): J. Math. Phys. **39**, 9
10. Canarutto, D.(2000): Acta Appl. Math. **62**, 2

The Spectrum of Endstates
of Spherical Gravitational Collapse

G. Magli

Abstract. The Cosmic Censorship conjecture in spherical symmetry is re-formulated as the problem of characterizing a set of functions in terms of the behaviour of an ordinary differential equation near a singular point. Existing results and perspectives of this approach are briefly reviewed.

The problem of predicting the final state of gravitational collapse in Einstein's theory is open. In fact, it is well known that stable, non singular states of superdense matter can exist only if the mass of the final object is less than a physical limit, namely the Chandrasekar limit (about $1.4 M_\odot$) in the case of white dwarfs or the neutron star limit (of the order of $3 M_\odot$) in the case of neutron stars. For collapsing objects which are unable to radiate away a sufficient amount of mass to fall below such limits, no final stable state is avaliable and therefore we have to admit that singularities are formed. Since we actually know that masses greater than the neutron star limit exist in nature (for instance in active galactic nuclei) it becomes urgent to understand the nature of such objects. The natural answer is, of course, that the end state of such a collapse process is a black hole. However, this "Cosmic Censorship conjecture" has never been rigourously proved. Therefore, it can well be the truth that very "embarassing" objects are going around in the universe: the naked singularities.

A familiar example of a black hole is the Schwarzschild solution, which is best visualized in the Kruskal plane. It is then seen that the singularity "lying at $r = 0$" is spacelike, namely, it is "a time placed in the future of any observer who crossed the event horizon". As a consequence, no observer can send information about "what happens at the singularity" to any other observer, not only to those remaining outside the horizon. As a consequence one says that this singularity, besides being "weakly" censored (as covered by the horizon) is also strongly censored.

A familiar example of a naked singularity is the Kerr solution. If the angular momentum per unit mass is less than the mass, an event horizon is present (so that the singularity is unvisible to far-away observers) but also a Cauchy horizon is present, and, as a consequence, the singularity "lying at $r = 0, \theta = \pi/2$" is timelike. It is at best visualized as a ring, and paths of timelike particles exist that avoid meeting the singularity and can receive signals emitted from "there": the singularity is naked. If the angular momentum per unit mass is greater than the mass, no horizon is present at all and the singularity is visible to far-away observers, i.e. it is globally naked.

Due to the possibility of extending the Kerr spacetime in the "$r < 0$ sheet", where closed timelike curves exist, it is widely believed that spacetimes with naked singularities always contain local or even global violations of causality. This is essentially the main reason for which many physicst are strongly convinced about the validity of the idea that nature should avoid formation of naked singularities. However, it is absolutely not obvious that all naked singularities should behave as the Kerr one. As a matter of fact, we don't know how a naked singularity looks like, and only very recently there have been attempts of calculating, for istance, the emission of gravitational waves from such objects. On the other hand, no general proof of a Censorship theorem is avaliable, while many examples of spacetimes containing naked singularities are known.

In such a scenario, a reasonable first task is to understand fully the most simple case, namely, that of spherically symmetric objects, where the Birkhoff theorem fixes uniquely the vacuum field to be a portion of Schwarzschild. Surprisingly enough, our knowledge even of such a simple case is still very far from being complete. We are going to present here the main results of a research carried out in the past few years, devoted to construct a mathematically consistent framework in which one could hopefully formulate and proof a cosmic censorship theorem in spherical symmetry. It is important to stress that this research has been developed on the basis of many fundamental contributions, which started with the pioonering papers by Eardley [1] and Christodoulou [2] and have received contributions by several authors during the last twenty years. It is impossible to cite them all here, so we refer the reader to [3] for a complete discussion and references.

The general, spherically symmetric, non-static line element in comoving coordinates t, r, θ, φ can be written in terms of three functions ν, η, Y of r and t

$$ds^2 = -e^{2\nu}dt^2 + \eta^{-1}dr^2 + Y^2(d\theta^2 + \sin^2\theta d\varphi^2). \tag{1}$$

Throughout this paper, we shall assume that the material composing the body is interpretable in terms of relativistic continuum mechanics, i.e. we assume the existence of a well-behaved equation of state (this will be the case, for instance, if barotropic perfect fluids are considered). To choose a specific material one has, therefore, to specify the *internal energy* ϵ as a function of the parameters characterizing the state of strain of the body. It can be shown that, using the comoving frame, the deformation can be described in terms of "purely gravitational" degrees of freedoom [4]. This means that the strain parameters can be identified with Y and η. Therefore, we introduce as *equation of state* of the material a positive function

$$\epsilon = \psi(r, Y, \eta),$$

where the explicit dependence on r takes into account possible inhomogeneities. The function ψ has to satisfy various requirements of physical reasonability. Taking into account regularity, weak energy condition, and local stability of matter one can construct a certain set Ψ of admissible functions of state (essentially, these functions are C^1 with respect to the strain parameters and admit a absolute minimum in the state of local relaxation of the material, see details in [5]).

The energy-momentum tensor can be readily calculated: the result is a diagonal tensor of the form $T_\nu^\mu = \mathrm{diag}(-\epsilon, \Sigma, \Pi, \Pi)$. The stress-strain relations (i.e. the relations giving the radial stress Σ and the tangential stress Π in terms of the constitutive function) are given by

$$\Sigma = 2\eta \frac{\partial \psi}{\partial \eta} - \psi, \quad \Pi = -\frac{1}{2} Y \frac{\partial \psi}{\partial Y} - \psi. \tag{2}$$

From the relations above one sees that different choices of the function ψ lead to materials having different physical properties. For instance, the case of barotropic perfect fluids corresponds to isotropic pressure that means $\Sigma = \Pi$ or $\psi(r, Y, \eta) = \tilde{\psi}(r, \rho)$ where $\rho = Y\sqrt{\eta}$. In particular, the pressures identically vanish if $\tilde{\psi}(r, \rho) = \rho$. This last choice, therefore, characterizes the dust model. As soon as one allows anisotropy to occur, other interesting models appear (see e.g. [6]). Recently, the case of vanishing radial stress has been studied in details [7,8]. For such a model, the condition $\Sigma = 0$ implies that the dependence of ψ on η must be a multiplicative dependence from $\sqrt{\eta}$ only. Therefore, materials with vanishing radial stresses can be characterized by equations of state of the form $\psi(r, \eta, Y) = \sqrt{\eta}h(r, Y)$ where h is a (positive) arbitrary function.

Once an equation of state has been chosen, the Einstein field equations become a closed system; in spherical symmetry there are three independent equations for the three variables v, η and Y. It is useful to re-write the field equations as a system of *four* differential equations. This is done by introducing the Misner-Sharp mass function $m(r, t)$ defined by $1 - 2m/Y = g^{\mu\nu}Y_{,\mu}Y_{,\nu}$. From the definition it is seen that the curve $t_h(r)$ defined implicitly by $Y(r, t) = 2m(r, t)$ gives the boundary of the region of trapped surfaces, i.e. the apparent horizon. In terms of m the field equations are

$$m' = 4\pi \epsilon Y^2 Y', \qquad \dot{m} = -4\pi \Sigma Y^2 \dot{Y}, \tag{3}$$

$$Y'\dot{\eta} = -2\eta(\dot{Y}' - \dot{Y}v'), \quad \Sigma' = -(\epsilon + \Sigma)v' - 2(\Sigma - \Pi)(Y'/Y). \tag{4}$$

A solution of the Einstein field equations describes the collapse of an initially regular distribution of matter only if the spacetime admits a spacelike hypersurface ($t = 0$) carrying *regular* initial data. This means that the metric, its inverse, and the second fundamental form are continuous at $t = 0$. It can be shown that the initial data for spherically symmetric collapse can be described in terms of two arbitrary functions F and f of the radial coordinate r, which carry the information about the initial velocity profile and the initial density profile of the collapsing matter (physical reasonability imposes some restrictions on both F and f, see [7] for details). Those systems for which f vanishes are called marginally bound and are usually the simplest to be analyzed.

Singularities of spherically symmetric matter filled spacetimes are easily recognized from the diverging behavior of the energy density and/or curvature scalars. These singularities can be of two different kinds: *shell crossing* singularities, at which Y' vanishes while Y is non-zero, and *shell focusing* singularities at which Y vanishes. The shell crossing singularities have been frequently considered as "weak" although no proof of extendibility is as yet avaliable in the literature. In any case, in most physically interesting situations such singularities do not occur, so that we shall concentrate attention here only on the shell focussing case.

The locus of the zeroes of the function $Y(r, t)$ defines a *singularity curve* $t_f(r)$ by the relation $Y(r, t_f(r)) = 0$. Physically, $t_f(r)$ is that comoving time at which the shell of matter labelled by r becomes singular. Usually, the singularity forming at $r = 0, t = t_0 = t_f(0)$ is called central. In many cases, this the unique singularity that can be naked (it can be shown, for instance, that non-central singularities cannot be naked if the radial pressure is positive).

The key idea on which the analysis of the causal structure of the central singularity is based is the following [9]. If the singularity is visible, at least one outgoing null geodesic must exist, that meets the singularity in the past. Such a geodesic will be a solution of the equation

$$\frac{dt(r)}{dr} = \Phi(r, t),$$

(where $\Phi(r, t) := \sqrt{-g_{rr}/g_{00}} = \eta^{-1/2}e^{-\nu}$) having a definite outgoing tangent at $r = 0, t = t_0$. Therefore, the above equation must have a solution of the kind $t = t_0 + x_0 r^\alpha + \cdots$ where $\alpha > 1$ and x_0 is a positive constant. This obviously means that, if existing, x_0 is defined as the limit of t/r^α. Using L'Hospital rule we finally obtain

$$x_0 = \lim_{r \to 0} \frac{t(r)}{r^\alpha} = \lim_{r \to 0} \frac{1}{\alpha r^{\alpha-1}} \frac{dt(r)}{dr} = \lim_{r \to 0} \frac{1}{\alpha r^{\alpha-1}} \Phi(r, t_0 + x_0 r^\alpha), \tag{5}$$

thus x_0 must be a solution of an *algebraic* equation, usually called *root equation*. Of course, the corresponding solution $t(r)$ must reach the point $(r, t(r))$ before the apparent horizon $t_h(r)$, i.e. the condition $t(r) < t_h(r)$ must be satisfied near the singularity (usually a locally naked singularity of this kind can also be extended to a globally naked one).

Once the material and the initial data have been chosen, Einstein field equations provide a unique evolution for such data. Thus, there is a one-to-one correspondence between spherically symmetric solutions and choices of triplets (s, say) of functions $s = \{F(r), f(r), \psi(r, \eta, Y)\}$ (we are, of course, identifying solutions *modulo* gauge transformations). We denote the set of solutions parameterized in this way by $\mathcal{S}$. Since in this parameterization we have already taken into account regularity as well as physical admissibility, the whole physical content of the cosmic censorship problem can now be translated in the mathematical terms of predicting the endstate of any choice of $s \in \mathcal{S}$. This can be formulated in terms of the following purely mathematical strategy:

1) Within the space $\mathcal{S}$, characterize the subset $\mathcal{N}$ of those functions that, via Einstein field equations, produce a root equation which has at least one positive solution.
2) State a "Cosmic Censorship theorem": a naked singularity will not form if and only if the data do not belong to $\mathcal{N}$ (generally speaking this sector will contain both black holes and globally regular solutions).
3) Come back to physics trying to understand physically the conditions that characterize the set $\mathcal{N}$.

We stress that the above program avoids – at least in principle – the complete integration of the field equations. In other words, what is required is to to find any mathematical tecnique that, given the data, "runs across" a system of p.d.e. (the Einstein field equations) and gives the behaviour near the singular point of an *ordinary* differential equation whose r.h.s is constructed out from two of the three unknowns of the system. It goes without saying, however, that understanding the mathematical structure of the subset $\mathcal{N}$ is not an easy task at all. Therefore, one has to switch again to less ambitiuous objectives. There are essentially two choices: one possibility is to study the singularities at fixed choice of the materials, i.e. one can study the nature of solutions of the kind $s_{\overline{\psi}} = \{F(r), f(r), \overline{\psi}(r, \eta, Y)\}$ where the equation of state $\overline{\psi}$ is held fixed and the initial data are varied; the second possibility is to fix the data and let the equation of state vary. These two possibilities correspond to two completely different "phylosophies" in the way the initial configurations are being prepared in a ideal laboratory: one can choose a matter ingredient (e.g. a barotropic fluid) and see what happens when it collapses starting from different velocities and density profiles, but one can also see what happens if different materials (for example, a dust cloud and a barotropic fluid) collapse from the same initial status. It is this second philosophy that we have recently proposed as a way to get some insight in the problem, as we shall now briefly recall.

There are only two sectors of the space $\mathcal{S}$ which is, as for now, are fully understood: that of dust materials, for which the explicit solution (the so called Tolman–Bondi metric) is avaliable (see [3] and references therein), and the so called "Einstein cluster" i.e. rotating but spherically symmetric clouds of particles [10,11]. This latter case will not be considered here because the presence of stresses is due to the rotation of the particles, and therefore the "state function" – essentially the relativistic analogue of the $l^2/2mr^2$ effective potential of Newtonian mechanics – being unbounded, does not belong to Ψ. As a guideline example, therefore, we shall use the dust case. Considering marginally bound solutions $s = \{F, 1, \rho\}$, the endstates can be uniquely characterized by the expansion of the function $F(r)$ at $r = 0$ or, and that is the same, by the expansion of the initial density profile $\epsilon = \epsilon_0 + \epsilon_n r^n + \cdots$. It turns out that the singularity is naked if $n = 1$ or $n = 2$ (i.e. if the initial profile is linear or parabolic). If $n = 3$ the root equation becomes the quartic: $2x^4 + x^3 + \xi(x-1) = 0$. This quartic has a real positive root (so that the singularity is naked) only if $\xi < \xi_c = -(26+15\sqrt{3})/2$. If $n > 3$, a blackhole always forms. The structure of the endstates can therefore be thought of as a spectrum dependent on the parameter n. These results can be extended to the general case of collapsing dust clouds, so that the final fate of the dust solutions $s = \{F, f, 1\}$ is completely known; the endstates depend on the values of a parameter in a similar fashion [12].

Each dust solution can be viewed as the limiting case of a sequence of solutions in the way outlined above: one chooses an initial density and velocity profile, and then varies the state function keeping fixed these profiles. It is easy to check, that the state function of a physically valid material can always be written in the form

$$\psi(r, \eta, Y) = \rho + \lambda \psi_1(r, Y, \eta),$$

where λ is a parameter to be considered as "small" in what follows. A crucial point is now that the equation of state remains well behaved everywhere, even if the material forms a singularity when it self-interacts with its own gravity. Therefore, one can use the expansion irrespectively how strong the gravity is, provided that quantum effects do not arise (when trying to understand singularities within General Relativity, one is always in some sense stretching the theory beyond its own boundary). Now, what occurs in special cases which have been analyzed so far, is that the contribution of a non-trivial equation of state to the r.h.s of the root equation stays finite at the singular point. This means that, in such cases, it is impossible to regularize a diverging behaviour in the root equation by adding any kind of – physically valid – stress. This observation led us [8] to a conjecture stating that all collapses that have *dust* initial data leading to black holes inevitably lead to black holes. There are as for now several indications of validity of this "stress cannot undress" conjecture. If definitively proved, it would give the answer on the final fate of a complete sector of the space S. What would remain to be analyzed is the possibility of a "dressing effect" due to addition of stresses to data sets leading to dust naked singularities.

References

1. Eardley, D.M. (1974): Phys. Rev. Lett. **33**, 442
2. Christodoulou, D. (1984): Commun. Math. Phys. **93**, 171
3. Joshi, P.S. (1993): Global aspects in gravitation and cosmology. Clarendon press, Oxford
4. Kijowski, J., Magli, G. (1998): Class. Quantum Gravity. **15**, 3891
5. Jhingan, S., Magli, G. (2000): Gravitational collapse of fluid bodies and cosmic censorship: analytic insights, in *Recent developments in General Relativity*, ed. by B. Casciaro, D. Fortunato, A. Masiello, M. Francaviglia , Springer-Verlag, Milano, pp. 307–322
6. Herrera, L., Santos, N. (1997): Phys. Rep. **286**, 2
7. Magli, G. (1997): Class. Quantum Grav. **14**, 1937
8. Magli, G. (1998): Class. Quantum Grav. **15**, 3215
9. Joshi, P.S., Dwivedi, I.H. (1992): Commun. Math. Phys. **146**, 333
10. Harada, T., Iguchi, H., Nakao, K. (1998): Phys. Rev. D **58**, R041502
11. Jhingan, S., Magli, G. (2000): Phys. Rev. D **61**, 124006
12. Singh, T.P., Joshi, P.S. (1996): Class. Quantum Grav. **13**, 559

Quantum Zeno Effect
and the Detection of Gravitomagnetism

A. Camacho

Abstract. In this work we introduce two experimental proposals that could shed some light upon the inertial properties of the intrinsic spin. In particular we will analyze the role that the gravitomagnetic field of the Earth could have on a quantum system with spin 1/2. We will first deduce the expression for the Rabi transitions, which depend explicitly, on the coupling between the spin of the quantum system and the gravitomagnetic field of the Earth. Then the continuous measurement of the energy of the spin 1/2 system is considered, and an expression for the emerging quantum Zeno effect is obtained. Thus, it will be proved that gravitomagnetism, in connection with spin 1/2 systems, could induce not only Rabi transitions but also a quantum Zeno effect.

1 Introduction

In more than three-quarters of a century that the theory of general relativity (GR) has achieved a great experimental triumph. Neverwithstanding, at this point it is also important to comment that all the current direct confirmations of GR are confirmations of weak field corrections to the Galilei–Newton mechanics [1]. We must also add that one of the most important, and yet undetected, predictions of GR is the so called gravitomagnetic field [1], sometimes also called Lense–Thirring effect [2], which is generated by mass-energy currents.

The first efforts in the detection of this gravitomagnetic field are quite old [3] and have already included many interesting proposals [4–6].

An additional topic in connection with gravitomagnetism is related to its coupling with intrinsic spin. This issue is of fundamental interest since it comprises the inertial properties of intrinsic spin. It is noteworthy to comment that this point is under constant analysis [7].

In this work we introduce two experimental proposals that could lead to the detection of the coupling between intrinsic spin and the gravitomagnetic field. We analyze the role that the gravitomagnetic field of the Earth could have on a quantum system with spin 1/2. In particular we deduce a Rabi formula, which depends on the coupling between the spin of the quantum system and the gravitomagnetic field of the Earth. Following this, the continuous measurement of the energy of the spin 1/2 system is considered, and a Zeno effect is obtained.

2 Rabi transitions and the gravitomagnetic field

Let us consider a spin 1/2 system immersed in the gravitational field of a rotating uncharged, idealized spherical body with mass M and angular momentum J. In the weak field and slow motion limit the metric, in the Boyer–Lindquist coordinates, reads [8]

$$ds^2 = -c^2 \left(1 - \frac{2GM}{c^2 r}\right) dt^2 + \left(1 - \frac{2GM}{c^2 r}\right)^{-1} dr^2$$
$$+ r^2 \left(d\theta^2 + \sin^2\theta d\varphi^2\right) - \frac{4GJ}{c^2 r} \sin^2\theta d\varphi dt. \tag{1}$$

The gravitomagnetic field in this case is approximately [1]

$$\boldsymbol{B} = 2\frac{G}{c^2}\frac{\boldsymbol{J} - 3(\boldsymbol{J}\cdot\hat{x})\hat{x}}{|\boldsymbol{x}|^3}. \tag{2}$$

We will assume that the expression that describes the precession of the orbital angular momentum immersed, for instance, in the gravitational field of the Earth, can be also used for the description of the dynamics in the case of intrinsic spin. This is a natural extension of general relativity [7].

Let us now denote the angular momentum of our spherical body by $\boldsymbol{J} = J\hat{z}$, being $\hat{z}$ the unit vector along the direction of the angular momentum. Our quantum particle is prepared such that $\boldsymbol{S} = S_z\hat{z}$, it has vanishing small velocity and acceleration, and it is located on the z-axis, with coordinate Z.

There is a formal analogy between the weak field and slow motion of the gravitomagnetic field in general relativity and the magnetic field in electromagnetism [1]. Following this analogy we may write down the interaction Hamiltonian (acting in the two-dimensional spin space of our spin $1/2$ system), which gives the coupling between $\boldsymbol{B}$ and the spin, $\boldsymbol{S}$, of our particle

$$H = -\boldsymbol{S}\cdot\boldsymbol{B}. \tag{3}$$

Introducing expression (2) we may rewrite the interaction Hamiltonian as follows

$$H = 2\frac{GJ\hbar}{c^2 Z^3}\Big[|+\rangle\,\langle+| - |-\rangle\,\langle-|\Big]. \tag{4}$$

Here $|+\rangle$ and $|-\rangle$ represent the eigenkets of S_z. Clearly, the introduction of the gravitomagnetic field renders two energy states

$$E_{(+)} = 2\frac{GJ\hbar}{c^2 Z^3}, \tag{5}$$

$$E_{(-)} = -2\frac{GJ\hbar}{c^2 Z^3}, \tag{6}$$

where $E_{(+)}$ $(E_{(-)})$ is the energy of the spin state $+\hbar/2$ $(-\hbar/2)$. Let us now define the frequency

$$\Omega = (E_{(+)} - E_{(-)})/\hbar = 4\frac{GJ}{c^2 Z^3}. \tag{7}$$

The present analogy allows us to consider the emergence of Rabi transitions [9]. In order to do this let us now introduce a rotating magnetic field, which, at the point where the particle is located, has the following form

$$\boldsymbol{b} = b\Big[\cos(wt)\hat{x} + \sin(wt)\hat{y}\Big],\tag{8}$$

where $\hat{x}$ and $\hat{y}$ are two unit vectors perpendicular to the z-axis, and b is a constant magnetic field.

Under these conditions the total Hamiltonian reads

$$\begin{aligned}
H_T = {}& 2\frac{GJ\hbar}{c^2 Z^3}\Big[|+\rangle\,\langle+| - |-\rangle\,\langle-|\Big]\\
& - \frac{eb\hbar}{2mc}\Big[\cos(wt)\Big(|+\rangle\,\langle-| + |-\rangle\,\langle+|\Big)\\
& + I\sin(wt)\Big(-|+\rangle\,\langle-| + |-\rangle\,\langle+|\Big)\Big].
\end{aligned}\tag{9}$$

Looking for a solution in the form $|\alpha\rangle = c_{(+)}(t)|+\rangle + c_{(-)}(t)|-\rangle$, we find the usual situation [9] (our quantum system has been initially prepared such that $c_{(-)}(0) = 1$ and $c_{(+)}(0) = 0$.)

$$\frac{|c_{(-)}(t)|^2}{|c_{(-)}(t)|^2 + |c_{(+)}(t)|^2} = \left[1 + \frac{(\frac{eb}{2mc\Gamma})^2 sin^2(\Gamma t)}{cos^2(\Gamma t) + \frac{(w-\Omega)^2}{4\Gamma^2}sin^2(\Gamma t)}\right]^{-1},\tag{10}$$

where $\Gamma = \sqrt{\left(\frac{eb}{2mc}\right)^2 + \frac{(w-\Omega)^2}{4}}$.

Clearly, the Rabi transitions depend upon the coupling between spin and the gravitomagnetic field.

$$\left(4\frac{GJ}{c^2 Z^3} - w\right)^2 = 4\left[\Gamma^2 - \left(\frac{eb}{2mc}\right)^2\right].\tag{11}$$

3   Quantum Zeno effect and gravitomagnetism

Let us now measure continuously, the energy of our spin $1/2$ system, such that E is the measurement output, and that this experiment lasts a time T. This kind of measuring process can be described by the so called effective Hamiltonian formalism [10], which is one of the models that exist in the topic of quantum measurement theory

[11]. In our case the corresponding effective Hamiltonian reads

$$
\begin{aligned}
H_{\text{eff}} = {} & 2\frac{GJ\hbar}{c^2 Z^3}\Big[|+\rangle\langle+| - |-\rangle\langle-|\Big] \\
& - \frac{eb\hbar}{2mc}\Big[\cos(wt)\Big(|+\rangle\langle-| + |-\rangle\langle+|\Big) \\
& + i\sin(wt)\Big(-|+\rangle\langle-| + |-\rangle\langle+|\Big)\Big] \\
& - i\frac{\hbar}{T\Delta E^2}\Big[E^2\Pi - 2E\frac{2GJ\hbar}{c^2 Z^3}\Big(|+\rangle\langle+| - |-\rangle\langle-|\Big) \\
& + (\frac{2GJ\hbar}{c^2 Z^3})^2\Big(|+\rangle\langle+| + |-\rangle\langle-|\Big)\Big],
\end{aligned}
\tag{12}
$$

where Π is the unit operator in the spin space of our particle. Consider now the following definitions $\tilde{\Gamma} = \frac{(w-\Omega)}{2} + \frac{i}{2T\Delta E^2}\Big[(E_{(+)} - E)^2 - (E_{(-)} - E)^2\Big]$, $\beta^2 = (\gamma/\hbar)^2 + \tilde{\Gamma}^2$, and finally $\gamma = -\frac{eb\hbar}{2mc}$.

Let us now suppose that the measurement output is the energy of the ground state, $E_{(-)}$, that we have a resonant perturbation, and that initially only the lowest energy state was populated, in other words, $E = E_{(-)}$, $\hbar w = E_{(+)} - E_{(-)}$, and $c_{(-)}(0) = 1$, $c_{(+)}(0) = 0$.

Hence assuming that $\frac{(E_{(+)} - E_{(-)})^4}{4T^2\Delta E^4} > \gamma^2/\hbar^2$, we find that

$$
P_{(-)}(t) = \left[1 + \frac{\sinh^2(\frac{\gamma}{\hbar}\tilde{\Omega}t)}{\tilde{\Omega}^2[\cosh(\frac{\gamma}{\hbar}\tilde{\Omega}t) + \frac{\hbar(E_{(+)} - E_{(-)})^2}{2T\gamma\tilde{\Omega}\Delta E^2}\sinh(\frac{\gamma}{\hbar}\tilde{\Omega}t)]^2}\right]^{-1},
\tag{13}
$$

where $\tilde{\Omega} = \sqrt{\frac{\hbar^2(E_{(+)} - E_{(-)})^4}{4T^2\gamma^2\Delta E^4} - 1}$, $\gamma = -\frac{eb\hbar}{2mc}$, and $P_{(-)}(t) = \frac{|c_{(-)}(t)|^2}{|c_{(-)}(t)|^2 + |c_{(+)}(t)|^2}$.

In the case $t \to \infty$ this last expression reduces to

$$
P_{(-)}^{(\infty)} = \left[1 + \left(\frac{c^2 Z^3}{4GJ\hbar}\right)^2\frac{ebT\Delta E^2}{mc}\right.
$$
$$
\left. \cdot\left(\sqrt{1 - \left(\frac{c^2 Z^3}{4GJ\hbar}\right)^4\left(\frac{ebT\Delta E^2}{mc}\right)^2} - 1\right)^{-2}\right]^{-1}.
\tag{14}
$$

Clearly, Rabi transitions are inhibited, and the asymptotic value that here appears depends explicitly upon the coupling between the intrinsic spin and the gravitomagnetic field, i.e., J emerges in the last expression.

Acknowledgements. The author would like to thank A.A. Cuevas-Sosa and A. Camacho-Galván for their help, and D.-E. Liebscher for the fruitful discussions on the subject.

The hospitality of the Astrophysikalisches Institut Potsdam is also kindly acknowledged. This work was supported by CONACYT (México) Posdoctoral Grant No. 983023.

References

1. Ciufolini, I. and Wheeler, J.A. (1995): Gravitation and Inertia. Princeton University Press, Princeton, NJ
2. Lense, J., Thirring, H. (1918): Phys. Z. **19**, 711–750
3. Friedländer, B., Friedländer, I. (1896): Absolute und relative Bewegung. Simion-Verlag, Berlin
4. Braginsky, V.B., Polnarev, A.G., Thorne, K.S. (1984): Phys. Rev Lett. **53**, 863–866
5. Vitale, S., Bonaldi, M., Falferi, P., Prodi, G.A., Cerdonio, M. (1989): Phys. Rev. B **39**, 11993–12002
6. Ciufolini, I., Pavlis, E., Chieppa, F., Fernandes-Vieira, E., Pérez-Mercader, J. (2000): Science **279**, 2100–2103
7. Mashhoon, B. (1999): Gen. Rel. Grav. **31**, 681–691
8. Boyer, R.H., Lindquist, R.W. (1967): J. Math. Phys. **8**, 265–281
9. Sakurai, J.J. (1995): Modern Quantum Mechanics. Addison-Wesley, Reading, MA
10. Mensky, M.B. (1993): Continuous Quantum Measurements and Path Integrals. IOP, Bristol and Philadelphia
11. Presilla, C., Onofrio, R., Tambini, U. (1996): Ann. Phys. **248**, 95–121

Sagnac, Gclock Effect and Gravitomagnetism

A. Tartaglia

Abstract. New techniques to evaluate the clock effect are described. They are based on the flatness of the cylindrical surface containing the world lines of the rays constrained to move on circular trajectories about a spinning mass. The effect of the angular momentum of the source is manifested in the fact that inertial observers must be replaced by local non rotating observers. Starting from this, exact formulas for circular trajectories are found. Numerical estimates for the Earth environment show that light would be a better probe than actual clocks to evidence the angular momentum influence. The advantages of light in connection with some principle experiments are shortly reviewed and a couple of specific solutions are suggested.

1 Introduction

The gravitomagnetic clock effect (synthetically gclock effect) is in a sense a phenomenon announced since the early days of relativity [1]. It is due to the fact that the proper time measured by a clock is indeed the four dimensional "length" of the world line of the clock between two given events: different allowed four dimensional paths between the same pair of events correspond to different proper time intervals and produce desynchronization of initially synchronous clocks.

When treating rotating systems (and clocks) a specific consequence of this fact is the Sagnac effect, known since 1913 or even before and named after G. Sagnac [2], who interpreted it with an antirelativistic and "etherial" (in the sense of the ether theory) attitude. The Sagnac effect earned a whole literature for itself during many decades [3]; the simplest and most complete description of it is indeed the geometrical one [4,5].

The introduction of nonspinning or spinning masses in the empty space time does not in principle modify the situation, but for computational complications. The latter case (spinning sources of gravity) is however the one to which the expression "gravitomagnetic clock effect" has been reserved. The name comes from the fact that, in weak field approximation, the gravitational field may be decomposed into a gravitoelectric (radial) field and a gravitomagnetic (solenoidal) field in analogy with electromagnetism: the gclock effect is determined by the gravitomagnetic component. This description and treatment has been particularly emphasized by Mashhoon and others [6,7], but a general treatment is in principle rather simple and can be given before any approximation [8].

The importance of the gclock effect is in the hope that it can be used to test the influence of the angular momentum of the source on the gravitational field. In fact the revolution time of a freely orbiting clock on a circular trajectory around a spinning mass is different according to whether the orbit is prograde or retrograde.

The difference between the two periods expressed as the times needed to recover a fixed azimuth in the inertial frame of a distant observer is [7]:

$$\delta t = 4\pi \frac{a}{c},$$
(1)

where a is a length given by the ratio between the angular momentum of the source J and its mass M multiplied by c: $a = \frac{J}{Mc}$. The result (1) is true for any axisymmetric metric with an a value independent from time [8].

Besides the actual numbers that should be measured (which, at least in the case of δt, are indeed big), a real experiment around the Earth encounters a series of difficulties not easy to be surmounted [7,9], and one of them is the stability and coincidence between opposite orbits. Considering this specific problem, electromagnetic waves come again on stage as probes more interesting than actual clocks. In fact a single ring of orbiting mirrors could be used to reflect at the same instant both prograde and retrograde "light" beams in a sort of Sagnac experiment in space. The influence of gravity on the issue of such an experiment can be worked out as a general relativistic correction of the classical effect [10], but the problem can be better solved in principle and in general using the method outlined in [8].

2   Geometric vision of the gclock effect

In order to describe our general method we move from the fact that the coordinate angular velocity $\omega = d\varphi/dt$ of an object along a circular trajectory is a helix, which becomes a straight line in the development of the cylindrical world tube in a plane.

The situation is graphically summarized in Fig. 1. The observer's world line is OAB. The figure is drawn considering two oppositely rotating light beams. Any pair of objects revolving in opposite directions with the same coordinate speed possess symmetric world lines with respect to a static observer at the origin (vertical straight line there) and cross each other again at the same event for such observer: the revolution times both in the objects proper times and in the observer's time are exactly the same for both. If however the observer is in turn steadily rotating his world line is an oblique straight line and the intersecting events with the world lines of the objects do not coincide any more: the revolution times measured by the observer differ from each other and so do the proper times of the two objects. The interval between A and B, in Fig. 1, is then proportional to the proper time lapsed (for the observer) between the completions of one revolution started simultaneously by the two probes. When the "objects" are light beams the AB interval is a measure of the Sagnac effect.

When the central body possesses an angular momentum, the equivalent in the vicinity of the static inertial observers of the previous case are the so called Locally Non Rotating Observers (LNRO) [11], i.e. the observers to whom radially infalling matter appears locally as non rotating. The LNRO's themselves are indeed rotating as seen by a distant inertial observer; their coordinate angular speed is:

$$\Omega = -g_{t\varphi}/g_{\varphi\varphi},$$
(2)

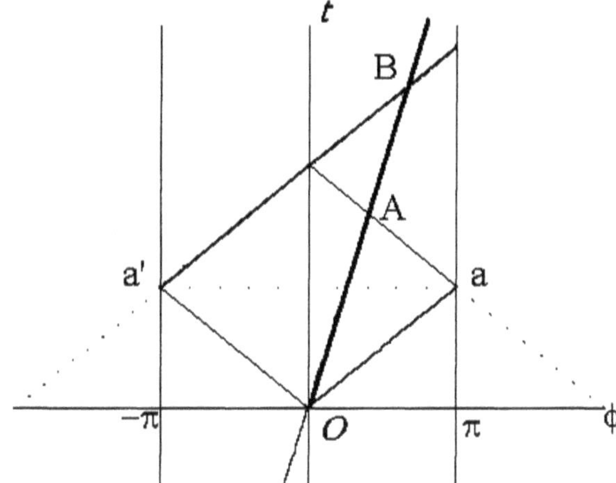

Fig. 1. Scheme of the opened world tube on whose surface the world lines of the observer and the "clocks" lie. $Oaa'B$ represents a prograde light ray; $Oa'aA$ is a retrograde light ray; OB is the world line of the observer. The segment AB visualizes the proper time delay which is indeed the Sagnacs effect

(the g's are elements of the metric and the use of polar coordinates is understood). In practice the considerations about the flatness of the cylindrical surface containing the world lines of constant r and θ remain unchanged, but now the world lines of co-rotating and counter-rotating light rays have a different inclination to cope with the role of an LNRO and his skew world line; in other words after leaving the LNRO at O the two beams must cross each other again on the (oblique) world line of the LNRO, in order it to be equivalent to the static observer of rotation free space times. Graphically the situation is the one now shown in Fig. 2.

Now the entity of the Sagnac effect is measured by the interval between A' and B' which is clearly different from the one between A and B in Fig. 1, depending on the value of the parameter a which is hidden in the $g_{t\varphi}$ element of the metric tensor.

3 Influence of the angular momentum on the Sagnac effect

What is visually determined by the simple graphic treatment of the previous section can be put in analytical form by the use of Minkowskian geometry in two dimensions. Let us restrict our study to light rays and the Sagnac effect in axisymmetric metrics. Events A and B in Fig. 1 (A' and B' in Fig. 2) are localized as the intersections of two straight lines: the world line of the observer and the world line of a light beam. When an angular momentum is absent the former's equation is $t = \varphi/\omega_o$, where ω_o is the observer's angular coordinate speed. The equation of the world line of a light ray constrained to move along a circular trajectory is in turn $t = (2\pi - \varphi)/\omega_{l0}$ for the counter-rotating beam and $t = (\varphi + 2\pi)/\omega_{l0}$ for the co-rotating one; the equations reproduce the lines respectively ascending to the left through a and ascending to

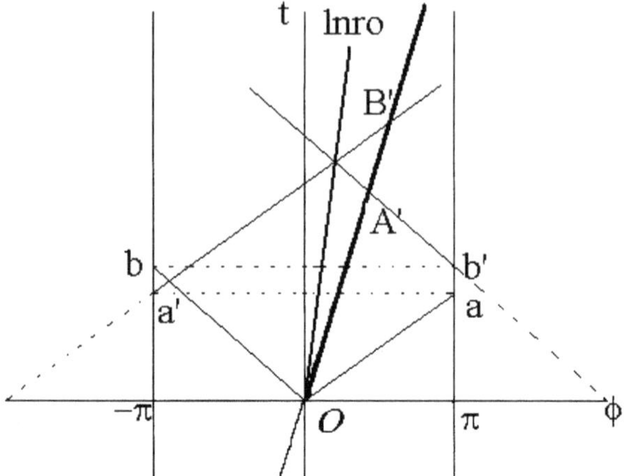

Fig. 2. The scheme is the same as in Fig. 1, but now the light rays are represented by different inclinations on the right and on the left of the origin. Events A' and B' delimitate the proper time span between one complete revolution of retrograde and prograde light rays. The rays meet each other at the world line of the LNRO passing through the origin event

the right through a' in Fig. 1. The coordinate angular speed of light for a circular path in a plane orthogonal to the symmetry axis (let us say it is the equatorial plane of the system) is $\omega_{l0} = \sqrt{-g_{tt}/g_{\varphi\varphi}}$. The intersections between the straight world lines of the observer and the beams identify points A (counter-rotating beam) and B (corotating beam) with their t and φ coordinates:

$$\begin{cases} \varphi_{A,B} = 2\pi \dfrac{\omega_o}{\omega_{l0} \pm \omega_o}, \\ t_{A,B} = \dfrac{2\pi}{\omega_{l0} \pm \omega_o}. \end{cases}$$

(3)

The $+$ $(-)$ sign corresponds to event A (B). The proper time corresponding to the interval OA (OB) in Fig. 1 is in practice the length (divided by c) of the hypotenuse of a right-angled triangle in a two dimensional flat Minkowski space time, whose other sides are $\sqrt{g_{tt}t_{A,B}^2}$ and $\sqrt{g_{\varphi\varphi}\varphi_{A,B}^2}$. The explicit result is:

$$\tau_{A,B} = \frac{2\pi}{c\left(\omega_{l0} \pm \omega_o\right)}\sqrt{g_{tt} + g_{\varphi\varphi}\omega_o^2},$$

and the Sagnac delay turns out to be

$$\delta\tau = \tau_B - \tau_A = 4\pi \frac{\omega_o}{c\left(\omega_{l0}^2 - \omega_o^2\right)}\sqrt{g_{tt} + g_{\varphi\varphi}\omega_o^2} = 4\pi \frac{\omega_o}{c}\sqrt{\frac{-g_{\varphi\varphi}}{\omega_{l0}^2 - \omega_o^2}}. \tag{4}$$

The g_{tt} term has been eliminated thanks to the explicit expression of ω_{l0}.

When $g_{t\varphi} \neq 0$ the coordinate angular speed of light on a circular path has two different values according to the fact that the light beam is co- or counter-rotating; the actual values are:

$$\omega_{l\pm} = \sqrt{\Omega^2 + \omega_{l0}^2} \pm \Omega,$$ (5)

($+$ is for prograde motion, $-$ is for retrograde motion).

Now the intersection method leads to:

$$\begin{cases} \varphi_{A',B'} = \dfrac{2\pi\omega_o}{\omega_{l\mp} \pm \omega_o}, \\ t_{A',B'} = \dfrac{2\pi}{\omega_{l\mp} \pm \omega_o}. \end{cases}$$ (6)

The proper observer's intervals of time are obtained as before:

$$\tau_{A',B'} = \dfrac{2\pi}{c\,(\omega_{l\mp} \pm \omega_o)} \sqrt{g_{tt} + 2g_{t\varphi}\omega_o + g_{\varphi\varphi}\omega_o^2}.$$

Finally the Sagnac delay is:

$$\begin{aligned} \delta\tau = \tau_{B'} - \tau_{A'} &= \dfrac{4\pi}{c} \dfrac{\omega_o - \Omega}{\omega_{l0}^2 + 2\Omega\omega_o - \omega_o^2} \sqrt{g_{tt} + 2g_{t\varphi}\omega_o + g_{\varphi\varphi}\omega_o^2} \\ &= \dfrac{4\pi}{c} \dfrac{-g_{\varphi\varphi}\omega_o - g_{t\varphi}}{\sqrt{g_{tt} + 2g_{t\varphi}\omega_o + g_{\varphi\varphi}\omega_o^2}}. \end{aligned}$$ (7)

The parameter Ω has been eliminated using the explicit expression for the coordinate angular speed of an LNRO.

If now we pose $\omega_o = \Omega$, we consistently verify that it is:

$$\delta\tau_{\text{LNRO}} = 0.$$

A truly inertial observer ($\omega_o = 0$) would find:

$$\delta t = -\dfrac{4\pi}{c} \dfrac{g_{t\varphi}}{\sqrt{g_{tt}}}.$$ (8)

Formula (7) treats ω_o and r as independent parameters; the situation changes when the observer is freely orbiting. For a circular geodesic trajectory in the equatorial plane there are two solutions for ω_o [8] (again $+$ is pro- and $-$ is retrograde motion of the observer):

$$\omega_{o\pm} = \dfrac{-g_{t\varphi,r} \pm \sqrt{g_{t\varphi,r}^2 - g_{\varphi\varphi,r}g_{tt,r}}}{g_{\varphi\varphi,r}}.$$ (9)

Commas denote ordinary partial differentiation. Correspondingly (7) becomes:

$$\delta\tau_\pm = \dfrac{4\pi}{c} \dfrac{-g_{\varphi\varphi}\omega_{o\pm} - g_{t\varphi}}{\sqrt{g_{tt} + 2g_{t\varphi}\omega_{o\pm} + g_{\varphi\varphi}\omega_{o\pm}^2}}.$$ (10)

4 Terrestrial environment

Considering the situation in the surroundings of our planet the appropriate metric is the one corresponding to a weak axisymmetric field. Actually this means that:

$$g_{tt} = c^2 \left(1 - 2G\frac{M}{c^2 r} \right),$$

$$g_{t\varphi} = 2aG\frac{M}{cr},$$

$$g_{\varphi\varphi} \simeq -r^2.$$

Consequently it is:

$$\Omega = 2aG\frac{M}{cr^3},$$

$$\omega_{l0} = \frac{c}{r}\sqrt{1 - 2G\frac{M}{c^2 r}}.$$

Now using (7) we obtain:

$$\delta\tau \simeq \frac{4\pi}{c^2} \frac{r^2\omega_o - 2aG\frac{M}{cr}}{\sqrt{1 - 2G\frac{M}{c^2 r} + 4aG\frac{M}{c^3 r}\omega_o - \frac{r^2\omega_o^2}{c^2}}}. \tag{11}$$

For not too high observer's velocities and radii of the order of the size of the Earth (11) becomes:

$$\delta\tau \simeq \frac{4\pi}{c}r \left(\frac{r\omega_o}{c} + \frac{r^3\omega_o^3}{2c^3} + G\frac{M}{c^3}\omega_o - 2G\frac{M}{c^2 r^2}a \right). \tag{12}$$

One easily recognizes, in the order, the traditional Sagnac term, then a purely special relativistic correction, a "Schwarzschild" correction and finally an angular momentum correction which is independent from the velocity of the observer.

When the observer is freely orbiting on a circular trajectory the starting point is formula (10); developing and keeping only the lowest order terms up to the one containing the angular momentum, the result is

$$\delta\tau_\pm \simeq \frac{4\pi}{c^2} \left(\pm\sqrt{GMr} \pm \frac{1}{2c^2}\sqrt{\frac{G^3 M^3}{r}} \pm G\frac{M}{c^2}\sqrt{\frac{GM}{r}} - 2a\frac{GM}{cr} \right).$$

Now summing the time differences for the two rotation directions we obtain

$$\Delta\left(\delta\tau\right) = \delta\tau_+ + \delta\tau_- = -16\pi\frac{GM}{c^3 r}a. \tag{13}$$

The order of magnitude of this quantity is $\sim 10^{-16}$ s.

5   Conclusions

We have given a general geometric treatment of the problem of the phase shift of electromagnetic waves moving in a closed circular path about a massive spinning body. Applying the general methodology to the terrestrial environment reproduces approximate results obtainable by various alternative techniques. In terms of time differences accumulated during round trips around the Earth the numbers remain extremely small and partly smaller than some values typical of the gclock effect. The idea of using electromagnetic waves in some orbital Sagnac type experiment deserves however some attention because of a few advantages light has.

To begin with, if a chain of orbiting mirrors or transponders are considered it is possible to use them simultaneously for prograde and retrograde rays, thus ensuring the coincidence of the two trajectories and the significance of the time of flight difference.

Next we can notice that, even for an "experimental apparatus" as big as the orbit of a satellite, the time of flight of light for a round trip is ~ 0.1 s. In that time the observer on a spacecraft could move for a few hundred meters at most and an observer on the surface of the Earth would move much less than that; this fact in principle allows for experiments at different values of the equivalent angular velocity ω_o, to be maintained for a few seconds at most.

Other possible solutions one can think of in order to experimentally verify the influence of the parameter a on physical phenomena, besides the observation of the Lense–Thirring effect on the orbit of satellites [12], could be the timing of pulsars when passing behind a spinning mass such as the Sun or Jupiter, or a ground based experiment reproducing in the laboratory a mini orbiting system with a central spinning massive sphere.

Pulsars are indeed good clocks in the sky and the difference in time of flight of their signals when passing on the right and on the left of a body such as the Sun or Jupiter would be in the order of 10^{-10}s and could be revealed comparing the signal pattern before and after the occultation.

The idea of a laboratory spinning mass with freely orbiting "satellites" magnetically suspended on a horizontal plane could be interesting, because a practicable value for a for a laboratory spinning body could be in the order of 10^{-7}m (for the whole Earth it is ~ 1 m). Consequently the time difference (1) would be $\sim 10^{-14}$ s which, converted in a phase shift for visible light, corresponds to a few fringes.

References

1. Einstein, A. (1905): Ann. der Physik **17**, translated by W. Perrett, G.B. Jeffery and reprinted in: The principle of relativity, Dover Publications, 1952 (New York)
2. Sagnac, G. (1913): Comptes Rendus, **157**, 708
3. Stedman, G.E. (1997): Rep. Prog. Phys., **60**, 615
4. Anandan, J. (1981): Phys. Rev. D **24**, 338
5. Rizzi, G., Tartaglia, A. (1998): Found. Phys. **28**, 1663
6. Cohen, J.M., Mashhoon, B. (1993): Phys. Lett. A **181**, 353

7. Mashhoon, B., Gronwald, F., Theiss, D.S. (1999): Ann. Phys. **8**, 135
8. Tartaglia, A. (2000): Gen. Rel. Grav. **32**, (9), 1745
9. Lichtenegger, H.I.M., Gronwald, F., Mashhoon, B. (1998): Advances in Space Research (in press); gr-qc/9808017
10. Tartaglia, A. (1998): Phys. Rev. D **58**, 064009
11. Straumann, N. (1984): General Relativity and Relativistic Astrophysics. Springer-Verlag, Berlin Heidelberg New York
12. Ciufolini, I., Chieppa, F., Lucchesi, D., Vespe, F. (1997): Class. Quantum Grav. **14**, 2710

Radiation from an Electric Charge

A. Harpaz

Abstract. The conditions in which an electromagnetic radiation is formed are discussed. It is found that the main condition for the emission of radiation by an electric charge is the existence of a relative acceleration between the charge and its electric field. Such a situation exists both for a charge accelerated in a free space, and for a charge supported at rest in a gravitational field. Hence, in such situations, the charge radiates. It is also shown that relating radiation to the relative acceleration between a charge and its electric field, solves several difficulties that existed in earlier approaches, like the "energy balance paradox", and the "relativistic" nature of the observation of the emitted radiation.

1 Introduction

A theory for electromagnetic radiation leans on the basic principle of the field theory – the principle that an electric field is an independent physical entity. It posseses inertia, it falls in a free fall in a gravitational field, and it is detached from the charge that induced it. When the charge is moving, the electric field is not carried with the charge, and since the charge continues to induce the field, the form of the field changes due to the fact that the inducing source, the charge, continuously changes its location. This is why we have to use the method of the "retarded potentials" to calculate the potentials of charges in motion. When the charge accelerates, the ratio between the charge velocity and the velocity at which the field expands in space is not constant, the field lines become curved, and a stress force is created. Later we shall show the curvature of the electric field of a uniformly accelerated charge, discussing the interaction of the stress force existing in the curved field with the charge. A similar curvature forms in the electric field of a static charge located in a gravitational field.

We discuss here several questions concerned with the theory of radiation:

- A very wide used answer to the question what is the source of the energy carried by the radiation, is that there exists a radiation reaction force, which contradicts the acceleration of the charge. A work should be done against this force, and this work is the source of the energy carried by the radiation. However, in many cases this radiation reaction force is very weak, and in the simplest case, the case of a constant acceleration at zero velocity, the radiation reaction force entirely vanishes.
  This problem was named "The Energy Balance Paradox", and many attempts were made to solve this paradox. We show later that there is indeed a reaction force, which is responsible for the work that creates the energy carried by the radiation. But this is not a reaction force created by the radiation. This reaction force is the stress force created in the curved electric field of the accelerated charge.

- The question who can observe the radiation concerns general relativity considerations.

  In general textbooks of electrodynamics [1, 2, 3], radiation is dealt with in the frame of special relativity, using flat systems of reference. In such systems, the existence of acceleration has an absolute nature, and it is clear how a radiation field is calculated, and how this field can be observed. When general relativity is considered, the three vector of acceleration becomes a relative property, which may exist in one system, and vanish in another.

  Rohrlich [4] and Boulware [5] argue that the radiation phenomenon also has a relative nature, and an observer co-accelerated with an accelerated charge, will not observe radiation from this charge, while other observers, relative to which the charge is accelerated, will observe electromagnetic radiation from the charge. According to this approach, observation of radiation depends on the relative acceleration between the charge and the observer.

  Ascribing on relativistic nature to radiation, raises serious questions concerning energy and angular momentum transfer, connected with emission and absorption of radiation. In the process of emission of radiation, the source loses energy and angular momentum, and when radiation is absorbed by some physical system, this system gains the energy and the angular momentum carried by the radiation. The excitation states of the systems involved in this process change, and these changes cannot be transformed away by a coordinate transformation.

  There is an objective measure when radiation is created, and this measure is the relative acceleration between the charge and its own electric field. This situation of a relative acceleration between a charge and its electric field does not depend on the observer, and is observed from any system of reference. When such an acceleration exists, the electric field of the charge is curved. The curvature creates a stress force that interacts with the charge as a reaction force. The external force that accelerates the charge has to perform an extra work to overcome the stress force, and this work is the source of the energy carried by the radiation.

- The common answer to the question whether a static charge in a gravitational field radiates is "The situation is static, no magnetic field exists, and hence the Poynting vector vanishes, and no radiation is created". We show that:
  - The absolute acceleration of a static charge in a gravitational field does not vanish.
  - The existence of a Poynting vector is not a covariant measure for the existence of radiation.
  - The electric field of the supported charge falls in a free fall in the gravitational field, and hence there exists a relative acceleration between the charge and its electric field.
  - The freely falling electric field is curved, and a stress force exists in this curved field, which gives rise to the creation of the energy carried by the radiation. Hence we conclude that a static charge, supported in a gravitational field, does radiate.

2   A uniformly accelerated charge

In analyzing the radiation from an accelerated charge we present the angular distribution of the radiation [1]:

$$\frac{dP}{d\Omega} = \frac{e^2 a^2}{4\pi c^3} \frac{\sin^2 \theta}{(1 - \beta \cos \theta)^5},$$

(1)

where θ is the angle measured from the direction of motion.

Integrating Eq. 1 over the angles, (with $\beta \to 0$) yields Larmor formula:

$$P = \frac{2}{3} \frac{e^2 a^2}{c^3}.$$

(2)

We find that for an acceleration at zero velocity – the radiation is emitted in a plane perpendicular to the direction of motion, and no radiation reaction force exists, and it cannot create the energy carried by the radiation. Below we show that the stress force created in the curved electric field of the accelerated charge, is the reaction force responsible for the creation of the radiation.

What is the source of the energy carried by the radiation?

An electric field is an independent physical entity, whose behaviour depends on the properties of space on which it is induced. It is not accelerated with the charge, and there is a relative acceleration between the charge and its field. The field is inertial with respect to the local (free) system of reference, characterized by the geodesics that cover this system.

For an accelerated motion, the electric field of the accelerated charge becomes curved. Fulton and Rohrlich [6] calculated the electromagnetic fields of a charge moving in a hyperbolic motion, with an acceleration a along the z axis. We use $\alpha = c^2/a$ as the location of the charge at $t = 0$. We present here the electric fields in a cylindrical coordinate system (ρ, z), at $t = 0$, and $z = \alpha$ (in a plane perpendicular to the direction of motion, at the moment the accelerated charge is at rest). We find:

$$E_\rho = \frac{e}{\rho^2 \left[1 + \frac{\rho^2}{4\alpha^2}\right]^{3/2}},$$

(3)

$$E_z = \frac{-e}{2\alpha\rho \left[1 + \frac{\rho^2}{4\alpha^2}\right]^{3/2}} = \frac{-ea}{2\rho c^2 \left[1 + \frac{\rho^2}{4\alpha^2}\right]^{3/2}}.$$

(4)

These equations show that for $\rho < \alpha$, E_ρ, which is the longitudinal field, is the Coulomb field that falls as $1/\rho^2$, and E_z, the transverse field is the radiation field that falls as $1/\rho$. This field is proportional to the acceleration.

The radius of curvature of the field lines is: $R_c = c^2/a \sin \theta$, where θ is the angle between the initial direction of the field line and the acceleration. Now, we follow in brief the calculations given in [7]. In a curved electric field a stress force exists, whose density, f_s, is given by:

$$f_s = \frac{E^2}{4\pi R_c},$$

(5)

where E is the electric field, and R_c is the radius of curvature of the field lines. In the immediate vicinity of the charge, the electric field can be taken as $E = e/r^2$. The stress force is perpendicular to the field lines, so that the component of this force along the acceleration is $-f_s \sin \varphi$, where φ is the angle between the local field line and the acceleration. In the immediate vicinity of the charge ($r \ll c^2/a$), $\varphi \sim \theta$, and we can write for the parallel component of the stress force:

$$-f_s \sin \varphi \simeq -f_s \sin \theta = -\frac{a \sin^2 \theta}{4\pi c^2} \frac{e^2}{r^4}. \tag{6}$$

In order to sum over f_s, we have to integrate over a sphere whose center is located on the charge. Naturally, such an integration involves a divergence (at the center). To avoid such a divergence, we take as the lower limit of the integration a small distance from the center, $r = c\Delta t$, (where Δt is infinitesimal). We calculate the work done by the stress force in the volume defined by $c\Delta t \leq r \leq r_{\text{up}}$, where r_{up} is some large distance from the charge, satisfying the demand: $c\Delta t \ll r_{\text{up}} \ll c^2/a$. These calculations are performed in a system of reference S, defined by the geodesics, and momentarily coincides with the frame of reference of the accelerated charge at time $t = 0$, at the charge location.

Integration of the stress force over a spherical shell extending from $r = c\Delta t$ to r_{up}, yields the total force due to stress, F_s:

$$F_s = 2\pi \int_{c\Delta t}^{r_{\text{up}}} r^2 dr \int_0^{\pi} \sin \theta d\theta [-f_s \sin \theta] = -\frac{2}{3} \frac{a}{c^2} \frac{e^2}{c\Delta t} \left(1 - \frac{c\Delta t}{r_{\text{up}}}\right). \tag{7}$$

Clearly the second term in the parenthesis can be neglected. The power supplied by the external force on acting against the electric stress is $P_s = -F_s v$, where v is the velocity of the charge in system S at time $t = \Delta t$, $v = a\Delta t$. Substituting this value for v we get for P_s:

$$P_s = \frac{2}{3} \frac{a^2 e^2}{c^3}. \tag{8}$$

This is the power radiated by an accelerated charged particle at zero velocity (Eq. (2)).

3   A charge supported in a gravitational field

The situation of a relative acceleration between a charge and its electric field also exists for a charge supported statically in a gravitational field. The electric field, which is not supported with the charge, falls in a free fall, and there exists a relative acceleration between the charge and its field. Prima facie, the situation seems static, but it is not. It is a steady state situation. It is correct that in the lab system (the system of the supported charge), the Poynting vector vanishes, but a Poynting vector is not an invariant, and cannot be taken as a covariant measure for the existence of radiation. We demand that a covariant measure for the existence of radiation, is the non-vanishing of the Poynting vector in the system of reference defined by the geodesics. Such a system is the one that falls freely parallel to the charge, and in this system, the supported charge is accelerated upward. A magnetic field exists in this system, and the Poynting vector does not vanish (see [8]). In this system, the electric field is curved, a stress force exists, and calculations similar to those carried for the accelerated charge, yield the power carried by the radiation. The power can be calculated according to Larmor formula, where we put g for the acceleration, namely:

$$P = \frac{2}{3} \frac{g^2 e^2}{c^3}. \tag{8a}$$

As a general measure for the radiation, we take the four acceleration vector A^μ. It is calculated by:

$$A^\mu = D_4 U^\mu = \frac{d}{dx^4} U^\mu + \left\{ {}_{4}{}^{\mu}{}_{\nu} \right\} U^\nu \tag{9}$$

For a static charge $U^\mu = \begin{pmatrix} 0 \\ 1 \end{pmatrix}$. The regular derivative in Eq. (9) vanishes, and we are left with:

$$A^\mu = \left\{ {}_{4}{}^{\mu}{}_{4} \right\} U^4 \tag{10}$$

and for Larmor formula we should use: $a^2 = g_{\mu\mu} A^\mu A^\mu$.

Let us examine two cases:

(1) A homogenous static gravitational field.
(2) A Schwarzschild field.

(1) Consider a homogenous gravitational field directed downward in the $z(= x^3)$ direction, that generates an acceleration g. This field is characterized by $u(z)$: $\frac{1}{u} = \cosh\sqrt{(1 - gz)^2 - 1}$, and $g_{33} = (u'/g)^2$ [8].

$$A^3 = \left\{ {}_{4}{}^{3}{}_{4} \right\} U^4 = g^2 \frac{u}{u'} \tag{11}$$

and

$$g_{33} A^3 A^3 = g^2 u^2, \tag{12}$$

which in the plane $(z = 0)$, yields g^2 to be introduced in Larmor formula.

(2) In a Schwarzschild field, the only relevant coordinate is $r(= x^1)$. We have:

$$A^1 = \left\{ {}_{4}{}^{1}{}_{4} \right\} U^4 = \frac{m}{r^2} \left(1 - \frac{2m}{r} \right). \tag{13}$$

By differentiation, we find that A^1 is maximal at $r = 3m$.

The charge has an outward acceleration of $g(1 - 2m/r)$. For Larmor formula we have to use $a^2 = g_{11}A^1A^1 = g^2(1 - 2m/r)$, and we find that for a static charge in a weak field, Larmor formula yields a power very close to that radiated by an accelerated charge with an acceleration g.

Let us calculate the electric field of a supported charge in a homogenous gravitational field. Using the equations given by [8], we have:

$$E_\rho = \frac{8e\alpha^3\rho u^2}{\left[(\alpha^2(1 - u^2) - \rho^2)^2 + 4\alpha^2\rho^2\right]^{3/2}}, \tag{14}$$

$$E_z = \frac{-4e\alpha^3 uu'}{\left[(\alpha^2(1 - u^2) - \rho^2)^2 + 4\alpha^2\rho^2\right]^{3/2}} \left(\alpha^2(1 - u^2) + \rho^2\right), \tag{15}$$

where $u(z)$ is given above, $u' = du/dz$, and here $\alpha = c^2/g$, is the characteristic radius of curvature of the electric field close to the location of the charge. To calculate the fields in a horizontal plane passing through the location of the charge, we take $z = 0$, which gives: $u \to 1$, $uu' \to 1/\alpha$. These substitutions yields for the fields:

$$E_\rho = \frac{e}{\rho^2} \frac{1}{\left[1 + \frac{\rho^2}{4\alpha^2}\right]^{3/2}}, \tag{14a}$$

$$E_z = \frac{-e}{2\alpha\rho} \frac{1}{\left[1 + \frac{\rho^2}{4\alpha^2}\right]^{3/2}} = \frac{-eg}{2\rho c^2\left[1 + \frac{\rho^2}{4\alpha^2}\right]^{3/2}}, \tag{16}$$

which are equal to Eqs. (3) and (4), that represent the fields of the charge accelerated uniformly in a free space. Certainly, a charge supported at rest in a gravitational field, and a charge accelerated uniformly in a free space, should be treated on the same footing.

What is the source of the energy carried by the radiation? The charge is supported by a solid object, which is static in the GF (Gravitational Field). This solid object is rigidly connected to the source of the GF. Otherwise, it will fall together with the "supported" charge. Actually, the supporting object is part of the object that creates the GF.

The charge is static and no work is done on the charge. However, the electric field of the charge is not static, and it falls in a free fall in the GF. With no interaction between the electric field and the charge, the field would follow a geodetic line and no work would be needed to keep it in a free fall. But the field is curved, and a stress force is implied. The interaction between the curved field and the supported charge creates a force that contradicts the free fall. In order to overcome this force and cause the electric field to follow the geodetic lines, a work is done on the electric field, and this work is done by the GF. This work is the source of the energy carried by the radiation. The energy carried away by the radiation is supplied by the GF, that loses this energy.

4   Conclusions

An accelerated charge does radiate, and also a charge supported at rest in a gravitational field does radiate. One should look for radiation whenever a relative acceleration exists between an electric charge and its electric field. We agree with the calculations of Fulton and Rohrlich [6, 8] (actually, we use these calculations), but we do not agree with their interpretation that the importatnt point is the relative acceleration between the observer and the charge. We find that the important issue is the relative acceleration between a charge and its electric field, because this acceleration gives rise to a curvature in the electric field, which creates the stress force, that acts as a reaction force on the accelerated charge.

The electric field which falls freely in the gravitational field is accelerated relative to the static charge. The field is curved, and the work done in overcoming the stress force created in the curved field, is the source of the energy carried by the radiation. This work is done by the gravitational field on the electric field, and the energy carried by the radiation is created at the expence of the gravitational energy of the system.

References

1. Jackson, J.D. (1975): Classical Electrodynamics. 2nd Edition, John Wiley & Sons, New York
2. Panofsky, W.K.H., Phillips, M. (1962): In: *Classical Electricity and Magnetism.* 2nd Edition, Addison-Weseley Pub. Co., MA
3. Parrott, S. (1987): In: *Relativistic Electrodynamics and Differential Geometry.* Springer-Verlag, New York
4. Rohrlich, F. (1965): In: *Classical Charged Particles.* Addison-Wesley Pub. Co., MA
5. Boulware, D.G. (1980): Annals of Physics **124**, 169
6. Fulton, A., Rohrlich, F. (1960): Annals of Physics **9**, 499
7. Harpaz, A., Soker, N. (1998): Gen. Rel. Grav. **30**, 1217
8. Rohrlich, F. (1963): Annals of Physics **22**, 169

Trajectories for Relativistic Particles in an Electromagnetic Field

E. Caponio, A. Masiello

Abstract. We present a variational theory for the trajectories of a massive, charged particle in a relativistic spacetime with an electromagnetic field. We state some result on the existence and the multiplicity of such trajectories joining two points, assuming that the spacetime is standard stationary and the electromagnetic field is constant with respect to the standard time coordinate. The results are obtained using critical point theory for functionals on infinite dimensional manifolds and the topological properties of the spacetime.

1 World lines of massive and charged particles

In General Relativity a gravitational field is described by a spacetime (M, g), where M is a connected, smooth 4-dimensional manifold and g is a metric tensor on M with segnature $(+, +, +, -)$. The physical properties of the gravitational field are related to the geometric properties of the space-time (M, g) by the Einstein equations. We refer to the classical books [1–3] for more details on space-times and Lorentzian manifolds.

In a space-time a massive particle is described by its *worldline*, that is a smooth curve $\gamma : I \longrightarrow M$ such that $g(\gamma(s))[\dot{\gamma}(s), \dot{\gamma}(s)] < 0$. Such curves are called *timelike*.

Let Y be a smooth vector field on the manifold M. We are interested in the trajectories of a relativistic particle when its dynamic is described by the action integral

$$F(z) = \frac{1}{2} \int_0^1 g(z)[\dot{z}, \dot{z}] \, \mathrm{d}s + \int_0^1 g(z)[Y(z), \dot{z}] \, \mathrm{d}s. \tag{1}$$

Fix two points p and $q \in M$, then a trajectory $z(s)$ for such a particle joining p and q is a stationary point for the functional F:

$$
\begin{aligned}
0 = F'(z)[\zeta] = {} & \int_0^1 g(z)[\dot{z}, \nabla_s \zeta] \, \mathrm{d}s \\
& + \int_0^1 g(z)[Y'(z)[\zeta], \dot{z}] \, \mathrm{d}s + \int_0^1 g(z)[Y(z), \nabla_s \zeta] \, \mathrm{d}s,
\end{aligned}
\tag{2}
$$

for any smooth vector field $\zeta(s)$ along $z(s)$ such that $\zeta(0) = 0$, $\zeta(1) = 0$. Here, $\nabla_s \zeta$ denotes denotes the covariant derivative along z of the vector field ζ, $\mathcal{X}(M)$ denotes the set of smooth vector fields on M, ∇ denotes the Levi–Civita connection associated to the metric g and $Y'\colon \mathcal{X}(M) \longrightarrow \mathcal{X}(M)$ is the differential of the vector field Y defined for any $X \in \mathcal{X}(M)$ by $Y'(X) = \nabla_X Y$. For any $p \in M$ the differential Y' defines a linear map $Y'(p)\colon T_p M \longrightarrow T_p M$ [2].

Whenever $Y \equiv 0$, Eq. (3) reduces to the classical *geodesic equation* (see [4] for results and references on geodesics).

Integrating by parts in (2) shows that a stationary curve joining the points p and q satisfies the system of nonlinear second order differential equations on the manifold M (representing the Euler–Lagrange equations for the functional F):

$$\nabla_s \dot{z} = \big(Y'(z)^* - Y'(z)\big)[\dot{z}], \tag{3}$$

where for any $p \in M$, $(Y'(p))^*$ is the adjoint operator of $Y'(p)$ on $T_p M$ with respect to the Lorentzian inner product $g(p)[\cdot\,,\cdot]$.

Multiplying by $\dot{z}$ in (3) shows that there exists a real constant E_z such that $E_z = g(z(s))[\dot{z}(s), \dot{z}(s)]$. Then, as for geodesics, the causal character of a curve solution of (3) is well defined and the timelike solutions represent the world-lines of particles whose action is given by (1).

The interest to study the action functional (1) (or equivalently Eqs. (3)) comes from relativistic electrodynamics. In General Relativity an electromagnetic field is described by a closed 2-form Ω on M. The equation $d\Omega = 0$. represent the Maxwell equations of the electromagnetism (see [5,6]). Whenever Ω is exact (it always occurs if second De Rham cohomology group $H^2(M)$ is trivial as in the Minkowski spacetime of Special Relativity), there exists a 1-form ω on M such that $\Omega = d\omega$. By the metric equivalence between vector fields and 1-forms on a space-time, there exists a vector field Y on M such that $\omega(p)[v] = g(p)[Y(p), v]$, for any $p \in M$ and $v \in T_p M$. The trajectories of a relativistic massive and charged particle under the action of the electromagnetic field Ω satisfies (whenever Ω is exact) the nonlinear system of second order differential equations (3). The curves solutions of (3) satisfy a variational principle since they can be characterized as the stationary points of the action functional (1).

We outline that the functional (1) (and Eqs. (3)) are *gauge invariant*, that is if $\Phi\colon M \longrightarrow \mathbf{R}$ is a smooth function on M and $\nabla\Phi$ is the gradient of Φ with respect to the metric g. Set $Y_\Phi = Y + \nabla\Phi$ and set

$$F_\Phi(z) = \frac{1}{2}\int_0^1 g(z)[\dot{z}, \dot{z}]\,ds + \int_0^1 g(z)[Y_\Phi(z), \dot{z}]\,ds.$$

Then the Euler–Lagrange equations of F_Φ are the same as F.

In [7] it is shown that if $\bar{z}(s)$ is a *timelike* solution of (3), after a change of variable (which is possible if the Lorentzian manifold admit a global time coordinate), a suitable reparameterization of $\bar{z}$ is a stationary point of the functional

$$\mathcal{S}(z) = \int_a^b \sqrt{-g(z(s))[\dot{z}(s), \dot{z}(s)]}\,ds + \int_a^b g(z(s))[\dot{z}(s), Y(s)]\,ds.$$

In the literature the functional $\mathcal{S}$ is often used instead of the functional (1) to describe the motion of a massive and charged particle (see for instance [5]). However the

Euler–Lagrange equations for the timelike solutions of S are the same (up to a reparameterization) as for the the functional (1). Therefore functionals S and F are equivalent in the search of timelike solutions. It is interesting to point out that the stationarity properties of the trajectories of the particle allow to define the *mass* and the *charge* of the particle. We refer to [7] for more details.

In this note we present some recent results on the existence and the multiplicity of timelike solutions of (3) obtained using *Variational Methods in Global Nonlinear Analysis*. We set $I = [0, 1]$. Let M be a smooth connected n-dimensional manifold. Let $H^{1,2}(I, M)$ be the infinite dimensional Sobolev manifold of the curves on M having a $H^{1,2}$ regularity, i.e. they are continuous and the tangent vector field is square integrable with respect to some Riemannian metric on M (for details see for instance [8]. Moreover, let $p, q \in M$ and consider the set $\Omega^{1,2}(p, q; M)$ of the curves $z \in H^{1,2}(I, M)$ such that $z(0) = p$ and $z(1) = q$. It is well known [8] that the space $\Omega^{1,2}(p, q; M)$ has a structure of infinite dimensional submanifold of $H^{1,2}(p, q; M)$.

We consider now a Lorentzian manifold (M, g) and a smooth vector field Y on M. Fix two points p and q on M and consider the functional $F(z)$ defined by (1) on the manifold $\Omega^{1,2}(p, q; M)$. Straightforward calculations show that F is smooth on $\Omega^{1,2}(p, q; M)$. For any $z \in \Omega^{1,2}(p, q; M)$ the first differential $F'(z)$: $T_z\Omega^{1,2}(p, q; M) \longrightarrow \mathbf{R}$ is given by (2).

Now, let $z \in \Omega^{1,2}(p, q; M)$ be a critical point of F, i.e. $F'(z) = 0$. An usual boot strap argument shows that z is a smooth curve. Integrating by parts in (2) gives that z is a solution of (3) and joining p and q. Moreover, multiplying by $\dot{z}$ we have that $g(z(s))[\dot{z}(s), \dot{z}(s)] = E_z$ is a real constant independent on s. Clearly only critical points z of F such that $E_z < 0$ (i.e. z is timelike) are physically interesting.

The search of timelike critical points of the functional F presents some difficulty. First of all, since the metric g is indefinite, the functional F is unbounded from below and from above, then minimization arguments cannot be directly applied to get a critical point for F. Indeed, any critical point of F is an infinite dimensional saddle point, in the sense that the Morse index and the Morse co-index are equal to $+\infty$. This fact makes difficult to apply also well-known results on the existence of critical points for unbounded functionals. Moreover, even if the causal character of a critical point z of F is well defined, it cannot be directly deduced by its critical value $F(z)$ (as in the geodesic problem). So in order to study the causal character of a solution, some estimates of the critical values of the functional are needed.

2 Standard stationary space-times

Recently, some results on the existence of timelike solutions of (3) have been obtained in [9] assuming that the space-time is *standard stationary* and the vector field Y is stationary with respect to the standard time coordinate. A space-time (M, g) is said *stationary* if it is endowed with a smooth, timelike *Killing* vector field W. A stationary space-time (M, g) is said to be *standard* if M is diffeomorphic to the product manifold $M_0 \times \mathbf{R}$ and the metric g has the following form:

$$g(z)[\zeta, \zeta'] = \langle \xi, \xi' \rangle_x + \langle \delta(x), \xi \rangle_x \tau' + \langle \delta(x), \xi' \rangle_x \tau - \beta(x)\tau\tau', \tag{4}$$

where $z = (x, t) \in M \equiv M_0 \times \mathbf{R}$, $\zeta = (\xi, \tau)$, $\zeta' = (\xi', \tau') \in T_z M \equiv T_x M_0 \times \mathbf{R}$, $\langle \cdot, \cdot \rangle_x$ is a Riemannian metric on M_0, δ is a smooth vector field on M_0 and β is a smooth positive scalar field on M_0. The vector field $W = (\mathbf{0}, 1)$ is a timelike Killing vector field on M. Moreover it is complete, because its integral curves are defined on the whole line $\mathbf{R}$. Notice that the smooth function $\mathcal{T}(x, t) = t$ is a time-function on M.

Now let Y be a smooth vector field on M and assume that Y is *stationary* with respect to W, that is $L_W Y = [W, Y] = 0$. Let $Y(z) \equiv Y(x, t) \equiv (A(x, t), \varphi(x, t))$, with $A(x, t) \in T_x M_0$ and $\varphi(x, t) \in \mathbf{R}$, then the condition $[W, Y] = 0$ is equivalent to require that $\partial_t A(x, t) = 0$ and $\partial_t \varphi(x, t) = 0$. So we have $Y(z) \equiv Y(x, t) \equiv Y(x) \equiv (A(x), \varphi(x))$. The spatial component $A(x)$ is called *vector potential*, while the time component $\varphi(x)$ is called *scalar potential*.

The functional F takes now the following form:

$$F(z) = F(x, t) = \int_0^1 F_1(x, \dot{x}) \, ds + \int_0^1 F_2(x, \dot{x}) \dot{t} \, ds - \frac{1}{2} \int_0^1 \beta(x) \dot{t}^2 \, ds, \quad (5)$$

where $F_1, F_2 : T M_0 \longrightarrow \mathbf{R}$ are defined in the following way ($T M_0$ denotes the tangent bundle of M_0): for any $(x, v) \in T_x M_0$,

$$F_1(x, v) = \frac{1}{2} \langle v, v \rangle + \langle A(x), v \rangle + \langle \delta(x), v \rangle \varphi(x),$$

$$F_2(x, v) = \langle \delta(x), v \rangle + \langle \delta(x), A(x) \rangle - \beta(x) \varphi(x).$$

We consider two arbitrary points $p = (x_0, t_0)$ and $q = (x_1, t_1)$ on M. Since the metric is stationary, it is not too restrictive to assume that $t_0 = 0$. We are interested to find timelike curves $z : [0, 1] \longrightarrow M$ joining p and q and solving equations (3). This problem has in general no solution. For instance, if p and q are *simultaneous* (i.e. $t_1 = 0$), there are no timelike curves joining p and q, because of the existence of a time function on a standard stationary Lorentzian manifold.

For this reason we consider the time coordinate of the end point q as a parameter. Let $T \in \mathbf{R}$ and consider the points $p = (x_0, 0)$ and $q_T = (x_1, T)$. The following result on the existence of a timelike curve joining p and q_T and satisfying (3) holds.

Theorem 1. *Assume that (M, g) is a standard stationary Lorentzian manifold and Y is a stationary vector field on M. Moreover assume that the Riemannian manifold $(M_0, \langle \cdot, \cdot \rangle_x)$ is complete and that there exist positive constants v, μ, λ, C, D such that*

$$0 < v \leq \beta(x) \leq \mu; \tag{6}$$

$$F_1(x, v) \geq \frac{1}{2} |v|^2 - \lambda |v|; \tag{7}$$

$$|F_2(x, v)| \leq C|v| + D, \tag{8}$$

for every $(x, v) \in T M_0$. Let $p = (x_0, 0)$ and $q_T = (x_1, T)$, then there exists $T_0 > 0$ such that for any $T \in \mathbf{R}$, $|T| > T_0$, there exists a timelike solution of (3) connecting $p = (x_0, 0)$ and $q_T = (x_1, T)$.

Assumptions (6)–(8) are satisfied for instance if M_0 is a compact manifold or more generally if the coefficients $\delta(x)$, $\beta(x)$, $A(x)$ and $\varphi(x)$ are bounded.

If the topology of the manifold is not trivial, the following multiplicity result holds.

Theorem 2. *Under the assumptions of Theorem 1, assume also that M is non contractible in itself. Let $N(T)$ the number of timelike solutions of (3) joining $p = (x_0, 0)$ and $q_T = (x_1, T)$, then*

$$\lim_{|T| \to \infty} N(T) = +\infty.$$

The theorems stated above are proved in [9]. The proofs are based on a *min–max reduction* which allows to study the solutions of (3) by means of a new functional which is bounded from below and satisfies the Palais–Smale condition under assumptions (6)–(8). Theorem 1 is proved by a minimum argument and by an estimate of the causal character of a minimum point depending on the parameter T. Theorem 2 is proved exploiting the topological richness of the manifold by means of a topological invariant, the *Ljusternik–Schnirelmann category* of the based loop space $\Omega(M)$ of the manifold M.

References

1. Beem, J.K., Ehrlich, P.E., Easley, K.L. (1996): Global Lorentzian Geometry. 2nd Edition. Marcel Dekker, New York–Basel
2. O'Neill, B. (1983): Semiriemannian Geometry with Applications to Relativity. Academic Press, New York
3. Hawking, S.W., Ellis, G.F.R. (1972): The Large Scale Structure of Space-Time. Cambridge University Press, Cambridge
4. Masiello, A. (2000): Applications of Calculus of Variations to General Relativity, in *Recent Developments in General Relativity, 13th Italian Conference on General Relativity and Gravitational Physics, Monopoli, Italy, September 18–22, 1998*, ed. by B. Casciaro, D. Fortunato, M. Francaviglia, A. Masiello. Springer-Verlag, Milano, pp. 173–195
5. Landau, L., Lifschitz, L. (1970): Théorie des Champs. Mir, Moscow
6. Thirring, W. (1992): Classical Mathematical Physics. Springer-Verlag, Berlin Heidelberg New York
7. Benci, V., Fortunato, D. (1998): Found. Phys. **26**, 333
8. Masiello, A. (1994): Variational Methods in Lorentzian Geometry. Pitman Research Notes in Mathematics **309** Longman, London
9. Caponio, E., Masiello, A. (2001): Non linear analysis, in press

Relativistic Astrophysics
and Cosmology

A Determination of any Physical Characteristics of Magnetic Monopole in Basis of Metrics of Curved Space-Time

H.L. Szőcs

Abstract. Using the Reissner–Nordström curved space-time metric type and the generalised Reissner–Nordström–Weyl metric, the radius, the mean density of Dirac's magnetic monopole and the ratio of the electron and magnetic monopole radii is evaluated and the stability of this particle is also studied.

1 The radius of magnetic monopole

We shall here calculate the charge and deduce the expression for the radius of the magnetic monopole [1,2] in function of different metrics. In this aim we used the following notations: e the charge of electron, g the charge of magnetic monopole, h the Planck's constant, $h*$ the relative uncertainty, G the gravitational constant, c the velocity of light in vacuum. At the same time we provide the transformation factor between the IS system, in which the previous values are given [3], and the Gauss-system. With this, we have:

$$e_{\text{cgs}} = e_{\text{is}}\frac{c}{10} \text{ and } g_{\text{cgs}} = g_{\text{is}}\frac{10^8}{c} \tag{1}$$

One passes from geometrized to physical units (indexed by "geom" and "ph", respectively) and conversely, by the formulae:

$$e_{\text{geom}} = e_{\text{ph}}\frac{G^{1/2}}{c^2} \text{ and } g_{\text{geom}} = g_{\text{ph}}\frac{G}{c^2} \tag{2}$$

So, using the Dirac's quantization relation [4]:

$$e \cdot g = n\frac{h}{2} \tag{3}$$

we obtain for $n = 1$ and for

$$e = 1.60221773335 \times 10^{-19}\,\text{A s} = 4.803206814 \times 10^{-10}\frac{\text{cm}^3 g}{s^2}$$
$$g = 2.067834616 \times 10^{-15}\text{V s} = 6.89755382 \times 10^{-18}(\text{cm}\cdot g)^{1/2} \tag{4}$$

The radii of the electron and the magnetic monopole have been obtained using the Reissner–Nordström metric [1,5]:

$$ds^2 = (1 - 2m/r + Q^2/r^2)dt^2$$
$$- (1 - 2m/r + Q^2/r^2)^{-1}dr^2 - r^2(d\theta^2 + \sin^2\theta d\Phi^2) \tag{5}$$

where r, θ, Φ are the spherical coordinates.

The components of the acceleration will be [5]:

$$a^r = 2(m/r^2 - Q^2/r^3),\, a^\theta = 0,\, a^\Phi = 0, \tag{6}$$

and the module of acceleration is:

$$a = |2(m/r^2 - Q^2/r^3)|(1 - 2m/r + Q^2/r^2)^{-1/2}, \tag{7}$$

where

$$Q^2 = Q_e^2 + Q_m^2. \tag{8}$$

Q_e being the electric, and Q_m the magnetic charge, respectively [5].

The elementary force (the "accretion") at surface is:

$$dF = a^r \cdot dm \tag{9}$$

where dm is an elementary mass.

This mass is in equilibrium (stable on the surface of spherical body) if $dF = 0$, namely $a^r = 0$, resulting

$$r = Q^2/m, \tag{10}$$

in geometrized units.

From (2) and (10) we deduce for the electron the radius:

$$r_e = \frac{e_{ph}^2/m_{ph}}{c^2} = 2.81794091 \times 10^{-13}\, \text{cm} \tag{11}$$

(the numerical value being very well into conformity with specific literature value [3])s, and for the magnetic monopole:

$$r_{g,cgs} = \frac{g_{ph}^2}{m_{ph}} \tag{12}$$

We have in the case of de Rújula-type of mass of monopole [2,6]:

$$m_g < 8.7\, n \cdot h\, \text{PeV}\, (10^6\ \text{GeV}) \tag{13}$$

(between $200 m_{proton}$ and m_{GUT}, respectively), the radius is the order of 10^{-19} cm.

We remark that the Planck-length $l_{Planck} = 1.616605(10) \times 10^{-33}$ cm is smaller than the radius of Planck-mass monopole only with three orders of magnitude, at the same time the mean-density of this monopole is smaller than the Planck-density of ten orders (the Planck-density being: $5.5174(57) \times 10^{93}$ g/cm^3).

2   The radius in basis of generalized Reissner–Nordström–Weyl (RNW) metric

The RNW metric we wrote in the following form [2,4,6]:

$$ds^2 = (1 - 2mG/r + Q^2G/r^2 + \beta Gr^2/3)dt^2$$
$$- (1 - 2mG/r + Q^2G/r^2 + \beta Gr^2/3)^{-1}dr^2 \qquad (14)$$
$$- r^2(d\theta^2 + \sin^2\theta d\Phi^2),$$

where β is proportional with the cosmological constant. In this case it results for radius of monopole the following equation:

$$\beta r^4 + 3mr - 3Q^2 = 0, \qquad (15)$$

where

$$\beta = \beta_{\rm ph}\left(\frac{(G/c^2)^3 \cdot h}{c}\right)^{1/2} = 7.42466 \times 10^{-58}\,{\rm cm}^{-2}$$

in geometrized units, with $\beta_{\rm ph} = 10^{-29}\,{\rm gcm}^{-3} = 10^{-47}\,{\rm GeV}$.

In this case computing the radius of monopole from (15) we come to the conclusion that the contribution of the cosmological constant is negligible (with relative error of order $10^{-115}\%$).

3   The black hole radius of magnetic monopole and this stability

The Schwarzschild-radius of a chargeless body is:

$$r_{Schw} = r_{grav} = 2m_{\rm geom} = 2m_{\rm ph}(G/c^2). \qquad (16)$$

Computing these radii in case of different masses for the monopole, we can establish that the Schwarzschild-radius is smaller than the radius of the particle, consequently the collapses is possible when we leave the charge out of consideration. But considering the electric and/or the magnetic charged particle, we obtain for the black hole radius in case of RN metric the following expression:

$$r_{\rm black-hole} = m_{\rm geom} + (m^2_{\rm geom} - Q^2_{\rm geom})^{1/2} < r_{\rm Schw}. \qquad (17)$$

Considering $r = Q^2_{\rm geom}/m_{\rm geom}$ we have:

$$r_{\rm black-hole} = m_{\rm geom} + m^{1/2}_{\rm geom}(m_{\rm geom} - r)^{1/2}, \qquad (18)$$

and the black hole type collapse is possible only:

$$m_{\rm geom} \geq r. \qquad (19)$$

For example considering the maximum mass of magnetic monopole as the Planck-mass, we have:

$$m^2_{\text{Planck}} = 2.611611138 \cdot 10^{-66} < 3.532187774 \cdot 10^{-64} = Q^2$$

and the black hole radius becomes imaginary and the black-hole type collapse is impossible.

It follows that the electric and/or magnetic charge impeds the black hole type collapse of a particle. In this case the black hole solution of problem does not exist.

Finally we analyse the radius of Schwinger-type dyon, resulting the following:

$$r_{\text{dyon}} = Q^2_{\text{geom}}/m_{\text{geom}} = (e^2_{\text{geom}} + g^2_{\text{geom}})/m_{\text{geom}} \approx g^2_{\text{geom}}/m_{\text{geom}} = r_{\text{mon}}$$

because of: $e^2_{\text{geom}} \ll g^2_{\text{geom}}$.

Consequently we conclude that the electric charge does not modify essentially the radius of dyon.

Appendix: The components of acceleration

The metric of space-time (outside) of spherically symmetrical central body can be written in generalized spherical coordinates as follows [4]:

$$ds^2 = g_{00}(dx^0)^2 + g_{11}(dx^1)^2 + g_{22}(dx^2)^2 + g_{33}(dx^3)^2, \tag{20}$$

where x^0 is the time-coordinate, while $x^1 = r, {}^2 = \theta, x^3 = \Phi$ are the space spherical coordinates. The three-dimensional components of the acceleration in geometrized units are:

$$a^i = -\Gamma^i_{00}/g_{00} \quad i = 1, 2, 3 \tag{21}$$

and the module of the acceleration is:

$$a^i = (a^i a^j \gamma^{ij})^{1/2}, \tag{22}$$

where:

$$\gamma_{ij} = -g_{ij} + g_{0i}g_{0j}/g_{00}, \quad i, j = 1, 2, 3. \tag{23}$$

are the compoents of three-dimensional metric tensor, and:

$$\Gamma^i_{nk} = g^{im}(g_{mk,n} + g_{mn,k} - g_{kn,m})/2 \tag{24}$$

are Cristoffel's symbols of the second kind [2,6]. In our case only the components:

$$\gamma_{11} = -g_{11}, \quad \gamma_{22} = -g_{22}, \quad \gamma_{33} = -g_{33}, \tag{25}$$

are nonzero, hence:

$$\Gamma^i_{00} = -\frac{g_{00,1}}{2g_{11}}. \tag{26}$$

In our case results the module of acceleration:

$$a = (a^r a^r \gamma_{11})^{1/2} = \frac{1}{2}(-g_{11})^{-1/2} \left| \frac{g_{00,1}}{g_{00}} \right|, \tag{27}$$

and from (21) and (25) it results:

$$a^i = -\frac{g_{00,1}}{2} = -\frac{1}{2}\frac{\partial g_{00}}{\partial r}, \tag{28}$$

and, finally:

$$a = |a^1 (g_{00})^{-1/2}|. \tag{29}$$

Acknowledgements. This work was sponsored by the Leybold Didactic GmbH Cologne-Budapest. Huba L. Szocs, University College J. Kodolanyi, 8000 Székesfe-hérvár, Hungary, Szabadságharcos Str. 59.

References

1. Szőcs, H.L. (1989): Univ. Napocensis, Fac. Math. Phys., Res. Sem. Preprint **5**, 275
2. Szőcs, H.L. (1993): Talk at International Láncos K Centenary Meeting. Székesfehérvár, Hungary
3. Taylor, B.N., Cohen, E.R. (1990): J. Res. National Inst. Standards Technol **95**, 497
4. Dirac, P.M. (1931): Proc. Roy. Soc. London A **133**
5. Sass, H.I.A. (1987): Univ. Napocensis, Fac. Math. Phys., Res. Sem. Preprint **3**, 49
6. Szőcs, H.L. (1994): Talk at International Conference on Chiral Perturbation Theory. Comenius University, Bratislava

Gravitational Waves
from Colored Spinning Cosmic Strings

R.J. Slagter

Abstract. We investigated the SU(2) Einstein–Yang–Mills system on a time-dependent non-diagonal cylindrical symmetric spacetime. From the numerical investigation, wave-like solutions are found, consistent with the familiar string-like features and dressed with gauge-charges.

1 Introduction

Cosmic defects, such as cosmic strings, are remnants of the symmetry-breaking process in grand unified theories in the very beginning of the universe. These topological defects could have played a substantial role in the structure formation, as is seen now on large cosmological scales. The observed anisotropy from COBE of the cosmic microwave sky, indicates that cosmic string-induced structures are preferred over the other defects. One also assumes that cosmological inflation, necessary to solve major problems in standard cosmological models, can be induced by cosmic defects. The accelerated expansion of the universe is driven by the energy of the false vacuum of the Higgs field. Furthermore, there is a profound connection between condensed matter physics and cosmic defects in the early universe [1]. For example, defects formation in second-order phase transitions in superfluid ^{4}He is described by a scalar field φ being the Bose condensate wave function and the Ginzburg–Landau model of superconductivity is described by a order-parameter φ representing the Cooperpair wave function. Especially the Aharonov–Bohm effect induced by spinning cosmic strings shows a striking analogy with the Iordanskii force acting on the vortex in superfluids. In elementary particles spontaneous symmetry breaking models, the φ-field is represented by the Higgs field. In cosmic string models, one usually starts with the Abelian Higgs model with Lagrangian $\mathcal{L} = \bar{D}_\mu \bar{\varphi} D^\mu \varphi - V(\varphi) - \frac{1}{4} F_{\mu\nu} F^{\mu\nu}$, where φ is a complex scalar field, $D_\mu = \partial_\mu - ieA_\mu$, $F_{\mu\nu} = \partial_\mu A_\nu - \partial_\nu A_\mu$ and A_μ a gauge vector field. In this model it is known that the space-time around (spinning) gauge strings could have, besides the usual angle deficit feature, exotic properties, such as the violation of causality, time-delay effects, helical structure of time and frame-dragging. Now it is believed that the Higgs fields describing spontaneous symmetry breaking are not fundamental and should be interpreted, instead, as phenomenological order parameters, like the magnetization vector for a ferromagnet or the Cooper pair wave function for a superconductor. A natural step is then to consider the non-Abelian Yang–Mills model and parameterize the YM potential by a 2D Abelian one-form, a 3D Higgs field and a matrix valued scalar. Taking the Yang–Mills model as the matter part of the equations of Einstein, one finds a rich spectrum of stationary regular and non-regular solutions. This in contrast with the vacuum solution and Einstein–Maxwell (EM) solutions, where the Schwarzschild and the Reissner–Nordström (RN) solutions are the only static black hole solutions. In the Einstein–Yang–Mills (EYM) theory one finds not only the embedding of the Abelian RN black hole, but in addition there are genuine non-Abelian coloured black holes. Their co-existence gives rise to the violation of the no-hair conjecture, since they carry the same (magnetic) charge. After the discovery by Bartnik and McKinnon (BMK) [2] of these non-Abelian particle-like and later the non-RN (hairy) black hole solutions of the SU(2) EYM theory, thorough instability investigations where done. It turns out that under spherically symmetric perturbations the BMK solution is unstable. The nth BMK solution has in fact $2n$ unstable modes, comparable with the flat space-time sphaleron solution. The solutions mentioned above need not be spherically symmetric. Since the pioneering work of Stachel [3], we know that stationary cylindrical symmetric rotating models possess intriguing features. First, the evolution of the "classical" cylindrical gravitational wave solution in the vacuum situation can be described in terms of two effects: the usual cylindrical reflections and a rotation of the polarization vector of the waves, an effect comparable with the Faraday rotation [4]. Secondly, the time-dependent solutions can be related to the cosmic-string solution interacting with gravitational waves [5]. This is quite surprising, because usually the cosmic string solution is found in the U(1)-gauge Einstein–Higgs system. It is remarkable that the cylindrical rotating vacuum solution possesses the same conical structure. Moreover, it is asymptotically flat and free of singularities, in contrast with the Higgs-model where singular behavior is encountered at finite distance from the core [6,7]. Also in the Einstein–Maxwell model (electrovac) comparable string-like features are obtained.

2   The non-Abelian string solution

Consider the Lagrangian of the SU(2) EYM system

$$S = \int d^4x \sqrt{-g}\left[\frac{\mathcal{R}}{16\pi G} - \frac{1}{4}\mathcal{F}^a_{\mu\nu}\mathcal{F}^{\mu\nu a}\right], \tag{1}$$

with the YM field strength $\mathcal{F}^a_{\mu\nu} = \partial_\mu A^a_\nu - \partial_\nu A^a_\mu + g\epsilon^{abc}A^b_\mu A^c_\nu$, g the gauge coupling constant, G Newton's constant, A^a_μ the gauge potential, and $\mathcal{R}$ the curvature scalar. The field equations then become $G_{\mu\nu} = 8\pi G T_{\mu\nu}$, $\mathcal{D}_\mu \mathcal{F}^{\mu\nu a} = 0$, with T the energy-momentum tensor $T_{\mu\nu} = \mathcal{F}^a_{\mu\lambda}\mathcal{F}^{\lambda a}_\nu - \frac{1}{4}g_{\mu\nu}\mathcal{F}^a_{\alpha\beta}\mathcal{F}^{\alpha\beta a}$, and $\mathcal{D}$ the gauge-covariant derivative, $\mathcal{D}_\alpha \mathcal{F}^a_{\mu\nu} \equiv \nabla_\alpha \mathcal{F}^a_{\mu\nu} + g\epsilon^{abc}A^b_\alpha \mathcal{F}^c_{\mu\nu}$.

Let us consider the cylindrical space time

$$ds^2 = -\Omega^2(dt^2 - dr^2) + f(dz + \omega d\varphi)^2 + \frac{r^2}{f}d\varphi^2, \tag{2}$$

where Ω, f and ω are functions of t and r. We will use a (2+1)+1 reduction scheme in order to obtain a suitable parameterization for the Yang–Mills potential

$$A_\mu = \Phi(dz + \omega d\varphi) + \Psi d\varphi + \mathbf{a}_a dx^a, \tag{3}$$

where **a** is the dynamical part of the YM field, parameterized by a 2D complex Abelian one-form, Φ a 3D Higgs field and Ψ a matrix-valued scalar. The YM-part of the Lagrangian can then be written as

$$\mathcal{L}_{\mathrm{YM}} = \Omega^2 r \left[\left\{ f_{ab} f^{ab} + \frac{2f}{r^2} (D_a \Psi + \Phi \partial_a \omega - a'_a)(D^a \Psi + \Phi \partial^a \omega - a'^a) \right\} \right. $$
$$\left. + \frac{2}{f} \left\{ D_a \Phi D^a \Phi + \frac{f}{r^2} (\Phi' + g[\Phi, \Psi])^2 \right\} \right],$$

$$(4)$$

with $D_a \Psi = \partial_a \Psi + g[a_a, \Psi] - k_a \Psi'$, f_{ab} the 2D field strength $f_{ab} = \partial_a a_b - \partial_b a_a + g[a_a, a_b] + a'_a k_b - a'_b k_a + \Phi \omega_{ab} + \Psi k_{ab}$, and a prime denoting the partial derivative with respect to φ.

Let us consider the following parameterization: $a_1 = A_2(t, r)\tau_\varphi, a_2 = A_1(t, r)\tau_\varphi$, $\Psi = W_1(t, r)\tau_r + (W_2(t, r) - 1)\tau_z$ and $\Phi = \Phi_1(t, r)\tau_r + \Phi_2(t, r)\tau_z$, with τ_i the usual set of orthonormal vectors. From the condition $T_{tt} - T_{rr} = 0$, we obtain

$$\partial_t A_1 = \partial_r A_2, \qquad W_2 = \frac{\Phi_2 W_1}{\Phi_1} + \frac{g-1}{g}. \qquad (5)$$

From the Einstein equations and the YM equations we obtain the set of partial differential equations for Ω, f, ω, Φ_1, Φ_2 and W_1. It follows from a combination of the YM equations, that A_1 and A_2 can be expressed in Φ_1 and Φ_2:

$$A_1 = \frac{\Phi_1 \partial_r \Phi_2 - \Phi_2 \partial_r \Phi_1}{g(\Phi_1^2 + \Phi_2^2)}, \qquad A_2 = \frac{\Phi_1 \partial_t \Phi_2 - \Phi_2 \partial_t \Phi_1}{g(\Phi_1^2 + \Phi_2^2)} \qquad (6)$$

For the gauge $\partial_t A_2 = \partial_r A_1$ and hence $\partial_t^2 A_i - \partial_r^2 A_i = 0$, we can solve the coupled set of equations.

3 Numerical solution

In the vacuum situation, the behavior of the cylindrical wave solution is well established. One can have reflection of incoming into outgoing waves and vice versa with the same polarization. Furthermore, there is a non-linear effect describing the interaction between the two different polarization modes. If an outgoing cylindrical wave is linearly polarized, its polarization vector rotates as it propagates ("Faraday" rotation). When a "Rosen"-pulse is emitted it turns out that $\Omega^2 f$ changes of some amount, indicating that the mass per unit length decreases. In our EYM case we solved the equations for different initial values. We have used a 500×500 grid and applied the method of lines with cubic Hermite polynomials. The roundoff error remained below 0.001. In Figs. 1 and 2 we have plotted two typical solutions, with slightly different initial values of the angular momentum ω. We find that there is reflection of the outgoing gravitational- and YM-waves, just as in the vacuum case. The solution remains regular, even for long-time runs. The solution depends on the

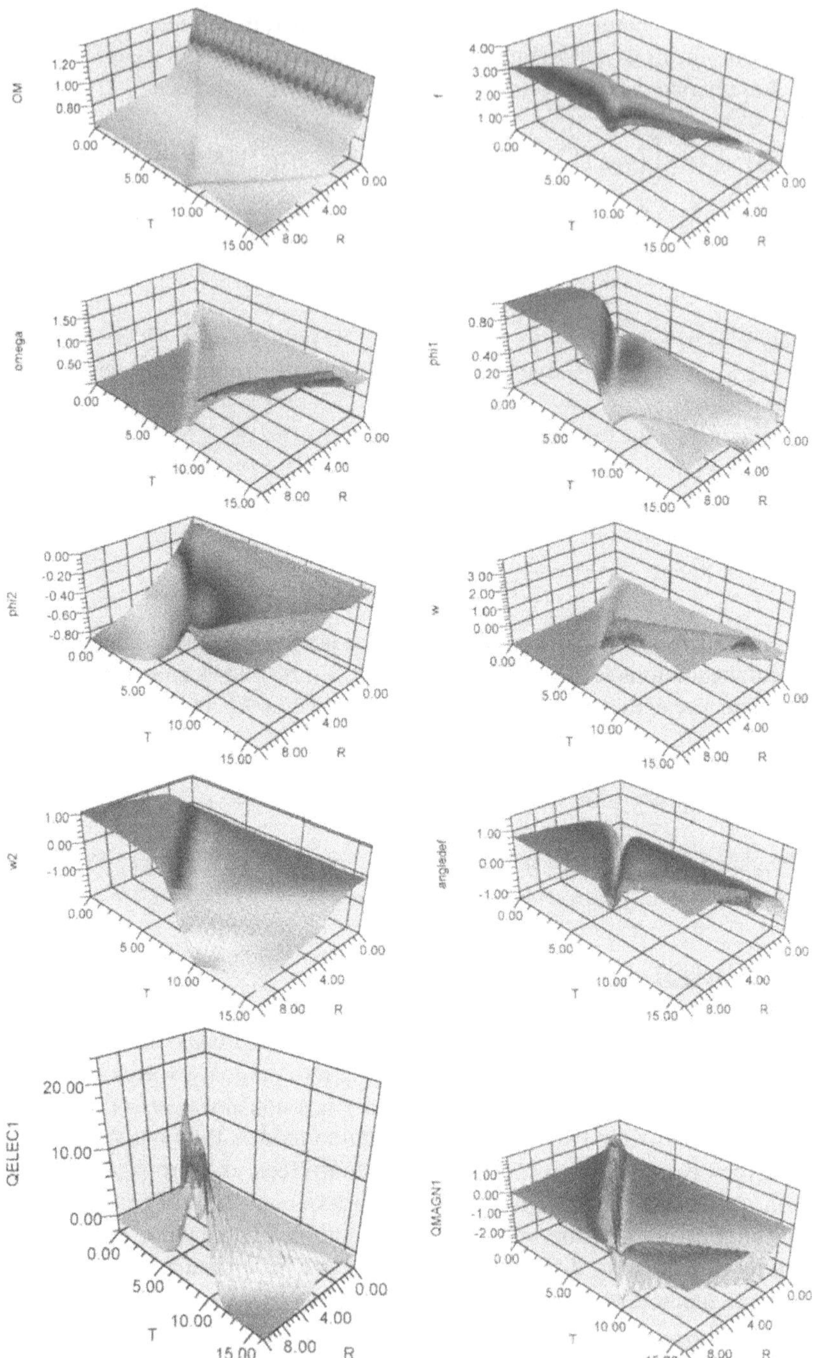

Fig. 1. Plot of a typical wave-like solution for some initial values. We also plotted the electric and magnetic charge integrant of Eq. (7) (first part)

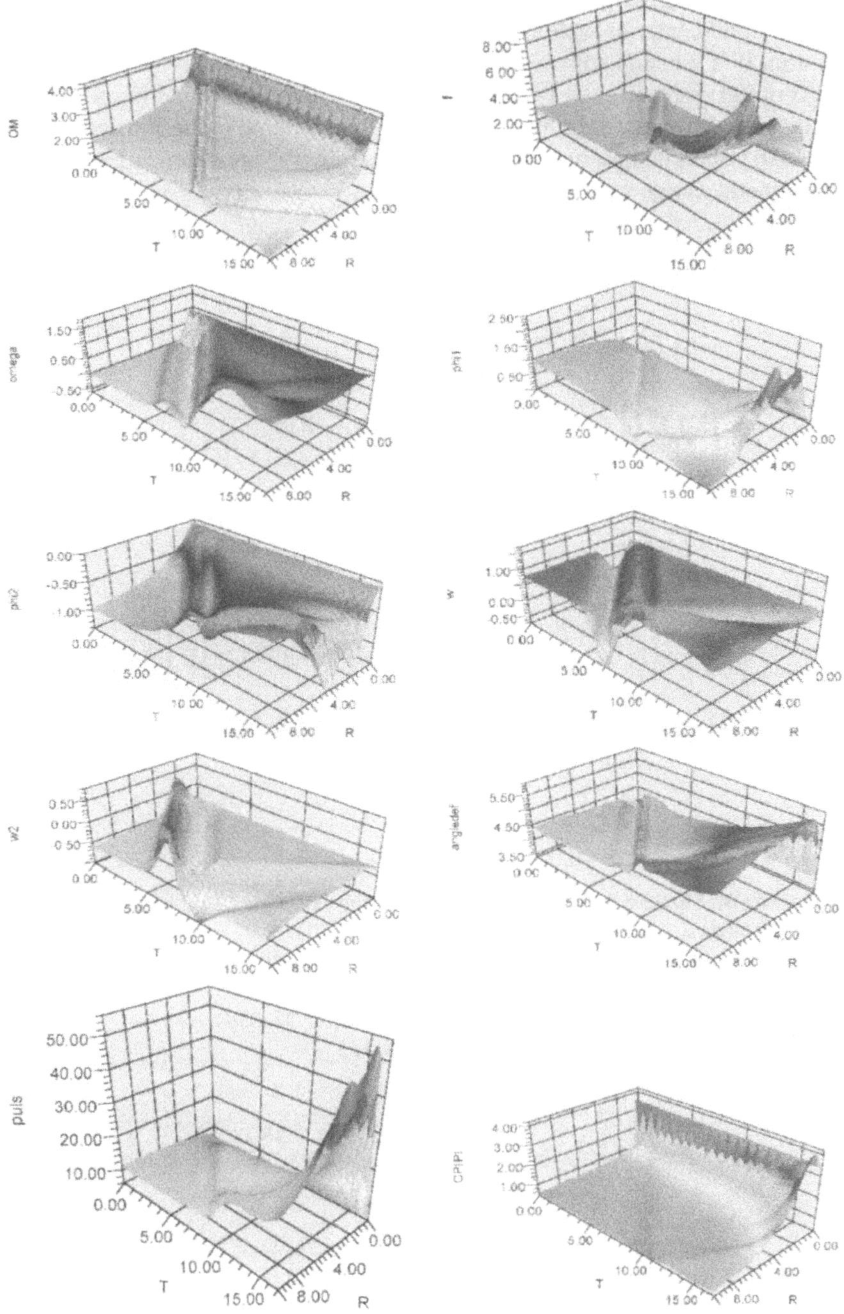

Fig. 2. As Fig. 1, but now for a slightly different initial ω. We also plotted the metric component $g_{\varphi\varphi}$ (second part)

initial mass per unit length of the string, just as in the vacuum case. The angle deficit changes significantly as well as the gravitational outgoing pulse, when we take for the initial W a function with an other node-number. So the Yang–Mills waves has an impact on the gravitational waves. The wave-like solutions found in this model are significantly different from the electrovac solutions found in the Einstein–Maxwell model. An important question is if the solution admits gauge-charges. For the electric and magnetic charge we obtain

$$Q_E = \int r \left\{ \left[\partial_t(W_1 + \omega\Phi_1) + \frac{gA_2\Phi_2}{\Phi_1}(W_1 + \omega\Phi_1) \right] \tau_r \right.$$
$$\left. + \left[\partial_t\left(\omega\Phi_2 + \frac{\Phi_2 W_1}{\Phi_1} \right) - gA_2(W_1 + \omega\Phi_1) \right] \tau_z \right\} dr.$$
$$Q_M = \int \frac{1}{r\Omega^2} \left\{ \left[\frac{(f^2\omega^2 + r^2)}{f}\left(\partial_r\Phi_1 + gA_1\Phi_2 \right) \right. \right.$$
$$\left. - f\omega\left(\partial_r(W_1 + \omega\Phi_1) + \frac{gA_1\Phi_2}{\Phi_1}(W_1 + \omega\Phi_1) \right) \right] \tau_r \tag{7}$$
$$+ \left[\frac{(f^2\omega^2 + r^2)}{f}\left(\partial_r\Phi_2 - gA_1\Phi_1 \right) \right.$$
$$\left. \left. - f\omega\left(\partial_r(\frac{W_1\Phi_2}{\Phi_1} + \omega\Phi_2) - gA_1(W_1 + \omega\Phi_1) \right) \right] \tau_z \right\} dr,$$

When we implement the integrant in the numerical code, we observe from Fig. 1 that the solution admits gauge charges.

References

1. Vilenkin, A., Shellard, E.P.S. (2000): Cosmic Strings and Other Topological Defects. Cambridge University Press, Cambridge
2. Bartnik, R., McKinnon, J. (1988): Phys. Rev. Lett. **61**, 141
3. Stachel, J. (1966): Math. Phys. **7**, 1321
4. Piran, T., Safier, P.N., Stark, R.F. (1985): Phys. Rev. D **32**, 3101
5. Xanthopoulos, B.C. (1986): Phys. Lett. B **178**, 163
6. Slagter, R.J. (1996): Phys. Rev. D **54**, 4873
7. Slagter, R.J. (1999): Phys. Rev. D **59**, 59025009

"Multiple Bending of Light Ray"
Can Create Many Images for One Galaxy:
In Our Dynamic Universe – A Computer Simulation⋆

S. Gupta

Abstract. The light from distant galaxies can take different paths. It can come into our solar system directly. It can go into free space. It can go grazingly about a star in our galaxy. In addition to this, the light can pass grazingly over a multiple number of stars in series, and finally reach our earth. Such a path need not be unique.

1 Introduction

The present work is an extension to the one presented in GRG 15 held in Pune by the author S. Gupta [1]. Here, in this computer simulation, two such paths are generated. Two rays start from a distant galaxy, and reach earth in different directions by following different paths. Both these paths will produce different images for the single galaxy at earth. These images are different in direction from that of the galaxy. Here it is simulated to say it is possible that the IMAGES are more than 90 degrees apart. Distance of the IMAGE 1 (Fig. 2) is more than 3 times and distance of the IMAGE 2 (Fig. 3) is more than 5 times the distance of the galaxy in their respective directions.

2 Process

Light is emitted in all directions from a galaxy. A thin cone of light may encounter a star in our galaxy. Many rays that bent near this star may go into space and some cone of light will reach some other star where it bends again and it will reach another star. This process continues till some light reaches our earth. This whole process is shown in Fig. 1: which we call as the process of multiple bending. Light bends from a straight path by a small angle α near a gravitational mass like the sun, as deduced by Einstein:

$$\alpha = (4gM)/(Rc^2), \tag{1}$$

where G is the gravitational constant. M is mass, R is radius of the star and c is the velocity of light. This gravitational deflection will happen always, for all those rays

---

⋆ This is a ground work to show the existence of two of such Subbarao paths. By using some real data one can detect some real images. Listing of stars in both the paths is available on request.

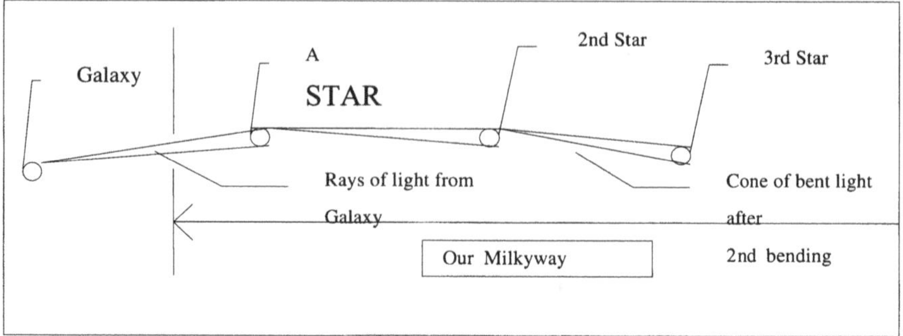

Fig. 1. Process of multiple bending

that pass tangential to a star or a heavy mass. The word "image" may mean it is not real, but I could not find any other word to replace it; it is the same galaxy seen in a different direction, seen in a different path. For an observer on earth, the light that comes to the earth without deflection will take a shorter time compared to light ray that bends at many stars and follows a longer path. The observer on earth sees both the images at the same time. Hence the image, that is formed due to the longer path must be an earlier photograph of the same galaxy. This path (Subbarao path) need not be unique. Many paths can co-exist together. In every path the observer on earth will be seeing a different image of the original galaxy. He can see these images in deferent directions simultaneously. All these are snapshots of the same galaxy taken at deferent times. They all will show a common structure of same galaxy.

3   Computer simulation

Our Milky-way is made of lumps of various dynamically moving masses of different sizes and parameters. We can not compare it with a model with a point mass located at the origin and other wise empty universe. There is no math model developed for our universe where we live in every day. The alternative is to simulate the Milky-way, part by part, using some standard model of the Milky-way like Caldwell and Ostriker [2].

In the first path, the angle of deflection was taken always towards the center of Milky-way. This resulted in stars that lie in a single plane. The second path was taken in three dimensions. We will find lots of dead stars, compact stars in the halo which do heavy bending for the second path.

Here in this simulation simple spherical coordinates are used. The center of Milky-way is taken as the origin. The direction of the sun from the center of galaxy is taken at a point co-ordinates with $\theta = 1.641111$ rad and $\varphi = -1.58244$ rad. The solar system assumed to be at a distance of 9176 pc. The dimensions of the Milky-way are taken as follows (as discussed by Caldwell and Ostriker [2]):

- Diameter of core $= 6\,\mathrm{pc}$;
- Distance of sun $= 9.1\,\mathrm{kpc}$;
- Diameter of halo $= 50.0\,\mathrm{kpc}$;
- Diameter of disk $= 60.0\,\mathrm{kpc}$;
- Thickness of disk $= 0.6\,\mathrm{kpc}$.
- The distance of the external galaxy is assumed to be $78\,\mathrm{Mpc}$; direction of this galaxy is assumed to be: $\theta = 2.677335\,\mathrm{rad}$ and $\varphi = 1.570796\,\mathrm{rad}$.

4 Algorithm

Let us take a ray of light starting from a distant galaxy which is tangential to the rim of the Milky-way. The stars in the path can be selected as follows.

Step 1 Take some incremental volume in spherical co ordinate $(dR, d\theta, d\varphi)$ from some base position $(R_{\min}, \theta_{\min}, \varphi_{\min})$.

Step 2 Find No. of stars by using a database of stars (created in such a way that density of stars in both disk and bulge decreases exponentially with increasing radius).

Step 3 No. of Stars to be simulated in this volume = Stars factor $\times$ density. Note a star factor $= 0.001$ was taken here. Total number of stars in Milky-way is approximated at 10^8.

Step 4 Simulated star location are calculated, so that stars are located in the incremental volume.

Step 5 Search for suitable stars in that location; if no suitable star is found go to Step 1 or go to Step 6.

Step 6 Simulate mass and radius of star. Mass $\leq 1.2 \times M_{\odot}$; and $R_{\odot} \leq$ star Radius $\leq 10 R_{\odot}$.

Table 1. Caldwell and Ostriker Model of our Milkyway

Component	Type of Stars	Density	Mass $= 10^{10}$ $\times$ sun	Radius of half Mass	Mass % within R_0	Mass % contribution	Approx overall radius
Disc	pop1	falls exponentially	6	7 kpc	44%	3.5%	30 kpc
halo	pop2	falls as R^{-3}	6	7.4 kpc	33%	3.5 %	25 kpc
corona	unknown	falls as R^{-2}	160	80 kpc	23%	93%	not considered

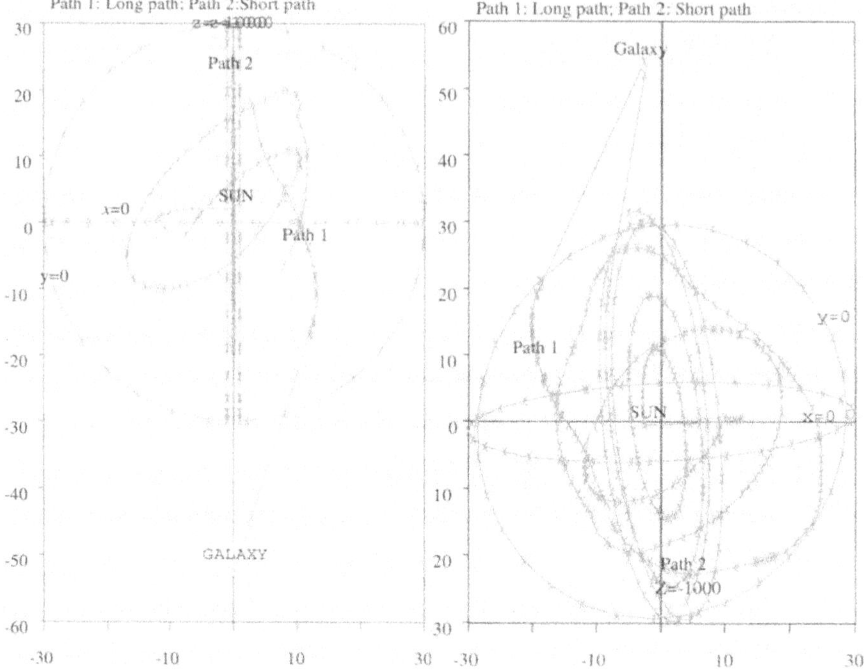

Fig. 2. Shows the paths in different projections. Path 1 is planar; Path 2 is 3-dimensional. **left:** 90 degree about y; **right:** another view. Axis in thousand parsecs

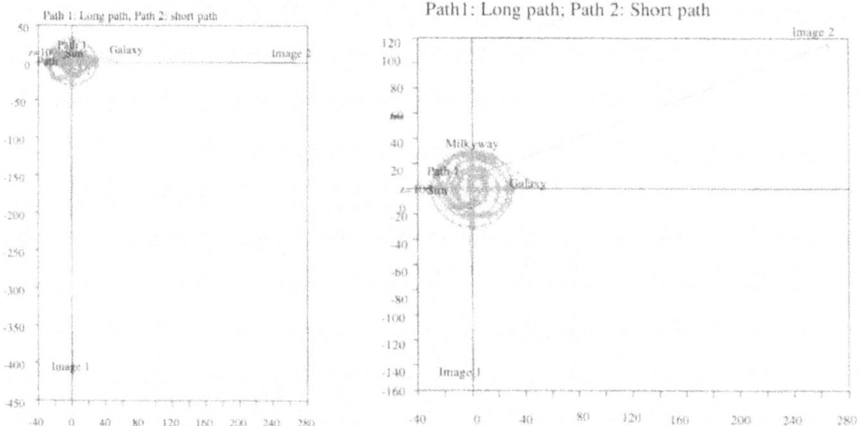

Fig. 3. Indicate the angle between the galaxy and its images . It is clearly visible that the angle so formed is larger than 90 degrees. Axis in thousand parsecs

Acknowledgements. This research work was continuously guided by THE ALMIGHTY VAK I acknowledge the continuous help provided by my spouse SAVITRI while writing this paper.

References

1. Gupta, S. (1997): Multiple bending of light ray in our dynamic universe: A computer simulation. Gr15: 15th International Conference on Gravitation and Relativity, 16–21 December 1995, Pune, India
2. Caldwell, S.A.R., Ostriker, J.P. (1981): Astroph. J. **251**, 61

Experimental Gravitation and Space Physics

Recent Improvements
on the EXPLORER Gravitational Wave Antenna

P. Astone, M. Bassan, P. Bonifazi, P. Carelli, M.G. Castellano, G. Cavallari,
E. Coccia, C. Cosmelli, S. D'Antonio, V. Fafone, Y. Minenkov, G. Modestino,
I. Modena, A. Moleti, G. Pizzella, G.V. Pallottino, R. Terenzi, G. Torrioli, M. Visco

Abstract. The EXPLORER gravitational wave detector has been recently improved. The antenna has been equipped with a new read-out. The use of a new transducer, characterized by a very small gap, and a dc-Squid with a high coupling, led to a better sensitivity and a larger bandwidth. In the first six months of the year 2000, an effective temperature of 3 mK and a bandwidth of almost 8 Hz were reached. Further improvements are expected. The new experimental configuration appears also very promising for use in the ultra-cryogenic antenna NAUTILUS.

1 Introduction

EXPLORER [1], installed at CERN Laboratories, is one of the two resonant gravitational detectors of the Rome group. The antenna is made of an high Q alloy Al 5056, has a mass $M = 2200$ kg and can be cooled in a cryostat by superfluid liquid helium at a temperature around 2.5 K. The antenna, equipped with a capacitive resonant transducer (mushroom shaped) and a planar geometry multiloop dc-SQUID as amplifier, had been working along the nineties (see Fig. 1), in coincidence with other detectors, with a good duty cycle. [2–5]. At the beginning of 1999, we decided to modify the detection apparatus to improve its sensitivity.

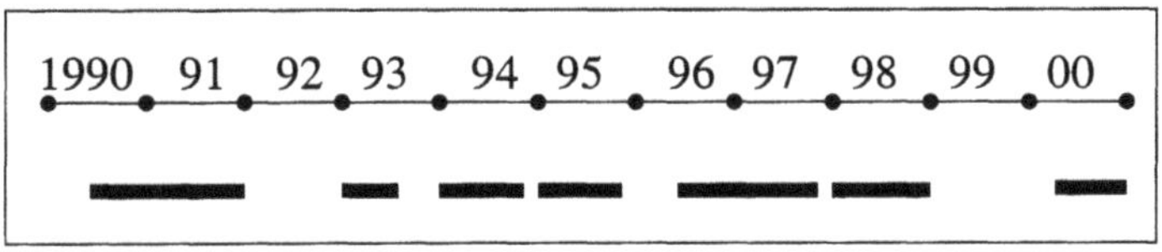

1990 91 92 93 94 95 96 97 98 99 00

Fig. 1. Operation during the past 10 years, most of the interruptions are due to cryogenic maintenance

2 Sensitivity of the detector

The two fundamental sources of noise that affect a gravitational wave (GW) detector and limit its sensitivity are the thermal noise associated with dissipation in the antenna and the electronic noise of the amplifier.

In the past years many efforts were made in reducing the termal noise in the bar, and new detectors, cooled at a thermodynamic temperature close to 100 mK [6,7], have begun operation. The most significant progress can be now achieved by improving the transduction-read-out scheme.

The sensitivity of bar detectors is usually expressed by means of the function of the frequency $S_h(\omega)$ that represents the contribution of the different sources of noise referred to the input of the detector as if they were a GW spectral density. If *δ-pulses* of gravitational waves are considered, the sensitivity can be calculated integrating $S_h(\omega)$ over the spectrum and the performances of a detector can be conveniently expressed by the so called effective temperature T_{eff}. This quantity, multiplied by the Boltzman constant, gives the minimum change of energy detectable in the antenna in case of burst signals.

The effective possibility of detecting a gravitational signal depends, of course, also on the quantity of energy released on the antenna by the gravitational wave. This energy is function only of the characteristics of the wave, direction, polarization and intensity, and of the geometrical parameters of the bar (mass distribution).

Let us consider an antenna resonating at a frequency $\omega_0/2\pi$ cooled at temperature T and with a mechanical quality factor Q, whose read-out circuit consists of a capacitive resonant transducer coupled, by means of a superconductor transformer, to a dc-SQUID, used as amplifier (Fig. 2). Under quite general conditions the effective temperature can be expressed in terms of the circuit parameters [8]:

$$T_{\text{eff}} \propto \frac{\Phi_n}{\alpha}\sqrt{\frac{\omega_0}{Q}}\sqrt{m_t \cdot T}, \tag{1}$$

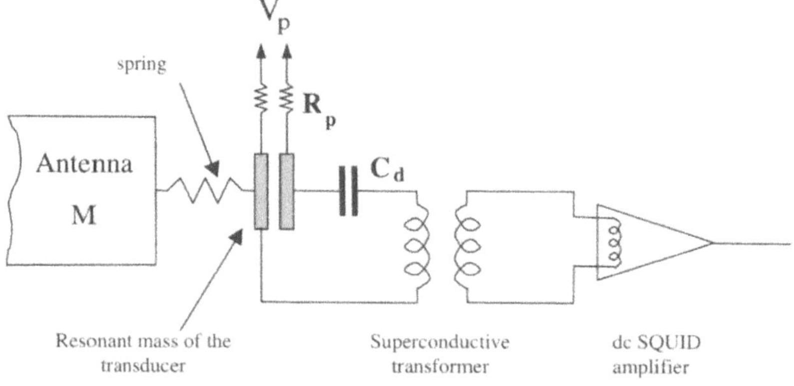

Fig. 2. Read-out circuit

where Φ_n is the SQUID noise, m_t the transducer mass, and α is a coupling term due both to transducer and to SQUID, that can be written as:

$$\alpha \propto C_t E N M_s, \tag{2}$$

where C_t is the transducer capacitance, N the transformer ratio, E the electrical field in the transducer and M_s the mutual inductance of the SQUID.

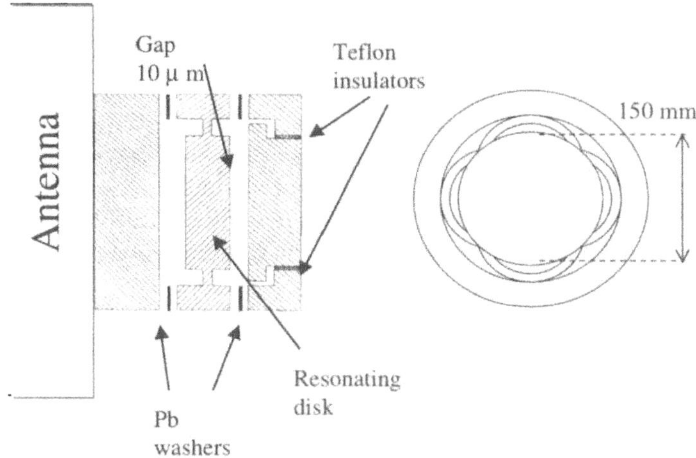

Fig. 3. New resonant capacitive transducer

With given values of the thermodynamic temperature T and mechanical Q, a lower effective temperature yields a larger band-width:

$$\frac{\Delta\omega}{\omega_0} = \frac{4}{Q} \frac{T}{T_{\text{eff}}}. \tag{3}$$

3 New experimental configuration

dc-SQUIDs have been used for many years as amplifiers in gravitational wave experiments. Nowadays dc-SQUID are available with a noise very close to quantum limit, at frequencies around 1 kHz. Much progress can be done improving the electromechanical transducer and its coupling to the SQUID.

Our group has developed a new capacitive transducer, with a very small gap, that has been tested for the first time on a detector. The new transducer [9] has a gap as small as $10\,\mu$m and a capacitance $C_t = 11\,$nF that is more then 3 times larger then in the transducers used in the past.

The dc-SQUID is a commercial device produced by Quantum Design, it has an input flux noise Φ_n comparable to that measured with the SQUID previously used, but the input coil mutual inductance $M_s = 10\,\text{nH}$ is about 3 times better then in the past.

The overall expected improvement of the coupling calculated using (2) is about 10. This corresponds to an effective temperature of the order of 1 mK and a bandwidth of the order of 10 Hz. This target was considered particularly challenging, as it would represent a tenfold increase in bandwidth with respect to all existing resonant detectors.

During the detector stop, needed to modify the read-out system, some significant improvements were also made on the vibration insolation apparatus in order not to be limited by environmental noise. Both the room temperature part and the low temperature part of the mechanical suspension were modified.

The low temperature suspension consists of a double stage with two rings as intermediate masses. The four separate titanium cables that suspended one intermediate mass were substituted by two U shaped cables wrapped around it, that led to a better distribution of the loads and consequently to a better attenuation. The antenna is suspended, like in the past, to the intermediate ring by a titanium cable wrapped around the bar central section.

The room temperature filters are composed of three stages, obtained by metallic discs separated by rubber pads. The new adopted design aims at giving a better stability and an higher attenuation than before.

4 Performance during year 2000 and perspectives

The EXPLORER new run has started at the beginning of year 2000 and several months were necessary to fully understand the new features and tune the detector.

During the first months of operation, the antenna was affected by extra mechanical noise, probably due to the boil-off of the cryogenic liquids and for most of the time the performances were strongly compromised (from the end of November much more stable performances were obtained by modifying the cryogenic conditions).

In Fig. 4 the plot of the $S_h(\omega)$ relative to 2 hours of data on June 2000 is reported. The frequencies of the two modes of oscillation of the antenna-transducer system, with a biasing field $E = 3\,\text{MV/m}$, are $\nu_- = 905.79\,\text{Hz}$ and $\nu_+ = 929.02\,\text{Hz}$. The mechanical quality factors are respectively $Q_- = 4 \times 10^5$ and $Q_+ = 2 \times 10^5$.

The bandwidth is about 8 Hz with an improvement of about one order of magnitude with respect to the previous value.

The performances obtained with the new transduction scheme on EXPLORER are encouraging, better results can be reached increasing the biasing field in the transducer to $7 \div 9\,\text{MV/m}$.

Further improvements in sensitivity are expected once a similar experimental configuration is used with our ultra-low temperature detector NAUTILUS.

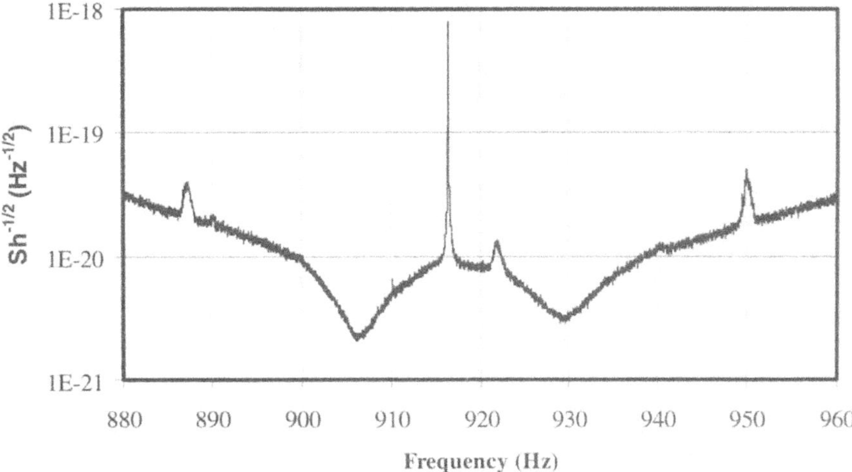

Fig. 4. $\sqrt{S_h}$ calculated over 2 hours of data on June 1, 2000. The large peak at 917 Hz is due to a monochromatic signal introduced to continuously monitor the electrical gain

References

1. Astone, P., Bassan, M. et al. (1993): Phys. Rev. D **47**, 362
2. Amaldi, E., Aguiar, O., Bassan, M. et al. (1989): Astron. Astrophys. **216**, 325
3. Astone, P., Bassan, M., Bonifazi, P. et al. (1999): Phys. Rev. D **59**, 122001
4. Astone, P., Bassan, M., Bonifazi, P. (1999): Astron. Astrophys. **351**, 811
5. Allen, Z.A., Astone, P., Baggio, L. (international Gravitational Event Collabroration (2000): Phys. Rev. Lett. **85**, 5046
6. Astone, P., Bassan, M., Bonifazi, P. (1997): Astroparticle Phys. **7**, 231
7. Prodi, G.A., Conti, L., Mezzena, R. (1998): Initial operation of the gravitational wave detector AURIGA, in *Second Edoardo Amaldi Conference on Gravitational Wave Experiments*, Geneve, Switzerland, 1997, ed. by E. Coccia, G. Veneziano, G. Pizzella, World Scientific, Singapore, pp. 148–158
8. Bassan, M., Pizzella, G.: Sensitivity of a Capacitive Transducer for Resonant Gravitational Wave Antennas. INFN Internal report LNF-95 / 064
9. Bassan, M., Minenkov, Y., Simonetti, R. (1997): Advances in Linear Transducers for Resonant Gravitational Wave Antennas, in *Proceedings of the Virgo Conference on Gravitational Waves: Sources and Detectors, Cascina*, March 1996, ed. by I. Ciufolini, F. Fidecaro, World Scientific, Singapore, pp. 225–228

NAUTILUS Recent Results

V. Fafone, P. Astone, M. Bassan, P. Bonifazi, P. Carelli, E. Coccia, S. D'Antonio,
G. Federici, A. Marini, Y. Minenkov, I. Modena, G. Modestino, A. Moleti,
G.V. Pallottino, G. Pizzella, L. Quintieri, A. Rocchi, F. Ronga, R. Terenzi,
M. Visco, L. Votano

Abstract. The resonant-mass gravitational wave detector NAUTILUS, operated by the ROG collaboration at the INFN Frascati National Laboratories, is in continuous data taking since June 1998 with peak sensitivity of $\tilde{h} \simeq 3 \cdot 10^{-22}\,\mathrm{Hz}^{-1/2}$. It has recently proven to be capable of recording signals due to the passage of cosmic rays. The results of the latest coincidence analyses between the NAUTILUS data and the cosmic ray signals measured by detectors located above and below NAUTILUS are reported.

The ultracryogenic gravitational wave (GW) detector NAUTILUS [1] has been operating in continuous data taking since June 1998. The detector consists of an aluminium alloy 2300-kg bar cooled at about 0.13 K, below the material superconducting transition temperature of 0.92 K. The read-out system is made by a resonant capacitive transducer, followed by a superconducting transformer and a dc SQUID. The bar and the resonant transducer form a coupled oscillator system with two resonant modes at 906.40 Hz and 921.95 Hz. The NAUTILUS data, recorded with a sampling time of 4.54 ms, are processed by a filter [2] optimized to detect impulsive signals applied to the bar.

The measured strain sensitivity at the two resonances is $\tilde{h} \simeq 3 \cdot 10^{-22}\,\mathrm{Hz}^{-1/2}$, with a bandwidth of about 1 Hz, and the best pulse sensitivity is of the order of $h \simeq 4 \cdot 10^{-19}$, with a duty cycle only limited by cryogenic operations.

NAUTILUS is also equipped with a cosmic ray (c.r.) detector system consisting of seven layers of streamer tubes for a total of 116 counters [3]. Three superimposed layers, each one with area of $36\,\mathrm{m}^2$, are located over the cryostat. Four superimposed layers are under the cryostat, each one with area of $16.5\,\mathrm{m}^2$. Each counter measures the charge, which is proportional to the number of particles. The detector is able to measure particle density up to $5000\,\mathrm{particles/m}^2$ without large saturation effects and it gives a rate of showers in good agreement with the expected number [3,4], as verified here using the up particle density, which is not affected by the interaction in the NAUTILUS detector.

The work initially done by Beron and Hofstander [5,6] Strini and Tagliaferri [7] and refined calculations by several authors [8–12] estimated the possible acoustic effects due to the passage of particles in a metallic bar. All these models agree in predicting, for the vibrational energy E of the excited fundamental mode of an aluminium cylindrical bar like NAUTILUS, the following formula:

$$E = 7.64 \cdot 10^{-9} W^2 f, \tag{1}$$

where E is expressed in kelvin degrees, W in GeV, is the energy delivered by the particle to the bar and f is a geometrical factor of the order of unity. The adopted mechanism assumes that the mechanical vibrations originate from the local thermal expansion caused by the warming up due to the energy lost by the particles crossing the material. The above formula has been recently verified with an experiment at room temperature [13], using a small aluminium cylinder and an electron beam. We notice that the GW bar used as particle detector has characteristics very different from the usual particle detectors, because the usual detectors are sensitive only to ionisation losses.

The data regarding the vibrational energy of the bar have been correlated with the data obtained by the cosmic ray (c.r.) detector in the period October 1998 to January 1999.

We have performed two different analyses, looking for both low-energy and high-energy c.r. events. In the first analysis [14] we have made use of NAUTILUS data selection based on the theoretical expectations, proceeding as follows: a) for each c.r. event we have used 20.000 samples (for a total time of 90.8 s) centered at the time when the number of particles (due to the c.r. event) crossing the lower detector exceeded $M = 10^4$; b) the data stretches with noise temperature T_{eff} (obtained by averaging the filtered data over 6 minutes included the time of the c.r. event) larger than 5 mK were rejected, in order to select periods when the detector was properly working and the noise was of the order of the expected signals.

In this way we selected 93 stretches for $M \geq 10^4$ during a total time of 47.7 days. The analysis showed the presence of small signals correlated (above twenty standard deviations) with c.r. showers. These signals were correlated with the shower multiplicity, thus confirming, within a factor of three the experimental data and the calculations on the c.r. effect both in terms of rate of occurrance and in terms of amplitude of the c.r. interaction with a resonant detector.

After having performed this analysis we noted a very large NAUTILUS signal (a few Kelvin) in coincidence with a c.r. shower, so we decided to proceed to a new analysis on the following different criteria, in order to study the large signals. We consider antenna events defined as follows: we apply to the filtered data a threshold corresponding to signal to noise ratio SNR > 19.5, and for each threshold crossing we take the maximum value above threshold and its time of occurrence. These two quantities define the event of the GW detector. We wish to stress that we here consider only events with energy greater than about twenty times the noise, a procedure entirely different from that adopted in [14]. With this selection we collected 26466 NAUTILUS events and 94775 c.r. shower events for a total observation time of 83.4 days. The events produced by NAUTILUS were already posted on the WEB within the IGEC collaboration among the groups that operate resonant GW detectors [15].

We have determined a) the number of coincidences, using a time window of 0.5 s, as a function of the particle density of the c.r. events; b) the corresponding background of accidental coincidences estimated by performing one hundred time shifts of the NAUTILUS event times, in steps of 2 seconds. The result of the analysis, i.e. the number n_c of observed coincidences and the estimated number n of accidental coincidences versus the particle density is given in Fig. 1.

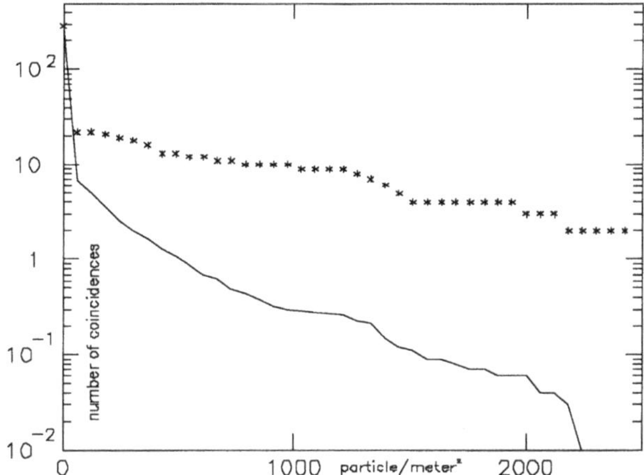

Fig. 1. Coincidences between the GW detector NAUTILUS and the c.r. detector. The asterisks show the integral number of observed coincidences versus the particle density observed by the c.r. counters located under the NAUTILUS cryostat. The continuous line shows the estimated number of accidental coincidences

Clear coincidence excess above background is found, when the showers have particle density large enough to give a signal in the bar. In particular, we noticed an unexpected extremely large NAUTILUS event in coincidence with a c.r. event, with energy $E = 57.89\,\mathrm{K}$. Both the up and down particle density of the c.r. detector are the largest ones in this case (2036 particles/m^2 in the upper detectors and 3556 particles/m^2 in the lower ones).

Using Eq. (1) we find that this NAUTILUS event requires that $W = 87\,\mathrm{TeV}$ of energy be released by the shower to the bar.

Let us discuss some important points:

- Using the down particle density we can calculate the energy of the NAUTILUS signals that we expect under the hypothesis the shower consists of electrons. In the previous work [14], finalised to the study of small signals, we had found that this energy is given by $E = \Lambda^2 \cdot 4.7 \cdot 10^{-10}\,\mathrm{K}$, where Λ is the number of particles in the bar. For the biggest event the above formula gives $E = 0.019\,\mathrm{K}$, that is more than three orders of magnitude smaller than the recorded 58 K. In the same way we calculate energies much smaller than those reported for most of the coincident events. Thus we conclude that most of the observed NAUTILUS events are not due to electromagnetic showers. On the contrary, when using the NAUTILUS measurements at zero time delay with energy of the order or below the noise and add them up at the cosmic ray trigger time, as done in the previous analysis [14], we find that the electromagnetic showers account for the energy observations

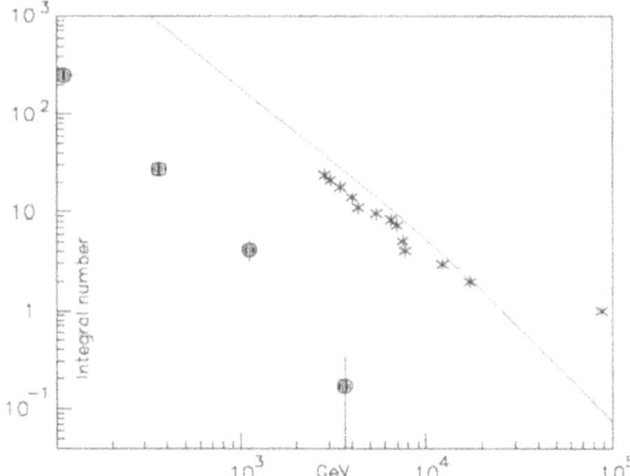

Fig. 2. Comparison between calculations and measurements. The asterisks indicate the integrated number of coincident events versus the energy delivered by the c.r. to the bar, expressed in GeV units, to be compared with the points having error bars, which give the number of events due to hadrons, we expect in the NAUTILUS bar. The dashed line is the experimental integral spectrum for the hadronic component of the showers, for the 83.4 days of observation

within a factor of three. For the previous result [14] the energy of the small signals is correlated with the c.r. particle density. Instead no correlation with the lower particle density is found for these large signals. This confirms the idea that the observed large events are not due to electromagnetic showers. In conclusions, the NAUTILUS signals are associated to two distinct families of c.r. showers. In one family the signals can be interpreted as due to the electromagnetic component of the showers, in the other family the known c.r. particles in the shower do not justify the amplitude or the rate of the observed signals.

- One must consider the possibility that the large events are due to the contribution of hadrons in the showers. Previous calculations have been made [4,16] on the frequency of both hadrons and multihadrons showers. The calculated values appear to disagree with our observation by more than an order of magnitude. Recently we have estimated the expected rate of hadronic events in the bar by means of new Monte Carlo calculations, using the CORSIKA package [17] with the QGSJET model for the hadronic interaction and simulating the NAUTILUS detector with the GEANT package. This is compared with the integrated number of the eighteen highest down particle density coincidences versus the NAUTILUS event energy. (The covered time periods are different for the various energy thresholds, which vary during the observations, depending on the noise. We have normalised the number of detected events to the total time of 83.4 days). Using Eq. (1) we can express the integral number in terms of the energy W delivered to the NAUTILUS bar by the cosmic rays. The result is shown in Fig. 2 where we report also recent measurements of the hadronic components of extensive air showers [18]. These measurements prove that the Monte Carlo calculations have been done correctly, since, because of the small diameter of the bar, we expect that only a few percent of the hadronic energy is absorbed by the bar, just as shown in Fig. 2. An immediate finding is that the highest energy event occurs in a time period more than one hundred times shorter than estimated under the hypothesis that the signals in the bar are due to hadrons. This big specific event could be explained as due to a large fluctuation, but we also notice a large disagreement between predicted and observed rates for all other events. Thus our observations exceed the expectation by one or two orders of magnitude.

- Some unexpected behaviour of NAUTILUS due to its superconducting state and to the transition to the normal state along the particle trajectories can be considered. These effects have been estimated [8,9] for type I superconductors (as aluminium). They are very small and cannot explain our observations if the shower includes only electromagnetic and hadronic particles. Thus, the unexpected behaviour of NAUTILUS should be relevant for the field of particle detectors based on the superconducting transitions. Another possibility is that the impact of a particle could trigger non-elastic audiofrequency vibrational modes with a much larger energy release. This has been already suggested [19,20] for the case of the interaction with gravitational waves, to explain cross-sections possibly higher than calculated. However, in this case, the agreement we have found for the small signals between experiment and calculation using formula 1 requires that the breaking of the model occur rather infrequently.

- Other possibilities to explain our observations must be considered, as anomalous composition of cosmic rays due to exotic particles (like nuclearities [21]). Previous search of these particles with resonant GW detectors gave upper limits [12,22] not inconsistent with the present experimental data. We remark that a resonant GW detector is able to observe particles with a mechanism different from the usual c.r. detectors.

Finally we remark that the presence of signals due to c.r. does not jeopardise a coincidence experiment with two or more GW detectors. Even without the use of veto systems employing c.r. detectors, the few dozen of events in a file, which includes thousand events, does not appreciably affect the number of accidental coincidences.

References

1. Astone, P. et al. (1997): Astropart. Phys. **7**, 231
2. Astone, P., Buttiglione, C., Frasca, S., Pallottino, G.V., Pizzella, G. (1997): Il Nuovo Cimento **20**, 9
3. Coccia, E., Marini, A., Mazzitelli, G , Modestino, G., Ricci, F., Ronga, F., Votano, L. (1995): Nucl. Instrum. Methods Phys. Res. Sect. A **335**, 624
4. Cocconi, G. (1961): In: *Encyclopedia of Physics*, ed. by S. Flugge, Vol. **46**, No. 1, p. 228
5. Beron, B.L., Hofstander, R. (1969): Phys. Rev. Lett. **23**, 184
6. Beron, B.L., Boughn, S.P., Hamilton, W.O., Hofstander, R., Tartin, T.W. (1970): IEEE Trans. Nucl. Sci. **17**, 65

7. Grassi Strini, A.M., Strini, G., Tagliaferri, G. (1980): J. Appl. Phys. **51**, 849
8. Allega, A.M., Cabibbo, N. (1983): Lett. Nuovo Cimento **83**, 263
9. Bernard, C., De Rujula, A., Lautrup, B. (1984): Nucl. Phys. B **242**, 93
10. De Rujula, A., Glashow, S.L. (1984): Nature **312**, 734
11. Amaldi, E., Pizzella, G. (1986): Nuovo Cimento **9**, 612
12. Liu, B., Barish, G. (1988): Phys. Rev. Lett. **61**, 271
13. van Albada, G.D. et al. (2000): Rev Sci Instrum. **71**,1345
14. Astone, P. et al. (2000): Phys. Rev. Lett. **84**, 14
15. Prodi, G. et al. (1999): In: Proc. of 4th Gravitational Wave Data Analysis Workshop (GWDAW 99), Rome, Italy, 2–4 December,
16. Chiang, J., Michelson, P., Price, J. (1992): Nucl. Instrum. Methods A **311**, 363
17. Heck, D. et al. (1998): In: Report FZKA 6019, Forschungszentrum Karlsruhe
18. Horandel, L.R. et al. (1999): In: ICRC Cosmic Ray Conference, Salt Lake City, Vol. **1**, p. 337
19. Desalvo, R. (1997): In: Proc. Amaldi Conference on Gravitational Waves, CERN, July 1997, ed. by E. Coccia, G. Pizzella, G. Veneziano, World Scientific
20. Fitzgerald, E.R. et al. (1974): Nature **252**, 638
21. Witten, E. (1984): Phys. Rev. D **30**, 272
22. Astone, P. et al. (1993): Phys. Rev. D **47**, 4770

The Low Frequency Facility,
R&D Experiment of the Virgo Project

A. Di Virgilio

Abstract. The Low Frequency Facility(LFF) experiment plans to make a measurement of the displacement thermal noise of two mirrors, suspended to the last stage of the VIRGO suspension. The expected displacement target sensitivity is 10^{-18} m/$\sqrt{\text{Hz}}$ at around 10 Hz, enough to measure outside resonance the thermal noise spectrum. We give the basic principle of the experiment and its progress report.

1 Introduction

Laser interferometric gravitational wave detectors aim to detect signals from several astrophysical sources in a band of frequency from around 10 Hz up to 10 kHz. The arm length of the laser interferometers range from 300 m up to 4 km, and the major projects are GEO (German-British), LIGO (USA), TAMA (Japanese) and VIRGO (French-Italian) [1]. A passing gravitational wave has the effect of changing the distance between the freely suspended end mirrors and the beam splitter, in phase opposition, thereby inducing a phase shift between the interfering beams in the two arms of the interferometer [2,3]. This causes a change in the output intensity of light, which has to be detected in the presence of several sources of noise among which are the intrinsic shot noise of the laser light, the radiation pressure induced fluctuation of the mirrors, the suspension noise, and the thermal noise of the mirror. The displacement sensitivity for the interferometers under construction is close to the state of the art in interferometry, that is roughly speaking around 10^{-19} m/$\sqrt{\text{Hz}}$. One of the major efforts at VIRGO is the improvement of the low frequency performance of the antenna, which would increase the sensitivity to spinning neutron stars. A large community is [4] working hard to improve the sensitivity of the antennas. LIGO plans to make an upgrade in five years from now. The effort is mainly concentrated on improving the suspension and keeping as low as possible the thermal noise of the pendulum and mirror substrate. With the aim of reducing thermal noise, several methods have been proposed, like cryogenic cooling, choice of low loss material for the test mass, as sapphire or silicon, and novel cancellation techniques. Since the low frequency noise of VIRGO [5], below 900 Hz and above 10 Hz, is almost entirely dominated by the suspension and test mass thermal noise, the improvement below 900 Hz implies changes in the suspension. In order to demonstrate that a new choice, once applied to the antenna, makes a real improvement in the sensitivity of the whole antenna it is necessary to make tests in conditions as close as possible to the real one. The Low Frequency Facility (LFF) [6,7], has been conceived in order to measure the noise of test masses attached to the SA, with a sensitivity at 10 Hz close to the Virgo one: 10^{-18} m/$\sqrt{\text{Hz}}$. The goal is to make a measurement of the thermal noise of the Virgo suspension, while the system is under active control of the low frequency motion. The thermal noise [8] comes from the suspension as well as from the mirror itself; but the sensitivity of the LFF at the beginning will be enough only to measure the thermal noise of the Virgo pendulum.

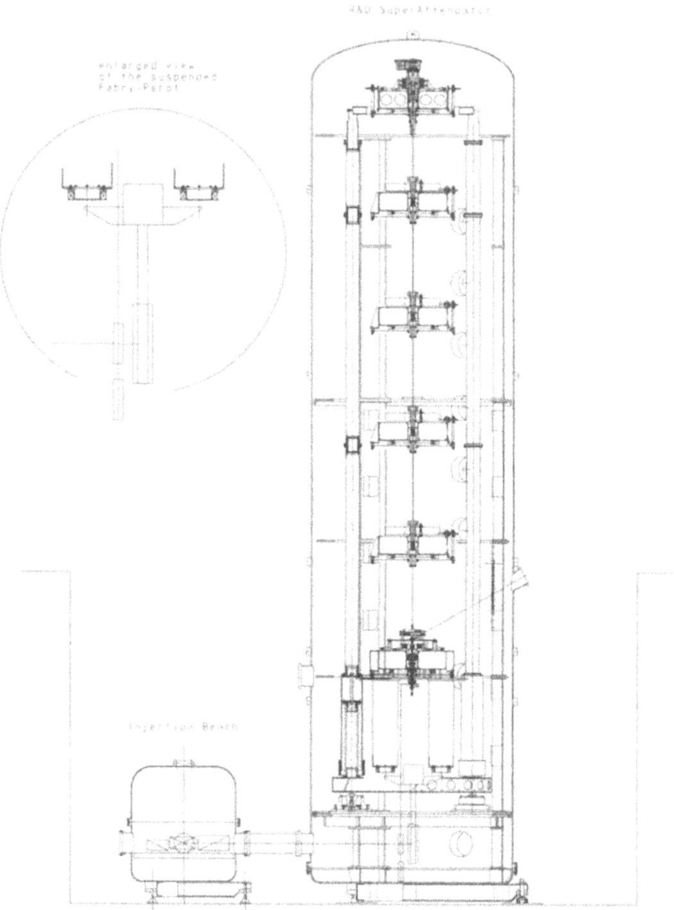

Fig. 1. The R&D SA and the injection table of the LFF, inside vacuum tanks; the enlarged picture shows the two mirrors forming a Fabry–Perot cavity

In this way, it will be possible to characterize the suspension in an environment close to the real one. We describe in Sect. 2 the experiment, in Sect. 3 the actual status and in Sect. 4 the conclusions.

2 Description of the experiment

As shown in Fig. 1, LFF consists of a SA, called R&D SA, which holds two mirrors instead of one only, the two mirrors define a Fabry–Perot cavity, the rest of the optical circuit lays on the optical table; the whole experiment is contained inside two vacuum tanks. Of the two suspended mirrors, one is the standard Virgo mirror, with a mass above 20 kg, while the other [9], usually called (AX), will be different, and can be changed and loaded with a variable mass in order to test the noise in different conditions, reduce the displacement noise due to low frequency radiation pressure and have evidence of creep events, increasing the suspension wire tension. The Fabry–Perot cavity will act as a displacement transducer and allow the measurement of the combined thermal noise of the two mirrors, and related suspension. The use of a Fabry–Perot cavity as a displacement transducer relies on the excellent frequency stabilization of the laser beam. A frequency stabilization, for a Nd-YAG laser, of the order of $10^{-1} - 10^{-2}$ Hz/$\sqrt{\text{Hz}}$ at 10 Hz has been obtain, for example by the Virgo team [10]. A copy of this circuit provides the frequency stabilization for the LFF, it is already operational, and has obtained a closed loop behavior similar to the Virgo prestabilization circuit. The technique to reduce the influence of power fluctuation is to extract signal from the Fabry–Perot with the well known Pound-Drever-Hall, where the injected light is phase modulated at high frequency (usually around 10 MHz, where the laser is shot noise limited). The signal is extracted in reflection, the noise is due to the shot noise, and the signal to noise ratio SNR for the cavity detuning $\delta \nu$ is:

$$\text{SNR} = \frac{P_{\text{in}} 8 J_0(m) J_1(m) F (1 - \frac{A_{\text{loss}} F}{\pi}) \frac{2L}{c} \delta \nu}{2\sqrt{\eta P_{DC} h \nu_0}}, \tag{1}$$

where P_{in} represents the incident power on the cavity, A_{loss} the total loss in the cavity during one round trip, P_{DC} is the DC term of the power reflected from the cavity, $\sqrt{2\eta P_{DC} h \nu}$ is the shot noise expression with η the quantum efficiency of the photodiode, h the Plank constant, ν_0 is the laser frequency, L the cavity length, F the finesse, c the speed of light, and J_0 and J_1 the Bessel functions, functions of the modulation depth m. The LFF cavity parameters are: Finesse 3000, $P_0 = 30$ mW, cavity length $L = 1$ cm. With the above parameters, the sensitivity at 10 Hz is limited by the frequency fluctuation, and a power spectrum density of 10^{-1} Hz/$\sqrt{\text{Hz}}$ is necessary in order to reach the displacement sensitivity of 10^{-18} m/$\sqrt{\text{Hz}}$ at 10 Hz.

Let us say few words about the suspension. Roughly speaking, above few Hz, the seismic noise is function of frequency ν and goes as ν^{-2}. In our site the measured spectrum is about $10^{-6} - 10^{-7}/\nu^2$ m/$\sqrt{\text{Hz}}$, in all degrees of freedom; this imply a factor at least 10^{10} of noise reduction at 10 Hz. The suspension is a complex multipendulum [11]. The suspension takes care of the seismic isolation and of the test mass control as well. All the control forces are done by means of coils and magnets acting between different stages of the SA, in order not to reintroduce seismic noise through the actuators. The control has two main parts: The damping and the mirror control [12,13]. The damping is a local control, which reduces the large motion of the

SA, the feed-backs are produced using accelerometer signals and electromagnetic actuators [12]. The mirror control acts directly on the test mass, through a special stage called marionetta, for this loop the interferometer signal is used. Due to the very high importance of the control, and due to the fact that the controls are active in frequency region of interest for the detection, the LFF has been designed keeping as much as possible the Virgo SA control tools.

It is expected that a relative measurement between the Virgo thermal noise spectrum and the AX thermal noise can lead to an understanding of dissipation processes in the AX suspension as well as in the VIRGO suspension, since the noise spectra simply add up. We have described in [9] a framework to carry out the computation of the thermal noise spectrum of the double suspension. The motion of the wire due to thermal noise (due to internal dissipation mechanisms like for example thermo-elastic damping) induces a motion in the AX, by virtue of the fact that the wires create stress forces and moments at the point where they are clamped to the masses. We use the well known fluctuation-dissipation theorem, which requires us to find the admittance between an applied force and the induced motion.

The parameters of the two suspended mirrors are the following: The Virgo mirror weight is about 25 kg, and it is done by a 2.5 cm diameter curved mirror (radius of curvature 10 cm) supported by a steel cylindrical frame, the AX mirror is a plane mirror of about 200 gr, diameter 10 cm and 2 cm width. The AX is loaded by a variable mass, ranging from 0.5 kg up to 5 kg. The sensitivity will be similar to the Virgo one: 10^{-18} m/$\sqrt{\text{Hz}}$; later on an improvement of a factor 100 will be necessary, to become the test bench area of future suspension improvement. We plan to reduce the cavity length to 1 mm, in fact since the sensitivity is dominated by the frequency jitters of the beam, a reduction by a factor ten of the cavity length means a factor ten gain in displacement sensitivity. In this case the cavity has two plane mirrors, and is not stable. We have studied this kind of cavity [14], and we are developing a very high frequency (100 Mhz) Pound-Drever-Hall detection scheme (work done in Naple by S. Solimeno and A. Porzio).

3   LFF status report

A prototype of the SA has been assembled in Pisa and has allowed to test the assembly procedure and to develop the control loop. It has been used to identify the low frequency resonance of the multipendulum and to give the first measurement of the seismic noise transfer function [15,16]. It has been used to study and set up the inertial damping, which has been already implemented in all the Virgo towers [12]. The test mass position with respect to the ground is measured by an array of 8 sensors called LVDT. They exhibit a large linearity region [17] (it depends on the design, but they can exhibit 1% linearity regions over 2 cm, with a displacement sensitivity better than 10^{-8} m/$\sqrt{\text{Hz}}$; in the present test the linearity region was about 2 mm). The LVDT array are positioned in front of the test mass, in a way to allow the spatial reconstruction of the position test mass center of mass. It has been used so far to measure transfer functions of the below 10 Hz and tune and check the suspension

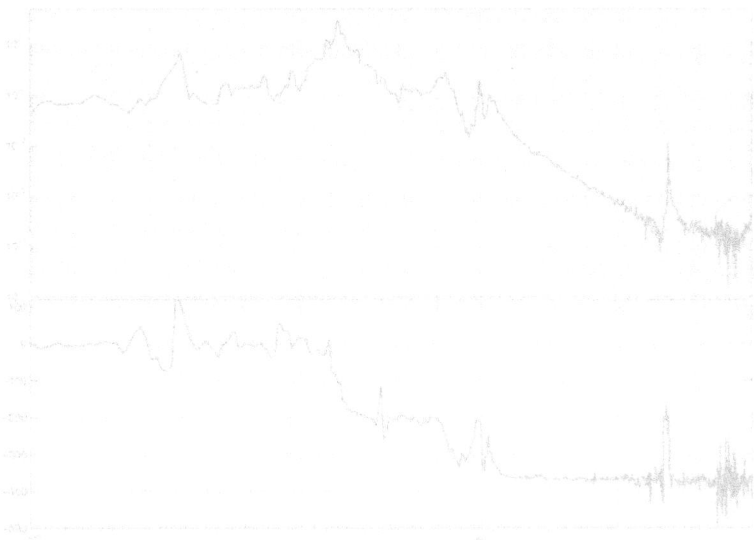

Fig. 2. Transfer function for the translation along the beam axis, magnitude and phase

simulation program. In Fig. 2, the transfer function current in the marionetta coils versus the translation along the beam axis is shown [18]. The test mass control from the last stage is a very important task, especially because we are interested on the low frequency region. Using the LVDT signals we have performed a test to lock the test mass, with respect to the ground, in the two most important degrees of freedom for LFF: The translation along optical beam, fundamental to keep tuned the cavity, and the rotation around the vertical axis, which gives the larger motion of the test mass [19]. The lock was done in the worse condition, with the suspension in air and without inertial damping, which reducing the r.m.s. motion of the test mass reduces the feedback required force, reducing the risk of saturation in the feedback. Figures 3 and 4 show the test mass displacement, with respect to the LVDT array, with and without feedback, for the degrees of freedom under examination.

All the above tests have been done with the only Virgo mirror mounted on the suspension. The AX suspension and the new LVDT array, with improved sensors, vacuum compatible and remotely adjustable, are going to be mounted. As far as the optics is concerned, the full optical lay-out is mounted on the optical table, with a geometry close to the real one, important to correctly tune the optical beam parameters. The frequency stabilization circuit is running, and tests of the control of the 1 cm cavity have been done using the final electronics, i.e. the Virgo digital electronics, based on a DSP. The injection vacuum tank and the injection table are ready; the table is a 1 m diameter steel disk, supported by vertical spring done by maragin blades, as the vertical spring of the SA.

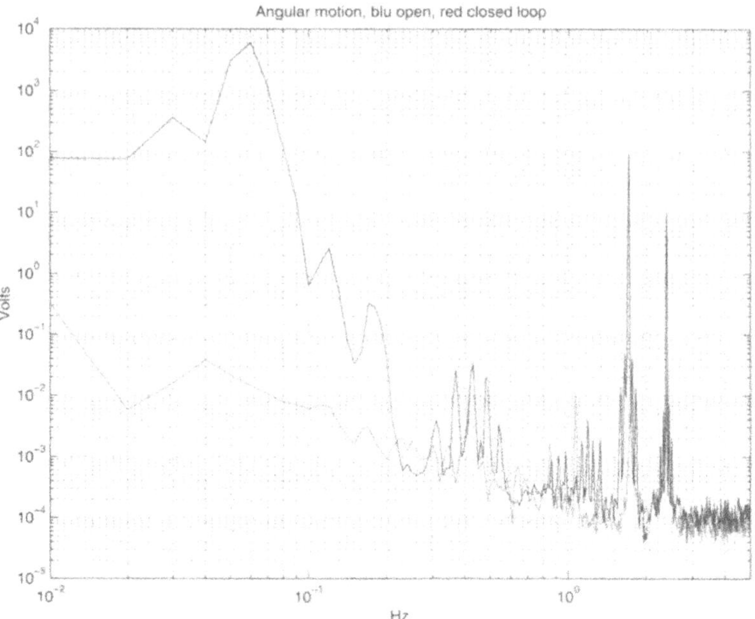

Fig. 3. Closed loop and open loop motion for the rotation around the vertical axis

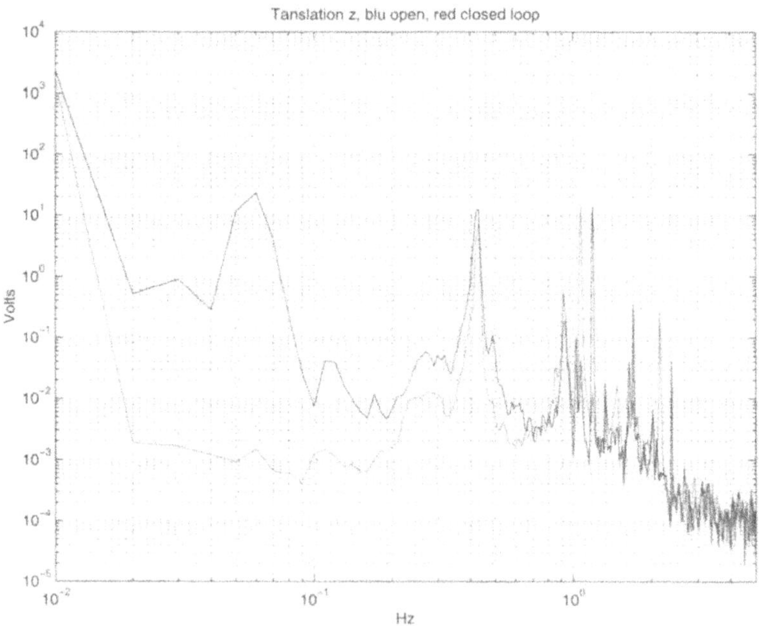

Fig. 4. Closed loop and open loop motion for the translation along the beam axis

4   Conclusions

The LFF aims the measure the thermal noise of pendulum and mirrors attached to a suspension called R&D SA, similar to the Virgo suspension. The two main parts of the experiment have been separately tested: the optical lay out has been assembled and the frequency stabilization circuit is ready, the R&D SA main controls have been developed. The experiment is going to be fully assembled before spring 2001, in this early phase only the R&D SA will be under vacuum.

References

1. Meshkov, S. (1999): Gravitational Waves, 3rd Edoardo Amaldi Conference, Caltec, July 1999, World Scientific, Singapore
2. Hello, P. (1998): Progress in Optics. **XXXVIII**, ed. by E. Wolf, Elsevier Science B.V.
3. Saulson, P. (1994): Fundamentals of interferometric gravitational wave detectors. World Scientific, Singapore,
4. Meshkov, S. (1999): Aspen winter Conference on gravitational waves and their detection, http://www.ligo.caltech.edu/
5. Gamaitoni, L., Kovalik, J., Punturo, M. (1997): The VIRGO sensitivity curve. VIRGO internal note, NOT-PER-1390-84, 30–3
6. Di Virgilio, A. et al. (1997): Virgo Note, VIR-NOT-PIS-8000-101, 15/04/97; Bernardini, M. et al. (1998): Phys. Lett. A **243**, 187–194
7. Benabid, F. et al. (2000): The low Frequency Facility, R&D experiment of the Virgo Project. J. Opt. B, Quantum Semiclass. Opt. **2**, 172–178
8. Bondu, F., Hello, P., Vinet, J.-Y. (1998): Phys Lett A **246**, 227–236; Levin, Yu (1998): PRD **57**, 659–663
9. Cella, G. et al. (2000): Suspension for the Low Frequency Facility. Phys. Lett. A, in press
10. Bondu, F. et al. (1996): Opt. Lett. **14**, 582
11. Braccini, S. et al. (1996): Rev.Sci.Instrum **67**, (8), 2899–2902; Beccaria, M. et al. (1997): Nucl. Instrum. Methods A **394**, 397–408; Braccini, S. et al. (1997): Rev. Sci. Instrum. **68** (10), 3904–3906; De Salvo, R. et al. (1999): Nucl.Instr. and Methos A **420**, 316–335; Losurdo, G. et al. (1999): Rev. Sci. Instrum. **70**, 5, 2507–2515
12. Losurdo, G. (2000): Active control of interferometric detectors of gravitational waves: Inertial damping of the Super-Attenuator. Urbino Scool, Italy, September 1999, World Scientific; gr-qc/00020006
13. Bernardini, A. et al. (1999): Rev. Sci. Instrum. **70**, 8, 3463–3471
14. La Penna, P. et al. (1999): Trasmittivity profile of high finesse plane parallel Fabry–Perot cavities illuminated by Gaussian Beams. Opt. Com. **162**, 267–279
15. Braccini, S. et al. (2001): Measurement of the Virgo SA performances for seismic noise suppression. Rev. Sci. Instrum., in press
16. Di Virgilio, A., Pasotti, M. (1999): Alignment method of the Filter7-Marionetta System. VIR-NOT-PIS-1390-150
17. Di Virgilio, A., Nucita, A. (1999): Design of the LVDT displacement transducers. VIR-NOT-PIS-1390-142
18. Vicere, A. (1999): Introduction to the mechanical simulation of seismic isolation systems. Urbino School, Italy, September
19. Di Virgilio, A. et al. (2001): Two degrees of freedom control of the R&D SA test mass. VIR-NOT-PIS-1390-167

The Newton's Gravitational Law

S. Focardi

Abstract. A geophysical experiment to check Newton's inverse square law is described. As a consequence, the value of the constant G at a 50 m effective distance is obtained. A project to measure G at a 1 m distance using a method different from those normally employed is also presented.

1 Introduction

The validity of Newton's inverse square law has been checked in the last fifteen years at distances of tens of meters. All these experiments had as an objective the search for particles other than the graviton coupling to masses. They were spurred by a paper of Fischbach et al. [1] consisting of a re-analysis of the Eötwös experiment [2]. In that paper, the existence of one or more non-Newtonian gravitational terms was suggested. We performed an experiment [3,4], which we will briefly describe in the next section and obtained a result which seemed to indicate the existence of a gravitational mediator having a mass different from zero. Recently, we re-analysed the previous result and reached the conclusion that Newton's inverse square law is verified at the 0.17 % level at a 50 m effective distance. Such a re-analysis is described in Sect. 3. The G value obtained by us at a 50 m distance is in agreement with the CODATA value [5]. In the last section the improvements of the superconducting gravimeter [6] used for the measurements are described. Such changes are essential to accomplish our programme which consists of measuring G at a 1 m distance and at a precision level comparable with the best existing results. Our method, which uses a superconducting gravimeter, is different from those traditionally employed and which make use of a torsion balance or torsion pendulum.

2 The previous experiment

Our experiment has two important characteristics:

- it is practically independent of the knowledge of the Earth's gravitational field
- it uses a superconducting gravimeter, which is the most precise instrument to measure changes of the gravity acceleration g.

  The g changes produced by the motion of large water masses were measured. In practice, the gravimeter was installed in a tunnel below a lake whose level changes during the day. This lake (Brasimone, 845 m above sea level, midway between Bologna and Florence, in Italy) is used by ENEL, together with Suviana lake as a power storage system. The experiment required very accurate measurements and the control of many physical parameters influencing the final result. The interested reader is referred to [3] for further details. Here I shall limit myself to a concise description. An aerophotogrammetric and a terrestrial photogrammetric survey of the lake shore were done when the lake was at its minimum level. This allowed us to know the Newtonian effect on the gravimeter with an accuracy of 1.4×10^{-4}. We continuously measured during two runs, for a total of 327 days, the value of g in the tunnel, the lake level, lake water temperature, atmospheric pressure, air temperature and humidity. The g changes measured in the tunnel are due to five different contributions: changing water masses, Moon and Sun tides, atmospheric pressure changes, instrument drifts and vertical gravimeter displacement in the Earth's gravitational field (produced by load changes).

Two auxiliary measurements were also needed. The tidal signal depends both on the direct Moon and Sun effects and on the Earth's elastic response. The first contribution is known and can be calculated in a given terrestrial point for every Moon and Sun position; the second one must be measured. For this reason, the gravimeter was run, for a period of about five months, in a laboratory in the same area but far away enough from the lake so as not to be influenced by its level variations. The analysis of the output signal permitted us to calculate, for every theoretical frequency component A_i of the tide, a gravimetric factor δ_i such that $A_{i,\exp} = \delta_i A_i$, which takes into account the Earth's elastic response. The tidal model so obtained [7] was used to compute the Moon and Sun effects in the tunnel. This run allowed us also to obtain the instrumental drift (2.709 ± 0.001 μ gal/day) and the pressure admittance (-0.201 ± 0.007 μ gal/mbar). In the same laboratory we calibrated the gravimeter. Such an operation was made by moving a precisely known annular mass (273.40 ± 0.01 kg) up and down around the instrument. The maximum Δg variation ($\Delta g_{\max} = 6.731$ μ gal) produces an output voltage change of about 100 mV. This two numbers were used to derive the calibration coefficient C, a parameter which multiplied by the voltage variations gives the corresponding acceleration changes expressed in μ gal. Since it was impossible, for reasons of space, to directly obtain the calibration coefficient in the tunnel, we adopted a transport procedure. This consisted in measuring both in the laboratory and in the tunnel the output voltage change induced by a known electrostatic force applied to the gravimeter sensor, obtaining as a result an electrostatic calibration coefficient. In the laboratory the relationship giving the calibration coefficient versus the electrostatic one was obtained experimentally. As a consequence the calibration coefficient in the tunnel was derived.

The data analysis consisted in comparing the theoretical (Newtonian) and the experimental effects associated with 10 cm lake level variations. Such changes of level, which occur in a period of the order of 15 minutes, allowed us to disregard any effect due to water temperature changes, to water table variations and to the lack of knowledge of the long period tidal components. For a water level variation of 7.376 m, the ratio R between the experimental effect and the theoretical one resulted $R = 1.0127 \pm 0.0013$. This value takes into account all corrections to the experimental data, in particular gravity changes due to the subsidence and uplift produced by the load variations on the gravimeter station in the tunnel. Such result could be explained only with the existence of a non Newtonian term.

3 The re-analysis

The above result convinced us of the importance of making any possible effort to check the conclusions achieved in the previous experiment. On having realised that the critical point was the calibration transport, we planned to repeat the experiment [4] by placing into the tunnel an absolute gravimeter [8]. The impossibility of having at our disposal such an instrument for the long time needed forced us to modify the original program.

The absolute gravimeter FG5 [8] of the Institut de Physique du Globe of Strasburg was installed in the Brasimone laboratory in order to calibrate our superconducting instrument. The FG5 was operated in runs of 25 drops with a drop every 15 seconds and 4 runs in one hour [9]. The two data sets coming from the absolute gravimeter and the superconducting one were compared after removing the drift. A linear regression between them allowed us to obtain the calibration coefficient of the superconducting gravimeter with an accuracy of 1.4×10^{-3}.

We used the signal output acquired by the superconducting gravimeter in the six month period centred around the FG5 calibration week in order to obtain amplitudes, gravimetric factors and phases for the main tidal species of the Tamura 1987 catalogue [10]. In particular we obtained the gravimetric factor δO_1 relative to the diurnal wave O_1 which has the important characteristic of being practically independent of the geographical position. We obtained, after correction for the oceanic loading, $\delta O_1 = 1.153(7)$ which is to be compared with the values of other European stations which range in the interval $1.152 \div 1.154$.

The set of phases so obtained was used to compute the tidal signal in other periods. Data collected over periods of six months were compared with the computed tide: a linear regression of the experimental data versus computed tide gave the best calibration coefficient for each data run. This coefficient permitted us to obtain for each period an amplitudes and phase data set together with the gravitational factor δO_1. The constancy of δO_1 in time is a good check for the method.

In the tunnel, the tidal signal was derived from that obtained in the laboratory for comparison with the FG5 instrument, taking into account the different position. Other components of the signal recorded in the tunnel are due to atmospheric pressure, lake effect, gravimeter drift and noise. The two important effects produced by lake and tides were separated by using a recursive method consisting of the following steps:

- an arbitrary calibration coefficient C is chosen;
- tide and pressure effect are subtracted by the signal;
- the residual noise is minimised by changing the lake admittance A ($A = \Delta g / \Delta h$) and the gravimeter drift;
- the process is repeated from the beginning, with a new C value.

At the end, the couple A, C which gives the minimum residual noise is chosen.

Starting from the C values relative to the two runs in the tunnel, the relative tidal analysis were performed from which we obtained the gravimetric factors δO_1. The obtained values agree perfectly with their values determined in the laboratory. We repeated the same data analysis of the previous experiment using as calibration coefficients the new C values but retaining all other data. We obtained for the ratio R the new value $R = 1.0123 \pm 0.0017$ confirming the correctness of the Newton's law.

Because of the particular shape of the lake, the admittance A is practically independent of the water level allowing us to obtain an independent determination of R. Starting from the admittance value, after correction for thermal variations of water density, we obtained for R the same result as reported above.

As a consequence from our experiment the value of the constant G is derived. It is $G = (6.688 \pm 0.011)10^{-11}\ \mathrm{Nm^2/kg^2}$.

4   The constant G

As pointed out by Gillies [11], due to the extreme weakness of gravity, the value of G is not known with the precision we would like a fundamental constant to have. Several recent measurements in this field are characterised by a good precision. However, some of them present such large deviations from the accepted mean value to suggest the increase by a factor of twelve of the error of the last CODATA [5] determination with respect of the previous one.

We are evidently in presence of systematic errors. Therefore, measurements of G based on methods different from those normally used can usefully contribute to the determination of this important coupling constant. Our method consists in measuring the gravity variations obtained by moving up and down around the superconducting gravimeter the well known annular mass used in the previous experiment and deriving the G value from these measurements. Obviously this requires a preliminary gravimeter calibration. Using our data we verified that the calibration coefficient C (the factor which multiplied by the output change in volts gives the corresponding value in μgals) and the factor δO_1 are strongly correlated: the relative variations of both parameters are equal. This conclusion was achieved by artificially changing the C value and computing the corresponding δO_1 factor. Practically this characteristic permits us to use δO_1 in order to derive C. As reported above, δO_1 is almost constant in Europe. Combining our δO_1 value with those measured in various European stations we get $\delta O_1 = 1.1537 \pm 0.00016$. The result shows an excellent agreement between data coming from several gravimeters, each of them subjected to various calibrations.

The annular mass will be moved up and down around the gravimeter at a frequency of the order of 3 cycles per hour. In this frequency region the instrumental noise is 0.135 ngal [12]. Since the gravity signal produced by our mass is 6.731 μgal, the ratio between the noise and the signal produced by the moving mass is 2×10^{-5}. As a consequence, the precision of this measurement is fixed by the knowledge of the gravitational effect of the annular mass, i. e. 9.5×10^{-5}.

The gravitational effect of the mass, thanks to the cylindrical symmetry of the annular system, is easily obtained on the axis. We have calculated the gravity changes versus the gravimeter sensor distance from the axis. If this distance is less than 1 mm the relative gravity change is less than 10^{-5}. We have also evaluated effects due to misalignement between the system axis and the vertical. For angles smaller than 0.3 mrad the relative change in gravity is smaller than 1.7×10^{-5}. Angles of this order can be easily controlled by means of bubble adjustment. The only condition is to use a very stiff guide-rail, which is also useful to avoid vertical and horizontal mass oscillations which in turn could produce oscillations in the gravimeter output signal.

The analysis reported above suggested us the changes needed for our gravimeter. The instrument is now at GWR in California where it was built for a complete renovation. In particular:

- Sensor change. The new generation GWR gravimeters are characterised by a drift of some μgal/year. We need as small a drift as possible to improve the quality level of the measurements.
- Addition of a system to detect with 1 millimetre precision, from outside and with the instrument at liquid helium temperature, of the horizontal position of the sensor.
- Displacement of the "verticality" controls on the instrument axis, in order to avoid influence of the mass position on the "verticality".

We estimate that G should be obtained with a precision of 1.7×10^{-4}, which is one order of magnitude better than the actual CODATA value.

References

1. Fischbach, E., Sudarsky, D., Szafer, A., Talmadge, C., Aronson, S.H. (1986): Phys. Rev. Lett. **56**, 3
2. Eötwös, R., Pekar, D., Fekete, E. (1922): Ann. Phys. **68**, 11
3. Achilli, V., Baldi, P., Casula, P., Errani, M., Focardi, S., Palmonari, F., Pedrielli, F. (1997): Nuovo Cim. B **112**, 775
4. Focardi, S. (2000): Testing Newton's Inverse Square Law, in *Recent Developments in General Relativity*, ed. by B. Casciaro, D. Fortunato, M. Francaviglia, A. Masiello, Springer, Milano, pp. 513–521
5. Particle Data Group (2000): Eur. Phys. Jour. C **15**, 73
6. Goodkind, J.M. (1999): Rev. Sci. Instr. **70**, 4131
7. Baldi, P., Casula, G., Focardi, S., Palmonari, F. (1995): Ann. Geofis. **38**, 161
8. Niebauer, T.M., Sasagawa, G.S., Faller, J.E., Hilt, R., Klopping, F. (1995): Metrologia **32**, 159
9. Baldi, P., Casula, G., Hinderer, J., Amalvict, M. (2000): Cahiers Centre Europeenne Geodynamique Seismologie **17**, 137
10. Tamura, Y. (1987): Bull. Inf. Marees Terrestres **99**, 6813
11. Gillies, G.T. (1997): Rep. Prog. Phys. **60**, 151
12. Richter, B., Wenzel, H.G., Zürn, W., Klopping, F. (1995): Phys. of the Earth and Plan. Interiors **91**, 131

Measurement of Gravitational Interaction on mm Scale

A.M. Panin

Abstract. An experiment with a Cavendish balance and $Dxh = 19x1.6$ mm lead disc masses withsurfaces separated by 0.2–0.3 mm revealed no noticeable deviation from the Newtonian inverse square law. The experiment rules out at least significant deviations from the inverse square law on a 1-mm scale.

1 Introduction

Recently [1–4] there were proposals in connection with the Planck scale hierarchy problem, that gravitons may actually propagate in $5 + 1$ (instead $3 + 1$) dimensions thus the inverse square law for gravitational interaction may not hold on the distances of order of 1 mm and less (it should switch to inverse 4th power). A preliminary experiment to test the inverse square law on 1-mm distance scale has been conducted. The results of the experiment and comparison with calculations are described below.

2 Experiment

The experimental set consisted of the Cavendish balance with 250 mm bronze tape suspension (tape cross section $10 \times 150\,\mu$m) with $D = 19$ mm lead discs of the thickness $h = 1.6$ mm and the mass $m = 5.1$ g each on $l = 100$ mm aluminum rod. The source masses were also similar lead discs $D = 19 \times 1.6$ mm placed at variable distance from the balance masses (see Fig. 1). The Cavendish balance had a flat mirror to monitor the angular position of the balance masses. A solid-state laser light is reflected by the balance mirror, and then back by non-moving mirror. After a second reflection by the balance mirror the light produces a 10-mm spot 20.4 m away, position of which was monitored with the accuracy ±1.5 mm thus giving the angular sensitivity of the balance in the 15 microradian range although long-term drift limited overall accuracy to $20 \div 30\,\mu$rad, (which is equivalent to $2 \div 3$ pN of gravitational force per each pair of masses).

The balance was operating in the air inside a glass/metal enclosure to protect from air currents and static. All metal parts including masses themselves were grounded. Ambient temperature was monitored during experiment to so that temperature changes were within $\pm\,0.3\,^\circ$C level.

The value of the gravitational force F was calculated by measuring the angle of turn of the balance. The torsion constant k of the bronze tape were measured by measuring the period of small torsion oscillations of the balance ($T = 320$ s). Although unharmonism was small, the angular amplitude of oscillations during measurements of the period was kept close to the amplitude of the turn caused by actual gravitational attraction between masses.

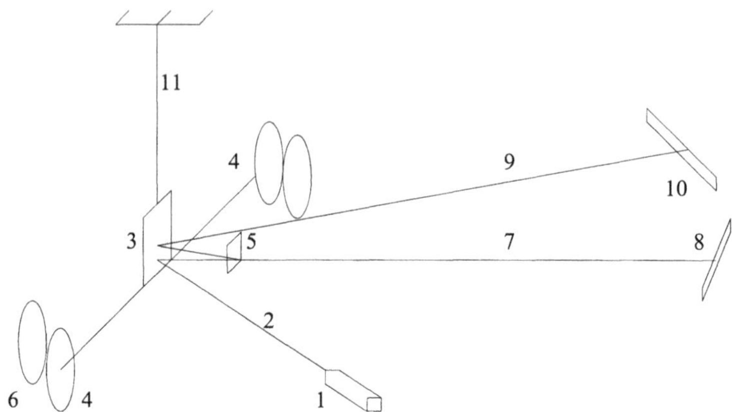

Fig. 1. Experimental set-up. 1. Laser (solid-state, wavelength 0.65 μm). 2. Laser beam. 3. Cavendish balance mirror. 4. Balance lead discs ($D = 19$ mm, $h = 1.6$ mm, $m = 5.1$ g). 5. Outer non-moving mirror (glass plate 12 mm away from the balance mirror). 6. Source masses ($D = 19$ mm, $h = 1.6$ mm, $m = 5.1$ g). 7. Single-reflected laser beam. 8. Monitoring scale 20.4 m away from balance. 9. Double-reflected laser beam. 10. Monitoring scale for double reflected beam (also 20.4 m away from the balance). 10. $0.01 \times 0.15 \times 250$ mm bronze suspension tape

3   Theory

To compare with the measurements, the gravitational force between two objects (1 and 2) was numerically calculated using the equation:

$$F = -G\rho_1\rho_2 \int\limits_{V_1} \int\limits_{V_2} \frac{dV_1\,dV_2}{r^2}, \tag{1}$$

where $G = 6.67 \times 10^{-11}$ Nm^2/kg^2 is the gravitational constant, ρ_1, ρ_2 are the densities of the source and balance masses. The main contribution ($> 96\%$) was from the gravitational attraction between the lead disc masses themselves. Nevertheless, the contribution from supporting Al rod was calculated as well.

For two parallel discs with the radius R and thickness h each separated by the distance d the force F_{z1} along their common axis was calculated as

$$F_{z1} = G\rho_1\rho_2 \int_d^{d+h} \int_{-\sqrt{R^2-x_1^2}}^{\sqrt{R^2-x_1^2}} \int_{-R}^{R} \int_{-h}^{0} \int_{-\sqrt{R^2-x^2}}^{\sqrt{R^2-x^2}} \int_{-R}^{R}$$

$$\frac{(z_1 - z)\, dx\, dy\, dz\, dx_1\, dy_1\, dz_1}{\left((x_1 - x)^2 + (y_1 - y)^2 + (z_1 - y)^2\right)^{3/2}} \tag{2}$$

The second (about 2% by magnitude) contribution was from the supporting Al bar:

$$F_{z2} = G\frac{2\rho_1\rho_2}{l} \int_d^{d+h} \int_{-\sqrt{R^2-x_1^2}}^{\sqrt{R^2-x_1^2}} \int_{-R}^{R} \int_{-\sqrt{R^2-y^2}}^{\sqrt{R^2-x^2}} \int_{-R}^{R_1} \int_0^{l}$$

$$\frac{x(z_1 - z)\, dx\, dy < dz\, dx_1\, dy_1\, dz_1}{\left((x_1 - x)^2 + (y_1 - y)^2 + (z_1 - y)^2\right)^{3/2}} \tag{3}$$

Here F_{zi} is the normal to the discs and supporting bar component of the gravitational force, ρ_i the densities of interacting bodies (2.69 g/cm^3 for Al and 11.4 g/cm^3 for Pb), $l = 100$ mm is the Al support bar length, $R_1 = 1.25$ mm is the bar radius.

Contribution from other parts was negligible (less than 1%).

4 Results of measurements and calculations

The torsion constant

$$k = \frac{\tau}{\Theta} = \frac{Fl}{\Theta}, \tag{4}$$

can be expressed via measured period of torsion oscillations

$$T = 2\pi\sqrt{\frac{I}{k}}, \tag{5}$$

where τ is torque of gravitational force F, l – balance bar length (equal to the distance between the centers of discs), I – the moment of inertia of the balance, Θ – angle of turn of the balance under the gravitational attraction. Substituting k from (4) into (5) and solving for F, we get:

$$F = \frac{4\pi^2 I\Theta}{T^2 l}. \tag{6}$$

On Fig. 2 the force F is plotted as a function of the distance between surfaces of lead discs. Solid curve corresponds to the calculated force $F_{\text{calc}} = F_{z1} + F_{z2}$.

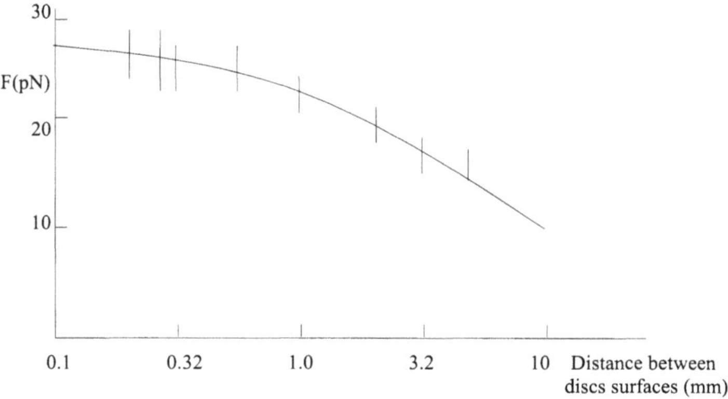

Fig. 2. Force versus distance between discs

5   Conclusions

The result of the experiment shows that there is no significant difference between calculated and measured gravitational force down to sub-mm distance between masses. At the minimal measured distance $d = 0.2$ mm about a quarter of disc masses is within 1 mm from each other. The overall accuracy of measurement is about $10 \div 15\%$, thus at least significant $(30 \div 40\%)$ deviations from the inverse square law could be ruled out on 1-mm distance.

References

1. Hamed, N.A., Dimopolous, S., Dvali, G. (1998): Phys. Lett. B **429**, 263–272
2. Hamed, N.A., Kaloper, N., Sundrum, R. (2000): Phys. Lett. B **480**, 193–199
3. Hamed, N.A., Dimopolous, S., Dvali, G., Kaloper, N. (2000): Phys. Rev. Lett. **84**, 586–589
4. Randall, L., Sundrum, R.: LANLonline ArXive; hep-ph/9905221, hep-th/9906064

Quantum Relativity
and Strings

Constraint Supersymmetry Breaking and Non-Perturbative Effects in String Theory

C. Kokorelis

Abstract. We discuss supersymmetry breaking mechanisms at the level of low energy $\mathcal{N} = 1$ effective heterotic superstring actions that exhibit $SL(2, Z)_T$ target space modular duality or $SL(2, Z)_S$ strong-weak coupling duality. The allowed superpotential forms use the assumption that the source of non-perturbative effects is not specified and as a result represent the most general parametrization of non-perturbative effects. The minimum values of the limits on the parameters in the superpotential may correspond to vacua with vanishing cosmological constant.

One of the biggest problems that heterotic string theory, and its "equivalents", e.g. type II, I, have to face today is the question of four dimensional $\mathcal{N} = 1$ space-time supersymmetry breaking. The breaking, due to the presence of the gravitino, that determines the scale of supersymmetry breaking, in the effective action, must be spontaneous and not explicit. Several mechanisms have been used in recent years to break consistently supersymmetry. They can distinguished as to when they are at work at the string theory level or at the effective superstring action level. The first category of mechanisms includes the tree level coordinate dependent compactification mechanism [1], the magnetized tori approach [2], the type I brane breaking [3–5], the partial breaking [6] while the latter category includes approaches that use target space duality e.g. [7–10] or S-duality [11,12] at the level of effective superstring action, related to gaugino condensation [13], to constrain the allowed superpotential forms. The main problem of the approaches is the creation of an appropriate potential for the moduli and the dilaton that can fix their vacuum expectation values. Because heterotic string theory has target space duality as one of its properties, we can use modular forms to parametrize the unknown non-perturbative dynamics [10], practically to parametrize the unknown non-perturbative contributions to the gauge kinetic function f.

The purpose of this paper is to reexamine the issue of constructing superpotentials W that affect supersymmetry breaking at the level of $\mathcal{N} = 1$ effective heterotic superstring actions when the source of non-perturbative effects, is not specified.

The modification of the $\mathcal{N} = 1$ heterotic effective action that we examine in this work amounts to modifying the superpotential when T-duality or S-duality non-perturbative effects are included. Moreover we want to break supersymmetry dynamically, rather than geometrically, thus we make use of gaugino condensation in supergravity [14].

In string theories physical quantities like masses of matter, Higgs fields depend on the moduli fields Φ. The latter fields have flat potential to all orders of string perturbation theory, so their vacuum expectation values remains undetermined.

In general there are two different approaches in describing gaugino condensation. These are the effective lagrangian approach [7,9], where we can use a gauge singlet bilinear superfield U as a dynamical degree of freedom [14,15] and the effective superpotential approach [8,10,16–18]. In the latter formalism the gauge singlet bilinear superfield is integrated out through its equation of motion.

The $\mathcal{N} = 1$ effective supergravity theory coming from superstrings is described by the knowledge of three functions, the Kähler potential K, the superpotential W and the gauge kinetic function f, that all depend on the moduli fields.

In general, duality stabilizes the potentials with local minima at the points T, $S = 1$, ρ. The minima in the cases considered in [10,11] are either at the self-dual points giving unbroken space-time supersymmetry in the T, S field sector or in the general case supersymmetry breaking minima with negative cosmological constant. In the latter case the minima occur at the boundary of the moduli space.

While considerable progess has been made to the understanding of non-perturbative effects based on D-branes, the problem of determining the moduli values still remains. In the absence of such a non-perturbative mechanism, we choose to break $\mathcal{N} = 1$ local SUSY, via gaugino condensation. We choose a four dimensional heterotic orbifold [19] with Kähler potential $K = -3\log(T + \bar{T})$, where T the complex structure modulus.

For (2, 2) heterotic string compactifications with $\mathcal{N} = 1$ supersymmetry there is at least one complex modulus T which we denote by $T = R^2 + ib$, where R is the breathing mode of the six dimensional internal space and b is the internal axion θ. The T-field corresponds, when T large, to the globally defined (1, 1) Kähler form. Here we will restrict our study to the simplest (2, 2) models where there is a single overall modulus, by freezing all other T-moduli.

The four dimensional string effective supergravity action is invariant to all orders of perturbation theory under the target space duality transformations

$$T \to \frac{AT - iB}{iCT + D}, \quad AD - BC = 1, \tag{1}$$

since $G = K + \log Y'W'^2$ has to remain invariant, while the superpotential transforms wit modular weight -3, $W \to (iCT + D)^{-3}W$. Local supersymmetry is spontaneously broken if the auxiliary fields $h^i = Y'W'^{K/2}(K^i + W^i/W)$ gets non-vanishing vacuum expectation values. If we choose the superpotential W in the, factorized, form $W(T, S) = \Omega(T)K(S)\eta^{-6}$, $K(S) = e^{-3S/2b}$ that is equivalent to defining the gauge kinetic function into the form

$$f = S - \frac{|G_i|}{|G|}(b_a^{N=2}\log[\eta^4(T)(T + \bar{T})] - \frac{1}{16\pi^2}\mathrm{Re}\{\partial_T \partial_U h^{(1)}(T, U)$$
$$- 2\log((j(T) - j(U))\}) + (b_a/3)\log|\Omega(T)|^2 + \mathcal{O}(e^{-S}), \tag{2}$$

where $h^{(1)}$ is the one-loop prepotential[1], G_i the orders of the subgroup G which leaves the i-complex plane unrotated and we have included the one-loop Green–Schwarz term. Because $\Omega(T)$ parametrizes our ignorance of non-perturbative corrections(from target space duality) for the T-modulus, it was assumed [10] that the T-modulus dependent part of the superpotential takes the form

$$W(T) = (j - 1728)^{m/2} j^{n/3} [\eta(T)]^{2r} \mathcal{P}(j(T)),$$
$$W(T) = \Omega(T)[\eta(T)]^{2r} \mathcal{P}(j(T)), \tag{3}$$
$$\Omega(T) = (j - 1728)^{m/2} j^{n/3},$$

where m, n positive integers and G_6, G_4 are the Eisenstein functions of modular weight six and four and $\mathcal{P}(j(T))$ an arbitrary polynomial of the absolutely $SL(2, Z)$ modular invariant $j(T)$. Analyzing the supergravity scalar potential corresponding to (3) we found negative cosmological constant at its minimum. The minima on the cases considered in [10,11] are either at the self-dual points giving unbroken space-time supersymmetry in the T, S field sector respectively or in the general case supersymmetry breaking minima with negative cosmological constant. In the latter case the minima occur at the boundary of the moduli space.

In addition, depending on the value of $(S + \bar{S})\partial K(S)/\partial S$ the dilaton minimum can be at weak or strong coupling [10]. However, because W is a section on a flat holomorphic bundle over the moduli space, the constraints

$$n \quad \text{mod } 3, \quad m \quad \text{mod } 2, \tag{4}$$

that have to be included apriori [12] in (3) under the dilatational transformations $T \to T + 1$ are missing. The construction of the superpotentials that incorporates apriori the constraints (5) is performed in [23] and treats not the η-invariant as a fundamental quantity, as in (3), bur rather the cusp forms $\Delta(T)$ instead. As a result, by writing down for a moduli field Φ the most general weight zero modular form

$$j^a(\Phi)(j - 1728)^b(\Phi), \tag{5}$$

where a, b are integers and j is the $SL(2, Z)$ modular function, (3) is generalized[2] and the following new T-modulus superpotentials are both allowed, namely

$$W(T) = \frac{\tilde{\Sigma}(T)}{\eta^6(T)} \mathcal{P}(j), \tag{6}$$

$$W(T) = \frac{1}{\eta^6(T)} \frac{1}{\tilde{\Sigma}(T)} \mathcal{P}(j). \tag{7}$$

---

[1] The one loop $\mathcal{N} = 2$ four dimensional vector multiplet prepotential $h^{(1)}$ was calculated as an ansatz solution to a differential equation involving one loop corrections to gauge coupling constant in [20]. However, its exact general form for any four dimensional compactification of the heterotic string was calculated in [21]. Higher derivatives of $h^{(1)}$ were calculated in [22].

[2] See for example [24].

Defining the candidate T-dual superpotentials into the forms (6), (7) is equivalent to demanding that do not allow or that we do allow poles in the upper half plane respectively. That happens because modular functions which are allowed to have poles in upper half plane are exactly rational functions in j (quotients of polynomials in j) whereas modular functions which are not allowed to have such poles are polynomials in j. Including the dilaton part $K(S)$ in (6), (7) results in the construction of the full superpotential. The superpotentials (6), (7) can be further constrained by examining stability constraints on the scaler potential at the self-dual points $T = 1$, $T = \rho = e^{i\pi/6}$. By employing the latter method we derive minimum values on the a, b parameters,

$$(a; b) = (1, 2, 3, \dots; 2, 3, \dots). \tag{8}$$

It is worth noticing that the stable minimum values of the a, b parameters correspond to vacua with vanishing cosmological constant at a generic point in the moduli space. The new non-perturbative superpotentials may test the nature of non-perturbative effects in the 4D orbifold constructions of the heterotic string.

References

1. Scherk, J., Schwarz, J.H. (1979): Phys. Lett. B **82**, 60;
   Ferrara, S., Kouns, C., Lüsz, P., Zwigner, F. (1992): Nucl. Phys. B **372**, 14
2. Bachas, C. (2000): hep-th/9503030, hep-th/9509067;
   Bianchi, M., Stanev, M. (1999): Nucl. Phys. B **553**, 133, hep-th/9711069
3. Antoniadis, I., Dudas, E., Sagnotti, A. (1999): Nucl. Phys. B **544**, 469, hep-th/9807011;
   Antoniadis, I., D'Appollonio, G., Dudas, E., Sagnotti, A. (1999): Nucl. Phys. B **553**, 133, hep-th/9812118
4. Angelantonj, C. (1988): Phys. Lett. B **444**, 309, hep-th/9810214;
   Blumenhagen, R., Font, A., Lüst, D. (1999): Nucl. Phys. B **558**, 159, hep-th/9904069
5. Altazabal, G., Uranga, A.M. (1999): JHEP 9910 024;
   Altazabal, G., Ibáñez, L.E., Quevedo, F., Uranga, A.M. (2000): JHEP **0008**, 002, hep-th/0005067
6. Antoniadis, I., Partouche, H., Taylor, T.R. (1996): Phys. Lett. B **372**, 92;
   Taylor, T.R., Vafa, C. (2000): Phys. Lett. B **474**, 130
7. Ferrara, S., Magnoli, N., Taylor, T.R., Veneziano, G. (1990): Phys. Lett. B **245**, 409
8. Font, A., Ibáñez, L.E., Lüst, D., Quevedo, F. (1990): Phys. Lett. B **245**, 401
9. Binetry, P., Gailard, M.R. (1991): Phys. Lett. B **253**, 119
10. Cvetic, M., Font, A., Ibáñez, L.E., Lüst, D., Quevedo, F. (1991): Nucl. Phys. B **361**, 194
11. Font, A., Ibáñez, L., Lüst, D., Quevedo, F. (1990): Phys. Lett. B **249**, 35
12. Horne, J.H., Moore, G. (1994): Nucl. Phys. B **432**, 109
13. Veneziano, G., Yankielowicz, S. (1982): Phys. Lett. B **113**, 231
14. Taylor, T.R. (1990): Phys. Lett. B **252**, 59
15. Casas, J.A., Lalak, Z., Munoz, C., Ross, G. (1990): Nucl. Phys. B **347**, 243
16. Lüst, D., Taylor, T.R. (1991): Phys. Lett. B **253**, 335
17. Lüst, D., Munoz, C. (1992): Phys. Lett. B **279**, 272
18. De Carlos, B., Casas, J.A., Munoz, C. (1993): Nucl. Phys. B **399**, 623

19. Dixon, L., Harvey, J., Vafa, C., Witten, E. (1986): Nucl. Phys. B **278**, 769; (1986): Nucl Phys. B **274**, 285
20. Harvey, J., Moore, G. (1996): Nucl. Phys. B **463**, 315
21. Kokorelis, C. (1999): Nucl. Phys. B **542**; 89, hep-th/9812068, for an extended version see hep-th/9802099
22. De Wit, B., Kaplunovsky, V., Louis, J., Lüst, D. (1995): Nucl. Phys. B **451**, 53; Ferrara, S., Antoniadis, I., Gava, E., Narain, K.S., Taylor, T.R. (1995): Nucl. Phys. B **447**, 35; Foerger, K., Stieberger, S. (1998): Nucl. Phys. B **514**, 135
23. Kokorelis, C. (2000): Constraint supersymmetry breaking and non-perturbative effects in string theory. hep-th/0006194
24. Koblitz, N. (1983): Introduction to Elliptic Curves and Modular Forms. Springer Verlag, Berlin Heidelberg New York

T-Dualization and Symmetry Breaking

P.-Y. Casteill

Abstract. We consider the most general σ-model defined on a group manifold $G_L \times G_R/G_D$. Its metric, invariant under the action of G_L, includes appropriate parameters describing all the possible breakings of the group G_R. A direct computation of the Ricci tensor shows that the one-loop renormalizability of its T-dualized σ-model is strictly equivalent to the one-loop renormalizability of the original model for any symmetry breaking.

1 Introduction

The subject of classical versus quantum equivalence of T-dualized σ-models has been intensely studied in recent years, and extensive reviews covering abelian, non-abelian dualities and their applications to string theory and statistical physics are available [1–3]. At the classical level the equivalence of the abelian [4,5] and the non-abelian [6–8] dualized theory is proved and the explicit form of the canonical transformation given.

Restricting ourselves to the non-abelian dualizations of Lie groups, the one-loop equivalence was established for the group manifold of $SU(2)$ in [9,10]. A further decisive progress was to generalize the one-loop equivalence to an arbitrary Lie group [11]. However, as pointed out in [12], Tyurin's analysis considers only models with explicit invariance under the left group action (whose existence is crucial for the dualization process) leaving aside the right action and the possible symmetry breaking schemes for it. The one-loop equivalence problem in this more general setting has been examined recently [12,13] for the group manifold $SU(2)_L \times SU(2)_R/SU(2)_D$, where $SU(2)_R$ is broken down to a $U(1)$. The renormalizability does survive despite the lowering of the right isometries. It is the purpose of the present paper to analyze the geometry of the dualized model for any $G_L \times G_R/G_D$, with arbitrary breaking of G_R. While in [11] supersymmetry considerations à la Busher [14,15] were of great help to derive the dualized geometry, we will show that a direct computation in local coordinates is fairly efficient to extract the Ricci tensor in the presence of symmetry breaking. A more complete description of these derivations can be found in [16].

2 The initial and dualized theories

Let us consider a target space with coordinates g, with a left and right action of some Lie group G. Its corresponding action can be written as

$$S = \frac{1}{2} \int d^2x \, B_{ij} \, \eta^{\mu\nu} J_\mu^i \, J_\nu^j, \tag{1}$$

where $g^{-1}\partial_\mu g = J^i_\mu T_i$. The matrices of the adjoint representation $(T_i)^k_j = -f^k_{ij}$ verify the Jacobi identity $[T_i, T_j] = f^k_{ij} T_k$ and the constant matrix B_{ij} is symmetric and invertible. The action (1) is invariant under G_L, while the matrix B_{ij} will generally break G_R down to some subgroup H (possibly trivial). Through the identification of the currents J^i_μ with the vielbein e^i, a straightforward and usual computation gives for the Ricci tensor:

$$\mathrm{ric}_{ij} = R^s_{i,sj} = -\omega^s_{i,t}\,\omega^t_{j,s} + \omega^t_{s,t}\,\omega^s_{i,j}.$$

where $\omega^i_j = \omega^i_{j,s} e^s$ is the spin-connection. The scalar curvature $R = (B^{-1})_{ij}\,\mathrm{ric}_{ij}$ is a constant.

For a simple group G, the most symmetric action is given by the bi-invariant metric $B_{ij} = \frac{1}{\rho}\mathrm{Tr}\,(T_i T_j)$. Such a metric is Einstein: $\mathrm{ric}_{ij} = -\frac{\rho}{4} B_{ij}$. This simple structure will have a counterpart in the dualized theory.

The essence of the dualization process [1,10] is to switch from the coordinates g to new coordinates ψ_i defined as the Lagrange multipliers of the Bianchi identities. Concretely this transformation is carried out starting from the action

$$S = \frac{1}{4}\int d^2x \left\{ B_{ij}\,\eta^{\mu\nu} J^i_\mu J^j_\nu - \epsilon^{\mu\nu}\psi_i(\partial_\mu J^i_\nu - \partial_\nu J^i_\mu + f^i_{st} J^s_\mu \wedge J^t_\nu) \right\}.$$

The field equations obtained from the variations with respect to the currents J^i_μ give, using Minkowskian coordinates on the worldsheet

$$J^{\mu i} = B^{ij}\,\epsilon^{\mu\nu}\left(\partial_\nu\psi_j - (A\cdot\psi)_{jk} J^k_\nu\right),$$

$$(A^s)_{ij} = (T_i)^s_j = -f^s_{ij}, \quad (A\cdot\psi)_{ij} = (A^s)_{ij}\psi_s.$$

Using this relation, one can write, up to total derivatives and in light-cone coordinates, the dual action:

$$S = \frac{1}{2}\int d^2x\, G_{ij}\,\partial_+\psi^i\,\partial_-\psi^j, \quad G_{ij} = (B + A\cdot\psi)^{-1}_{ij}. \tag{2}$$

The metric and torsion potential are given by:

$$g_{ij} = \frac{1}{2}(G_{ij} + G_{ji}), \qquad h_{ij} = \frac{1}{2}(G_{ij} - G_{ji}), \qquad G_{ij} = g_{ij} + h_{ij}.$$

Let us emphasize that the final action of the dualized theory is independent of the choice of the representation for the fields parametrizing G. One has then to compute the connection with torsion in order to obtain the Ricci tensor. It is then possible to prove the important relation:

$$\mathrm{Ric}_{ij} = -G_{is}\,\mathrm{ric}_{st}\,G_{tj} + D_j v_i, \qquad v_i = -2\,G_{it}\,f^s_{st} - \partial_i\ln(\sqrt{\det g}). \tag{3}$$

One can notice the appearance of the initial Ricci tensor in this tensorial equality.

Another useful relation appears in the calculation:

$$D_j w_i + \partial_{[i} W_{j]} = \frac{1}{2} G_{ij} + \frac{1}{2}(GBG)_{ij}, \quad w_i = g_{is}\psi^s, \quad W_i = \psi^s G_{si}. \tag{4}$$

When we dualize the bi-invariant metric on G, we get for the dual theory

$$\mathrm{Ric}_{ij} = \frac{\rho}{4}(GBG)_{ij} + D_j v_i,$$

$$D_j w_i = \frac{1}{2} G_{ij} + \frac{1}{2}(GBG)_{ij}, \quad w_i = W_i = (B^{-1})_{is}\psi^s,$$

from which we deduce

$$\mathrm{Ric}_{ij} = -\frac{\rho}{4} G_{ij} + D_j \mathcal{V}_i, \qquad \mathcal{V}_i = v_i + \frac{\rho}{2} w_i,$$

a relation which establishes the quasi-Einstein character of the bi-invariant torsionful dual metric.

3 One loop equivalence

We are now in position to discuss the quantum properties of the dualized models at the one loop level. Let us first consider the broken principal models with classical action (1). Its one loop counterterm is given by Friedan [17]:

$$\frac{1}{2\pi\epsilon} \frac{1}{2} \int d^2x \, \mathrm{ric}_{ij} \, \eta^{\mu\nu} J_\mu^i J_\nu^j, \qquad d = 2 - \epsilon,$$

where the Ricci components are computed in the vielbein basis.

Renormalizability in the strict field theoretic sense requires that these divergences have to be absorbed by (field independent) deformations of the coupling constants $\hat{\rho}_s$ hidden in the matrix B and possibly by a non-linear field renormalization. This last ingredient is not required here because the scalar curvature is a constant. Therefore the renormalizability of the theory (1) is ensured by

$$\mathrm{ric}_{ij} = \hat{\chi}_s(\rho)\frac{\partial}{\partial\hat{\rho}_s} B_{ij}. \tag{5}$$

The one loop renormalizability is clear for two extreme choices of metrics:

- The bi-invariant metric, which is Einstein.
- The maximally broken metric, for which the matrix B contains $\nu(\nu+1)/2$ independent coupling constants $\hat{\rho}_s$. Since the Ricci is also a symmetric matrix, it can always be absorbed by a deformation of the coupling constants.

For partial breakings of the group G_R, relation (5) may fail to hold and is indeed a constraint which mixes conditions involving the breaking matrix B and the algebra through its structure constants.

We will now assume that the initial theory is renormalisable at one loop order so that (5) always holds. In order to define cleanly the coupling constants and their renormalization we switch from the couplings $\{\hat{\rho}_i, \ i = 1, \ldots, c\}$ to (λ, ρ_i) through the relations

$$\hat{\rho}_1 = \frac{1}{\lambda}, \quad \hat{\rho}_{i+1} = \frac{\rho_i}{\lambda}, \quad i = 1, \ldots c - 1, \quad B_{ij} = \frac{1}{\lambda} S_{ij}(\rho_i), \tag{6}$$

where, for simplicity, the matrix S can be taken linear in the couplings ρ_i. Writing (5) with these new coupling constants, the full one loop action for the initial theory becomes

$$\frac{1}{\lambda} \frac{1}{2} \int d^2 x \left[\left(1 + \frac{\lambda \chi_\lambda}{2\pi \epsilon}\right) S_{ij}(\rho) + \frac{\lambda}{2\pi \epsilon} \sum_s \chi_s \frac{\partial}{\partial \rho_s} S_{ij}(\rho) \right] J_\mu^i J_\nu^j,$$

$$\epsilon = 2 - d, \tag{7}$$

from which we see that the divergences can be absorbed through coupling constant renormalizations:

$$\lambda_0 = \mu^\epsilon \lambda Z_\lambda, \quad Z_\lambda = 1 - \frac{\lambda}{2\pi \epsilon} \chi_\lambda, \quad \rho_i^{(0)} = \rho_i Z_i, \quad Z_i = 1 + \frac{\lambda}{2\pi \epsilon} \chi_i.$$

It follows that the corresponding beta functions are

$$\beta_\lambda = \mu \frac{\partial \lambda}{\partial \mu} = \lambda^2 \frac{\partial}{\partial \lambda} Z_\lambda^{(1)} = -\frac{\lambda^2}{2\pi} \chi_\lambda, \quad \beta_i = \mu \frac{\partial \rho_i}{\partial \mu} = \lambda \frac{\partial}{\partial \lambda} \left(\rho_i Z_i^{(1)}\right) = \frac{\lambda}{2\pi} \chi_i.$$

In order to compare to the renormalization properties of the dualized theory, let us recall that the most general conditions giving one loop renormalizability (we work with worldsheet components) are

$$\begin{cases} \text{Ric}_{(ij)} = \hat{\chi}_s \dfrac{\partial}{\partial \hat{\rho}_s} g_{ij} + D_{(i} v_{j)}, \\[2ex] \text{Ric}_{[ij]} = \hat{\chi}_s \dfrac{\partial}{\partial \hat{\rho}_s} h_{ij} + v_s T_{ij}^s + \partial_{[i} w_{j]} \end{cases}$$

$$\Longleftrightarrow \text{Ric}_{ij} = \hat{\chi}_s \frac{\partial}{\partial \hat{\rho}_s} G_{ij} + D_j v_i + \partial_{[i} (v + w)_{j]},$$

where the $\hat{\rho}_s$ are the coupling constants in the principal model we started from, appearing now in a non trivial way in the dualized model. The only constraint on the functions $\hat{\chi}_s$, is that they should be field independent.

In order to compute the divergences of the dualized theory in terms of the coupling constants defined in (6), we start from the dual classical action

$$\frac{1}{\lambda} G_{ij}(S, \tilde{\psi}) \partial_+ \tilde{\psi} \, \partial_- \tilde{\psi}^j, \quad G(B, \psi) = \lambda \, G(S, \tilde{\psi}), \quad \tilde{\psi}^i = \lambda \, \psi^i.$$

The one-loop counterterms follow from the ricci. We get from relation (3)

$$\mathrm{Ric}_{ij} = \lambda^2 \left\{ -G_{is}(S, \tilde{\psi}) \mathrm{ric}_{st} G_{tj}(S, \tilde{\psi}) + D_j v_i \right\}.$$

Using (4) and (5) appropriately scaled out, we get for the one loop action of the dualized theory

$$\frac{1}{\lambda} \int d^2x \left[\left(1 + \frac{\lambda \chi_\lambda}{2\pi\epsilon} + \frac{\lambda}{2\pi\epsilon} \sum_s \chi_s \frac{\partial}{\partial\rho_s} \right) G_{ij} \right.$$
$$\left. + \frac{\lambda}{2\pi\epsilon} \left(D_j \mathcal{V}_i - 2\chi_\lambda \partial_{[i} W_{j]} \right) \right] \partial_+ \psi^i \, \partial_- \psi^j, \quad \mathcal{V}_i = v_i - 2\chi_\lambda w_i.$$

$$(8)$$

Comparing this relation with (7) we conclude that, up to the non-linear field redefinition described by the vector $\mathcal{V}_i$ and the gauge transformation described by W_i, the coupling constants renormalize in exactly the same way than in the principal model we started from. We have thus established their equivalence, in the strict field theoretic sense, and proved that their beta functions do coincide. What is really new is that this equivalence survives even for the maximal breaking of G_R, i.e., even when the dualized theory has no isometries at all.

References

1. Alvarez, E., Alvarez-Gaumé, L., Barbón, J.L.F., Lozano, Y. (1994): Nucl. Phys. B **415**, 71
2. Alvarez, E., Alvarez-Gaumé, L., Lozano, Y. (1995): Nucl. Phys. Proc. Suppl. **41**, 1
3. Giveon, A., Porrati, M., Rabinovici, E. (1994): Phys. Rep. **244**, 77
4. Alvarez, E., Alvarez-Gaumé L., Lozano, Y. (1994): Phys. Lett. B **336**, 183
5. Giveon, A., Rabinovici, E., Veneziano, G. (1989): Nucl. Phys. B **322**, 167; cern-th. 5106/88
6. Curtright, T., Zachos, C. (1994): Phys. Rev. D **49**, 183
7. Lozano, Y. (1995): Phys. Lett. B **355**, 165
8. Sfetsos, K. (1996): Phys. Rev. D **54**, 1682
9. Fridling, B.E., Jevicki, A. (1984): Phys. Lett. B **134**, 70
10. Fradkin, E.S., Tseytlin, A.A. (1985): Ann. Phys. **162**, 31
11. Tyurin, E. (1995): Phys. Lett. B **348**, 386
12. Balázs, L.K., Balog, J., Forgács, P., Mohammedi, N., Palla, L., Schnittger, J. (1998): Phys. Rev. D **57**, 3585
13. Horváth, Z., Karp, R.L., Palla, L.: hep-th/9609198
14. Buscher, T. (1987): Phys. Lett. B **194**, 59
15. Buscher, T. (1988): Phys. Lett. B **201**, 466
16. Casteill, P.Y., Valent, G. (2000): Quantum structure of T-dualized models with symmetry breaking. Nucl. Phys. B **591**, 491; hep-th/0006186
17. Friedan, D. (1985): Ann. Phys. **163**, 1257

A Theory of Quantum Gravity from First Principles

G. Esposito

Abstract. When quantum fields are studied on manifolds with boundary, the corresponding one-loop quantum theory for bosonic gauge fields with linear covariant gauges needs the assignment of suitable boundary conditions for elliptic differential operators of Laplace type. There are however deep reasons to modify such a scheme and allow for pseudo-differential boundary-value problems. When the boundary operator is allowed to be pseudo-differential while remaining a projector, the conditions on its kernel leading to strong ellipticity of the boundary-value problem are studied in detail. This makes it possible to develop a theory of one-loop quantum gravity from first principles only, i.e. the physical principle of invariance under infinitesimal diffeomorphisms and the mathematical requirement of a strongly elliptic theory.

The space-time approach to quantum mechanics and quantum field theory has led to several profound developments in the understanding of quantum theory and space-time structure at very high energies [1,2]. In particular, we are here concerned with the choice of boundary conditions. On using path integrals, which lead, in principle, to the appropriate formulation of the ideas of Feynman, DeWitt and many other authors [2–5], the assignment of boundary conditions consists of two main steps:

(i)   Choice of Riemannian geometries and field configurations to be included in the path-integral representation of transition amplitudes;
(ii) Choice of boundary data to be imposed on the hypersurfaces Σ_1 and Σ_2 bounding the given space-time region.

The main object of our investigation is the second problem of such a list, when a one-loop approximation is studied for a bosonic gauge theory in linear covariant gauges. The well posed mathematical formulation relies on the "Euclidean approach", i.e., in geometric language, on the use of differentiable manifolds endowed with positive-definite metrics g, so that Lorentzian space-time is actually replaced by an m-dimensional Riemannian manifold (M, g). In particular, in Euclidean quantum gravity, mixed boundary conditions on metric perturbations h_{cd} occur naturally if one requires their complete invariance under infinitesimal diffeomorphisms, as is proved in detail in [6]. On denoting by N^a the inward-pointing unit normal to the boundary, by

$$q^a_{\ b} \equiv \delta^a_{\ b} - N^a N_b \tag{1}$$

the projector of tensor fields onto ∂M, with associated projection operator

$$\Pi_{ab}^{\ \ cd} \equiv q^c_{\ (a} \, q^d_{\ b)}, \tag{2}$$

the gauge-invariant boundary conditions for one-loop quantum gravity read [6]

$$\left[\Pi_{ab}{}^{cd}h_{cd}\right]_{\partial M} = 0, \tag{3}$$

$$\left[\Phi_a(h)\right]_{\partial M} = 0, \tag{4}$$

where Φ_a is the gauge-averaging functional necessary to obtain an invertible operator $P_{ab}{}^{cd}$ on metric perturbations. When $P_{ab}{}^{cd}$ is chosen to be of Laplace type, Φ_a reduces to the familiar de Donder term

$$\Phi_a(h) = \nabla^b\left(h_{ab} - \frac{1}{2}g_{ab}g^{cd}h_{cd}\right) = E_a^{bcd}\nabla_b h_{cd}, \tag{5}$$

where E^{abcd} is the DeWitt supermetric on the vector bundle of symmetric rank-two tensor fields over M (g being the metric on M):

$$E^{abcd} \equiv \frac{1}{2}\left(g^{ac}g^{bd} + g^{ad}g^{bc} - g^{ab}g^{cd}\right). \tag{6}$$

The boundary conditions (3) and (4) can then be cast in the Grubb–Gilkey–Smith form [7, 8]:

$$\begin{pmatrix} \Pi & 0 \\ \Lambda & I - \Pi \end{pmatrix}\begin{pmatrix} [\varphi]_{\partial M} \\ [\varphi_{;N}]_{\partial M} \end{pmatrix} = 0. \tag{7}$$

However, the work in [6] has shown that an operator of Laplace type on metric perturbations is then incompatible with the requirement of strong ellipticity of the boundary-value problem, because the operator Λ contains tangential derivatives of metric perturbations.

To take care of this serious drawback, the work in [9] has proposed to consider in the boundary condition (4) a gauge-averaging functional given by the de Donder term (5) plus an integro-differential operator on metric perturbations, i.e.

$$\Phi_a(h) \equiv E_a^{bcd}\nabla_b h_{cd} + \int_M \zeta_a^{cd}(x, x')h_{cd}(x')dV'. \tag{8}$$

We now point out that the resulting boundary conditions can be cast in the form

$$\begin{pmatrix} \Pi & 0 \\ \Lambda + \widetilde{\Lambda} & I - \Pi \end{pmatrix}\begin{pmatrix} [\varphi]_{\partial M} \\ [\varphi_{;N}]_{\partial M} \end{pmatrix} = 0, \tag{9}$$

where $\widetilde{\Lambda}$ reflects the occurrence of the integral over M in Eq. (8). It is convenient to work first in a general way and then consider the form taken by these operators in the gravitational case. On requiring that the resulting boundary operator

$$\mathcal{B} \equiv \begin{pmatrix} \Pi & 0 \\ \Lambda + \widetilde{\Lambda} & I - \Pi \end{pmatrix} \tag{10}$$

should remain a projector: $\mathcal{B}^2 = \mathcal{B}$, we find the condition

$$\Pi\widetilde{\Lambda} = \widetilde{\Lambda}\Pi, \tag{11}$$

by virtue of the property $\Pi\Lambda = \Lambda\Pi = 0$ considered in [6].

In Euclidean quantum gravity at one loop, on denoting by $W_{ab}{}^{cd}(x, x') = W_{ab}{}^{c'd'}$ the kernel associated to $\widetilde{\Lambda}$, i.e.

$$\widetilde{\Lambda}_{ab}{}^{cd} h_{cd}(x) \equiv \int_M W_{ab}{}^{cd}(x, x') h_{cd}(x') dV', \tag{12}$$

Eq. (11) leads to (cf. [10])

$$\int_M \left[\Pi_{ab}{}^{cd} W_{cd}{}^{e'f'} - W_{ab}{}^{c'd'} \Pi_{c'd'}{}^{e'f'} \right] h_{e'f'} dV' = 0. \tag{13}$$

Since this should hold for all $h_{e'f'} = h_{ef}(x')$, it eventually leads to the vanishing of the term in square brackets in the integrand, i.e.

$$\left[\Pi_{ab}{}^{cd} W_{cd}{}^{e'f'} - W_{ab}{}^{c'd'} \Pi_{c'd'}{}^{e'f'} \right] = 0. \tag{14}$$

We are now concerned with the issue of ellipticity of the boundary-value problem of one-loop quantum gravity. For this purpose, we begin by recalling what is known about ellipticity of the Laplacian (hereafter P) on a Riemannian manifold with smooth boundary. This concept is studied in terms of the leading symbol of P. It is indeed well known that the Fourier transform makes it possible to associate to a differential operator of order k a polynomial of degree k, called the characteristic polynomial or symbol. The leading symbol, σ_L, picks out the highest order part of this polynomial. For the Laplacian, it reads $\sigma_L(P; x, \xi) = |\xi|^2 I = g^{\mu\nu}\xi_\mu\xi_\nu I$. With a standard notation, (x, ξ) are local coordinates for $T^*(M)$, the cotangent bundle of M. The leading symbol of P is trivially elliptic in the interior of M, since $|\xi|^2$ is positive-definite, and one has

$$\det\left[\sigma_L(P; x, \xi) - \lambda \right] = (|\xi|^2 - \lambda)^{\dim V} \neq 0, \tag{15}$$

for all $\lambda \in C - \mathbf{R}_+$. In the presence of a boundary, however, one needs a more careful definition of ellipticity. First, for a manifold M of dimension m, the m coordinates x are split into $m - 1$ local coordinates on ∂M, hereafter denoted by $\{\hat{x}^k\}$, and r, the geodesic distance to the boundary. Moreover, the m coordinates ξ_μ are split into $m - 1$ coordinates $\{\zeta_j\}$ (with ζ a cotangent vector on the boundary), jointly with a real parameter $\omega \in T^*(\mathbf{R})$. At a deeper level, all this reflects the split

$$T^*(M) = T^*(\partial M) \oplus T^*(\mathbf{R}) \tag{16}$$

in a neighbourhood of the boundary [6, 11].

The ellipticity we are interested in requires now that σ_L should be elliptic in the interior of M, as specified before, and that strong ellipticity should hold. This means that a unique solution exists of the differential equation obtained from the leading symbol:

$$\left[\sigma_L\left(P;\left\{\hat{x}^k\right\},r=0,\left\{\zeta_j\right\},\omega\to-i\frac{\partial}{\partial r}\right)-\lambda\right]\varphi(r,\hat{x},\zeta;\lambda)=0,\tag{17}$$

subject to the boundary conditions

$$\sigma_g(B)\left(\left\{\hat{x}^k\right\},\left\{\zeta_j\right\}\right)\psi(\varphi)=\psi'(\varphi),\tag{18}$$

and to the asymptotic condition

$$\lim_{r\to\infty}\varphi(r,\hat{x},\zeta;\lambda)=0.\tag{19}$$

In Eq. (18), σ_g is the *graded leading symbol* of the boundary operator in the local coordinates $\left\{\hat{x}^k\right\},\left\{\zeta_j\right\}$, and is given by

$$\sigma_g(B)=\begin{pmatrix}\Pi & 0\\ i\Gamma^j\zeta_j & I-\Pi\end{pmatrix}.\tag{20}$$

Roughly speaking, the above construction uses Fourier transform and the inward geodesic flow to obtain the ordinary differential equation (17) from the Laplacian, with corresponding Fourier transform (18) of the original boundary conditions. The asymptotic condition (19) picks out the solutions of Eq. (17) which satisfy Eq. (18) with arbitrary boundary data $\psi'(\varphi)\in C^\infty(W',\partial M)$ for W' a vector bundle over the boundary, and vanish at infinite geodesic distance to the boundary. When all the above conditions are satisfied $\forall\zeta\in T^*(\partial M),\forall\lambda\in C-\mathbf{R}_+,\forall(\zeta,\lambda)\neq(0,0)$ and $\forall\psi'(\varphi)\in C^\infty(W',\partial M)$, the boundary-value problem (P,B) for the Laplacian is said to be strongly elliptic with respect to the cone $C-\mathbf{R}_+$.

However, when the gauge-averaging functional (8) is used in the boundary condition (4), the work in [9] has proved that the operator on metric perturbations takes the form of an operator of Laplace type $P_{ab}{}^{cd}$ plus an integral operator $G_{ab}{}^{cd}$. Explicitly, one finds [9] (with $R^a{}_{bcd}$ the Riemann curvature of the background geometry (M,g))

$$P_{ab}{}^{cd}=E_{ab}{}^{cd}(-\Box+R)-2E_{ab}{}^{qf}R^c{}_{qpf}g^{dp}-E_{ab}{}^{pd}R_p^{\ c}-E_{ab}{}^{cp}R_p^{\ d},\tag{21}$$

$$G_{ab}{}^{cd}=U_{ab}{}^{cd}+V_{ab}{}^{cd},\tag{22}$$

where (here $\zeta_b{}^{cq}(x,x')\equiv T^p\widetilde{\zeta}_{bp}{}^{cq}(x,x')$ for a suitable vector field T)

$$U_{ab}{}^{cd}h_{cd}(x)=-2E_{rsab}\nabla^r\int_M T^p(x)\widetilde{\zeta}^s_{\ p}{}^{cd}(x,x')h_{cd}(x')dV',\tag{23}$$

$$h^{ab}V_{ab}{}^{cd}h_{cd}(x)$$
$$=\int_{M^2}h^{ab}(x')T^q(x)\widetilde{\zeta}_{pqab}(x,x')T^r(x)\widetilde{\zeta}^p_{\ r}{}^{cd}(x,x'')h_{cd}(x'')dV'dV''.\tag{24}$$

We now assume that the operator on metric perturbations, which is so far an integro-differential operator defined by a kernel, is also pseudo-differential [11]. In the presence of pseudo-differential operators, both ellipticity in the interior of M and strong ellipticity of the boundary-value problem need a more involved formulation. In our paper, inspired by the flat-space analysis in [12], we make the following requirements [10].

(i) Ellipticity in the interior. Let U be an open subset with compact closure in M, and consider an open subset U_1 whose closure $\overline{U}_1$ is properly included into U: $\overline{U}_1 \subset U$. If p is a symbol of order d on U, it is said to be *elliptic* on U_1 if there exists an open set U_2 which contains $\overline{U}_1$ and positive constants C_0, C_1 so that

$$|p(x,\xi)|^{-1} \leq C_1(1+|\xi|)^{-d}, \tag{25}$$

for $|\xi| \geq C_0$ and $x \in U_2$, where $|\xi| \equiv \sqrt{g^{ab}(x)\xi_a\xi_b}$. The corresponding operator P is then elliptic.

(ii) Strong ellipticity in the absence of boundaries. Let us assume that the symbol under consideration is *polyhomogeneous*, in that it admits an asymptotic expansion of the form

$$p(x,\xi) \sim \sum_{l=0}^{\infty} p_{d-l}(x,\xi), \tag{26}$$

where each term p_{d-l} has the *homogeneity property*

$$p_{d-l}(x,t\xi) = t^{d-l}p_{d-l}(x,\xi) \quad \text{if} \quad t \geq 1 \quad \text{and} \quad |\xi| \geq 1. \tag{27}$$

The leading symbol is then, by definition, $p^0(x,\xi) \equiv p_d(x,\xi)$. Strong ellipticity in the absence of boundaries is formulated in terms of the leading symbol, and it requires that

$$\text{Re } p^0(x,\xi) \geq c(x)|\xi|^d, \tag{28}$$

where $x \in M$ and $|\xi| \geq 1$, c being a positive function on M. It can then be proved that the Gårding inequality holds, according to which, for any $\varepsilon > 0$,

$$\text{Re}(Pu,u) \geq b\|u\|^2_{\frac{d}{2}} - b_1\|u\|^2_{\frac{d}{2}-\varepsilon} \quad \text{for} \quad u \in H^{\frac{d}{2}}(M), \tag{29}$$

with $b > 0$.

(iii) Strong ellipticity in the presence of boundaries. The homogeneity property (27) only holds for $t \geq 1$ and $|\xi| \geq 1$. Consider now the case $l = 0$, for which one obtains the leading symbol which plays the key role in the definition of ellipticity. If $p^0(x,\xi) \equiv p_d(x,\xi) \equiv \sigma_L(P;x,\xi)$ is not a polynomial (which corresponds to the genuinely pseudo-differential case) while being a homogeneous function of ξ, it is irregular at $\xi = 0$. When $|\xi| \leq 1$, the only control over the leading symbol is provided by estimates of the form [12]

$$\left| (-i)^{\sum_{k=1}^{m}(\alpha_k+\beta_k)} \left(\frac{\partial}{\partial x_1}\right)^{\alpha_1} \cdots \left(\frac{\partial}{\partial x_m}\right)^{\alpha_m} \left(\frac{\partial}{\partial \xi_1}\right)^{\beta_1} \cdots \left(\frac{\partial}{\partial \xi_m}\right)^{\beta_m} p^0(x,\xi) \right| \tag{30}$$
$$\leq c(x)\langle\xi\rangle^{d-|\beta|}.$$

We therefore come to appreciate the problematic aspect of symbols of pseudo-differential operators [12]. The singularity at $\xi = 0$ can be dealt with either by modifying the leading symbol for small ξ to be a C^∞ function (at the price of loosing the homogeneity there), or by keeping the strict homogeneity and dealing with the singularity at $\xi = 0$ [12].

On the other hand, we are interested in a definition of strong ellipticity of pseudo-differential boundary-value problems that reduces to Eqs. (17)–(19) when both P and the boundary operator reduce to the form considered in [6]. For this purpose, and bearing in mind the occurrence of singularities in the leading symbols of P and of the boundary operator, we make the following requirements [10].

Let $(P + G)$ be a pseudo-differential operator subject to boundary conditions described by the pseudo-differential boundary operator $\mathcal{B}$ (the consideration of $(P + G)$ rather than only P is necessary to achieve self-adjointness, as is described in detail in [12] and [13]). The pseudo-differential boundary-value problem $((P + G), \mathcal{B})$ is strongly elliptic with respect to $\mathcal{C} - \mathbf{R}_+$ if:

(I)   The inequalities (25) and (28) hold.
(II)   There exists a unique solution of the equation

$$\left[\sigma_L\left((P + G); \left\{\hat{x}^k\right\}, r = 0, \{\zeta_j\}, \omega \to -i\frac{\partial}{\partial r} \right) - \lambda \right] \varphi(r, \hat{x}, \zeta; \lambda) = 0, \tag{31}$$

subject to the boundary conditions

$$\sigma_L(\mathcal{B}) \left(\left\{\hat{x}^k\right\}, \{\zeta_j\} \right) \psi(\varphi) = \psi'(\varphi) \tag{32}$$

and to the asymptotic condition (19). It should be stressed that, unlike the case of differential operators, Eq. (31) is not an ordinary differential equation in general, because $(P + G)$ is pseudo-differential.
(III)   The strictly homogeneous symbols associated to $(P + G)$ and $\mathcal{B}$ have limits for $|\zeta| \to 0$ in the respective leading symbol norms, with the limiting symbol restricted to the boundary which avoids the values $\lambda \notin \mathcal{C} - \mathbf{R}_+$ for all $\{\hat{x}\}$.

Condition (III) requires a last effort for a proper understanding. Given a pseudo-differential operator of order d with leading symbol $p^0(x, \xi)$, the associated strictly homogeneous symbol is defined by [12]

$$p^h(x, \xi) \equiv |\xi|^d p^0\left(x, \frac{\xi}{|\xi|}\right) \quad \text{for} \ \xi \neq 0. \tag{33}$$

This extends to a continuous function vanishing at $\xi = 0$ when $d > 0$. In the presence of boundaries, the boundary-value problem $((P+G), \mathcal{B})$ has a strictly homogeneous symbol on the boundary equal to (some indices are omitted for simplicity)

$$\begin{pmatrix} p^h\left(\{\hat{x}\}, r = 0, \{\zeta\}, -i\frac{\partial}{\partial r}\right) + g^h\left(\{\hat{x}\}, \{\zeta\}, -i\frac{\partial}{\partial r}\right) - \lambda \\ b^h\left(\{\hat{x}\}, \{\zeta\}, -i\frac{\partial}{\partial r}\right) \end{pmatrix},$$

where p^h, g^h and b^h are the strictly homogeneous symbols of P, G and $\mathcal{B}$ respectively, obtained from the corresponding leading symbols p^0, g^0 and b^0 via equations analogous to (33), after taking into account the split (16), and upon replacing ω by $-i\frac{\partial}{\partial r}$. The limiting symbol restricted to the boundary (also called limiting λ-dependent boundary symbol operator) and mentioned in condition (III) reads therefore [12]

$$a^h\left(\{\hat{x}\}, r = 0, \zeta = 0, -i\frac{\partial}{\partial r}\right)$$
$$= \begin{pmatrix} p^h\left(\{\hat{x}\}, r = 0, \zeta = 0, -i\frac{\partial}{\partial r}\right) + g^h\left(\{\hat{x}\}, \zeta = 0, -i\frac{\partial}{\partial r}\right) - \lambda \\ b^h\left(\{\hat{x}\}, \zeta = 0, -i\frac{\partial}{\partial r}\right) \end{pmatrix}, \tag{34}$$

where the singularity at $\xi = 0$ of the leading symbol in absence of boundaries is replaced by the singularity at $\zeta = 0$ of the leading symbols of P, G and $\mathcal{B}$ when a boundary occurs.

Let us now see how the previous conditions on the leading symbol of $(P + G)$ and on the leading symbol of the boundary operator can be used. The equation (31) is solved by a function φ depending on r, $\{\hat{x}^k\}$, $\{\zeta_j\}$ and, parametrically, on the eigenvalues λ. For simplicity, we write $\varphi = \varphi(r, \hat{x}, \zeta; \lambda)$, omitting indices. Since the leading symbol is no longer a polynomial when $(P + G)$ is genuinely pseudo-differential, we cannot make any further specification on φ at this stage, apart from requiring that it should reduce to (here $|\zeta|^2 \equiv \zeta_i\zeta^i$) $\chi(\hat{x}, \zeta)e^{-r\sqrt{|\zeta|^2-\lambda}}$ when $(P+G)$ reduces to a Laplacian.

The equation (32) involves the leading symbol of $\mathcal{B}$ and restriction to the boundary of the field and its covariant derivative along the normal direction. Such a restriction is obtained by setting to zero the geodesic distance r, and hence we write, in general form (here we denote again by Λ the full matrix element $\mathcal{B}_{21}$ in the boundary operator (10)),

$$\begin{pmatrix} \Pi & 0 \\ \sigma_L(\Lambda) & I - \Pi \end{pmatrix} \begin{pmatrix} \varphi(0, \hat{x}, \zeta; \lambda) \\ \varphi'(0, \hat{x}, \zeta; \lambda) \end{pmatrix} = \begin{pmatrix} \Pi\varrho(0, \hat{x}, \zeta; \lambda) \\ (I - \Pi)\varrho'(0, \hat{x}, \zeta; \lambda) \end{pmatrix}, \tag{35}$$

448   G. Esposito

where ϱ differs from φ, because Eq. (38) is written for $\psi(\varphi)$ and $\psi'(\varphi) \neq \psi(\varphi)$. Now Eq. (35) leads to

$$\Pi \varphi(0, \hat{x}, \zeta; \lambda) = \Pi \varrho(0, \hat{x}, \zeta; \lambda), \tag{36}$$

$$\sigma_L(\Lambda) \varphi(0, \hat{x}, \zeta; \lambda) + (I - \Pi) \varphi'(0, \hat{x}, \zeta; \lambda) = (I - \Pi) \varrho'(0, \hat{x}, \zeta; \lambda), \tag{37}$$

and we require that, for φ satisfying Eq. (31) and the asymptotic decay (19), with $\lambda \in \mathcal{C} - \mathbf{R}_+$, Eqs. (36) and (37) can be always solved with given values of $\varrho(0, \hat{x}, \zeta; \lambda)$ and $\varrho'(0, \hat{x}, \zeta; \lambda)$, whenever $(\zeta, \lambda) \neq (0, 0)$. The idea is now to relate, if possible, $\varphi'(0, \hat{x}, \zeta; \lambda)$ to $\varphi(0, \hat{x}, \zeta; \lambda)$ in such a way that Eq. (36) can be used to simplify Eq. (37). For this purpose, we consider the function f such that

$$\frac{\varphi'(0, \hat{x}, \zeta; \lambda)}{\varphi(0, \hat{x}, \zeta; \lambda)} = \frac{\varrho'(0, \hat{x}, \zeta; \lambda)}{\varrho(0, \hat{x}, \zeta; \lambda)} = f(\hat{x}, \zeta; \lambda), \tag{38}$$

$$\Pi(\hat{x}) f(\hat{x}, \zeta; \lambda) - f(\hat{x}, \zeta; \lambda) \Pi(\hat{x}) = C(\hat{x}, \zeta; \lambda). \tag{39}$$

The occurrence of C is a peculiar feature of the fully pseudo-differential framework. Equation (37) is then equivalent to (now we write explicitly also the independent variables in the leading symbol of Λ)

$$\left[(\sigma_L(\Lambda) - C)(\hat{x}, \zeta; \lambda) + f(\hat{x}, \zeta; \lambda)\right] \varphi(0, \hat{x}, \zeta; \lambda)$$
$$= \varrho'(0, \hat{x}, \zeta; \lambda) - C(\hat{x}, \zeta; \lambda) \varrho(0, \hat{x}, \zeta; \lambda). \tag{40}$$

On defining

$$\gamma(\hat{x}, \zeta; \lambda) \equiv \left[\sigma_L(\Lambda) - C\right](\hat{x}, \zeta; \lambda), \tag{41}$$

we therefore obtain the strong ellipticity condition

$$\det\left[\gamma(\hat{x}, \zeta; \lambda) + f(\hat{x}, \zeta; \lambda)\right] \neq 0 \quad \forall \lambda \in \mathcal{C} - \mathbf{R}_+, \tag{42}$$

which is satisfied if

$$\det\left[\mathrm{Re}^2 f(\hat{x}, \zeta; \lambda) - \mathrm{Im}^2 f(\hat{x}, \zeta; \lambda) - \gamma^2(\hat{x}, \zeta; \lambda)\right.$$
$$\left. + 2i \mathrm{Re} f(\hat{x}, \zeta; \lambda) \mathrm{Im} f(\hat{x}, \zeta; \lambda)\right] \neq 0. \tag{43}$$

We have therefore provided a complete characterization of the properties of the symbol of the boundary operator for which a set of boundary conditions completely invariant under infinitesimal diffeomorphisms are compatible with a strongly elliptic one-loop quantum theory. Our analysis is detailed but general, and hence has the merit (as far as we can see) of including all pseudo-differential boundary operators for which the sufficient conditions just derived can be imposed.

It would be now very interesting to prove that, by virtue of the pseudo-differential nature of $\mathcal{B}$ in (10), the quantum state of the universe in one-loop semiclassical theory can be made of surface-state type [14]. This would describe a wave function of the universe with exponential decay away from the boundary, which might provide a novel description of quantum physics at the Planck length. It therefore seems that by insisting on path-integral quantization, strong ellipticity of the Euclidean theory and invariance principles, new deep perspectives are in sight. These are in turn closer to what we may hope to test, i.e. the one-loop semiclassical approximation in quantum gravity. In the seventies, such calculations could provide a guiding principle for selecting couplings of matter fields to gravity in a unified field theory. Now they can lead instead to a deeper understanding of the interplay between non-local formulations [15–17], elliptic theory, gauge-invariant quantization [18] and a quantum theory of the very early universe [10].

References

1. DeWitt, B.S. (1965): Dynamical Theory of Groups and Fields. Gordon and Breach, New York
2. DeWitt, B.S. (1984): The Space-Time Approach to Quantum Field Theory, in *Relativity, Groups and Topology II*, ed. by B.S. DeWitt, R. Stora, North-Holland, Amsterdam pp. 381–738
3. Feynman, R.P. (1948): Rev. Mod. Phys. **20**, 367
4. Misner, C.W. (1957): Rev. Mod. Phys. **29**, 497
5. Hawking, S.W. (1979): The Path-Integral Approach to Quantum Gravity, in *General Relativity, an Einstein Centenary Survey*, ed. by S.W. Hawking, W. Israel, Cambridge University Press, Cambridge, pp. 746–789
6. Avramidi, I.G., Esposito, G. (1999): Commun. Math. Phys. **200**, 495
7. Grubb, G. (1974): Ann. Scuola Normale Superiore Pisa **1**, 1
8. Gilkey, P.B., Smith, L. (1983): J. Diff. Geom. **18**, 393
9. Esposito, G. (1999): Class. Quantum Grav. **16**, 3999
10. Esposito, G. (2000): Int. J. Mod. Phys. A **15**, 4539
11. Gilkey, P.B. (1995): Invariance Theory, the Heat Equation and the Atiyah–Singer Index theorem. Chemical Rubber Company, Boca Raton
12. Grubb, G.(1996): Functional Calculus of Pseudodifferential Boundary Problems. Birkhäuser, Boston
13. Esposito, G. (1999): Class. Quantum Grav. **16**, 1113
14. Schröder, M. (1989): Rep. Math. Phys. **27**, 259
15. Moffat, J.W. (1990): Phys. Rev. D **41**, 1177
16. Evens, D., Moffat, J.W., Kleppe, G., Woodard, R.P. (1991): Phys. Rev. D **43**, 499
17. Marachevsky, V.N., Vassilevich, D.V. (1996): Class. Quantum Grav. **13**, 645
18. Esposito, G., Stornaiolo, C. (2000): Int. J. Mod. Phys. A **15**, 449

Spin Models, TQFTs and Their Hierarchical Structure

G. Carbone, M. Carfora, A. Marzuoli

Abstract. State sum models based on the recoupling theory of $SU(2)$ angular momenta can be defined in any dimension d, and the corresponding hierarchy includes the Ponzano–Regge model in $d = 3$ and the Crane–Yetter invariant in $d = 4$. Here we establish an equivalence between each of these spin models and a corresponding BF theory by exploiting suitable d-dimensional generalizations of the Ooguri's approach in $d = 3$.

Topological Quantum Field Theories (see e.g. [1]) and discretized models based on decorated Piecewise–Linear manifolds are deeply connected to low dimensional quantum gravity models. Recently the authors have defined in [2] state sums representing $SU(2)$ spin network models in any dimension (and their associated PL invariants) arranged in the following multi-hierarchical structure

$$\text{dimension:} \quad 2 \quad 3 \quad 4 \quad \dots \quad d \quad d+1$$

$$Z_\chi^2 \quad Z_{PR}^3$$

$$Z_\chi^3 \quad Z_{CY}^4$$

$$Z_\chi^4 \quad \ddots$$

$$\ddots \quad \ddots$$

$$Z_\chi^d \quad Z^{d+1}$$

$$Z_\chi^{d+1}$$

The first diagonal displays invariants (related to the Euler number) built from colorings of the 1-skeleton of the dual triangulations, while the second diagonal, based on colorings of the dual 2-skeleton, embraces both known models (namely the Ponzano–Regge model [3] and the Crane–Yetter invariant [4]) as well as new combinatorial invariants in dimension greater than four.

Here we shall deal with the problem of the equivalence between the spin models in the second diagonal above and TQFTs of the BF type. In this way we shall generalize to any dimension the relation between the two classes of theories found in [5] and [6,7] in the three and four dimensional cases, respectively. Notice that the authors [8] and Oriti–Williams [9] have already proved – by using the discretization technique proposed by Freidel and Krasnow [10] – that the second diagonal of the PL hierarchy is actually in one-to-one correspondence with a BF hierarchy, in the sense that the corresponding partition functions are equivalent. However, to clarify the meaning of the resulting hierarchical arrangement on the BF side, and also to reach a better understanding of such equivalences, here we address the subject by employing a d-dimensional generalization of the original Ooguri approach [5] in dimension three. This method is complicated in many respects, and moreover is not completely generalized to d dimensional manifolds of arbitrary topology. On the other hand, it turns out to be particularly suitable in order to clarify the role of the auxiliary surfaces (introduced in [2]) which occur both in the definition of the state sums of the second diagonal and in the computation of the invariants associated with the first one. The proof will be carried out in three steps, following the scheme given in [5]. The first step consists in verifying the existence of an isomorphism between the Hilbert spaces of the two theories; then the structure of the isomorphism is improved by establishing the equivalence between the wave functions; finally, the identitification between the partition functions will be proved.

Let us start by considering a d-dimensional manifold M decomposed into three parts $M_1 \cup M_2 \cup N$, where N has the topology of $\Sigma \times [0, 1]$ and $\partial M_i = \Sigma$ for each $i = 1, 2$. Thus we can rewrite the partition function of the discretized model $Z^d[M^d] \equiv Z^d[M]$ according to

$$Z^d[M] = \sum_{\substack{c_1 \in C(\Delta_1) \\ c_2 \in C(\Delta_2)}} Z[M_1, \Delta_1](c_1) w_L^{-\Xi(\Delta_1)} P_{\Delta_1, \Delta_2}(c_1, c_2) \tag{1}$$

$$\cdot w_L^{-\Xi(\Delta_2)} Z[M_2, \Delta_2](c_2).$$

Here Δ_i $(i = 1, 2)$ denotes a fixed triangulation on each boundary ∂M_i; $C(\Delta_i)$ is the set of all admissible $SU(2)$ colorings on Δ_i; c_i is a particular coloring; w_L is a weight depending on the cut–off L and $\Xi(\Delta_i) = (N_0 - N_1 + N_2 + \cdots + (-1)^{d-3} N_{d-3})(\Delta_i)$, where N_k is the number of simplices of dimension k. Each factor $Z_{M_i, \Delta_i}(c_i)$ is given by the sum over all possible colorings on the $(d-2)$- and $(d-1)$-simplices in the interior of M_i, namely

$$Z[M_i, \Delta_i](c_i) = w_L^{(-1)^d \Xi^{\text{int}}} \prod_{\tilde{\sigma}^{d-2} \in \Delta_i} (-1)^{j_{\tilde{\sigma}^{d-2}}} (\sqrt{2 j_{\tilde{\sigma}^{d-2}} + 1})$$

$$\cdot \prod_{\tilde{\sigma}^{d-1} \in \Delta_i} \left(\prod_{D=1}^{d-3} (-1)^{J_D} (\sqrt{2 J_D + 1}) \right)_{\tilde{\sigma}^{d-1}}$$

$$\cdot \sum_{col} \prod_{\sigma^{d-2} \in int \, M_i} (-1)^{2 j_{\sigma^{d-2}}} (2 j_{\sigma^{d-2}} + 1)$$

$$\cdot \prod_{\sigma^{d-1} \in int \, M_i} \left(\prod_{C=1}^{d-3} (-1)^{2 J_C} (2 J_C + 1) \right)_{\sigma^{d-1}}$$

$$\cdot \prod_{\sigma^d \in M_i} \left\{ \frac{3}{2}(d - 2)(d + 1) j \right\}_{\sigma^d},$$

where we have used the expression found in [2] (the last symbol being the recoupling coefficient to be associated with each d-simplex σ^d). Notice also that in the above relation we keep fixed the coloring c_i on σ^{d-2} and σ^{d-1} lying in ∂M_i and that $\Xi^{\text{int}} = N_0^{\text{int}} - N_1^{\text{int}} + N_2^{\text{int}} + \cdots + (-1)^{d-2} N_{d-2}^{\text{int}}$. Similarly, $P_{\Delta_1,\Delta_2}(c_1, c_2)$ is given by a sum over all possible colorings on the internal σ^{d-2} and σ^{d-1} of N with fixed colorings c_1 and c_2 on $\partial N \cong \Sigma + \Sigma$ and it satisfies the following property with respect to compositions

$$\sum_{c_2 \in C(\Delta_2)} P_{\Delta_1,\Delta_2}(c_1, c_2) w_L^{-\Xi(\Delta_2)} P_{\Delta_2,\Delta_3}(c_2, c_3) = P_{\Delta_1,\Delta_3}(c_1, c_3). \tag{2}$$

If we introduce now an operator $\mathcal{P}$ acting on functionals $\varphi_\Delta(c)$ defined on (Σ, Δ) according to

$$\mathcal{P}[\varphi_\Delta](c) = \sum_{c' \in C(\Delta)} P_{\Delta,\Delta}(c, c') w_L^{-\Xi(\Delta)} \varphi_\Delta(c'), \tag{3}$$

then we can rewrite (1) as

$$\begin{aligned}
Z^d[M] = \sum_{c_1,c_2} &\mathcal{P}[Z[M_1, \Delta_1]](c_1') w_L^{-\Xi(\Delta_1)} P_{\Delta_1,\Delta_2}(c_1, c_2) \\
&\cdot w_L^{-\Xi(\Delta_2)} \mathcal{P}[Z[M_2, \Delta_2]](c_2').
\end{aligned} \tag{4}$$

Thus the physical Hilbert space $H(\Delta)$ for the triangulated $(d-1)$-dimensional surface Σ turns out to be the subspace projected out by $\mathcal{P}$, namely

$$\varphi_\Delta(c) \in H(\Delta) \iff \varphi_\Delta = \mathcal{P}[\varphi_\Delta]. \tag{5}$$

Moreover, by defining an inner product in $H(\Delta)$ according to

$$(\varphi_\Delta, \varphi_\Delta') \doteq \sum_{c,c' \in C(\Delta)} \varphi_\Delta(c) w_L^{-\Xi(\Delta)} P_{\Delta,\Delta}(c, c') w_L^{-\Xi(\Delta)} \varphi_\Delta'(c'), \tag{6}$$

we can rewrite the partition function simply as

$$Z^d[M] = (Z[M_1, \Delta], Z[M_2, \Delta]). \tag{7}$$

Recall that in order to deal with the Hilbert space of a pure d-dimensional BF theory (with a $\Sigma \times [0, 1]$ topology) we have first to integrate the classical action over the B field obtaining the constraint $F_{ij} = 0$ of flat connections. An element of H_{BF} is built up by means of Wilson-line operators $U_j(x, y)$ (x, y are point in Σ and $j = 0, \frac{1}{2}, 1, \ldots$), defined as $U_j(x, y) = P \exp(\int_x^y A^a t_j^a)$ with A a flat connection. Consider now a finite number of Wilson operators and take their tensor product $\otimes_{i=1}^n U_{j_i}(x_i, y_i)$. To make this quantity gauge-invariant, we need to contract the group indices of the U's with one of the following invariant tensors:

- Generalized Clebsch–Gordan (GCG) coefficients
  $\langle (j_1 \cdots j_{n-1})^A a j_n m_n | j_1 m_1 \cdots j_{n-1} m_{n-1} \rangle$ (cfr. [11] for the notations);
- the metric $g_{mm'}^{(j)} := \sqrt{2j+1} \langle jmm'|00 \rangle = (-1)^{j-m} \delta_{m+m',0}$.

A gauge-invariant function which arises in this way corresponds to a colored multivalent graph Y on Σ. It can be shown that such a graph can be decomposed into trivalent graphs $\hat{Y}$ by expanding the GCGs in terms of ordinary CG coefficients (associated with each vertex). Each of such colored trivalent graph $\hat{Y}$ on Σ identifies a function $\Psi_{\hat{Y}} \in H_{BF}$, and this space comes out to be freely generated by them. In general, if two graphs $\hat{Y}$ and $\hat{Y}'$ are homotopic, the corresponding functions $\Psi_{\hat{Y}}$ and $\Psi_{\hat{Y}'}$ have the same value on flat connections.

Conversely, with each colored trivalent graph $\hat{Y}$ one can always associate a colored triangulation and if two graphs $\hat{Y}$ and $\hat{Y}'$ are homotopically inequivalent, they correspond to distinct colored triangulations. In other words, with each colored triangulation Δ we can associate a physical wave function $\Psi_{\Delta,c}(A)$ in the BF framework. Thus the usual wave function $\Phi(A)$ (the explicit expression of which will be given in (12)) can be expanded in terms of such functionals as

$$\Phi(A) = \sum_{\Delta} \sum_{c \in \Delta} \varphi_{\Delta}(c) w_L^{-\Xi(\Delta)} \Psi_{\Delta,c}(A), \tag{8}$$

where in turn the coefficients are the functionals introduced in the discretized context. Moreover, the $\Psi_{\Delta,c}(A)$'s obey the set of relations

$$\Psi_{\Delta,c}(A) = \sum_{c' \in C(\Delta')} P_{\Delta,\Delta'}(c, c') w_L^{-\Xi(\Delta')} \Psi_{\Delta',c'}(A). \tag{9}$$

and it can be shown (see [8]) that these properties identify uniquely the set owing to the PL invariance of the underlying state sums proved in [2]. By substituting (9) into (8), we obtain

$$\Phi(A) = \sum_{c \in C(\Delta)} \varphi_{\Delta}(c) w_L^{-\Xi(\Delta)} \Psi_{\Delta,c}(A), \tag{10}$$

where $\varphi_{\Delta}(c)$ turns out to be defined as

$$\varphi_{\Delta}(c) = \sum_{\Delta'} \sum_{c' \in C(\Delta')} P_{\Delta,\Delta'}(c, c') w_L^{-\Xi(\Delta')} \varphi_{\Delta'}(c'),$$

for an arbitrary *fixed* triangulation Δ of Σ. Then it follows from (2) that $\varphi_{\Delta}(c)$ actually satisfies (5). Summing up, the relation (10) represents a one-to-one correspondence between $\{\varphi_{\Delta}\}$ and $\{\Phi(A)\}$, namely between the Hilbert spaces $H(\Delta)$ and H_{BF}. For completeness, recall that the natural inner product in H_{BF} is given by

$$(\Phi_1, \Phi_2)_{BF} = \int [dA]\delta(F_{ij})\Phi_1^*(A)\Phi_2(A). \tag{11}$$

For what concerns the second step of our program, we start by recalling that in a pure BF theory the physical wave function Φ_M for a d-dimensional manifold M (decomposed as indicated before) may be defined as

$$\Phi_M(A_{|\Sigma})\delta(F_{ij|\Sigma}) = \int [dB, dA]\exp(i \int_M B \wedge (dA + A \wedge A)), \tag{12}$$

where the integration over the A field is restricted by the condition that $A_{|\Sigma}$ is keeped fixed, while the integration over $B_{|\Sigma}$ give rise to the delta function $\delta(F_{ij|\Sigma})$. This ensures that the functional integrals in the right-hand side give a physical state of the BF theory.

In order to analyze the structure of $\Phi_M(A_{|\Sigma})$ we start by considering the exact sequence of homotopy groups for the d-dimensional pair $(M, \partial M)$, namely

$$\cdots \to \pi_2(M) \to \pi_2(M, \partial M) \to \pi_1(\partial M) \to$$
$$\to \pi_1(M) \to \pi_1(M, \partial M) \to \cdots \tag{13}$$

If G is a compact Lie group we may also apply the contravariant functor $\mathrm{Hom}(\cdot, G)$ to obtain another long exact sequence which reads

$$\cdots \to \mathrm{Hom}(\pi_1(M), G) \to \mathrm{Hom}(\pi_1(\partial M), G) \to$$
$$\to \mathrm{Hom}(\pi_2(M, \partial M), G) \to \cdots \tag{14}$$

From the exactness of this last sequence we get

$$\mu : \mathrm{Hom}(\pi_1(\partial M), G) \to \mathrm{Hom}(\pi_2(M, \partial M), G),$$

and

$$\varphi : \mathrm{Hom}(\pi_1(M), G) \to \mathrm{Hom}(\pi_1(\partial M), G),$$

and thus $\mathrm{Ker}\,\mu = \mathrm{Im}\varphi$. On the other hand, there exists an isomorphism between $\mathrm{Hom}(\pi_1(N), G)$ and the set of flat connections on N. We fix now the condition that the holonomy around any element of $\pi_2\ (M, \partial M)$ for a connection $A \in \mathrm{Hom}(\pi_1(\partial M), G)$ on ∂M is equal to zero (namely, A is an element of the Kernel of μ). Then A is the image under φ of a flat connection τ over M, namely τ is the preimage of A. In other words, the condition $\mathrm{Hol}(A) = 0$ over the group $\pi_2(M, \partial M)$ turns out to be *sufficient* to extend A to a flat connection on M. But it is also a *necessary* condition, since the holonomy of a flat connection around a contractible loop is zero.

From the above remarks it is natural to expect that the wave functions $\Phi_M(A_{|\Sigma})$ in (12) and $Z[M, \Delta](c)$ in (1) associated with the same d-manifold M are indeed related through

$$\Phi_M(A_{|\Sigma}) = D_b \sum_{c \in C(\Delta)} Z[M, \Delta](c) w_L^{-\Xi(\Delta)} \Psi_{\Delta,c}(A_{|\Sigma}), \tag{15}$$

where $A_{|\Sigma}$ is flat and D_b is a constant depending on $\mathrm{rank}(\pi_2(M, \partial M)) \equiv b$. To show that this is the case, notice first that our previous analysis implies the relation

$$\Phi_M(A_{|\Sigma}) = D_b' \prod_{a=1}^{b} \delta(U_{(a)} - \mathbf{1}), \tag{16}$$

where D_b' is a constant independent of $A_{|\Sigma}$, and $\delta(U - \mathbf{1})$ is a δ-function with respect to the Haar measure of $G = SU(2)$.

On the other hand, we need to prove that the sum over colorings in the right-hand side of equation (15) imposes the constraint $U_{(a)} = \mathbf{1}$ on $A_{|\Sigma}$, namely that the expression reduces to sums over the colorings of the contractible cycles according to

$$\sum_{c \in C(\Delta)} Z[M, \Delta](c) w_L^{-\Xi(\Delta)} \Psi_{\Delta,c}(A_{|\Sigma}) \prod_{a=1}^{b} \left[\sum_{j=0}^{\infty} (2j+1)\, Tr(U_{(a)}j) \right]. \quad (17)$$

To verify (17) note that since $A_{|\Sigma}$ is flat, we can use the invariance of (9) to remove one d-simplex at a time according to

$$\sum_{c \in C(\Delta)} Z[M, \Delta](c) w_L^{-\Xi(\Delta)} \Psi_{\Delta,c}(A_{|\Sigma})$$

$$= \sum_{c' \in C(\Delta')} Z[M, \Delta'](c) w_L^{-\Xi(\Delta')} \Psi_{\Delta',c'}(A_{|\Sigma}), \quad (18)$$

where Δ is the original triangulation of Σ, and Δ' is obtained by removing precisely that d-simplex. Since $Z[M, \Delta'](c)$ refers to a collection of $(n-1)$ d-simplices, by implementing this procedure we obtain a triangulation with a "tree" structure, in which each d-simplex is glued to the complementary triangulation through at most one $(d-1)$-subsimplices and/or some $(d-r)$-subsimplices, with $r \geq 2$. At the end of the procedure, the resulting state sum will contain only boundary contributions. At this point, as anticipated before, we have to restrict our analysis to particular topological types, and here we just address the case of the d-dimensional solid torus whose boundary has a d-torus topology. As shown in details in [8], the triangulation associated to a d-torus has exactly a triangulated 2-torus as its auxiliary 2-surface. Over this 2-torus we can operate as described in [5]: the resulting graph of the dual triangulation turns out to be simply the contractible loop, namely the generator of $\pi_2(M, \partial M)$. This proves the relation given in (15) and the equality $D_1 = D_1'$. Moreover we see that the auxiliary surface actually encodes all information on $\pi_2(M, \partial M)$.

Coming now to the equivalence between the partition functions (already proved in [8] and [9]), notice that both $Z_{BF}^d[M]$ and $Z^d[M]$ can be written in the form of scalar products between wave functions. More precisely, $Z_{BF}^d[M] = (\Phi_{M_1}, \Phi_{M_2})_{BF}$ (with the definition given in (11)), while the expression of $Z^d[M]$ is given in (7). Thus the equivalence is granted once we show that the isomorphism (8) preserves the inner products in the two Hilbert spaces, namely

$$(\Psi_{\Delta_1,c_1}, \Psi_{\Delta_2,c_2})_{BF} = D_b^2 \cdot P_{\Delta_1,\Delta_2}(c_1, c_2), \quad (19)$$

or, equivalently

$$\sum_{\substack{c_1 \in C(\Delta_1) \\ c_2 \in C(\Delta_2)}} \Psi_{\Delta_1,c_1}(A_1) w_L^{-\Xi(\Delta_1)} P_{\Delta_1,\Delta_2}(c_1, c_2) w_L^{-\Xi(\Delta_2)} \Psi_{\Delta_2,c_2}(A_2)$$

$$= D_b^{-2} \cdot K(A_1, A_2). \quad (20)$$

Here $K(A_1, A_2)$ is a kernel for the inner product. Since Ψ_{Δ_i, c_i} is evaluated on a flat connection A_i, we may use (9) to rewrite the left hand side of (20) as

$$\sum_{c \in C(\Delta)} w_L^{-\Xi(\Delta)} \Psi_{\Delta,c}(A_1) \Psi_{\Delta,c}(A_2). \tag{21}$$

The kernel $K(A_1, A_2)$ vanishes unless both A_1 and A_2 have flat extensions in $N = \Sigma \times [0, 1]$, and the flat extensions exist if and only if A_1 and A_2 are gauge-equivalent. Thus we need to show that the sum over colorings in (21) imposes the constraint $A_1 \simeq A_2$ For the d-torus case the simplicial decomposition of the auxiliary surface corresponds, in the dual lattice, to a Wilson line made by two holonomies U and V around the two homology cycles on Σ. Following as usual [5], we obtain that the wave functions $\Psi_{\Delta,c}$ can be cast into the form

$$\Psi_{\Delta;c}(U, V) = \sum_{m_i, m_i', m_i''} U_{j_1 m_1'}^{m_1} V_{j_2 m_2'}^{m_2} g_{j_1}^{m_1' m_1''} g_{j_2}^{m_2' m_2''} g_{j_3}^{m_3 m_3''}$$
$$\cdot \langle j_1 m_1 j_2 m_2 | j_3 m_3 \rangle \langle j_1 m_1'' j_2 m_2'' | j_3 m_3'' \rangle, \tag{22}$$

where j_1 and j_2 are the colorings associated with the holonomies U and V. These operators can be diagonalized simultaneously and it is easy to prove that the sum over the colors of (21) gives the desired result. So far we have found that the left hand side of (20) is equal to $K(A_1, A_2)$ up to a constant factor E_b, namely we can write

$$(\Psi_{\Delta_1, c_1}, \Psi_{\Delta_2, c_2})_{BF} = E_b^{-1} \cdot P_{\Delta_1, \Delta_2}(c_1, c_2). \tag{23}$$

By combining this last relation with (15) and by using the expressions of the partition functions in terms of pairings, we get

$$Z_{BF}^d[M] = D_b^2 E_b^{-1} Z^d[M] \tag{24}$$

Finally, it can be shown (see [8]) that this result does not actually depend on the topological type (namely on b, which is equal to 1 for the d-torus): as a consequence $D_b^2 E_b^{-1} = 1$ is always fulfilled and the equality holds.

References

1. Birmingham, D., Blau, ; Rakowski, M., Thomson, G. (1991): Phys. Rep. **209**, 129
2. Carbone, G., Carfora, M., Marzuoli, A. (2001): Nucl. Phys. **B 595 [PM]**, 654
3. Ponzano, G., Regge, T. (1968): Semiclassical limit of Racah coefficients, in *Spectroscopic and Group Theoretical Methods in Physics*, ed. by F. Bloch, S.G. Cohen, A. De Shalit, S. Sambursky, I. Talmi, North-Holland Publ. Co., Amsterdam, p. 1
4. Crane, L., Kauffman, L.H., Yetter, D.N.: State sum invariants of four manifolds; hep–th/9409167
5. Ooguri, H. (1992): Nucl. Phys. B **382**, 276

6. Broda, B. (1995): A gauge-field approach to 3- and 4-manifold invariants, in *Symplectic singularities and geometry of gauge field*, ed. by R. Budzynski, S. Janeczk, O.W. Kondracki, A.F. Kunzle, Polish Academy of Science, Warsaw, p. 201
7. Roberts, J. (1995): Topology **34**, 771
8. Carbone, G. (2000): Ph.D Thesis, S.I.S.S.A.-I.S.A.S., Trieste
9. Oriti, D., Williams, R.M. (2001): Phys. Rev. D **63**, 024022
10. Freidel, L., Krasnov, K. (1999): Adv. Theor. Math. Phys. **2**, 1183
11. Yutsis, A.P., Levinson, I.B., Vanagas, V.V. (1962): *The Mathematical Apparatus of the Theory of Angular Momentum*. Published for the National Science Foundation by the Israel Program for Scientific Translation, Jerusalem

Quantum Closed Timelike Curves in General Relativity

P.F. González-Díaz, L.J. Garay

Abstract. We review different spacetimes that contain nonchronal regions separated from the causal regions by chronology horizons and investigate their connection with some importants aspects one would expect to be present in a final theory of quantum gravity, including: stability to classical and quantum metric fluctuations, boundary conditions of the Universe and gravitational topological defects corresponding to spacetime kinks.

1 Introduction

The chronology protection conjecture [1] has been much debated in recent years and the idea has started to emerge that, quite the contrary to the spirit of the conjecture, closed timelike curves (CTC's) [2–4] (see also [5,6]) could play a rather decisive role in the future construction of the theory of quantum gravity. This work aims at considering: (i) some quantum mechanisms which may stabilize the spacetimes which are topological generalizations from Misner space, (ii) the noncausal boundary conditions describing the birth of the universe from itself, and (iii) the connections of CTC's with quantum gravitational kinks and the Euclidean formalism of quantum gravity.

2 The stability of spacetime holes

We shall restrict ourselves to consider those generalizations from Misner space which possess at least a chronology horizon (i.e. a closed surface separating the nonchronal region from the causal one [7]). The static metric of one such spacetimes can generally be written in the form

$$ds^2 = -e^{2\Phi(t,\ell)}dt^2 + F(\ell, \varphi_1, \varphi_2)d\ell^2 + d\Omega(\varphi_1, \varphi_2)^2, \tag{1}$$

where Φ is a generic function which depends on t and ℓ accoding to the particular topology of the considered manifold, ℓ is the proper radial distance of the transversal sections taken on the given manifold, and $d\Omega^2$ is the metric on the two-manifold defined by the angular coordinates φ_1 and φ_2 which set the topology. The function F becomes unity for the case of spherical symmetry (wormholes [3]) and takes on particular shapes $F \equiv F(\ell, \varphi_2)$ for orientable (ringholes [5]) or nonorientable (Klein bottleholes [6]) toroidal symmetries. Metric (1) can be converted into time machine by first setting one of the hole's mouths in motion relative to the other, and then identifying the two mouths. For the generalizations we are considering, the use of the point-splitting regularized Hadamard two-point function for a massless conformally coupled scalar field leads to diverging quantum vacuum polarization on each of the Nth-polarized hypersurfaces occurring at times [5,6,8]

$$T^{\pm}_{H_{Ni}} = \pm \frac{\xi^{-N} + 1}{\xi^{-N} - 1} b_i \left[1 \mp (-1)^i \cos \varphi_2 \right], \quad i = 1, 2, \tag{2}$$

where $\xi = \sqrt{(1 - v)/(1 + v)}$, with v the relative velocity between the hole mouths, and $b_i = A + B S_i\{\varphi/4\}$, in which $S_1\{x\} = \cos^2 x$, $S_2\{x\} = \sin^2 x$, and A and B are arbitrary parameters such that $A, B \neq 0$ for Klein bottlehole, and $A > 0$, $B = 0$ for ringhole and wormhole. All the Nth-polarized hypersurfaces are respectively nested in the corresponding chronology horizons $H_i^{\pm}$ occurring at times given by $\lim_{N \to \infty} T^{\pm}_{H_{Ni}}$. Divergence of the stress-tensor on the chronology horizons can however be avoided using generalizations [9] from a Misner space based on a quantization condition for time $T = (N + \alpha)T_0$, where $0 \leq \alpha \leq 1/2$ is an automorphic constant [10] and T_0 is a constant time of the order the Planck time. It follows that the condition for the existence of the Nth-polarized hypersurfaces should then imply

$$\frac{(1 + \xi^N)b_i}{(1 - \xi^N)(N + \alpha)} \left[1 \mp (-1)^i \cos \varphi_2 \right] = T_0. \tag{3}$$

Therefore, no Nth-polarized hypersurfaces could exist in any of the accelerating holes which are in this way rendered quantum-mechanically stable. However, these holes can only show CTC's at the Planck scale with chronology horizons whose minimum widths are also of the Planck scale [9]. These stable submicroscopic holes can be regarded as true virtual ingredients of the spacetime foam and are subject to the so-called quantum interest [11].

3   Stable Universe created from itself

Recent cosmological observations have led to the requirement that the boundary conditions of the Universe should predict an open or flat model. So far, the most popular initial conditions for cosmology are described in an Euclidean framework (including the Vilenkin's tunneling wave function [12] and the Hartle-Hawking's no-boundary proposal [13,14]) and predict the formation of a closed or even open universe from nothing. Nevertheless, a more natural and genuine condition for directly creating an open-inflationary process has recently been suggested by Gott and Li [15] who, basing on the periodicity properties of Misner space, assumed that the universe could have been created from itself. The existence of a nonchronal region in de Sitter space can be visualized by slicing the Schrödinger's five hyperboloid,

$$-v^2 + w^2 + x^2 + y^2 + z^2 = 3/\Lambda, \tag{4}$$

Λ being the positive cosmological constant, with embedding metric

$$ds^2 = -dv^2 + dw^2 + dx^2 + dy^2 + dz^2, \tag{5}$$

along the spacelike direction defined in terms of the proper time. These slices are negatively-curved surfaces describable by open cosmological solutions whose origin of time occupies any point on the equator of the resulting sphere and that contain CTC's in the region of the static space satisfying the Misner symmetry

$$(v, w, x, y, z) \Leftrightarrow \left(c \cosh(nb) + w \sinh(nb), w \cosh(nb) \right.$$
$$\left. + v \sinh(nb), x, y, z\right), \tag{6}$$

where n is any integer number and b is a dimensionless *a priori* arbitrary quantity. It turns out that this symmetry can only be satisfied in the region defined by $w > |v|$, covered by the static de Sitter metric. The multiply connected de Sitter space is just another classically stable [16] pathological generalization from Misner space, corresponding to the de Sitter symmetry. Let us then Wick rotate in the usual de Sitter-Kruskal metric and in the time identification $t \leftrightarrow t + nb\sqrt{3/\Lambda}$ to obtain

$$ds^2 = \frac{4\Lambda}{3\left(1 + \eta^2 + v^2\right)} \left(d\eta^2 + dv^2\right) + r^2 d\Omega_2^2, \tag{7}$$

$$v - i\eta = \sqrt{v^2 + \eta^2} \exp\left[i\left(nb + \tau\sqrt{\Lambda/3}\right)\right]. \tag{8}$$

Then, if we set (i) the Lorentzian time $t = -i\tau = (n + \alpha)t_0$, with t_0 about the Planck time, as implied by the modified Misner space, and (ii) $b = 2\pi t = 2\pi\alpha\sqrt{3/\Lambda}$, we see that quantum stability can be unambiguously restored, provided we accept restricting to nonchronal regions and CTC's on them to be at the Planck scale, i.e. $\sqrt{\Lambda} \equiv \ell_p^{-1}$.

4 Superluminal travels, CTC's and kinks

In what follows we shall show how the inner region of the two-dimensional Alcubierre space [17] (and most probably its four-dimensional counterpart) can be made nonchronal, while becoming still another quantum-mechanically stable generalization of the doubly-conical modified version of the Misner space. The comoving, manifestly static two-dimensional metric corresponding to Alcubierre spacetime can be written as [18]

$$ds^2 = -A(r)dt^2 + \frac{dr^2}{A(r)}, \tag{9}$$

where $A(r) = 1 - v_0\left[1 - f(r)\right]^2$, with v_0 the constant apparent velocity of the warp drive spaceship, and $f(r)$ is an arbitrary function just subjected to the boundary conditions that $f = 1$ at $r = 0$ (the location of the spaceship) and $f = 0$ at infinity. This dimensionally-reduced Alcubierre spacetime possesses an event horizon for the most interesting case where $v_0 > 1$, and can be visualized [19] as a three-hyperboloid

$$-v^2 + w^2 + x^2 = v_0^{-2}, \tag{10}$$

embedded in E^3, with an embedding metric

$$ds^2 = -dv^2 + dw^2 + dx^2, \tag{11}$$

corresponding to topology $R \times S^2$ and invariance group $SO(2, 1)$, using the coordinate transformations $v = v_0^{-1}\sqrt{A(r)}\sinh(v_0 t)$, $v = v_0^{-1}\sqrt{A(r)}\cosh(v_0 t)$, $x = F(r)$, where

$$\left[\frac{dF(r)}{dr}\right]^2 = -\left[\frac{\left(\frac{d(A-1)}{dr}\right)^2 - 4v_0^2}{4v_0^2 A}\right].$$

For $v_0 > 1$, one can convert the two-dimensional Alcubierre space into multiply connected space by adding the coordinate identifications $(t, r) \leftrightarrow (t + nb/v_0, r)$. Thus, the region defined by $w > |v|$ will be filled with CTC's and separated from the causal exterior by the surface at r_0 (now a chronology horizon) defined by $f(r_0) = 1 - v_0^{-1}$. Using first the coordinate re-definitions, $v = \frac{1}{2}(e^V + Te^{-V})$, $w = \frac{1}{2}(e^V - Te^{-V})$, $x = x$, and then new coordinates $V = Y + Z$, $T - \int T dV = Y - Z$, we can transform metric (11) into the maximally extended three-dimensional Misner metric $ds^2 = -dY^2 + dZ^2 + dx^2$, and hence we get a positive definite line element by using the continuation $Y = i\xi$. Upon rotating back to the Lorentzian sector, this ultimately leads to a period of the closed spacelike coordinate given by $2\pi T$ and finally, repeating the same procedure as in Sects. 2 and 3, to a quantum-mechanically stable superluminal Alcubierre wrap drive with CTC's at the Planck scale.

We note that metric (9) can be converted into the form

$$ds^2 = -A(r)\left[d\theta - \frac{v_0(1 - f(r))}{A(r)}dr\right]^2 + \frac{dr^2}{A(r)} \tag{12}$$

by re-defining the proper time so that $dt = d\theta + v_0[1 - f(r)]dr/A(r)$. Metric (12) can then be regarded as the kinked line element [20] that corresponds to the static metric (9), with apparent horizon at r_0. This can be seen by first re-defining the coordinates so that $d\theta' = A^{\frac{1}{4}}d\theta$, $dr' = A^{-\frac{1}{4}}dr$, and then take $A(r) = \cos^2(2\beta)$, with β the tilt angle of the light cones tipping over the hypersurfaces. Metric (12) thereby becomes that of a gravitational topological defect [20]

$$ds^2 = -\cos(2\beta)\left[(d\theta')^2 - (dr')^2\right] \pm 2\sin(2\beta)d\theta'dr', \tag{13}$$

with the choice of sign in the second term depending on whether a positive (upper sign) or negative (lower sign) topological charge is considered. Since $\sin\beta$ cannot exceed unity, for a complete description of a one-kink (β monotonically varying from 0 to π), one need two coordinate patches which are identified only on surface $r = 0$. However, in order to complete geodesic closed paths on the Kruskal diagrams, one would require additional identifications of the two coordinate patches also on the surfaces at $r = r_0$. Such identifications, and hence complete CTC's, are then made possible if, and only if we allow the existence of propagating quantum fields that lead to the process of thermal radiation at temperature $dA(r)/dr|_{r=r_0}/4\pi$ [18], usually described in the corresponding Euclideanized spacetime where any coherent process giving rise to an observable causality violation is destroyed. Our conclusion (which can readily be generalized to any spacetime with an apparent event horizon) is then that making any such Lorentzian spacetimes multiply connected by imposing Misner symmetry somehow makes thermal radiation to appear, so that CTC's become essential ingredients for any future theory of quantum gravity.

References

1. Hawking, S.W. (1992): Phys. Rev. D **46**, 603
2. Gödel, K. (1949): Rev. Mod. Phys. **21**, 447
3. Morris, M.S., Thorne, Yurtsever, U. (1988): Phys. Rev. Lett. **61**, 1446
4. Gott, J.R. (1991): Phys. Rev. Lett. **66**, 1126
5. González-Díaz, P.F. (1996): Phys. Rev. D **54**, 6122
6. González-Díaz, P.F., Garay, L.J. (1999): Phys. Rev. D **59**, 064026
7. Visser, M. (1996): Lorentzian Wormholes. AIP, Woodbury, NY
8. Kim, S.-W., Thorne, K.S. (1991): Phys. Rev. D **43**, 3929
9. González-Díaz, P.F. (1998): Phys. Rev. D **58**, 124011
10. Krasnikov, S.V. (1996): Phys. Rev. D **54**, 7322
11. Ford, L.H., Roman, T.A. (1999): Phys. Rev. D **60**, 104018
12. Vilenkin, A. (1982): Phys. Lett. B **117**, 25
13. Hartle, J.B., Hawking, S.W. (1983): Phys. Rev. D **28**, 2960
14. Hawking, S.W., Turok, N. (1998): Phys. Lett. B **425**, 25
15. Gott, J.R., Li, Li-Xin (1998): Phys. Rev. D **58**, 023501
16. González-Díaz, P.F. (1999): Phys. Rev. D **59**, 123513
17. Alcubierre, M. (1994): Class. Quant. Grav. **11**, L73
18. Hiscock, W.A. (1997): Class. Quant. Grav. **14**, L183
19. González-Díaz, P.F. (2000): Phys. Rev. D **62**, 044005
20. Finkestein, D., McCollum, G. (1975): J. Math. Phys. **16**, 2250

GPSR Compliance
The European Union's (EU) General Product Safety Regulation (GPSR) is a set
of rules that requires consumer products to be safe and our obligations to
ensure this.

If you have any concerns about our products, you can contact us on

ProductSafety@springernature.com

In case Publisher is established outside the EU, the EU authorized
representative is:

Springer Nature Customer Service Center GmbH
Europaplatz 3
69115 Heidelberg, Germany